CHAPTER 4

10. $\dfrac{d}{dx}e^x = e^x, \quad \dfrac{d}{dx}e^{f(x)} = f'(x)e^{f(x)}$

11. $\dfrac{d}{dx}\ln x = \dfrac{1}{x}, \quad x > 0;$

$\quad\;\; \dfrac{d}{dx}\ln f(x) = f'(x) \cdot \dfrac{1}{f(x)}, \quad f(x) > 0$

12. $\dfrac{d}{dx}\ln |x| = \dfrac{1}{x}, \quad x < 0;$

$\quad\;\; \dfrac{d}{dx}\ln |f(x)| = f'(x) \cdot \dfrac{1}{f(x)}, \quad f(x) < 0$

13. If $\dfrac{dP}{dt} = kP$, then $P(t) = P_0 e^{kt}$.

14. If $\dfrac{dP}{dt} = -kP$, then $P(t) = P_0 e^{-kt}$.

15. $\dfrac{d}{dx}a^x = (\ln a)a^x$

16. $\dfrac{d}{dx}\log_a x = \dfrac{1}{\ln a} \cdot \dfrac{1}{x}, \quad x > 0;$

$\quad\;\; \dfrac{d}{dx}\log_a |x| = \dfrac{1}{\ln a} \cdot \dfrac{1}{x}, \quad x < 0$

17. $a^x = e^{x(\ln a)}$

Calculus

AND ITS APPLICATIONS

SEVENTH EDITION

Calculus

AND ITS APPLICATIONS

SEVENTH EDITION

MARVIN L. BITTINGER

Indiana University—Purdue University at Indianapolis

 ADDISON-WESLEY

An imprint of Addison Wesley Longman, Inc.

Reading, Massachusetts • Menlo Park, California • New York • Harlow, England
Don Mills, Ontario • Sydney • Mexico City • Madrid • Amsterdam

Football is like calculus.
Someone has to show you how to do it,
but once you've got er' figgered, why, she's easy.

Bert Jones, Former quarterback, *Baltimore Colts*
(Bert Jones majored in mathematics in college.)

Sponsoring Editor	Laurie Rosatone
Senior Project Manager	Christine O'Brien
Assistant Editor	Ellen Keohane
Managing Editor	Karen Guardino
Production Supervisor	Rebecca Malone
Marketing Manager	Carter Fenton
Production Services	Martha Morong, Quadrata, Inc.
Text Design and Art Coordination	Geri Davis, The Davis Group, Inc.
Cover Design and Art Direction	Barbara T. Atkinson
Cover Photograph	© Dieter Hessel Photographer
Senior PrePress Supervisor	Caroline Fell
Manufacturing Buyer	Evelyn Beaton
Composition	The Beacon Group
Technical Art Illustration	Tech-Graphics Corp.

Photo credits appear on page 570.

Library of Congress Cataloging-in-Publication Data

Bittinger, Marvin L.
  Calculus and its applications / Marvin L. Bittinger—7th ed.
    p.   cm.
  Rev. ed. of: Calculus, 6th ed. c1996.
  Includes index.
  ISBN 0-201-33864-5
  1. Calculus. I. Bittinger, Marvin L. Calculus. II. Title.
QA303.B645   1999
515—dc21   99–047368
                                                               CIP

1 2 3 4 5 6 7 8 9 10—RNT—02010099

Contents

CHAPTER 3

Applications of Differentiation   *185*

CHAPTER 4

Exponential and Logarithmic Functions   *285*

CHAPTER 5

Integration 363

CHAPTER 6

Applications of Integration 433

CHAPTER 7

Functions of Several Variables   *495*

Preface

Appropriate for a one-term course, this text is an introduction to calculus as applied to business, economics, the life and physical sciences, the social sciences, and many general areas of interest to all students. A course in intermediate algebra is a prerequisite for the text, although the appendix, *Review of Basic Algebra*, together with Chapter 1 provides a sufficient foundation to unify the diverse backgrounds of most students.

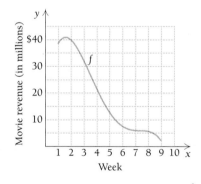

What's New in the Seventh Edition?

Functions Earlier

To minimize the amount of time needed for review, we have moved much of the algebra review material from Chapter 1 to an appendix. This allows us to begin Chapter 1 with coverage of functions, graphs, and models and to incorporate an earlier use of technology. New to this edition is Section 1.3, "Finding Domain and Range." This early and expanded function coverage allows for many uses of functions such as the expansion of graphing applications.

Mathematical Modeling, Curve Fitting, and Regression

This edition continues its emphasis on mathematical modeling, including the advantages of technology as appropriate. For example, the use of the grapher for modeling, as an optional topic, is introduced in an all-new Section 1.6 and then reinforced many times throughout the text. (In this text, we use the term "grapher" to refer to all graphing calculators and graphing software.)

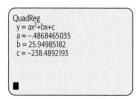
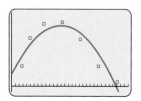

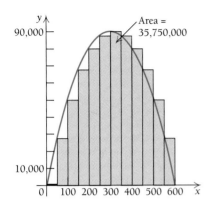

Integration Reorganized

The structure of the material on integration in Chapter 5 has been significantly reorganized. Section 5.1 introduces antiderivatives and indefinite integrals. The topics of area and definite integrals are then connected in Section 5.2. Section 5.3 develops limits of sums and accumulations on an intuitive level and is now placed earlier in the chapter. In effect, what was covered in Sections 5.2, 5.3, 5.4, and 5.8 in the preceding edition is now reorganized and covered in Sections 5.2, 5.3, and 5.4.

Enhanced Technology

Though still optional, the use of technology has been expanded and enhanced.

TECHNOLOGY CONNECTION FEATURES.   Technology Connection features (102 in all) are included throughout the text to illustrate the use of technology. Whenever appropriate, art that simulates graphs and tables generated using a grapher is included as well.

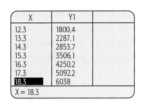

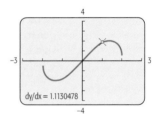

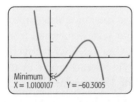

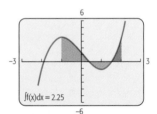

There are now four types of Technology Connections for students and instructors to use for exploring key ideas in business calculus.

- *Lesson/Teaching.* These provide students with an example, followed by exercises to work.
- *Checking.* These tell the students how to verify a solution by using a grapher.
- *Exercises.* These are simply exercises for students to work using a grapher.
- *Exploration/Investigation.* These provide questions to guide students through an investigation.

TECHNOLOGY CONNECTION EXERCISES.   Most exercise sets contain technology-based exercises that are indicated either with a [icon] icon or under the heading "Technology Connection." These exercises also appear in the Summary and Reviews and the

Chapter Tests. The Printed Test Bank supplement includes technology-based exercises as well.

Hallmark Features

Student Friendly

Calculus and Its Applications, Seventh Edition, is the most student-oriented applied calculus text on the market. The level of vocabulary and terminology is accessible and used in a consistent manner, allowing the content to be easily understood by students. There is a plentiful supply of examples with numerous and carefully placed art pieces to help students visualize topics.

Use of Color

The text uses full color in an extremely functional way, and not merely as window dressing. Color usage has been carried out in a methodical and precise manner so that it enhances the readability of the text for the student and the instructor. The following illustrates.

For example, when two curves are graphed using the same set of axes, one is usually red and the other blue with the red graph the curve of major importance. This is exemplified in the following graphs from Chapter 1 (pp. 60 and 61). Note that the equation labels are the same color as the curve. When the instructions say "Graph," the dots match the color of the curve.

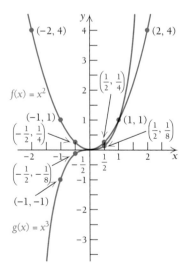
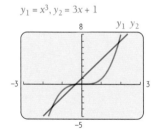

In the following figure from Chapter 2 (p. 127), we see how color is used to distinguish between secant and tangent lines. Throughout the text, blue is used for secant lines and red for tangent lines.

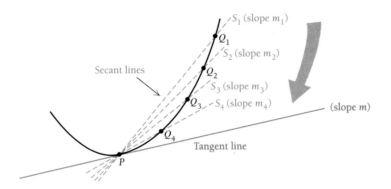

In the following graph and text development from Chapter 3 (pp. 205–206), the color red denotes substitution in equations and blue highlights maximum and minimum values (second coordinates). This specific use of color is carried out further, as shown in the figure that follows. Note that when dots are used for emphasis other than just merely plotting, they are black.

We then find second coordinates by substituting in the original function:

$$f(-3) = (-3)^3 + 3(-3)^2 - 9(-3) - 13 = 14;$$
$$f(1) = (1)^3 + 3(1)^2 - 9(1) - 13 = -18.$$

Thus there is a relative maximum at $(-3, 14)$ and a relative minimum at $(1, -18)$. The relative extrema are shown in the graph below.

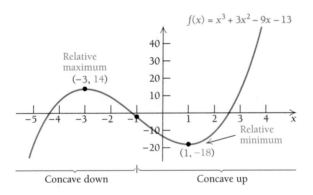

Beginning with integration in Chapter 5, the color amber is used to identify the area, unless certain special concepts are considered in the initial learning process.

The figure on the left below from Chapter 5 (p. 406) illustrates the vivid use of blue and red for the curves and labels, and amber for the area. In Chapter 7 (p. 513), we use extra colors for clarity. The top of the surface, as shown in the figure on the right below, is blue and the bottom of the surface is rose. This is carried out in all examples in Section 7.4.

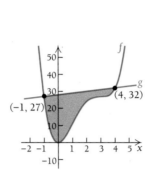

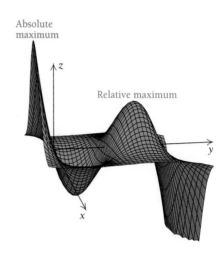

Intuitive Approach

Although the word "intuitive" has many meanings and interpretations, its use here means "experience based." Throughout the text, when a particular concept is discussed, its presentation is designed so that the students' learning process is based on their earlier mathematical experience or a new experience presented by the author before the concept is formalized. This is illustrated by the following situations.

• The definition of the derivative in Chapter 2 is presented in the context of a discussion of average rates of change (see p. 116). This presentation is more accessible and realistic than the strictly geometric idea of slope.

• When maximum problems involving volume are introduced (see pp. 250–251), a function is derived that is to be maximized. Instead of forging ahead with the standard calculus solution, the student is first asked to stop, make a table of function values, graph the function, and then estimate the maximum value. This experience provides students with more insight into the problem. They recognize that not only do different dimensions yield different volumes, but also that the dimensions yielding the maximum volume may be conjectured or estimated as a result of the calculations.

• Relative maxima and minima (Sections 3.1–3.3) and absolute maxima and minima (Section 3.4) are covered in separate sections in Chapter 3, so that students obtain a gradual buildup of these topics as they consider graphing using calculus concepts (see pp. 186–249).

• The understanding behind the definition of the number e in Chapter 4 is explained both graphically and through a discussion of continuously compounded interest (see pp. 291–292).

Applications

Relevant and factual applications drawn from a broad background of fields are integrated throughout the text as applied problems and exercises, and are also featured in separate application sections. These have been updated and expanded in this edition, to include even more applications using real data. In addition, each chapter opener includes an application that serves as a preview of what students will learn in the chapter.

The applications in the exercise sets are grouped under headings that identify them as reflecting real-life situations:

Business and Economics

Life and Physical Sciences

Social Science

General Interest

This organization allows the instructor to gear the assigned exercises to a particular student and also allows the student to know whether a particular exercise applies to his or her major.

The following illustrate some examples of the applications.

- An early discussion in Chapter 1 of compound interest, total revenue, cost and profit, as well as supply and demand functions sets the stage for subsequent applications that permeate the book (see pp. 49, 153, and 254).
- The concepts of average rate of change and instantaneous rate of change are introduced with new business applications in Chapter 2. This is prior to the introduction of applications of differentiation in Chapter 3, which includes maximum–minimum applications (see pp. 249–265) and implicit differentiation and related rates (see pp. 273–279).
- When the exponential model is studied in Chapter 4, other applications, such as continuously compounded interest and the demand for natural resources, are also considered (see pp. 317–344). Growth and decay are covered in separate sections in Chapter 4 to allow room for the many worthwhile applications that relate to these concepts.
- Applications of integration are covered in a separate chapter and include a thorough coverage of continuous probability (see pp. 433–494).
- The Extended Technology Application feature includes interesting and topical situations such as the ecological effect of global warming (see pp. 91–93), predicting the distance a home run will travel (see pp. 181–183), and using total box-office revenue data to determine when videotape sales should begin (see pp. 360–362).

Pedagogical Features

Chapter Openers

Each chapter opener includes an application to whet the student's appetite for the chapter material. These applications also provide an intuitive introduction to a key calculus topic.

Section Objectives

As each new section begins, its objectives are stated in the margin. These can be spotted easily by the student, and they provide the answer to the typical question, "What material am I responsible for?"

Variety of Exercises

There are over 3300 exercises in this edition. The exercise sets are enhanced not only by the inclusion of business and economics applications, detailed art pieces, and extra graphs, but also by the following features.

*   **TECHNOLOGY CONNECTION EXERCISES.**   These exercises appear in the Technology Connection features (see pp. 34, 136, and 307) and in the exercise sets (see pp. 108, 234, and 400).

*   **SYNTHESIS EXERCISES.**   Synthesis exercises are included at the ends of most exercise sets and all Summary and Reviews and Chapter Tests. They require students to go beyond the immediate objectives of the section or chapter and are designed both to challenge students and to make them think about what they are learning (see pp. 174, 248, 332, and 407).

*   **THINKING AND WRITING EXERCISES.**   These exercises appear both in the exercise sets and in the synthesis section at the end of most exercise sets (see pp. 123, 234, and 386). They are denoted by the symbol tw . These ask students to explain mathematical concepts in their own words, thereby strengthening their understanding. The answers to these exercises are given in the *Instructor's Solutions Manual*.

*   **APPLICATIONS.**   A section of applied problems is included in most exercise sets. The problems are grouped under headings that identify them as business and economics, life and physical sciences, social sciences, or general interest. Each problem is accompanied by a brief description of its subject matter (see pp. 155–158, 349, and 397–398).

Tests and Reviews

*   **SUMMARY AND REVIEW.**   At the end of each chapter is a summary and review. These are designed to provide students with all the material they need for successful review. Answers are at the back of the book, together with section references so that students can easily find the correct material to restudy if they have difficulty with a particular exercise. In each summary and review there is a list of *Terms to Know*, accompanied by page references to help students key in on important concepts (see p. 279).

*   **TESTS.**   Each chapter ends with a chapter test that includes synthesis and technology questions. There is also a Cumulative Review at the end of the text that can also serve as a final examination. The answers, with section references to the chapter tests and the Cumulative Review, are at the back of the book. Six additional forms of each of the chapter tests and the final examination with answer keys appear, ready for classroom use, in the *Instructor's Manual/Printed Test Bank*.

Supplements for the Student

Student's Solutions Manual
ISBN 0-201-33865-3

Written by Judith A. Penna, this manual provides complete worked-out solutions for all odd-numbered exercises in the exercise sets (with the exception of the Thinking and Writing exercises). This supplement is available to instructors and is for sale to students.

Graphing Calculator Manual
ISBN 0-201-65863-1

The *Graphing Calculator Manual* by Judith A. Penna, with the assistance of Daphne Bell, contains keystroke level instruction for the Texas Instruments TI-83®, TI-83 Plus®, TI-85®, TI-86®, and TI-89® models. This manual is available to instructors and is for sale to students. Contact your Addison Wesley Longman sales consultant for a separate HP38G® module.

The *Graphing Calculator Manual* uses actual examples and exercises from *Calculus and Its Applications,* Seventh Edition, to teach students to use a graphing calculator. The order of topics in the *Graphing Calculator Manual* mirrors that of the text, providing a just-in-time mode of instruction.

Web Site

The Web site for this text provides additional resources for both students and instructors.

Supplements for the Instructor

Instructor's Solutions Manual
ISBN 0-201-66977-3

The *Instructor's Solutions Manual* by Judith A. Penna contains worked-out solutions to *all* exercises in the exercise sets. This supplement is available to instructors.

Instructor's Manual/Printed Test Bank
ISBN 0-201-66976-5

The *Instructor's Manual/Printed Test Bank* (IM/PTB) by Laurie Hurley contains six alternate test forms for each chapter test and six comprehensive final examinations with answers. The IM/PTB also has answers for the even-numbered exercises in the exercise sets. These can be easily copied and handed out to students. The answers to the odd-numbered exercises are at the back of the text.

TestGen-EQ with QuizMaster EQ
Windows and Macintosh CD ISBN 0-201-61842-7

TestGen-EQ's friendly graphical interface enables instructors to easily view, edit, and add questions, transfer questions to tests, and print tests in a variety of fonts and forms. Search and sort features let the instructors quickly locate questions and arrange them in a preferred order. Six question formats are available, including short-answer, true–false, multiple-choice, essay, matching, and bimodal formats. A built-in question editor gives the user power to create graphs, import graphics, insert mathematical symbols and templates, and insert variable numbers or text. Computerized testbanks include algorithmically defined problems organized according to each textbook. An "Export to HTML" feature lets instructors create practice tests for the Web.

QuizMaster-EQ enables instructors to create and save tests using *TestGen-EQ* so students can take them for either practice or a grade on a computer network. Instructors can set preferences for how and when tests are administered. *QuizMaster-EQ* automatically grades exams, stores results on disk, and allows the instructor to view or print a variety of reports for individual students, classes, or courses.

Consult your Addison-Wesley representative for details.

Supplementary Chapters
ISBN 0-201-65864-X

The trigonometry chapter, Chapter 8 in the preceding edition, has been deleted from this text. It is available as a supplement to this edition. Also available are supplemental chapters, written by Marvin L. Bittinger and Michael A. Penna of Indiana University—Purdue University at Indianapolis, that present material on sequences and series and expand the book's presentation of differential equations. Many applications, along with optional technology material, are included. All three chapters are available together as a single paperback supplement.

Acknowledgments

Your author has taken many steps to ensure the accuracy of the manuscript. The graphic art pieces have been computer-generated. A number of devoted individuals comprised the team that was responsible for monitoring each step of the production process in such careful detail. In particular, I would like to thank Judith A. Beecher and Judith A. Penna for their helpful suggestions, proofreading, and checking of the art. Their efforts above and beyond the call of duty deserve infinite praise. I also wish to express my appreciation to my students for providing suggestions and criticisms so willingly during the preceding editions; to Mike Penna of IUPUI for his help with the computer graphics; and to Daphne Bell, Barbara Johnson, Judy Penna, Bill Saler, and Keith Schwingendorf for their precise checking and proofreading of the manuscript for mathematical accuracy. In addition, I would like to thank Scott Mortensen and Ken Hurley for their contributions to the technology material.

Finally, the following reviewers provided thoughtful and insightful comments that helped immeasurably in the revision of this text.

Viola Lee Bean, *Boise State University*

Franco Fedele, *University of West Florida*

Ian Gladwell, *Southern Methodist University*

Alex A. Himonas, *University of Notre Dame*

Alec Ingraham, *New Hampshire College*

Larry S. Johnson, *Metropolitan State College of Denver*

Mark S. Korlie, *Montclair State University*
Raimundo M. Kovac, *University of Minnesota, Duluth*
Thomas Lada, *North Carolina State University*
Frances J. Lane, *Virginia State University*
Patrick O'Brien, *Mesa Community College*
Richard D. Porter, *Northeastern University*
William T. Sledd, *Michigan State University*
James Thomas, *Colorado State University*
Bruce R. Wenner, *University of Missouri, Kansas City*

M.L.B.

Index of Applications

Index of Technology Connections

The following index lists, by topic, 64 optional Technology Connections. These features incorporate in-text comments, including guided explorations and exercises, that explain and encourage the use of technology to explore problem situations. The Technology Connection Features, along with the Technology Connection Exercises in the exercise sets, and the Extended Technology Applications, described in the Preface, bolster the optional integration of technology into the course. New to this edition is a *Graphing Calculator Manual,* available for students, featuring keystroke instructions for several calculator models.

Calculus

AND ITS APPLICATIONS

SEVENTH EDITION

1

Functions, Graphs, and Models

INTRODUCTION

This chapter introduces functions together with their graphs and applications. Also presented are many topics that we will consider often throughout the text: supply and demand, total cost, total revenue, and total profit, and the concepts of a mathematical model and curve fitting.

Skills in using a graphing calculator or graphing software (henceforth referred to as graphers) are also introduced in optional Technology Connections. Details on keystrokes are given in the Graphing Calculator Manual (GCM).

Those needing some algebra review might wish to study the appendix along with this chapter.

AN APPLICATION
Hearing-Impaired Americans

The graph below approximates the number N, in millions, of hearing-impaired Americans as a function of age x (*Source*: American Speech-Language Hearing Association). An equation for this graph is the polynomial function $N(x) = -0.00006x^3 + 0.006x^2 - 0.1x + 1.9$. Use the graph to determine the domain.

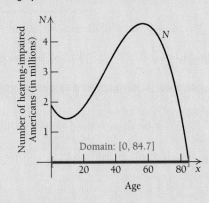

Domain: [0, 84.7]

Number of hearing-impaired Americans (in millions)

Age

Since a person's age cannot be negative, we do not consider negative inputs. Negative outputs would not make sense in this problem either. Thus the domain is the interval [0, 84.7].

This problem appears as Exercise 54 in Exercise Set 1.3.

1.1

OBJECTIVES

➤ Graph equations.
➤ Use graphs as mathematical models to make predictions.
➤ Carry out calculations involving compound interest.

Graphs and Equations

What Is Calculus?

What is calculus? This is a typical question when beginning a course like this. Let's consider a simplified answer for now. A more complete answer will evolve throughout the course.

Consider the advertising billboard shown at left. The following is an algebra problem that might be related to this billboard. Try to solve it. (If you need some algebra review, consult the appendix.)

Algebra Problem

A standard rectangular billboard seen along highways has a perimeter of 124 ft. The length is 6 ft more than three times the height. Find the dimensions of the billboard.

The answer is a length of 48 ft and a height of 14 ft.

The following is a calculus problem that might be related to this billboard. Try to solve it.

Calculus Problem

A billboard is to be built with a perimeter of 124 ft in such a way that the area of the billboard is as large as possible. What dimensions will allow the maximum area?

One way to solve this problem might be to choose several sets of dimensions for the billboard, compute the resulting areas, and determine which is the largest. If you have access to a computer spreadsheet, you might create one and expand the table at left. We let l = length and h = height. Then

$$2l + 2h = 124,$$
so $$l + h = 62,$$
and $$A = lh.$$

Dimensions That Allow a Perimeter of 124 ft		
Length, l	Height, h	Area, $A = lh$
34 ft	28 ft	952 ft^2
32 ft	30 ft	960 ft^2
31.5 ft	30.5 ft	960.75 ft^2
31 ft	31 ft	961 ft^2 ← Largest?
30.4 ft	31.6 ft	960.64 ft^2

From the data in the table, we might conclude that the largest area is 961 ft^2 for square dimensions of 31 ft by 31 ft. But how do we know for sure that there are no other dimensions that yield a larger area? We need the tools of calculus to answer this. We will study such maximum–minimum problems in more detail in Chapter 3.

Other topics we will consider in calculus are the slope of a curve at a point, rates of change, area under a curve, accumulations of quantities, and many statistical applications.

Graphs

We see graphs of all kinds in magazines and newspapers. Some examples are shown below. The study of graphs is an essential aspect of calculus. A graph offers the opportunity to visualize relationships. For instance, the graph showing Netscape Communications' stock price could be used to show how the price has changed over time. One topic that we consider in calculus is how a change in one quantity affects the change in another.

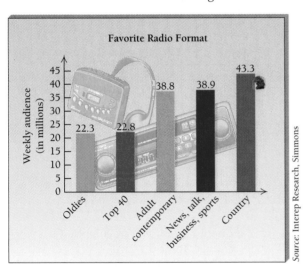

Favorite Radio Format

Source: Interep Research, Simmons

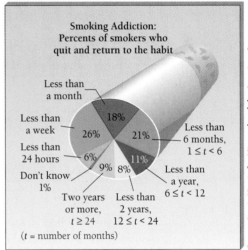

Smoking Addiction: Percents of smokers who quit and return to the habit

Less than a month — 18%
Less than a week — 26%
Less than 24 hours — 6%
Don't know 1%
Two years or more, $t \geq 24$ — 9%
Less than 2 years, $12 \leq t < 24$ — 8%
Less than 6 months, $1 \leq t < 6$ — 21%
Less than a year, $6 \leq t < 12$ — 11%

(t = number of months)

Source: Louis Harris for Nicotrol and the American Lung Association

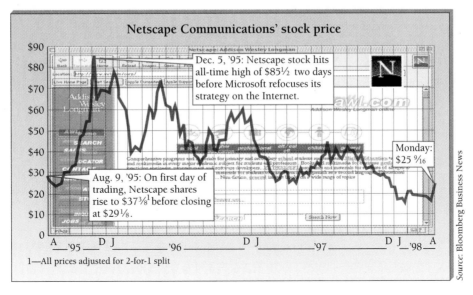

Netscape Communications' stock price

Dec. 5, '95: Netscape stock hits all-time high of $85½ two days before Microsoft refocuses its strategy on the Internet.

Aug. 9, '95: On first day of trading, Netscape shares rise to $37⅜[1] before closing at $29⅛.

Monday: $25 ⁹⁄₁₆

1—All prices adjusted for 2-for-1 split

Source: Bloomberg Business News

Ordered Pairs and Graphs

Each point in a plane corresponds to an ordered pair of numbers. Note in the figure below that the point corresponding to the pair (2, 5) is different from the point corresponding to the pair (5, 2). This is why we call a pair like (2, 5) an **ordered pair.** The first member, 2, is called the **first coordinate** of the point, and the second member, 5, is called the **second coordinate.** Together these are called the *coordinates of the point.* The vertical line is often called the *y-axis,* and the horizontal line is often called the *x-axis.*

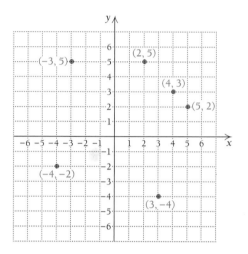

Graphs of Equations

A **solution** of an equation in two variables is an ordered pair of numbers that, when substituted for the variables, gives a true sentence. If not directed otherwise, we usually take the variables in *alphabetical* order. For example, $(-1, 2)$ is a solution of the equation $3x^2 + y = 5$, because when we substitute -1 for x and 2 for y, we get a true sentence:

$$\begin{array}{rcl} 3x^2 + y & = & 5 \\ \hline 3(-1)^2 + 2 & \overset{?}{} & 5 \\ 3 + 2 & & \\ 5 & & \text{\scriptsize TRUE} \end{array}$$

> **Definition**
>
> The **graph** of an equation is a drawing that represents all ordered pairs that are solutions of the equation.

We obtain the graph of an equation by plotting enough ordered pairs (that are solutions) to see a pattern. The graph could be a line, a curve (or curves), or some other configuration.

Example 1   Graph: $y = 2x + 1$.

Solution   We first find some ordered pairs that are solutions and arrange them in a table. To find an ordered pair, we can choose *any* number for x and then determine y. For example, if we choose -2 for x, then $y = 2(-2) + 1 = -4 + 1 = -3$. We substituted -2 for x in the equation $y = 2x + 1$. For balance, we make some negative choices for x, as well as some positive choices. If a number takes us off the graph paper, we usually omit the pair from the graph.

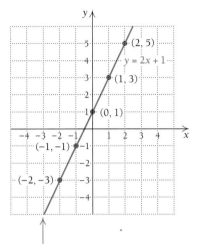

x	y	(x, y)
-2	-3	$(-2, -3)$
-1	-1	$(-1, -1)$
0	1	$(0, 1)$
1	3	$(1, 3)$
2	5	$(2, 5)$

(1)   Choose any x.
(2)   Compute y.
(3)   Form the pair (x, y).
(4)   Plot the points.

After we plot the points, we look for a pattern in the graph. If we had enough points, they would make a solid line. We can draw the line with a ruler and label it $y = 2x + 1$.   ◁

Example 2   Graph: $3x + 5y = 10$.

Solution   We could choose x-values, substitute, and solve for y-values, but we first solve for y to ease the arithmetic calculations.*

$$3x + 5y = 10$$

$$3x + 5y - 3x = 10 - 3x \qquad \text{Subtracting } 3x$$

$$5y = 10 - 3x \qquad \text{Simplifying}$$

$$\tfrac{1}{5} \cdot 5y = \tfrac{1}{5} \cdot (10 - 3x) \qquad \text{Multiplying by } \tfrac{1}{5}, \text{ or dividing by 5}$$

$$y = \tfrac{1}{5} \cdot (10) - \tfrac{1}{5} \cdot (3x) \qquad \text{Using the distributive law}$$

$$= 2 - \tfrac{3}{5}x$$

$$= -\tfrac{3}{5}x + 2$$

---

*We need the skill of solving for y when we use a grapher.

We use the equation $y = -\frac{3}{5}x + 2$ to find three ordered pairs, choosing multiples of 5 for x to avoid fractions.

x	y	(x, y)
0	2	$(0, 2)$
5	−1	$(5, -1)$
−5	5	$(-5, 5)$

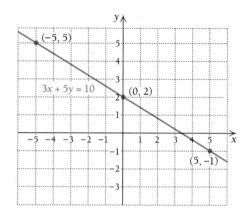

We plot the points, draw the line, and label the graph as shown.   ◄

Graphs and properties of linear equations (graphs that are straight lines) will be considered in greater detail in Section 1.4.

Example 3   Graph: $y = x^2 - 1$.

Solution

x	y	(x, y)
−2	3	$(-2, 3)$
−1	0	$(-1, 0)$
0	−1	$(0, -1)$
1	0	$(1, 0)$
2	3	$(2, 3)$

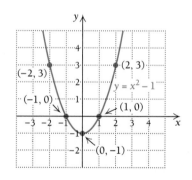

This time the pattern of the points is a curve called a *parabola*. We fill in the pattern and obtain the graph. Note that we must plot enough points to see a pattern.   ◄

Example 4   Graph: $x = y^2$.

Solution   In this case, x is expressed in terms of the variable y. Thus we first choose numbers for y and then compute x.

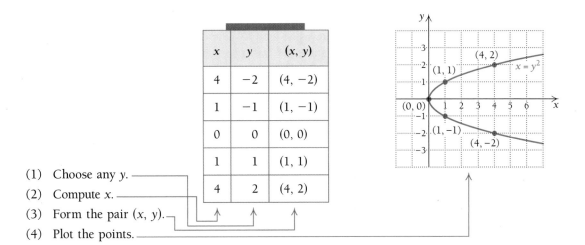

x	y	(x, y)
4	−2	(4, −2)
1	−1	(1, −1)
0	0	(0, 0)
1	1	(1, 1)
4	2	(4, 2)

(1)   Choose any *y*.
(2)   Compute *x*.
(3)   Form the pair (*x*, *y*).
(4)   Plot the points.

We plot these points, keeping in mind that *x* is still the first coordinate and *y* the second. We look for a pattern and complete the graph.

TECHNOLOGY   CONNECTION

Introduction to the Use of a Graphing Calculator: Windows and Graphs

Viewing Windows

With this first of the optional Technology Connections, we begin to create graphs using a graphing calculator and computer graphing software, referred to simply as **graphers.** Most of the coverage will refer to a TI-83 graphing calculator (and with adaptation to a TI-82) but in a somewhat generic manner, discussing features common to virtually all graphers. Although some reference to keystrokes will be mentioned, in general, exact details on keystrokes will be covered in the manual for your particular grapher or in the Graphing Calculator Manual (henceforth referred to as the GCM) that accompanies this text.

One feature common to all graphers is the **viewing window.** This refers to the rectangular screen in which a graph appears. Windows are described by four numbers, [**L, R, B, T**], which represent the **L**eft and **R**ight endpoints of the *x*-axis and the **B**ottom and **T**op endpoints of the *y*-axis. A WINDOW feature can be used to set these dimensions. Below is a window setting of [−20, 20, −5, 5] with axis scaling denoted as

Xscl = 5 and Yscl = 1, which means that there are 5 units between tick marks on the *x*-axis and 1 unit between tick marks on the *y*-axis. Graphs are made up of black rectangular dots called **pixels.** Roughly speaking, the notation Xres = 1 is an indicator of the number of pixels used in making a graph.* We will usually leave it at 1 and not refer to it unless needed.

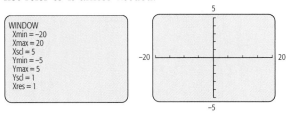

Axis scaling must be chosen with care, because tick marks become blurred and indistinguishable when too many appear. On some graphers, a setting of [−10, 10, −10, 10], Xscl = 1, Yscl = 1, Xres = 1 is considered **standard.**

---

*Xres sets pixel resolution at 1 through 8 for graphs of equations. At Xres = 1, equations are evaluated and graphed at each pixel on the *x*-axis. At Xres = 8, equations are evaluated and graphed at every eighth pixel on the *x*-axis. The resolution is better for smaller Xres values than for larger values.

Graphs

The primary use for a grapher is to graph equations. For example, let's graph the equation $y = x^3 - 5x + 1$. The equation can be entered using the notation $y = x \wedge 3 - 5x + 1$. Some software uses Basic notation, in which case the equation might be entered as $y = x \wedge 3 - 5 * x + 1$. We obtain the following graph in the standard viewing window.

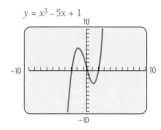

It is often necessary to change viewing windows in order to best reveal the curvature of a graph. For example, each of the following is a graph of $y = 3x^5 - 20x^3$, but with a different viewing window. Which do you think best displays the curvature of the graph?

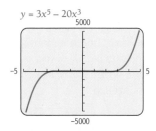

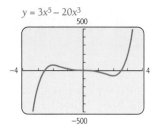

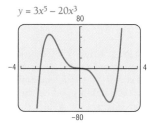

In general, choosing a window that best reveals a graph's characteristics involves some trial and error and, in some cases, some knowledge about the shape of that graph. We will learn more about the shape of graphs as we continue through the text.

To graph an equation like $3x + 5y = 10$, most graphers require that the equation be solved for y, that is, "$y = ...$" Thus we must rewrite and enter the equation as

$$y = \frac{-3x + 10}{5}, \quad \text{or} \quad y = -\frac{3}{5}x + 2.$$

(See Example 2.) Its graph is shown below in the standard window.

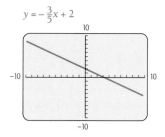

For the equation $x = y^2$, we would first obtain $y = \pm\sqrt{x}$ and then graph the individual equations $y_1 = \sqrt{x}$ and $y_2 = -\sqrt{x}$.

Exercises
Use a grapher to graph each of the following equations. Select the standard window $[-10, 10, -10, 10]$, with axis scaling Xscl $= 1$ and Yscl $= 1$.

1. $y = 2x - 2$ 2. $y = -3x + 1$
3. $y = \frac{2}{5}x + 4$ 4. $y = -\frac{3}{5}x - 1$
5. $y = 2.085x + 15.08$ 6. $y = -\frac{4}{5}x + \frac{13}{7}$
7. $2x - 3y = 18$ 8. $5y + 3x = 4$
9. $y = x^2$ 10. $y = (x + 4)^2$
11. $y = 8 - x^2$ 12. $y = 4 - 3x - x^2$
13. $y + 10 = 5x^2 - 3x$ 14. $y - 2 = x^3$
15. $y = x^3 - 7x - 2$ 16. $y = x^4 - 3x^2 + x$
17. $y = |x|$ (On most graphers, this is entered as $y = \text{abs}(x)$.)
18. $y = |x - 5|$
19. $y = |x| - 5$
20. $y = 9 - |x|$

Mathematical Models

When the essential parts of a problem are described in mathematical language, we say that we have a **mathematical model.** For example, the arithmetic of the natural numbers constitutes a mathematical model for situations in which counting is the essential ingredient. Situations in which calculus can be brought to bear often require the use of equations and functions (Section 1.2), and typically there is concern with the way a change in one variable affects a change in another.

Mathematical models are abstracted from real-world situations (see the figure below). Procedures within the mathematical model then give results that allow one to predict what will happen in that real-world situation. To the extent that these predictions are inaccurate or the results of experimentation do not conform to the model, the model is in need of modification.

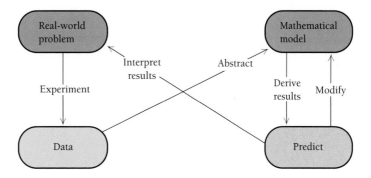

Equations as Mathematical Models

Mathematical modeling is an ongoing, possibly everchanging, process. For example, finding a mathematical model that will enable accurate prediction of population growth is not a simple problem. Surely any population model one might devise will need to be altered as further relevant information is acquired.

Although models can reveal worthwhile information, one must always be cautious when using them. An interesting case in point is a study* showing that world records in *any* running race can be modeled by a linear function. In particular, for the mile run,

$$R = -0.00582x + 15.3476,$$

where R is the world record, in minutes, and x is the year. Roger Bannister shocked the world in 1954 by breaking the 4-minute mile. Had people been aware of this model, they would not have been shocked, for when we substitute 1954 for x, we get

$$R = -0.00582(1954) + 15.3476$$
$$= 3.97532$$
$$\approx 3{:}58.5.$$

---

*Ryder, H. W., H. J. Carr, and P. Herget, "Future Performance in Footracing," *Scientific American* **234** (June 1976): 109–119.

The actual record was 3:59.4. Although this model will continue for 40 to 50 years to be worthwhile in predicting the world record in the mile run, we see that we cannot get meaningful answers to some questions. For example, we could use the model to find when the 1-minute mile will be broken. We set $R = 1$ and solve for x:

$$1 = -0.00582x + 15.3476$$
$$2465 = x.$$

Most track athletes would assure us that the 1-minute mile is beyond human capability. In fact, at the time of this writing, experienced runners think the record will never reach 3:40.0, the current world record being 3:44.39. Going to an even further extreme, we see that the model predicts that the 0-minute mile will be run in 2637. In conclusion, one must be careful in the use of any model. (You will develop this model in Exercise Set 7.5.)

Compound Interest

One model that we will use frequently throughout this course involves **compound interest.** Suppose that we invest P dollars at interest rate i, compounded annually. The amount A_1 in the account at the end of one year is given by

$$A_1 = P + Pi$$
$$= P(1 + i) = Pr,$$

where, for convenience, we let

$$r = 1 + i.$$

Going into the second year, we have Pr dollars, so by the end of the second year, we would have the amount A_2 given by

$$A_2 = A_1 \cdot r = (Pr)r$$
$$= Pr^2 = P(1 + i)^2.$$

Going into the third year, we have Pr^2 dollars, so by the end of the third year, we would have the amount A_3 given by

$$A_3 = A_2 \cdot r = (Pr^2)r$$
$$= Pr^3 = P(1 + i)^3.$$

In general, we have the following.

Theorem 1

If an amount P is invested at interest rate i, compounded annually, in t years it will grow to the amount A given by

$$A = P(1 + i)^t.$$

Example 5 Business: Compound Interest. Suppose that $1000 is invested at 8%, compounded annually. How much is in the account at the end of 2 yr?

Solution We substitute 1000 for P, 0.08 for i, and 2 for t into the equation $A = P(1 + i)^t$ and get

$$A = 1000(1 + 0.08)^2$$
$$= 1000(1.08)^2$$
$$= 1000(1.1664)$$
$$= \$1166.40.$$

There is \$1166.40 in the account after 2 yr.

For interest that is compounded quarterly, we can find a formula like the one above, as illustrated in the following diagram.

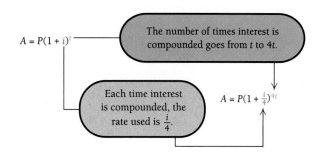

In general, the following theorem applies.

> **Theorem 2**
>
> If a principal P is invested at interest rate i, compounded n times a year, in t years it will grow to an amount A given by
>
> $$A = P\left(1 + \frac{i}{n}\right)^{nt}.$$

Example 6 *Business: Compound Interest.* Suppose that \$1000 is invested at 8%, compounded quarterly. How much is in the account at the end of 2 yr?

Solution We use the equation $A = P(1 + i/n)^{nt}$, substituting 1000 for P, 0.08 for i, 4 for n (compounding quarterly), and 2 for t. Then we get

$$A = 1000\left(1 + \frac{0.08}{4}\right)^{4 \times 2}$$
$$= 1000(1 + 0.02)^8$$
$$= 1000(1.02)^8$$
$$= 1000(1.171659381) \qquad \text{Using a calculator to approximate } (1.02)^8$$
$$= 1171.659381$$
$$\approx \$1171.66.$$

There is \$1171.66 in the account after 2 years.

A calculator with a $\boxed{y^x}$ or a $\boxed{\wedge}$ key and a ten-digit readout was used to find $(1.02)^8$ in Example 6. The number of places on a calculator may affect the accuracy of the answer. Thus you may occasionally find that your answers do not agree with those at the back of the book, which were found on a calculator with a ten-digit readout. In general, when using a calculator, do all your computations, and round only at the end, as in Example 6. Usually, your answer will agree to at least four digits. It might be wise to consult with your instructor on the accuracy required.

Graphs as Mathematical Models

Graphs can prove very useful as mathematical models, as shown in the following example.

Example 7   *Life Science: Incidence of Breast Cancer.* The following graph approximates the incidence of breast cancer y, per 100,000 women, as a function of age x, where x represents ages 25 to 100. No equation is given. Use the graph to answer the following.

a) What is the incidence of breast cancer in women of age 40?

b) For what ages is the incidence of breast cancer about 300 per 100,000 women?

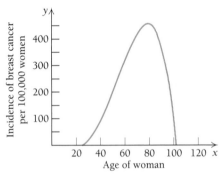

Source: National Cancer Institute

Solution

a) To estimate the incidence of breast cancer in women of age 40, we locate 40 on the horizontal axis and go directly up until we reach the graph (see the figure on the left below). Then we move left until we reach the value shown on the vertical axis. We estimate that value to be about 90.

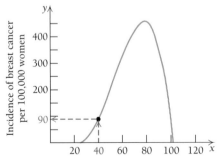

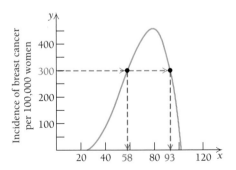

Source: National Cancer Institute

b) We locate 300 on the vertical axis and move horizontally across the graph, where we note that two x-values correspond to 300 (see the figure on the right above). They are ages 58 and 93, approximately.

EXERCISE SET 1.1

Exercises designated by the symbol **tw** are Thinking and Writing Exercises. They should be answered using one or two English sentences. Because answers to many such exercises will vary, solutions are not given at the back of the book.

Graph.

1. $y = x - 1$
2. $y = x$
3. $y = -3x$
4. $y = -\frac{1}{4}x$
5. $y = \frac{2}{3}x - 4$
6. $y = -\frac{5}{3}x - 2$
7. $x + y = -4$
8. $x - y = 5$
9. $8y - 2x = 4$
10. $6x + 3y = -9$
11. $y = 3 - x^2$
12. $y = x^2 - 3$
13. $x = 2 - y^2$
14. $x = y^2 + 2$
15. $y = |x|$
16. $y = |4 - x|$

Applications

BUSINESS AND ECONOMICS

17. *Compound Interest.* Suppose that $1000 is invested at 6%. How much is in the account at the end of 1 yr, if interest is compounded:
    a) annually?
    b) semiannually?
    c) quarterly?
    d) daily?
    e) hourly?

18. *Compound Interest.* Suppose that $1000 is invested at $8\frac{1}{2}$%. How much is in the account at the end of

1 yr, if interest is compounded:
    a) annually?   b) semiannually?
    c) quarterly?   d) daily?
    e) hourly?

Determining Monthly Payments on a Loan. If P dollars are borrowed, the monthly payment M, made at the end of each month for n months, is given by

$$M = P\left[\frac{\dfrac{i}{12}\left(1 + \dfrac{i}{12}\right)^n}{\left(1 + \dfrac{i}{12}\right)^n - 1}\right],$$

where i is the annual interest rate and n is the total number of monthly payments.

19. A car loan is $18,000, the interest rate is $9\frac{3}{4}$%, and the loan period is 3 yr. What is the monthly payment?

20. The mortgage on a house is $100,000, the interest rate is $7\frac{1}{2}$%, and the loan period is 30 yr. What is the monthly payment?

LIFE AND PHYSICAL SCIENCES

21. *Incidence of Breast Cancer.* Refer to the graph shown in Example 7.
    a) What is the incidence of breast cancer in women of age 60?
    b) For what ages is the incidence of breast cancer about 400 per 100,000 women?
    c) Examine the graph and try to determine the age at which the largest incidence of breast cancer occurs.
    tw d) What difficulties do you have making this determination?

22. *Hearing-Impaired Americans.* The number N, in millions, of hearing-impaired Americans of age x can be approximated by the graph shown below. (***Source:*** American Speech-Language Hearing Association).

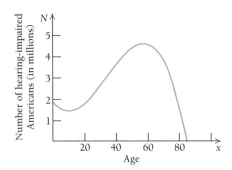

Use the graph to answer the following.

a) Approximate the number of hearing-impaired Americans of ages 20, 40, 50, and 60.

b) For what ages is the number of hearing-impaired Americans approximately 4 million?

c) Examine the graph and try to determine the age at which the greatest number of Americans is hearing-impaired.

d) What difficulties do you have in making this determination?

 TECHNOLOGY   CONNECTION

Graph.

23. $y = x + 200$

24. $y = 9 - |x|$

25. $y = x^3 + 2x^2 - 4x - 13$

26. $y = \sqrt{23 - 7x}$

27. $9.6x + 4.2y = -100$

28. $y = -2.3x^2 + 4.8x - 9$

29. $x = 4 + y^2$

30. $x = 8 - y^2$

1.2

Functions and Models

Identifying Functions

The idea of a *function* is one of the most important concepts in mathematics. In much the same way that ordered pairs form correspondences between first and second coordinates, a function is a special kind of correspondence from one set to another. Let's look at the following.

To each employee in a company there corresponds his or her salary.

To each CD in a music store there corresponds its price.

To each real number there corresponds the cube of that number.

In each of these examples, the first set is called the **domain** and the second set is called the **range.** Given a member of the domain, there is *exactly one* member of the range to which it corresponds. This type of correspondence is called a **function.**

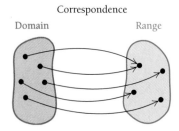

> **Definition**
> A **function** is a correspondence between a first set, called the **domain,** and a second set, called the **range,** such that each member of the domain corresponds to *exactly one* member of the range.

Example 1 Determine whether or not the correspondence is a function.

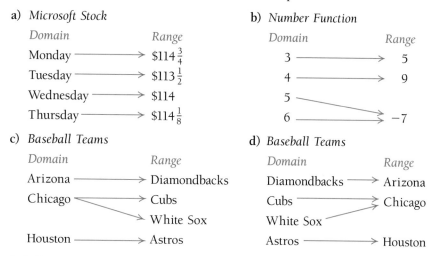

a) *Microsoft Stock*

Domain	Range
Monday	$114\frac{3}{4}$
Tuesday	$113\frac{1}{2}$
Wednesday	$114
Thursday	$114\frac{1}{8}$

b) *Number Function*

Domain	Range
3	5
4	9
5	
6	−7

c) *Baseball Teams*

Domain	Range
Arizona	Diamondbacks
Chicago	Cubs
	White Sox
Houston	Astros

d) *Baseball Teams*

Domain	Range
Diamondbacks	Arizona
Cubs	Chicago
White Sox	
Astros	Houston

Solution

a) The correspondence is a function because each member of the domain corresponds (is matched) to only one member of the range.

b) The correspondence is a function because each member of the domain corresponds to only one member of the range, even though two members of the domain correspond to −7.

c) The correspondence is not a function because one member of the domain, Chicago, corresponds to two members of the range, the Cubs and the White Sox.

d) The correspondence is a function because each member of the domain corresponds to only one member of the range, even though two members of the domain correspond to Chicago.   ◄

Example 2 Determine whether or not the correspondence is a function.

	Domain	Correspondence	Range
a)	A family	Each person's weight	A set of positive numbers
b)	The integers	Each number's square	A set of all nonnegative integers
c)	The set of all states	Each state's members of the U.S. Senate	The set of all U.S. senators

Solution

a) The correspondence *is* a function because each person has *only one* weight.

b) The correspondence *is* a function because each integer has *only one* square.

c) The correspondence *is not* a function because each state has two U.S. Senators.

When a correspondence between two sets is not a function, it is still an example of a **relation.**

> **Definition**
>
> A **relation** is a correspondence between a first set, called the **domain,** and a second set, called the **range,** such that each member of the domain corresponds to *at least one* member of the range.

Thus, although the correspondences of Examples 1 and 2 are not all functions, they *are* all relations. A function is a special type of relation—one in which each member of the domain is paired with *exactly one* member of the range.

Finding Function Values

Most functions considered in mathematics are described by equations like $y = 2x + 3$ or $y = 4 - x^2$. For example, we graph the function $y = 2x + 3$ by first performing calculations like the following:

for $x = 4$, $y = 2x + 3 = 2 \cdot 4 + 3 = 11$;

for $x = -5$, $y = 2x + 3 = 2 \cdot (-5) + 3 = -7$;

for $x = 0$, $y = 2x + 3 = 2 \cdot 0 + 3 = 3$; and so on.

For $y = 2x + 3$, the **inputs** (members of the domain) are values of x substituted into the equation. The **outputs** (members of the range) are the resulting values of y. If we call the function f, we can use x to represent an arbitrary *input* and $f(x)$ — read "f of x," or "f at x," or "the value of f at x"—to represent the corresponding *output*. In this notation, the function given by $y = 2x + 3$ is written as $f(x) = 2x + 3$ and the calculations above can be written more concisely as

$f(4) = 2 \cdot 4 + 3 = 11$;

$f(-5) = 2 \cdot (-5) + 3 = -7$;

$f(0) = 2 \cdot 0 + 3 = 3$; and so on.

Thus, instead of writing "when $x = 4$, the value of y is 11," we can simply write "$f(4) = 11$," which can also be read as "f of 4 is 11" or "for the input 4, the output of f is 11."

It helps to think of a function as a machine. Think of $f(4) = 11$ as putting a member of the domain (an input), 4, into the machine. The machine knows the correspondence $f(x) = 2x + 3$, multiplies 4 by 2 and adds 3, and gives out a member of the range (the output), 11.

Function: $f(x) = 2x + 3$	
Input	Output
4	11
-5	-7
0	3
a	$2a + 3$
$a + h$	$2(a + h) + 3$

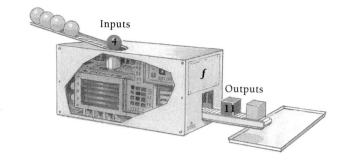

Be careful to note that the notation $f(x)$ *does not mean* "f times x" and should not be read that way.

Example 3 The squaring function f is given by

$$f(x) = x^2.$$

Find $f(-3), f(1), f(k), f(\sqrt{k}), f(1 + t)$, and $f(x + h)$.

Solution We have

$$f(-3) = (-3)^2 = 9;$$
$$f(1) = 1^2 = 1;$$
$$f(k) = k^2;$$
$$f(\sqrt{k}) = (\sqrt{k})^2 = k;$$
$$f(1 + t) = (1 + t)^2 = 1 + 2t + t^2;$$
$$f(x + h) = (x + h)^2 = x^2 + 2xh + h^2.$$

To find $f(x + h)$, remember what the function does: It squares the input. Thus, $f(x + h) = (x + h)^2 = x^2 + 2xh + h^2$. This amounts to replacing x on both sides of $f(x) = x^2$ with $x + h$. ◁

Example 4 A function f is given by $f(x) = 3x^2 - 2x + 8$. Find $f(0), f(1), f(-5)$, and $f(7a)$.

Solution One way to find function values when a formula is given is to think of the formula with blanks, or placeholders, as follows:

$$f(\blacksquare) = 3\blacksquare^2 - 2\blacksquare + 8.$$

To find an output for a given input, we think: "Whatever goes in the blank on the left goes in the blank(s) on the right." With this in mind, let's complete the example.

$$f(0) = 3 \cdot 0^2 - 2 \cdot 0 + 8 = 8;$$
$$f(1) = 3 \cdot 1^2 - 2 \cdot 1 + 8 = 3 \cdot 1 - 2 + 8 = 3 - 2 + 8 = 9;$$
$$f(-5) = 3(-5)^2 - 2 \cdot (-5) + 8 = 3 \cdot 25 + 10 + 8 = 75 + 10 + 8 = 93;$$
$$f(7a) = 3(7a)^2 - 2(7a) + 8 = 3 \cdot 49a^2 - 14a + 8 = 147a^2 - 14a + 8 \quad ◁$$

T E C H N O L O G Y   C O N N E C T I O N

The TABLE Feature

One way to find ordered pairs of inputs and outputs of functions makes use of the **TABLE** feature on a grapher. Let's consider the equation or function $f(x) = x^3 - 5x + 1$. We enter it into the grapher as $y_1 = x^3 - 5x + 1$. To use the **TABLE** feature, we access the **TABLE SETUP** screen and choose the x-value at which the table will start and an increment for the x-value. For this equation, let's set TblStart = 0.3 and ΔTbl = 1. (Other values can be chosen.) This means that the table's x-values will start at 0.3 and increase by 1.

```
TABLE  SETUP
  TblStart = .3
  ΔTbl = 1 ■
  Indpnt:  Auto  Ask
  Depend:  Auto  Ask
```

We also set Indpnt and Depend to Auto. The table that results when accessing the **TABLE** feature is shown below.

X	Y1	
.3	−.473	
1.3	−3.303	
2.3	1.667	
3.3	20.437	
4.3	59.007	
5.3	123.38	
6.3	219.55	
X = .3		

The arrow keys allow us to scroll up and down the table and extend it to other values not initially shown.

X	Y1	
12.3	1800.4	
13.3	2287.1	
14.3	2853.7	
15.3	3506.1	
16.3	4250.2	
17.3	5092.2	
18.3	6038	
X = 18.3		

Exercises

Use the function $f(x) = x^3 - 5x + 1$ for Exercises 1 and 2.

1. Use the **TABLE** feature to construct a table starting with $x = 10$ and ΔTbl = 5. Find the value of y when x is 10. Then find the value of y when x is 35.

2. Adjust the table settings to Indpnt: Ask. How does the table change? Enter a number of your choice and see what happens. Use this setting to find the value of y when x is 28.

Example 5   A function f subtracts the square of an input from the input. A description of f is given by

$$f(x) = x - x^2.$$

Find $f(4)$ and $f(x + h)$.

Solution   We replace the x's on both sides with the inputs. Thus,

$$f(4) = 4 - 4^2 = 4 - 16 = -12;$$
$$f(x + h) = (x + h) - (x + h)^2$$
$$= x + h - (x^2 + 2xh + h^2)$$
$$= x + h - x^2 - 2xh - h^2.$$

Consider the reciprocal function $f(x) = 1/(x - 3)$. A table for this function is shown below. Note that $\Delta\text{Tbl} = 0.25$.

X	Y1
2	−1
2.25	−1.333
2.5	−2
2.75	−4
3	ERROR
3.25	4
3.5	2
3.75	1.3333
4	1

Exercises

1. Make a table for the function $f(x) = 1/(x^2 - 4)$ from $x = -3$ to $x = 3$ and with $\Delta\text{Tbl} = 0.5$.
2. Create an input–output table for the function given in Example 6.

When a function is given by a formula, and nothing is said about the domain, its domain is understood to be the set of all numbers that can be substituted into the formula. For example, consider the reciprocal function

$$f(x) = \frac{1}{x}.$$

The only number that cannot be substituted into the formula is 0. We say that f *is not defined at* 0, or $f(0)$ *does not exist*. The domain consists of all nonzero real numbers.

Taking square roots *is not* a function, because an input can have more than one output. For example, the input 4 has two outputs, 2 and −2. Taking principal square roots (nonnegative roots) is a function. Let g be this function. Then g can be described as $g(x) = \sqrt{x}$. Recall from algebra that the symbol $\sqrt{a}$ represents the nonnegative square root of a for $a \geq 0$. There is only one such real-number root.

Example 6   Consider the principal square-root function
$$g(x) = \sqrt{x}.$$

a) Find the domain of this function.

b) Find $g(0)$, $g(2)$, $g(a)$, $g(16)$, $g(t + h)$, and $g(-3)$.

Solution

a) The domain consists of numbers that can be substituted into the formula. We can take the principal square root of any nonnegative number. The principal square root of a negative number is not a real number. (Taking square roots of negative numbers would require us to consider complex numbers, which we will not cover in this text.) Thus the domain consists of all nonnegative numbers.

b) We have
$$g(0) = \sqrt{0} = 0;$$
$$g(2) = \sqrt{2};$$
$$g(a) = \sqrt{a};$$
$$g(16) = \sqrt{16} = 4;$$
$$g(t + h) = \sqrt{t + h};$$
$$g(-3) \text{ does not exist as a real number.}$$

Functions are often implicit (the variable is not isolated) in certain equations. For example, consider
$$xy = 2.$$

For any nonzero x, there is a unique number y that satisfies the equation. This yields a function that is given explicitly by

$$y = f(x) = \frac{2}{x}.$$

On the other hand, consider the equation $x = y^2$. A positive number x would be related to two values of y, namely $\sqrt{x}$ and $-\sqrt{x}$. Thus this equation is not an implicit description of a function that maps inputs x to outputs y.

Graphs of Functions

Consider again the squaring function. The input 3 is associated with the output 9. The input–output pair $(3, 9)$ is one point on the *graph* of this function.

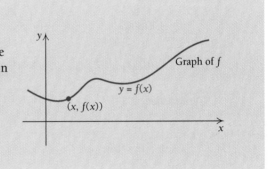

Definition

The **graph** of a function f is a drawing that represents all the input–output pairs $(x, f(x))$. In cases where the function is given by an equation, the graph of a function is the graph of the equation $y = f(x)$.

It is customary to locate input values (the domain) on the horizontal axis and output values (the range) on the vertical axis.

Example 7   Graph: $f(x) = x^2 - 1$.

Solution

x	$f(x)$	$(x, f(x))$
-2	3	$(-2, 3)$
-1	0	$(-1, 0)$
0	-1	$(0, -1)$
1	0	$(1, 0)$
2	3	$(2, 3)$

(1) Choose any x.
(2) Compute y.
(3) Form the pair (x, y).
(4) Plot the points.

We plot the input–output pairs from the table and, in this case, draw a curve to complete the graph.

TECHNOLOGY CONNECTION

Graphs and Function Values

We discussed graphing equations on a grapher in the Technology Connection of Section 1.1. Graphing a function makes use of the same procedure. We just change the "$f(x) =$" notation to "$y =$".

Consider the function $f(x) = 2x^2 + x$. We enter it into the grapher as $y_1 = 2x^2 + x$ and graph it in the standard viewing window.

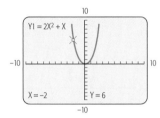

There are at least three ways in which to find function values on a grapher. The first is to use the TABLE feature, which we have already discussed. The second is to use the VALUE feature. We access the CALC menu and choose VALUE. To evaluate the function $f(x) = 2x^2 + x$ at $x = -2$ — that is, to find $f(-2)$ — we enter -2 for x. The function value, or y-value, $y = 6$, appears together with a trace indicator showing the point $(-2, 6)$.

Function values can also be found using the Y-VARS feature. Consult the manual for your particular grapher or the GCM for details.

Exercises

1. Graph $f(x) = x^2 + 3x - 4$. Then find $f(-5)$, $f(-4.7)$, $f(11)$, and $f(2/3)$. (*Hint*: To find $f(11)$, be sure that the window dimensions for the x-values include $x = 11$.)
2. Graph $f(x) = 3.7 - x^2$. Then find $f(-5)$, $f(-4.7)$, $f(11)$, and $f(2/3)$.
3. Graph $f(x) = 4 - 1.2x - 3.4x^2$. Then find $f(-5)$, $f(-4.7)$, $f(11)$, and $f(2/3)$.

The Vertical-Line Test

Let's now determine how we can look at a graph and decide whether it is a graph of a function. We already know that

$$x = y^2$$

does not yield a function that maps a number x to a unique number y. In its graph, which follows, note that the point x_1 has two outputs. This means that there is a vertical line that intersects the graph in more than one place.

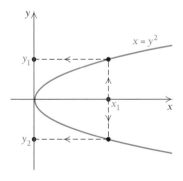

The Vertical-Line Test

A graph represents a function if it is impossible to draw a vertical line that intersects the graph more than once.

Example 8   Determine whether each of the following is the graph of a function.

a)

b)

c)
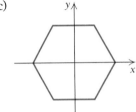

d)
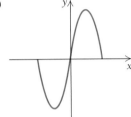

Solution

a) The graph is that of a function. It is impossible to draw a vertical line that intersects the graph more than once.

b) The graph is not that of a function. A vertical line (in fact, many) intersects the graph more than once.

c) The graph is not that of a function.

d) The graph is that of a function.

Functions Defined Piecewise

Sometimes functions are defined piecewise. That is, there are different output formulas for different parts of the domain.

Example 9   Graph the function defined as follows.

$$f(x) = \begin{cases} 4, & \text{for } x \leq 0 \\ & \text{(This means that for any input } x \text{ less than or equal} \\ & \text{to 0, the output is 4.)} \\ 4 - x^2, & \text{for } 0 < x \leq 2 \\ & \text{(This means that for any input } x \text{ greater than 0 and} \\ & \text{less than or equal to 2, the output is } 4 - x^2.) \\ 2x - 6, & \text{for } x > 2 \\ & \text{(This means that for any input } x \text{ greater than 2, the} \\ & \text{output is } 2x - 6.) \end{cases}$$

TECHNOLOGY CONNECTION

Graphing Functions Defined Piecewise

Graphing functions defined piecewise on a grapher generally involves the use of inequality symbols. On the TI-83, the **TEST** menu is used. The function in Example 9 is entered as follows:

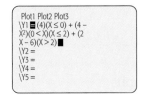

```
Plot1 Plot2 Plot3
\Y1 ■ (4)(X ≤ 0) + (4 −
X²)(0 < X)(X ≤ 2) + (2
X − 6)(X > 2)■
\Y2 =
\Y3 =
\Y4 =
\Y5 =
```

The graph is shown below in **DOT** mode.

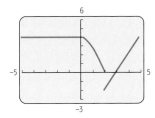

Exercises
Graph.

1. $f(x) = \begin{cases} -x - 2, & \text{for } x < -2, \\ \sqrt{4 - x^2}, & \text{for } -2 \le x < 2 \\ x + 3, & \text{for } x \ge 2 \end{cases}$

2. $f(x) = \begin{cases} x^2 - 2, & \text{for } x \le 3 \\ 1, & \text{for } x > 3 \end{cases}$

3. $f(x) = \begin{cases} x + 3, & \text{for } x \le -2, \\ 1, & \text{for } -2 < x \le 3, \\ x^2 - 10, & \text{for } x > 3 \end{cases}$

Solution   The graph of this function follows.

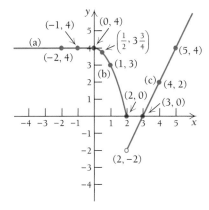

a) We graph $f(x) = 4$ for inputs less than or equal to 0 (that is, $x \le 0$).

For $f(x) = 4$,

$$f(-2) = 4,$$
$$f(-1) = 4, \quad \text{and}$$
$$f(0) = 4.$$

x	$f(x)$	$(x, f(x))$
-2	4	$(-2, 4)$
-1	4	$(-1, 4)$
0	4	$(0, 4)$

b) We graph $f(x) = 4 - x^2$ for inputs greater than 0 and less than or equal to 2 (that is, $0 < x \le 2$).

For $f(x) = 4 - x^2$,

$$f\left(\tfrac{1}{2}\right) = 4 - \left(\tfrac{1}{2}\right)^2 = 3\tfrac{3}{4},$$
$$f(1) = 4 - 1^2 = 3, \quad \text{and}$$
$$f(2) = 4 - 2^2 = 0.$$

x	$f(x)$	$(x, f(x))$
$\frac{1}{2}$	$3\frac{3}{4}$	$\left(\frac{1}{2}, 3\frac{3}{4}\right)$
1	3	$(1, 3)$
2	0	$(2, 0)$

The solid circle indicates that the point $(2, 0)$ is part of the graph.

c) We graph $f(x) = 2x - 6$ for inputs greater than 2 (that is, $x > 2$).

For $f(x) = 2x - 6$,

$$f(3) = 2(3) - 6 = 0,$$
$$f(4) = 2(4) - 6 = 2, \text{ and}$$
$$f(5) = 2(5) - 6 = 4.$$

x	$f(x)$	$(x, f(x))$
3	0	$(3, 0)$
4	2	$(4, 2)$
5	4	$(5, 4)$

The open circle at $(2, -2)$ indicates that the point is not part of the graph. ≺

Applications of Functions and Their Graphs

Regardless of whether an equation is given, functions are often described by graphs. Note that each point on a graph represents a pair of input–output values.

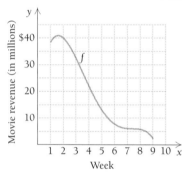

Example 10 *Business: Movie Revenue.* The graph at left approximates the weekly revenue, in millions of dollars, from the movie *Air Force One* (**Source:** Exhibitor Relations Co., Inc.). The revenue is a function f of the number of weeks since the movie was released. No equation is given for the function (we will consider it later).

Use the graph to answer the following.

a) What was the movie revenue for week 1? That is, find $f(1)$.

b) In what week was the movie revenue approximately \$13 million? That is, find any inputs x for which $f(x) = 13$.

Solution

a) To estimate the revenue for week 1, we locate 1 on the horizontal axis and move directly up until we reach the graph. Then we move across to the vertical axis. We estimate that value to be about \$38 million — that is, $f(1) = 38$.

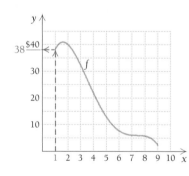 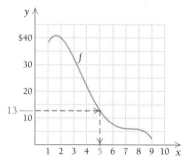

b) We locate 13 on the vertical axis and move horizontally across to the graph and note that one input corresponds to 13. We estimate that value to be about 5 — that is, $f(5) = 13$. The revenue is approximately \$13 million in the fifth week.

≺

TECHNOLOGY CONNECTION

The TRACE Feature

The TRACE feature can be used to determine the coordinates of points on a graph. When this feature is activated, a cursor appears on the line (or curve) that has been graphed and the coordinates at that point are displayed. Using the left and right arrow keys moves the cursor along the graph.

Let's consider the function $f(x) = x^3 - 5x + 1$ graphed in the window $[-5, 5, -10, 10]$. We select TRACE.

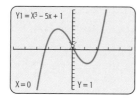

The coordinates at the bottom indicate that the cursor is at the point with coordinates $(0, 1)$. By using the left and right arrow keys, we can obtain coordinates of other points. For example, if we press the left arrow key ◁ seven times, we move the cursor to the location shown below, obtaining a point on the graph with coordinates $(-0.7446809, 4.3104418)$.

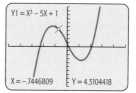

Exercises

1. Use the TRACE feature to find five different ordered-pair solutions of the equation $f(x) = x^3 - 5x + 1$.

Some Final Remarks

Almost all the functions in this text can be described by equations. Some functions, however, cannot. For example, there will be a function that assigns grades to students in this course, but that function will most likely not have a formula.

We sometimes use the terminology *y is a function of x.* This means that *x* is an input and *y* is an output. We often refer to *x* as the **independent variable** when it represents inputs and *y* as the **dependent variable** when it represents outputs. We may refer to "a function $y = x^2$," without naming it with a letter *f*. We may simply refer to x^2 (alone) as a function.

In calculus we will be studying how outputs of a function change when the inputs change.

EXERCISE SET 1.2

Determine whether the correspondence is a function.

1. *Domain*   *Range*

2 ——→ 9
5 ——→ 8
19

2. *Domain*   *Range*

5 ——→ 3
−3 ——→ 7
7
−7

3. *Domain*   *Range*

6 ——→ −6
7 ——→ −7
3 ——→ −3

4. *Domain*   *Range*

−5 ——→ 1
5
8

5. *Sales of Sunglasses*

Domain	Range
1993	$1.9 million
1994	$2.0 million
1995	$2.3 million
1996	$2.6 million
1997	

Source: Sunglass Association of America

6. *Average Price of Adult Admission at Disney World*

Domain	Range
1994	$36.00
1995	$37.00
1996	$40.81
1997	$42.14
1998	$44.52

Source: *Amusement Business*

Determine whether each of the following is a function.

	Domain	Correspondence	Range
7.	A math class	Each person's seat number	A set of numbers
8.	A set of numbers	Square each number and then add 8.	A set of positive numbers greater than or equal to 8
9.	A set of shapes	The perimeter of each shape	A set of positive numbers
10.	A family	Each person's height	A set of positive numbers
11.	A textbook	An even-numbered page in the book	A set of pages
12.	A set of avenues	An intersecting road	A set of cross streets

13. A function f is given by

$$f(x) = 2x + 3.$$

This function takes a number x, multiplies it by 2, and adds 3.

a) Complete this table.

Input	Output
4.1	
4.01	
4.001	
4	

b) Find $f(5)$, $f(-1)$, $f(k)$, $f(1 + t)$, and $f(x + h)$.

14. A function f is given by

$$f(x) = \frac{4}{x - 3}.$$

This function takes a number x, subtracts 3, and divides 4 by the result.

a) Complete this table.

Input	Output
5.1	
5.01	
5.001	
5	

b) Find $f(4)$, $f(3)$, $f(-2)$, $f(k)$, $f(1 + t)$, and $f(x + h)$.

15. A function g is given by

$$g(x) = x^2 - 3.$$

This function takes a number x, squares it, and subtracts 3. Find $g(-1)$, $g(0)$, $g(1)$, $g(5)$, $g(u)$, $g(a + h)$, and $g(1 - h)$.

16. A function g is given by

$$g(x) = x^2 + 4.$$

This function takes a number x, squares it, and adds 4. Find $g(-3)$, $g(0)$, $g(-1)$, $g(7)$, $g(v)$, $g(a + h)$, and $g(1 - t)$.

17. A function f is given by

$$f(x) = \frac{1}{(x + 3)^2}.$$

This function takes a number x, adds 3, squares the result, and takes the reciprocal of the result.

a) Find $f(4)$, $f(-3)$, $f(0)$, $f(a)$, $f(t + 1)$, $f(t + 3)$, and $f(x + h)$.

b) Note that f could also be given by

$$f(x) = \frac{1}{x^2 + 6x + 9}.$$

Explain what this does to an input number x.

18. A function f is given by

$$f(x) = (x + 4)^2.$$

This function takes a number x, adds 4, and squares the result.

a) Find $f(3)$, $f(-6)$, $f(0)$, $f(k)$, $f(t - 1)$, $f(t - 4)$, and $f(x + h)$.

b) Note that f could also be given by

$$f(x) = x^2 + 8x + 16.$$

Explain what this does to an input number x.

Graph the function.

19. $f(x) = 2x + 3$ **20.** $f(x) = 3x - 1$

21. $g(x) = -4x$ **22.** $g(x) = -2x$

23. $f(x) = x^2 - 1$ **24.** $f(x) = x^2 + 4$

25. $g(x) = x^3$ **26.** $g(x) = \frac{1}{2}x^3$

Use the vertical-line test to determine whether the graph is that of a function. (In Exercises 35–38, the vertical dashed lines are not part of the graph.)

27.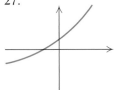

28.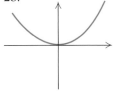

29.

30.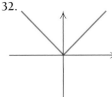

31.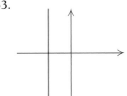

32.

33.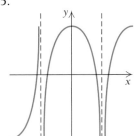

34.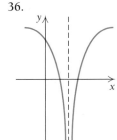

35.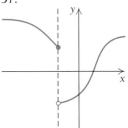

36.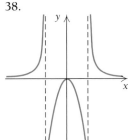

37.

38.

39. a) Graph $x = y^2 - 1$.
   b) Is this a function?

40. a) Graph $x = y^2 - 3$.
   b) Is this a function?

41. For $f(x) = x^2 - 3x$, find $f(x + h)$.

42. For $f(x) = x^2 + 4x$, find $f(x + h)$.

Graph.

43. $f(x) = \begin{cases} 1, & \text{for } x < 0, \\ -1, & \text{for } x \geq 0 \end{cases}$

44. $f(x) = \begin{cases} 2, & \text{for } x \text{ an integer}, \\ -2, & \text{for } x \text{ not an integer} \end{cases}$

45. $f(x) = \begin{cases} -3, & \text{for } x = -2, \\ x^2, & \text{for } x \neq -2 \end{cases}$

46. $f(x) = \begin{cases} -2x - 6, & \text{for } x \leq -2, \\ 2 - x^2, & \text{for } -2 < x < 2, \\ 2x - 6, & \text{for } x \geq 2 \end{cases}$

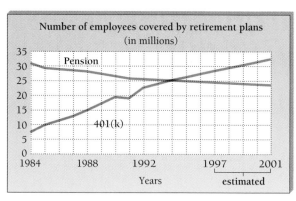

Source: Cerulli Associates

e) In what year was the number of pension plans the greatest?

f) In what year is the number of 401(k) plans estimated to be the greatest?

Applications

BUSINESS AND ECONOMICS

47. *Total Revenue.* Raggs, Ltd., a clothing firm, determines that its total revenue from the sale of x suits is given by the function

$$R(x) = 200x + 50,$$

where $R(x)$ is the revenue, in dollars, from the sale of x suits. Find $R(10)$ and $R(100)$.

48. *Compound Interest.* The amount of money in a savings account at 6%, compounded annually, depends on the initial investment x and is given by the function

$$A(x) = x + 6\%x,$$

where $A(x)$ is the amount in the account at the end of 1 yr. Find $A(100)$ and $A(1000)$.

49. *Movie Revenue.* Refer to the graph in Example 10.
   a) What was the movie revenue for week 2?
   b) In what week was the movie revenue approximately $9 million?
   c) In what week was the movie revenue approximately $30 million?

50. *Retirement Plans.* The following graph shows the number of employees, in millions, who are covered by two types of retirement plans: pension and 401(k).
   a) Is each graph that of a function?
   b) Estimate the number of pension plans in 1992 and the number estimated in 2001.
   c) Estimate the number of 401(k) plans in 1997 and the number estimated in 2001.
   d) In what year were the number of pension plans and the number of 401(k) plans the same?

LIFE AND PHYSICAL SCIENCES

51. *Scaling Stress Factors.* In psychology a process called *scaling* is used to attach numerical ratings to a group of life experiences. In the following table, various events have been rated from 1 to 100 according to their stress levels.

Event	Scale of Impact
Death of spouse	100
Divorce	73
Jail term	63
Marriage	50
Lost job	47
Pregnancy	40
Death of close friend	37
Loan over $10,000	31
Child leaving home	29
Change in schools	20
Loan less than $10,000	17
Christmas	12

Source: Thomas H. Holmes, University of Washington School of Medicine

   a) Does the table constitute a function? Why or why not?
   b) What are the inputs? What are the outputs?

Synthesis ··

Solve for *y* in terms of *x*. Decide whether the resulting equation represents a function.

52. $2x + y - 16 = 4 - 3y + 2x$

53. $2y^2 + 3x = 4x + 5$

54. $(4y^{2/3})^3 = 64x$ **55.** $(3y^{3/2})^2 = 72x$

 56. Suppose you were discussing the idea of a function with a friend. How would you explain it?

 57. Explain the special aspects of the graph of a function.

 TECHNOLOGY CONNECTION

In Exercises 58 and 59, use the **TABLE** feature to construct a table for the function under the given conditions.

58. $f(x) = x^3 + 2x^2 - 4x - 13$;
TblStart $= -3$; ΔTbl $= 2$

59. $f(x) = \dfrac{3}{x^2 - 4}$;
TblStart $= -3$; ΔTbl $= 1$

60. A function f is given by
$$f(x) = |x - 2| + |x + 1| - 5.$$
Find $f(-3)$, $f(-2)$, $f(0)$, and $f(4)$.

61. Graph the function in each of Exercises 43–46.

62. Use the **TRACE** feature to find several ordered-pair solutions of the function $f(x) = \sqrt{10 - x^2}$.

1.3

Finding Domain and Range

Set Notation

A **set** is a collection of objects. The set we consider most in calculus is the set of **real numbers.** There is a real number for every point on the number line.

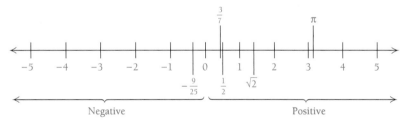

The set containing the numbers $-\frac{9}{25}$, 0, and $\sqrt{2}$ can be named $\left\{-\frac{9}{25}, 0, \sqrt{2}\right\}$. This method of describing sets is known as the **roster method.** It lists all the members of the set. To describe larger sets, we often use **set-builder notation** by specifying conditions under which an object is in a set. For example, the set of all real numbers less than 4 can be described as follows:

$$\{x \mid x \text{ is a real number less than } 4\}.$$

The set of ←

all numbers *x* ←

such that ←

x is a real number less than 4. ←

Interval Notation

We can also describe sets using **interval notation.** If a and b are real numbers such that $a < b$, we define the interval (a, b) as the set of all numbers between but not including a and b—that is, the set of all x for which $a < x < b$. Thus,

$$(a, b) = \{x \mid a < x < b\}.$$

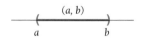

The points a and b are the **endpoints** of the interval. The parentheses indicate that the endpoints are *not* included in the graph.

The interval $[a, b]$ is defined as the set of all numbers x for which $a \le x \le b$. Thus,

$$[a, b] = \{x \mid a \le x \le b\}.$$

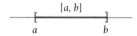

The brackets indicate that the endpoints *are* included in the graph.*

Be careful not to confuse the *interval* (a, b) with the *ordered pair* (a, b) used in connection with an equation in two variables in the plane, as in Section 1.1. The context in which the notation appears usually makes the meaning clear.

The following intervals include one endpoint and exclude the other:

$$(a, b] = \{x \mid a < x \le b\}. \quad \text{The graph excludes } a \text{ and includes } b.$$

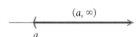

$$[a, b) = \{x \mid a \le x < b\}. \quad \text{The graph includes } a \text{ and excludes } b.$$

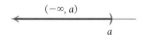

Some intervals extend without bound in one or both directions. We use the symbols ∞, read "infinity," and $-\infty$, read "negative infinity," to name these intervals. The notation (a, ∞) represents the set of all numbers greater than a—that is,

$$(a, \infty) = \{x \mid x > a\}.$$

Similarly, the notation $(-\infty, a)$ represents the set of all numbers less than a—that is,

$$(-\infty, a) = \{x \mid x < a\}.$$

---

*Some books use the representations ━●━━●━ and ━┤━━┝━ instead of, respectively,
  a b a b
━(━━)━ and ━[━━]━ .
  a b a b

The notations $[a, \infty)$ and $(-\infty, a]$ are used when we want to include the endpoints. The interval $(-\infty, \infty)$ names the set of all real numbers.

$$(-\infty, \infty) = \{x \mid x \text{ is a real number}\}$$

Interval notation is summarized in the following table.

Intervals: Notation and Graphs

Interval Notation	Set Notation	Graph
(a, b)	$\{x \mid a < x < b\}$	
$[a, b]$	$\{x \mid a \leq x \leq b\}$	
$[a, b)$	$\{x \mid a \leq x < b\}$	
$(a, b]$	$\{x \mid a < x \leq b\}$	
(a, ∞)	$\{x \mid x > a\}$	
$[a, \infty)$	$\{x \mid x \geq a\}$	
$(-\infty, b)$	$\{x \mid x < b\}$	
$(-\infty, b]$	$\{x \mid x \leq b\}$	
$(-\infty, \infty)$	$\{x \mid x \text{ is a real number}\}$	

Examples   Write interval notation for the given set or graph.

1. $\{x \mid -4 < x < 5\} = (-4, 5)$

2. $\{x \mid x \geq -2\} = [-2, \infty)$

3.

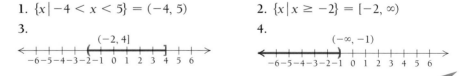

4.

Finding Domain and Range

A set of ordered pairs is a **relation.** The solutions of an equation in two variables consist of a set of ordered pairs and therefore form a relation. When a set of ordered pairs is such that no two different pairs share a common first coordinate, we have a function. The **domain** is the set of all first coordinates and the **range** is the set of all second coordinates.

Example 5   Find the domain and the range of the function f whose graph is shown below.

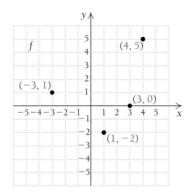

Solution   This is a rather simple function. Its graph contains just four ordered pairs and it can be written as

$$\{(-3,\ 1),\ (1,\ -2),\ (3,\ 0),\ (4,\ 5)\}.$$

We can determine the domain and the range by reading the x- and the y-values directly from the graph.

The domain is the set of all first coordinates, $\{-3,\ 1,\ 3,\ 4\}$. The range is the set of all second coordinates, $\{1,\ -2,\ 0,\ 5\}$.

Example 6   For the function f whose graph is shown below, determine each of the following.

a) The number in the range that is paired with 1 (from the domain). That is, find $f(1)$.

b) The domain of f

c) The numbers in the domain that are paired with 1 (from the range). That is, find all x such that $f(x) = 1$.

d) The range of f

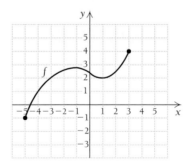

Solution

a) To determine which number in the range is paired with 1 in the domain, we locate 1 on the horizontal axis. Next, we find the point on the graph of f for which 1 is the first coordinate. From that point, we can look to the vertical axis to find the corresponding y-coordinate, 2. The input 1 has the output 2—that is, $f(1) = 2$.

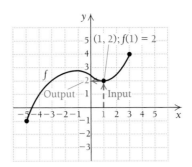

b) The domain of the function is the set of all *x*-values, or inputs, of the points on the graph. These extend from -5 to 3 and can be viewed as the curve's shadow, or projection, onto the *x*-axis. Thus the domain is the set $\{x \mid -5 \leq x \leq 3\}$, or, in interval notation, $[-5, 3]$.

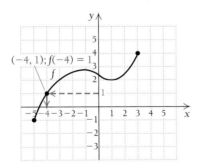

c) To determine which numbers in the domain are paired with 1 in the range, we locate 1 on the vertical axis (see the figure on the left below). From there, we look left and right to the graph of *f* to find any points (inputs) for which 1 is the second coordinate (output). One such point exists, $(-4, 1)$. For this function, we note that $x = -4$ is the only member of the domain paired with 1. For other functions, there might be more than one member of the domain paired with a member of the range.

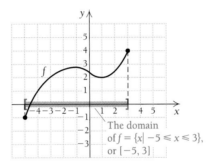

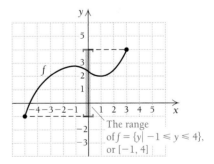

d) The range of the function is the set of all *y*-values, or outputs, of the points on the graph (see the figure on the right above). These extend from -1 to 4 and can be viewed as the curve's shadow, or projection, onto the *y*-axis. Thus the range is the set $\{y \mid -1 \leq y \leq 4\}$, or, in interval notation, $[-1, 4]$.

T E C H N O L O G Y   C O N N E C T I O N

Determining Domain and Range

Use a grapher to graph the function in the given viewing window. Then determine the domain and the range.

a) $f(x) = 3 - |x|$, $[-10, 10, -10, 10]$

b) $f(x) = x^3 - x$, $[-3, 3, -4, 4]$

c) $f(x) = \dfrac{12}{x}$, or $12x^{-1}$, $[-14, 14, -14, 14]$

d) $f(x) = x^4 - 2x^2 - 3$, $[-4, 4, -6, 6]$

We have the following.

a) $y = 3 - |x|$

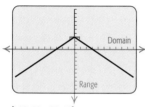

$[-10, 10, -10, 10]$

Domain = all real numbers;
range = $(-\infty, 3]$

b) $y = x^3 - x$

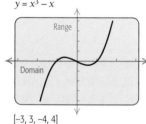

$[-3, 3, -4, 4]$

Domain = all real numbers;
range = all real numbers

c) $y = \frac{12}{x}$, or $12x^{-1}$

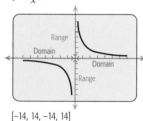

$[-14, 14, -14, 14]$

The number 0 is excluded as an input.
Domain = $\{x \mid x$ is a real number $and\ x \neq 0\}$,
or $(-\infty, 0) \cup (0, \infty)$;
range = $\{y \mid y$ is a real number $and\ y \neq 0\}$, or
$(-\infty, 0) \cup (0, \infty)$

d) $y = x^4 - 2x^2 - 3$

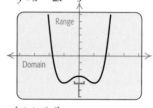

$[-4, 4, -6, 6]$

Domain = all real numbers;
range = $[-4, \infty)$

We can confirm our results using the **TRACE** feature, moving the cursor from left to right along the curve. We can also use the **TABLE** feature. In Example (d), it might not appear as though the domain is all real numbers because the graph seems "thin," but reexamining the formula shows that we can indeed substitute any real number.

Exercises

Use a grapher to graph the function in the given viewing window. Then determine the domain and the range.

1. $f(x) = |x| - 7$, $[-10, 10, -10, 10]$
2. $f(x) = 2 + 3x - x^3$, $[-5, 5, -5, 5]$
3. $f(x) = \dfrac{-16}{x}$, or $-16x^{-1}$, $[-20, 20, -20, 20]$
4. $f(x) = x^4 - 2x^2 - 7$, $[-4, 4, -9, 9]$
5. $f(x) = \sqrt{x + 4}$, $[-8, 8, -8, 8]$
6. $f(x) = \sqrt{9 - x^2}$, $[-5, 5, -5, 5]$
7. $f(x) = -\sqrt{9 - x^2}$, $[-5, 5, -5, 5]$
8. $f(x) = x^3 - 5x^2 + x - 4$,
   $[-10, 10, -20, 10]$

When a function is given by an equation or formula, the domain is understood to be the largest set of real numbers (inputs) for which function values (outputs) can be calculated. That is, the domain is the set of all allowable inputs into the formula. To find the domain, think, "What can we substitute?"

Example 7   Find the domain: $f(x) = |x|$.

Solution   We ask, "What can we substitute?" Is there any number x for which we cannot calculate $|x|$? The answer is no. Thus the domain of f is the set of all real numbers. ◄

Example 8   Find the domain: $f(x) = \dfrac{3}{2x - 5}$.

Solution   We ask, "What can we substitute?" Is there any number x for which we cannot calculate $3/(2x - 5)$? Since $3/(2x - 5)$ cannot be calculated when the denominator $2x - 5$ is 0, we solve the following equation to find those real numbers that must be excluded from the domain of f:

$$2x - 5 = 0 \qquad \text{Setting the denominator equal to 0}$$
$$2x = 5 \qquad \text{Adding 5}$$
$$x = \tfrac{5}{2}. \qquad \text{Dividing by 2}$$

Thus $\frac{5}{2}$ is not in the domain, whereas all other real numbers are.

The domain of f is $\left\{x \,\middle|\, x \text{ is a real number } and\ x \neq \tfrac{5}{2}\right\}$, or, in interval notation, $\left(-\infty, \tfrac{5}{2}\right) \cup \left(\tfrac{5}{2}, \infty\right)$. ◄

Example 9   Find the domain: $f(x) = \sqrt{4 + 3x}$.

Solution   We ask, "What can we substitute?" Is there any number x for which we cannot calculate $\sqrt{4 + 3x}$? Since $\sqrt{4 + 3x}$ is not a real number when the radicand $4 + 3x$ is negative, the domain is all real numbers for which $4 + 3x \geq 0$. We find them by solving the inequality. (See the appendix for a review of inequality solving.)

$$4 + 3x \geq 0$$
$$3x \geq -4 \qquad \text{Simplifying}$$
$$x \geq -\tfrac{4}{3} \qquad \text{Dividing by 3}$$

The domain is $\left[-\tfrac{4}{3}, \infty\right)$. ◄

Domains of Functions in Applications

Sometimes the domain (and, indeed, the range) of a function given by a formula is affected by the context of an application. Let's look again at Example 5 in Section 1.1.

Example 10   *Business: Compound Interest.* Suppose that $10,000 is invested at interest rate i, compounded annually, for 8 yr. From Theorem 1 in Section 1.1, we know that the amount in the account is given by

$$A = P(1 + i)^t$$
$$= 10{,}000(1 + i)^8.$$

The amount A is a function of the interest rate i and is given by

$$A(i) = 10{,}000(1 + i)^8.$$

Determine the domain.

Solution   We can substitute any real number for i into the formula, but neither negative nor zero interest rates are meaningful. The context of the application excludes nonpositive numbers. Thus the domain is the set of all positive numbers, $(0, \infty)$. However, large inputs for i may not be very realistic.

Example 11   *Life Science: Incidence of Breast Cancer.*   The following graph (considered in Section 1.1 without an equation) approximates the incidence of breast cancer I, per 100,000 women, as a function of age x. The equation for this graph is the polynomial function

$$I(x) = -0.0000554x^4 + 0.0067x^3 - 0.0997x^2 - 0.84x - 0.25.$$

Use the graph to find the domain.

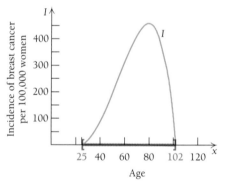

Source: National Cancer Institute

Solution   Inputs that yield negative outputs have no meaning in this case. From the graph, we see that function values (outputs) are nonnegative for values of x between 25 and 102. It seems reasonable that we consider the incidence of breast cancer in women of these ages. Thus the domain is [25, 102].*

The task of determining the domain and the range of a function is one that we will return to several times as we consider other types of functions in this book.

---

*The range is actually a set of integers between 0 and about 450, but we approximate this with the nonnegative real numbers.

The following is a review of the function concepts considered in Sections 1.1–1.3.

Function Concepts   *Graph*

- Formula for f: $f(x) = x^2 - 7$
- For every input of f, there is exactly one output.
- For the input 1, -6 is the output.
- $f(1) = -6$
- $(1, -6)$ is on the graph.
- Domain = The set of all inputs
           = The set of all real numbers
- Range = The set of all outputs
         = $[-7, \infty)$

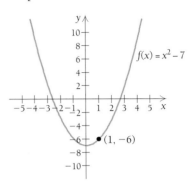

EXERCISE SET 1.3

In Exercises 1–8, write interval notation for the graph.

1.

2.
(number line from −1 to 6, bracket at −1 to bracket at 2)

3.
(number line from −10 to −3, bracket at −9 to parenthesis at −4)

4.
(number line from −10 to −3, parenthesis at −9 to bracket at −5)

5.

(x .... $x + h$)

6.
(x .... $x + h$)

7.
(parenthesis at p extending right)

8.
(bracket at q extending left)

Write interval notation for each of the following.

9. The set of all numbers x such that $-3 \le x \le 3$

10. The set of all numbers x such that $-4 < x < 4$

11. $\{x \mid -14 \le x < -11\}$

12. $\{x \mid 6 < x \le 20\}$

13. $\{x \mid x \le -4\}$

14. $\{x \mid x > -5\}$

In Exercises 15–26, the graph is that of a function. Determine for each one **(a)** $f(1)$; **(b)** the domain; **(c)** all x-values such that $f(x) = 2$; and **(d)** the range.

15.   16.

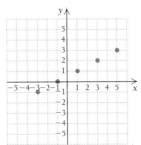   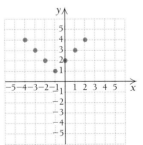

17.
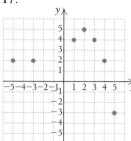

18.
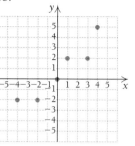

25.
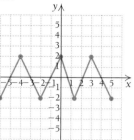

26.
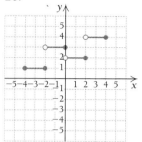

19.
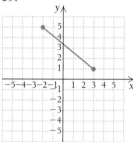

20.
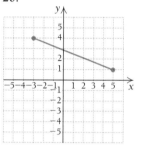

Find the domain.

27. $f(x) = \dfrac{7}{5 - x}$

28. $f(x) = \dfrac{2}{x + 3}$

29. $f(x) = 4 - 5x$

30. $f(x) = 2x + 1$

31. $f(x) = x^2 - 2x + 3$

32. $f(x) = x^2 + 3$

33. $f(x) = \dfrac{x - 2}{3x + 4}$

34. $f(x) = \dfrac{8}{5x - 14}$

35. $f(x) = |x - 4|$

36. $f(x) = |x| - 4$

37. $f(x) = \dfrac{x^2 - 3x}{|4x - 7|}$

38. $f(x) = \dfrac{4}{|2x - 3|}$

39. $g(x) = \dfrac{-11}{4 + x}$

40. $g(x) = \dfrac{1}{x - 1}$

41. $g(x) = 8 - x^2$

42. $g(x) = x^2 - 2x + 1$

43. $g(x) = 4x^3 + 5x^2 - 2x$

44. $g(x) = x^3 - 1$

45. $g(x) = \dfrac{2x - 3}{6x - 12}$

46. $g(x) = \dfrac{7}{20 - 8x}$

21.
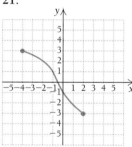

22.
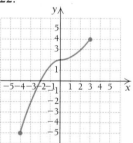

23.
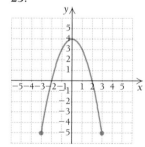

24.
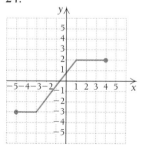

47. $g(x) = |x| + 1$

48. $g(x) = |x + 7|$

49. $g(x) = \dfrac{x^2 + 2x}{|10x - 20|}$

50. $g(x) = \dfrac{-2}{|4x + 5|}$

51. For the function f whose graph is shown below, find $f(-1)$, $f(0)$, and $f(1)$.

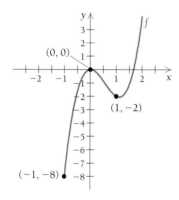

52. For the function g whose graph is shown below, find all the x-values for which $g(x) = 1$.

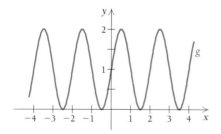

Applications

BUSINESS AND ECONOMICS

53. *Compound Interest.* Suppose that $10,000 is invested at interest rate i, compounded semiannually, for 8 yr.

   a) The amount A in the account is a function of the interest rate i. Find an equation for the function.

   b) Determine the domain.

LIFE AND PHYSICAL SCIENCES

54. *Hearing-Impaired Americans.* The following graph (considered in Exercise Set 1.1) approximates the number N, in millions, of hearing-impaired Americans as a function of age x (**Source:** American Speech-Language Hearing Association). The equation for this graph is the polynomial function

$$N(x) = -0.00006x^3 + 0.006x^2 - 0.1x + 1.9.$$

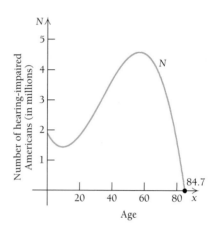

   a) Use the graph to determine the domain.

   tw b) Explain why the number -20 is not in the domain.

   tw c) Explain why the number 90 is not in the domain.

Synthesis

tw 55. For a given function, $f(2) = -5$. Give as many interpretations of this fact as you can.

tw 56. Explain the difference between the domain and the range of a function.

tw 57. Give an example of a function for which the number 3 is not in the domain and explain why it is not.

   TECHNOLOGY   CONNECTION

58. Determine the range of each of the functions in Exercises 28, 31, 35, 36, and 43.

59. Determine the range of each of the functions in Exercises 29, 32, 44, 47, and 48.

1.4

OBJECTIVES

➢ Graph equations of the type $x = a$ and $y = f(x) = b$.
➢ Graph linear functions.
➢ Find an equation of a line when given its slope and one point contained on the line and when given two points contained on the line.
➢ Solve applied problems involving slope and linear functions.

Slope and Linear Functions

Horizontal and Vertical Lines

Let's consider graphs of equations $y = b$ and $x = a$.

Example 1

a) Graph $y = 4$.

b) Decide whether the relation is a function.

Solution

a) The graph consists of all ordered pairs whose second coordinate is 4. To see how a pair such as $(-2, 4)$ could be a solution of $y = 4$, we can consider the equation above in the form

$$y = 0x + 4.$$

Then $(-2, 4)$ is a solution because

$$4 = 0(-2) + 4$$

is true.

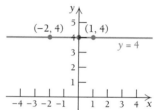

b) The vertical-line test holds. Thus this relation is a function.

Example 2

a) Graph $x = -3$.

b) Decide whether the relation is a function.

Solution

a) The graph consists of all ordered pairs whose first coordinate is -3. To see how a pair such as $(-3, 4)$ could be a solution of $x = -3$, we can consider the equation above in the form

$$x + 0y = -3, \quad \text{or} \quad x = 0y - 3.$$

Then $(-3, 4)$ is a solution because

$$(-3) + 0(4) = -3$$

is true.

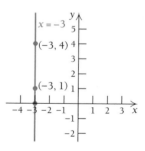

b) This relation is *not* a function because it fails the vertical-line test. The line itself meets the graph more than once—in fact, infinitely many times.

In general, we have the following.

> **Theorem 3**
>
> The graph of $y = b$, or $f(x) = b$, a horizontal line, is the graph of a function.
> The graph of $x = a$, a vertical line, is not the graph of a function.

TECHNOLOGY CONNECTION

Visualizing Slope

Exploratory

Squaring a Viewing Window. Consider the $[-10, 10, -10, 10]$ viewing window on the left below. Note that the distance between units is not visually the same on both axes. In this case, the length of the interval shown on the y-axis is about two-thirds of the length of the interval on the x-axis. If we change the dimensions of the window to $[-6, 6, -4, 4]$, we get a graph for which the units are visually about the same on both axes. Creating such a window is called **squaring the window**. On a TI-83 grapher, there is a ZSQUARE feature for automatic window squaring. This feature alters the standard window dimensions to $[-15.1613, 15.1613, -10, 10]$.

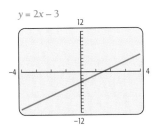

Each of the following is a graph of the line $y = 2x - 3$, but the viewing windows are different. When the window is square, as shown on the right, we get an accurate representation of the *slope* of the line.

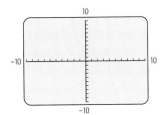
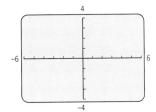
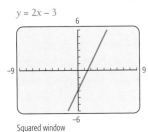

Squared window

Exercises

Use a square viewing window for each of these exercises.

1. Graph $y = x + 1$, $y = 2x + 1$, $y = 3x + 1$, and $y = 10x + 1$. What do you think the graph of $y = 247x + 1$ will look like?

2. Graph $y = x$, $y = \frac{7}{8}x$, $y = 0.47x$, and $y = \frac{2}{31}x$. What do you think the graph of $y = 0.000018x$ will look like?

3. Graph $y = -x$, $y = -2x$, $y = -5x$, and $y = -10x$. What do you think the graph of $y = -247x$ will look like?

4. Graph $y = -x - 1$, $y = -\frac{3}{4}x - 1$, $y = -0.38x - 1$, and $y = -\frac{5}{32}x - 1$. What do you think the graph of $y = -0.000043x - 1$ will look like?

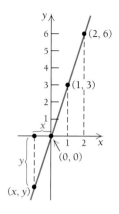

The Equation $y = mx$

Consider the following table of numbers and look for a pattern.

x	1	-1	$-\frac{1}{2}$	2	-2	3	-7	5
y	3	-3	$-\frac{3}{2}$	6	-6	9	-21	15

Note that the ratio of the bottom number to the top one is 3. That is,

$$\frac{y}{x} = 3, \quad \text{or} \quad y = 3x.$$

Ordered pairs from the table can be used to graph the equation $y = 3x$ (see the figure at left). Note that this is a function.

Theorem 4

The graph of the function given by

$$y = mx \quad \text{or} \quad f(x) = mx$$

is the straight line through the origin $(0, 0)$ and the point $(1, m)$. The constant m is called the **slope** of the line.

TECHNOLOGY CONNECTION

Exploring b

We can use a grapher to explore the effect of b when we are graphing $f(x) = mx + b$.

Exercises

1. Graph $y_1 = x$. Then, using the same viewing window, graph $y_2 = x + 3$ followed by $y_3 = x - 4$. Compare the graphs of y_2 and y_3 with that of y_1. Then without drawing them, compare the graphs of $y = x$ and $y = x - 5$.
2. Use the TABLE feature to compare the values of y_1, y_2, and y_3 when $x = 0$. Then scroll through other values and see if you can determine a pattern.

Various graphs of $y = mx$ for positive values of m are shown below. Note that such graphs slant up from left to right. A line with large positive slope rises faster than a line with smaller positive slope.

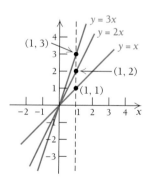

When $m = 0$, $y = 0x$, or $y = 0$. On the left at the top of the following page is a graph of $y = 0$. Note that this is both the x-axis and a horizontal line.

Lines of various slopes.

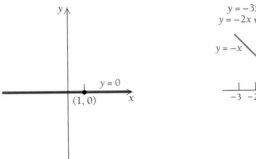

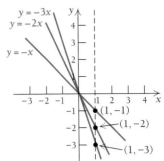

Graphs of $y = mx$ for negative values of m are shown on the right above. Note that such graphs slant down from left to right.

Direct Variation

There are many applications involving equations like $y = mx$, where m is some positive number. In such situations, we say that we have **direct variation,** and m (the slope) is called the **variation constant,** or **constant of proportionality.** Generally, only positive values of x and y are considered.

> ### Definition
> The variable y **varies directly** as x if there is some positive constant m such that $y = mx$. We also say that y is **directly proportional** to x.

Example 3 *Life Science: Weight on Earth and the Moon.* The weight M, in pounds, of an object on the moon is directly proportional to the weight E of that object on Earth. An astronaut who weighs 180 lb on Earth will weigh 28.8 lb on the moon.

a) Find an equation of variation.

b) An astronaut weighs 19.2 lb on the moon. How much will the astronaut weigh on Earth?

Solution

a) Since $M = mE$, then $28.8 = m(180)$ so $0.16 = m$. Thus, $M = 0.16E$.

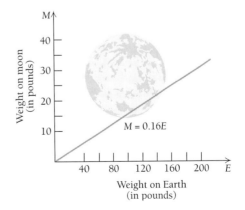

b) To find the weight on Earth of an astronaut who weighs 19.2 lb on the moon, we solve the equation of variation

$$19.2 = 0.16E \qquad \text{Substituting 19.2 for } M$$

and get

$$120 = E.$$

Thus an astronaut who weighs 19.2 lb on the moon weighs 120 lb on Earth.

The Equation $y = mx + b$

Compare the graphs of the equations

$$y = 3x \quad \text{and} \quad y = 3x - 2$$

(see the following figure). Note that the graph of $y = 3x - 2$ is a shift 2 units down of the graph of $y = 3x$, and that $y = 3x - 2$ has y-intercept $(0, -2)$. Note also that the graph of $y = 3x - 2$ is the graph of a function.

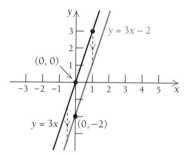

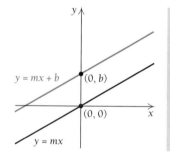

> ### Definition
> A **linear function** is given by
>
> $$y = mx + b \quad \text{or} \quad f(x) = mx + b$$
>
> and has a graph that is the straight line parallel to $y = mx$ with y-intercept $(0, b)$. The constant m is called the **slope.** (See the figure at left.)

When $m = 0$, $y = 0x + b = b$, and we have what is known as a **constant function** (see Theorem 3 at the beginning of the section). The graph of such a function is a horizontal line.

The Slope–Intercept Equation

Any nonvertical line l is uniquely determined by its slope m and its y-intercept $(0, b)$. In other words, the slope describes the "slant" of the line, and the y-intercept is the point at which the line crosses the y-axis. Thus we have the following definition.

> **Definition**
>
> $y = mx + b$ is called the **slope–intercept equation** of a line.

Example 4   Find the slope and the y-intercept of $2x - 4y - 7 = 0$.

Solution   We solve for y:

$$-4y = -2x + 7$$
$$y = \tfrac{1}{2}x - \tfrac{7}{4}$$

Slope: $\tfrac{1}{2}$ y-intercept: $\left(0, -\tfrac{7}{4}\right)$

The Point–Slope Equation

Suppose that we know the slope of a line and some point on the line other than the y-intercept. We can still find an equation of the line.

Example 5   Find an equation of the line with slope 3 containing the point $(-1, -5)$.

Solution   The slope is given as $m = 3$. From the slope–intercept equation, we have

$$y = 3x + b,$$

so we must determine b. Since $(-1, -5)$ is on the line, it follows that

$$-5 = 3(-1) + b,$$

so

$$-2 = b \quad \text{and} \quad y = 3x - 2.$$

If a point (x_1, y_1) is on the line

$$y = mx + b, \tag{1}$$

it must follow that

$$y_1 = mx_1 + b. \tag{2}$$

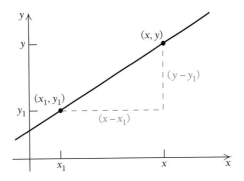

Subtracting equation (2) from equation (1) eliminates the b's, and we have

$$y - y_1 = (mx + b) - (mx_1 + b)$$
$$= mx + b - mx_1 - b$$
$$= mx - mx_1$$
$$= m(x - x_1).$$

Definition
$y - y_1 = m(x - x_1)$ is called the **point–slope equation** of a line.

This definition allows us to write an equation of a line given its slope and the coordinates of *any* point on it.

Example 6   Find an equation of the line with slope $\frac{2}{3}$ containing the point $(-1, -5)$.

Solution   Substituting in

$$y - y_1 = m(x - x_1),$$

we get

$$y - (-5) = \tfrac{2}{3}[x - (-1)]$$
$$y + 5 = \tfrac{2}{3}(x + 1)$$
$$y + 5 = \tfrac{2}{3}x + \tfrac{2}{3}$$
$$y = \tfrac{2}{3}x + \tfrac{2}{3} - 5$$
$$= \tfrac{2}{3}x + \tfrac{2}{3} - \tfrac{15}{3}$$
$$= \tfrac{2}{3}x - \tfrac{13}{3}.$$

Computing Slope

We now determine a method of computing the slope of a line when we know the coordinates of two of its points. Suppose that (x_1, y_1) and (x_2, y_2) are the coordinates of two different points, P_1 and P_2, respectively, on a line that is not vertical. Consider a right triangle with legs parallel to the axes, as shown in the following figure.

Which lines have the same slope?

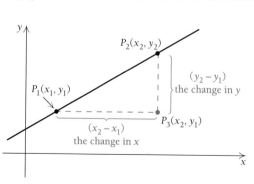

The point P_3 with coordinates (x_2, y_1) is the third vertex of the triangle. As we move from P_1 to P_2, y changes from y_1 to y_2. The change in y is $y_2 - y_1$. Similarly, the change in x is $x_2 - x_1$. The ratio of these changes is the slope. To see this, consider the point–slope equation,

$$y - y_1 = m(x - x_1).$$

Since (x_2, y_2) is on the line, it must follow that

$$y_2 - y_1 = m(x_2 - x_1).$$

Since the line is not vertical, the two x-coordinates must be different, so $x_2 - x_1$ is nonzero and we can divide by it to get the following theorem.

Theorem 5

$$m = \frac{y_2 - y_1}{x_2 - x_1} = \frac{\text{change in } y}{\text{change in } x} = \begin{array}{l}\text{slope of line containing points} \\ (x_1, y_1) \text{ and } (x_2, y_2)\end{array}$$

Example 7   Find the slope of the line containing the points $(-2, 6)$ and $(-4, 9)$.

Solution   We have

$$m = \frac{y_2 - y_1}{x_2 - x_1} = \frac{6 - 9}{-2 - (-4)}$$

$$= \frac{-3}{2} = -\frac{3}{2}.$$

Note that it does not matter which point is taken first, so long as we subtract the coordinates in the same order. In this example, we can also find m as follows:

$$m = \frac{9 - 6}{-4 - (-2)} = \frac{3}{-2} = -\frac{3}{2}.$$ ◄

If a line is horizontal, the change in y for any two points is 0. Thus a horizontal line has slope 0. If a line is vertical, the change in x for any two points is 0. Thus the slope is *not defined* because we cannot divide by 0. A vertical line has no slope. Thus "0 slope" and "no slope" are two very distinct concepts.

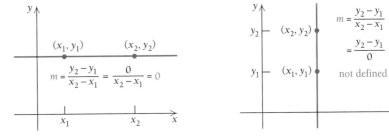
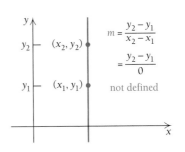

Applications of Slope

Slope has many real-world applications. For example, numbers like 2%, 3%, and 6% are often used to represent the *grade* of a road, a measure of how steep a road on a hill or mountain is. A 3% grade $\left(3\% = \frac{3}{100}\right)$ means that for every horizontal distance of 100 ft, the road rises 3 ft, and a -3% grade means that for every horizontal distance of 100 ft, the road drops 3 ft. An athlete might change the grade of a treadmill during a workout. An escape ramp on an airliner might have a slope of about -0.6.

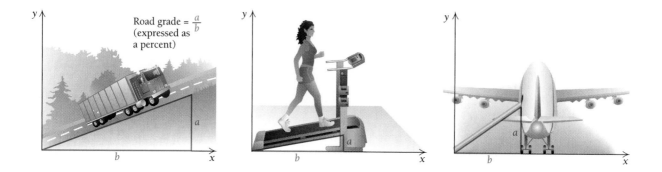

Architects and carpenters use slope when designing and building stairs, ramps, or roof pitches. Another application occurs in hydrology. When a river flows, the strength or force of the river depends on how far the river falls vertically compared to how far it flows horizontally. Slope can also be considered as an **average rate of change.**

Example 8 *Life Science: Amount Spent on Cancer Research.* The amount spent on cancer research has increased steadily over the years, as shown in the following graph. Find the average rate of change of that amount.

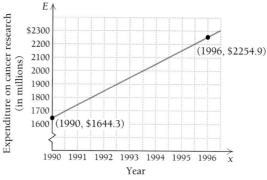

Source: *The New England Journal of Medicine*

Solution   First, we determine the coordinates of two points on the graph. In this case, they are given as (1990, $1644.3) and (1996, $2254.9). Then we compute the slope, or rate of change, as follows:

$$\text{Slope} = \text{Average rate of change} = \frac{\text{change in } y}{\text{change in } x}$$

$$= \frac{\$2254.9 - \$1644.3}{1996 - 1990} = \frac{\$610.6}{6} \approx 101.8 \; \frac{\$}{\text{yr}}.$$

This result tells us that each year the amount spent on cancer research has increased by about \$101.8 million.

Applications of Linear Functions

Many applications are modeled by linear functions.

Example 9 *Business: Total Cost.* Raggs, Ltd., a clothing firm, has **fixed costs** of \$10,000 per year. These costs, such as rent, maintenance, and so on, must be paid no matter how much the company produces. To produce x units of a certain kind of suit, it costs \$20 per suit (unit) in addition to the fixed costs. That is, the **variable costs** for producing x of these suits is $20x$ dollars. These are costs that are directly related to production, such as material, wages, fuel, and so on. Then the **total cost** $C(x)$ of producing x suits in a year is given by a function C:

$$C(x) = (\text{Variable costs}) + (\text{Fixed costs}) = 20x + 10{,}000.$$

a) Graph the variable-cost, the fixed-cost, and the total-cost functions.

b) What is the total cost of producing 100 suits? 400 suits?

c) How much more does it cost to produce 400 suits than 100 suits?

Solution

a) The variable-cost and fixed-cost functions appear in the graph on the left below. The total-cost function is shown in the graph on the right. From a practical standpoint, the domains of these functions are nonnegative integers 0, 1, 2, 3, and so on, since it does not make sense to make either a negative number or a fractional number of suits. It is common practice to draw the graphs as though the domains were the entire set of nonnegative real numbers.

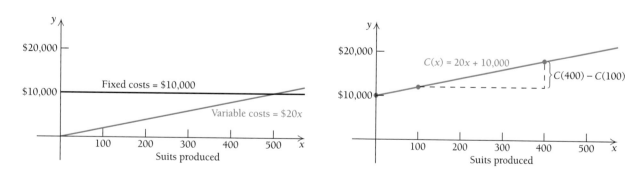

b) The total cost of producing 100 suits is

$$C(100) = 20 \cdot 100 + 10{,}000 = \$12{,}000.$$

The total cost of producing 400 suits is

$$C(400) = 20 \cdot 400 + 10{,}000$$
$$= \$18{,}000.$$

c) The cost of producing 400 suits exceeds the cost of producing 100 suits by

$$C(400) - C(100) = \$18{,}000 - \$12{,}000$$
$$= \$6000.$$

Example 10 *Business: Profit-and-Loss Analysis.* Refer to Example 9. Raggs, Ltd., determines that its total revenue from the sale of x suits is \$80 per suit. That is, the total revenue $R(x)$ is given by the function

$$R(x) = (\text{Unit price}) \times (\text{Quantity sold}) = 80x.$$

a) Graph $R(x)$ and $C(x)$ using the same set of axes.

b) The total profit $P(x)$ is given by a function P:

$$P(x) = (\text{Total revenue}) - (\text{Total costs}) = R(x) - C(x).$$

Determine $P(x)$ and draw its graph using the same set of axes as those used for the graph in part (a).

c) The company will *break even* at that value of x for which $P(x) = 0$ (that is, no profit and no loss). This is the point at which $R(x) = C(x)$. Find the **break-even value** of x.

Solution

a) The graphs of $R(x) = 80x$ and $C(x) = 20x + 10{,}000$ are shown below. When $C(x)$ is above $R(x)$, a loss will occur. This is shown by the color-shaded region. When $R(x)$ is above $C(x)$, a gain will occur. This is shown by the gray-shaded region.

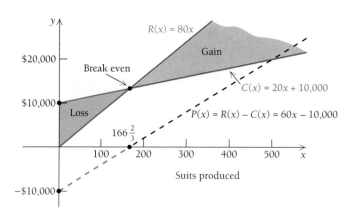

b) We see that

$$P(x) = R(x) - C(x) = 80x - (20x + 10{,}000)$$
$$= 60x - 10{,}000.$$

The graph of $P(x)$ is shown by the dashed line. The color dashed line shows a "negative" profit, or loss. The black dashed line shows a "positive" profit, or gain.

c) To find the break-even value, we solve $R(x) = C(x)$:

$$R(x) = C(x)$$
$$80x = 20x + 10{,}000$$
$$60x = 10{,}000$$
$$x = 166\tfrac{2}{3}.$$

How do we interpret the fractional answer, since it is not possible to produce $\frac{2}{3}$ of a suit? We simply round to 167. Estimates of break-even values are usually sufficient since companies want to operate well away from break-even values in order to maximize profit.

EXERCISE SET 1.4

Graph.

1. $y = -4$
2. $y = -3.5$
3. $x = 4.5$
4. $x = 10$

Graph. Find the slope and the y-intercept.

5. $y = -3x$
6. $y = -0.5x$
7. $y = 0.5x$
8. $y = 3x$
9. $y = -2x + 3$
10. $y = -x + 4$
11. $y = -x - 2$
12. $y = -3x + 2$

Find the slope and the y-intercept.

13. $2x + y - 2 = 0$
14. $2x - y + 3 = 0$
15. $2x + 2y + 5 = 0$
16. $3x - 3y + 6 = 0$

Find an equation of the line:

17. with $m = -5$, containing $(1, -5)$.
18. with $m = 7$, containing $(1, 7)$.
19. with $m = -2$, containing $(2, 3)$.
20. with $m = -3$, containing $(5, -2)$.
21. with y-intercept $(0, -6)$ and slope $\frac{1}{2}$.
22. with y-intercept $(0, 7)$ and slope $\frac{4}{3}$.
23. with slope 0, containing $(2, 3)$.
24. with slope 0, containing $(4, 8)$.

Find the slope of the line containing the given pair of points, if it exists.

25. $(-4, -2)$ and $(-2, 1)$
26. $(-2, 1)$ and $(6, 3)$
27. $\left(\frac{2}{5}, \frac{1}{2}\right)$ and $\left(-3, \frac{4}{5}\right)$
28. $\left(-\frac{3}{4}, \frac{5}{8}\right)$ and $\left(-\frac{1}{2}, -\frac{3}{16}\right)$
29. $(3, -7)$ and $(3, -9)$
30. $(-4, 2)$ and $(-4, 10)$
31. $(2, 3)$ and $(-1, 3)$
32. $\left(-6, \frac{1}{2}\right)$ and $\left(-7, \frac{1}{2}\right)$
33. $(x, 3x)$ and $(x + h, 3(x + h))$
34. $(x, 4x)$ and $(x + h, 4(x + h))$
35. $(x, 2x + 3)$ and $(x + h, 2(x + h) + 3)$
36. $(x, 3x - 1)$ and $(x + h, 3(x + h) - 1)$

37.–48. Find an equation of the line containing the pair of points in each of Exercises 25–36.

49. Find the slope (or grade) of the treadmill.

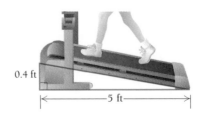

0.4 ft

|←——— 5 ft ———→|

50. Find the slope (or pitch) of the roof.

51. Find the slope (or head) of the river.

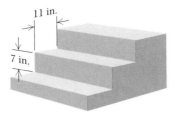

52. Public buildings regularly include steps with 7-in. risers and 11-in. treads. Find the grade of such a stairway.

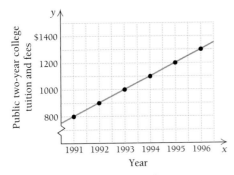

53. Find the average rate of change of the tuition and fees at public two-year colleges.

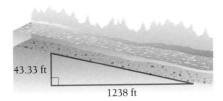

Year

Source: Statistical Abstract of the United States

54. Find the average rate of change of the cost of a formal wedding.

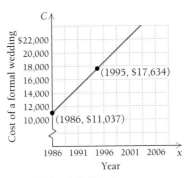

Year

Source: Modern Bride Magazine

Applications ....

BUSINESS AND ECONOMICS

55. *Investment.* A person makes an investment of P dollars at 8%. After 1 yr, it grows to an amount A.
  a) Show that A is directly proportional to P.
  b) Find A when $P = \$100$.
  c) Find P when $A = \$259.20$.

56. *Profit-and-Loss Analysis.* A ski manufacturer is planning a new line of skis. For the first year, the fixed costs for setting up the new production line are $22,500. The variable costs for producing each pair of skis are estimated at $40. The sales department projects that 3000 pairs can be sold during the first year at a price of $85 per pair.

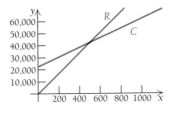

  a) Formulate a function $C(x)$ for the total cost of producing x pairs of skis.
  b) Formulate a function $R(x)$ for the total revenue from the sale of x pairs of skis.
  c) Formulate a function $P(x)$ for the total profit from the production and sale of x pairs of skis.
  d) What profit or loss will the company realize if the expected sales of 3000 pairs occurs?
  e) How many pairs must the company sell in order to break even?

57. *Profit-and-Loss Analysis.* Boxowitz, Inc., a computer firm, is planning to sell a new graphing calculator. For the first year, the fixed costs for setting up the new production line are $100,000. The variable costs for producing each calculator are estimated at $20. The sales department projects that 150,000 calculators can be sold during the first year at a price of $45 each.

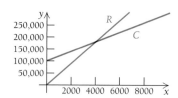

a) Formulate a function $C(x)$ for the total cost of producing x calculators.
b) Formulate a function $R(x)$ for the total revenue from the sale of x calculators.
c) Formulate a function $P(x)$ for the total profit from the production and sale of x calculators.
d) What profit or loss will the firm realize if the expected sales of 150,000 calculators occurs?
e) How many calculators must the firm sell in order to break even?

58. *Straight-Line Depreciation.* A company buys an office machine for $5200 on January 1 of a given year. The machine is expected to last for 8 years, at the end of which time its *trade-in value*, or *salvage value*, will be $1100. If the company figures the decline in value to be the same each year, then the *book value*, or *salvage value*, after t years, $0 \leq t \leq 8$, is given by the linear function

$$V(t) = C - t\left(\frac{C - S}{N}\right),$$

where C is the original cost of the item ($5200), N is the number of years of expected life (8), and S is the salvage value ($1100).

a) Find the linear function for the straight-line depreciation of the office machine.
b) Find the salvage value after 0 years, 1 year, 2 years, 3 years, 4 years, 7 years, and 8 years.

59. *Profit-and-Loss Analysis.* A college student decides to mow lawns in the summer. The initial cost of the lawnmower is $250. Gasoline and maintenance costs are $1 per lawn.

a) Formulate a function $C(x)$ for the total cost of mowing x lawns.

b) The student determines that the total-profit function for the lawnmowing business is given by $P(x) = 9x - 250$. Find a function for the total revenue from mowing x lawns. How much does the student charge per lawn?
c) How many lawns must the student mow before making a profit?

60. *Salvage Value.* Tyline Electric uses the function $S(t) = -700t + 3500$ to determine the *salvage value* $S(t)$, in dollars, of a photocopier t years after its purchase.

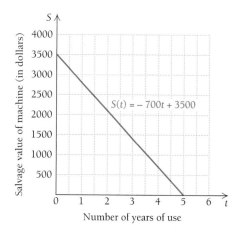

a) What do the numbers -700 and 3500 signify?
b) How long will it take the copier to *depreciate* completely?
tw c) What is the domain of S? Explain.

LIFE AND PHYSICAL SCIENCES

61. *Energy Conservation.* The R-factor of home insulation is directly proportional to its thickness T.
a) Find an equation of variation if $R = 12.51$ when $T = 3$ in.
b) What is the R-factor for insulation that is 6 in. thick?

62. *Nerve Impulse Speed.* Impulses in nerve fibers travel at a speed of 293 ft/sec. The distance D traveled in t seconds is given by $D = 293t$. How long would it take an impulse to travel from the brain to the toes of a person who is 6 ft tall?

63. *Brain Weight.* The weight B of a human's brain is directly proportional to his or her body weight W.
a) It is known that a person who weighs 200 lb

has a brain that weighs 5 lb. Find an equation of variation expressing B as a function of W.

b) Express the variation constant as a percent and interpret the resulting equation.

c) What is the weight of the brain of a person who weighs 120 lb?

64. *Muscle Weight.* The weight M of the muscles in a human is directly proportional to his or her body weight W.

a) It is known that a person who weighs 200 lb has 80 lb of muscles. Find an equation of variation expressing M as a function of W.

b) Express the variation constant as a percent and interpret the resulting equation.

c) What is the muscle weight of a person who weighs 120 lb?

Muscle weight is directly proportional to body weight.

65. *Stopping Distance on Glare Ice.* The stopping distance (at some fixed speed) of regular tires on glare ice is given by a linear function of the air temperature F,

$$D(F) = 2F + 115,$$

where $D(F)$ is the stopping distance, in feet, when the air temperature is F, in degrees Fahrenheit.

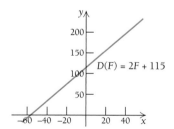

a) Find $D(0°)$, $D(-20°)$, $D(10°)$, and $D(32°)$.

tw b) Explain why the domain should be restricted to the interval $[-57.5°, 32°]$.

66. *Reaction Time.* While driving a car, you see a child suddenly cross the street unattended. Your brain registers the emergency and sends a signal to your foot to hit the brake. The car travels a distance D, in feet, during this time, where D is a function of the speed r, in miles per hour, that the car is traveling when you see the child. That reaction distance is a linear function given by

$$D(r) = \frac{11r + 5}{10}.$$

a) Find $D(5)$, $D(10)$, $D(20)$, $D(50)$, and $D(65)$.

b) Graph $D(r)$.

tw c) What is the domain of the function? Explain.

67. *Estimating Heights.* An anthropologist can use certain linear functions to estimate the height of a male or female, given the length of certain bones. The *humerus* is the bone from the elbow to the shoulder. Let $x =$ the length of the humerus, in centimeters. Then the height, in centimeters, of a male with a humerus of length x is given by

$$M(x) = 2.89x + 70.64.$$

The height, in centimeters, of a female with a humerus of length x is given by

$$F(x) = 2.75x + 71.48.$$

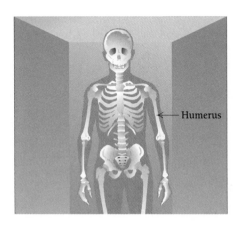

A 26-cm humerus was uncovered in a ruins.

a) If we assume it was from a male, how tall was he?

b) If we assume it was from a female, how tall was she?

SOCIAL SCIENCES

68. *Urban Population.* The population of a town is P. After a growth of 2%, its new population is N.

   a) Assuming that N is directly proportional to P, find an equation of variation.
   b) Find N when $P = 200{,}000$.
   c) Find P when $N = 367{,}200$.

69. *Median Age of Women at First Marriage.* In general, our society is marrying at a later age. The median age of women at first marriage can be approximated by the linear function

$$A(t) = 0.08t + 19.7,$$

   where $A(t)$ is the median age of women at first marriage the tth year after 1950. Thus, $A(0)$ is the median age of women at first marriage in the year 1950, $A(50)$ is the median age in 2000, and so on.

   a) Find $A(0)$, $A(1)$, $A(10)$, $A(30)$, and $A(50)$.
   b) What will be the median age of women at first marriage in 2003?
   c) Graph $A(t)$.

Synthesis ··

tw 70. Explain and compare the situations in which you would use the slope–intercept equation rather than the point–slope equation.

tw 71. Discuss and relate the concepts of fixed cost, total cost, total revenue, and total profit.

72. *Business: Daily Sales.* Match each sentence with the most appropriate graph below.

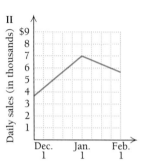

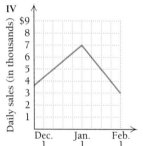

a) After January 1, daily sales continued to rise, but at a slower rate.

b) After January 1, sales decreased faster than they ever grew.

c) The rate of growth in daily sales doubled after January 1.

d) After January 1, daily sales decreased at half the rate that they grew in December.

   TECHNOLOGY   CONNECTION

73. Graph some of the total-revenue, total-cost, and total-profit functions in this exercise set using the same set of axes. Identify regions of profit and loss.

1.5

Other Types of Functions

Quadratic Functions

> **Definition**
>
> A **quadratic function** f is given by
>
> $$f(x) = ax^2 + bx + c, \quad \text{where } a \neq 0.$$

We have already considered some quadratic functions—for example, $f(x) = x^2$ and $g(x) = x^2 - 1$. We can create hand-drawn graphs of quadratic functions using the following information.

The graph of a quadratic function $f(x) = ax^2 + bx + c$ is called a **parabola.**

a) It is always a cup-shaped curve, like those in Examples 1 and 2 that follow.
b) It has a turning point, or **vertex,** at a point whose first coordinate is given by

$$x = -\frac{b}{2a}.$$

c) It has the vertical line $x = -b/2a$ as a line of symmetry (not part of the graph).
d) It opens up if $a > 0$ or opens down if $a < 0$.

Example 1   Graph: $f(x) = x^2 - 2x - 3$.

Solution   Let's first find the vertex, or turning point. The x-coordinate of the vertex is

$$x = -\frac{b}{2a}$$

$$= -\frac{-2}{2(1)} = 1.$$

Substituting 1 for x in the equation, we find the second coordinate of the vertex:

$$y = f(1) = 1^2 - 2(1) - 3$$
$$= 1 - 2 - 3$$
$$= -4.$$

The vertex is $(1, -4)$. The vertical line $x = 1$ is the line of symmetry of the graph. We choose some x-values on each side of the vertex, compute y-values, plot the points, and graph the parabola:

$$f(x) = x^2 - 2x - 3.$$

x	$f(x)$	
1	-4	← Vertex
0	-3	
2	-3	
3	0	
4	5	
-1	0	
-2	5	

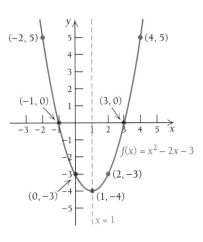

Exercises
Using the procedure of Examples 1 and 2, graph each of the following functions. Use the **TABLE** feature to create an input–output table for each function. Then check the graph with a grapher.

1. $f(x) = x^2 - 6x + 4$
2. $f(x) = -2x^2 + 4x + 1$

Example 2   Graph: $f(x) = -2x^2 + 10x - 7$.

Solution   Let's first find the vertex, or turning point. The x-coordinate of the vertex is

$$x = -\frac{b}{2a}$$

$$= -\frac{10}{2(-2)} = \frac{5}{2}.$$

Substituting $\frac{5}{2}$ for x in the equation, we find the second coordinate of the vertex:

$$y = f\left(\tfrac{5}{2}\right) = -2\left(\tfrac{5}{2}\right)^2 + 10\left(\tfrac{5}{2}\right) - 7$$

$$= -2\left(\tfrac{25}{4}\right) + 25 - 7$$

$$= \tfrac{11}{2}.$$

The vertex is $\left(\frac{5}{2}, \frac{11}{2}\right)$, and the line of symmetry is $x = \frac{5}{2}$. We choose some x-values on each side of the vertex, compute y-values, plot the points, and graph the parabola:

$$f(x) = -2x^2 + 10x - 7.$$

x	$f(x)$	
$\frac{5}{2}$	$\frac{11}{2}$	← Vertex
0	-7	
1	1	
2	5	
3	5	
4	1	
5	-7	

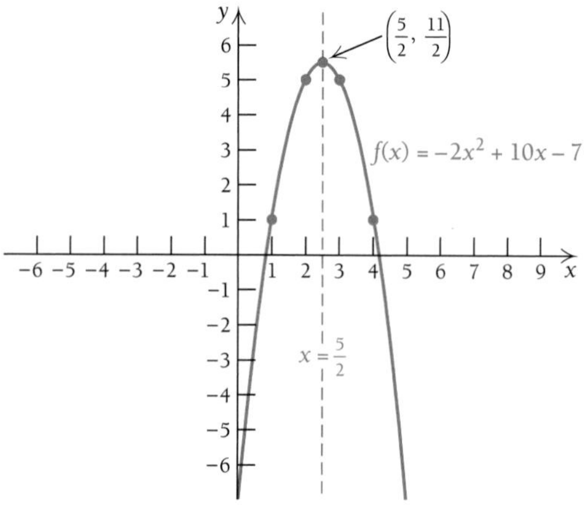

First coordinates of points at which a quadratic function intersects the x-axis (x-intercepts), if they exist, can be found by solving the quadratic equation $ax^2 + bx + c = 0$. If real-number solutions exist, they can be found using the *quadratic formula*.

Theorem 6

The Quadratic Formula

The solutions of any quadratic equation $ax^2 + bx + c = 0$, $a \neq 0$, are given by

$$x = \frac{-b \pm \sqrt{b^2 - 4ac}}{2a}.$$

When solving a quadratic equation, $ax^2 + bx + c = 0$, $a \neq 0$, first try to factor and then use the principle of zero products (see Appendix A). When factoring is not possible or seems difficult, try the quadratic formula. It will always give the solutions. When $b^2 - 4ac < 0$, there are no real-number solutions, but there are solutions in an expanded number system called the *complex numbers*. In this text, we will be considering only real-number solutions.

Example 3   Solve: $3x^2 - 4x = 2$.

Solution   We first find standard form $ax^2 + bx + c = 0$, and then determine a, b, and c:

$$3x^2 - 4x - 2 = 0,$$
$$a = 3, \quad b = -4, \quad c = -2.$$

We then use the quadratic formula:

$$x = \frac{-b \pm \sqrt{b^2 - 4ac}}{2a}$$

$$= \frac{-(-4) \pm \sqrt{(-4)^2 - 4(3)(-2)}}{2 \cdot 3}$$

$$= \frac{4 \pm \sqrt{16 + 24}}{6} = \frac{4 \pm \sqrt{40}}{6}$$

$$= \frac{4 \pm \sqrt{4 \cdot 10}}{6} = \frac{4 \pm 2\sqrt{10}}{6}$$

$$= \frac{2(2 \pm \sqrt{10})}{2 \cdot 3} = \frac{2 \pm \sqrt{10}}{3}.$$

The solutions are $(2 + \sqrt{10})/3$ and $(2 - \sqrt{10})/3$.

Algebraic–Graphical Connection

Let's make an algebraic–graphical connection between the solutions of a quadratic equation and the x-intercepts of a quadratic function.

We just considered the graph of a quadratic function $f(x) = ax^2 + bx + c$, $a \neq 0$. Let's look at the graph of the function $f(x) = x^2 + 6x + 8$ and its x-intercepts, shown below. The **x-intercepts**, $(-4, 0)$ and $(-2, 0)$, are the points at which the graph crosses the x-axis. These pairs are also the points of intersection of the graphs of $f(x) = x^2 + 6x + 8$ and $g(x) = 0$ (the x-axis).

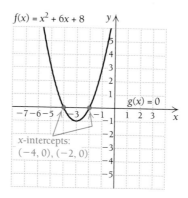

$f(x) = x^2 + 6x + 8$

$g(x) = 0$

x-intercepts:
$(-4, 0), (-2, 0)$

TECHNOLOGY CONNECTION

Exercise

1. **a)** Below is the graph of

$$f(x) = x^2 - 6x + 8.$$

Using *only* the graph, find the x-intercepts of the graph.

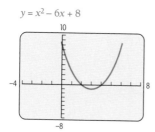

$y = x^2 - 6x + 8$

b) Using *only* the graph, find the solutions of $x^2 - 6x + 8 = 0$.

c) Compare the answers to (a) and (b).

Now let's consider solving the quadratic equation $x^2 + 6x + 8 = 0$. We use factoring although we could use the quadratic formula instead.

$$x^2 + 6x + 8 = 0$$
$$(x + 4)(x + 2) = 0 \qquad \text{Factoring}$$
$$x + 4 = 0 \quad or \quad x + 2 = 0 \qquad \text{Principle of Zero Products}$$
$$x = -4 \quad or \qquad x = -2.$$

We see that the solutions of $0 = x^2 + 6x + 8$, -4 and -2, are the first coordinates of the x-intercepts, $(-4, 0)$ and $(-2, 0)$, of the graph of $f(x) = x^2 + 6x + 8$.

Polynomial Functions

Linear and quadratic functions are part of a general class of *polynomial functions*.

> *Definition*
>
> A **polynomial function** f is given by
>
> $$f(x) = a_n x^n + a_{n-1} x^{n-1} + \cdots + a_2 x^2 + a_1 x^1 + a_0,$$
>
> where n is a nonnegative integer and $a_n, a_{n-1}, \ldots, a_1, a_0$ are real numbers, called the **coefficients** of the polynomial.

The following are examples of polynomial functions:

$f(x) = -5$,   (A constant function)
$f(x) = 4x + 3$,   (A linear function)
$f(x) = -x^2 + 2x + 3$,   (A quadratic function)
$f(x) = 2x^3 - 4x^2 + x + 1$.   (A cubic function)

In general, creating graphs of polynomial functions other than linear and quadratic functions is difficult unless we use a grapher. We use calculus to sketch such graphs in Chapter 3. Some **power functions,** such as

$$f(x) = ax^n,$$

are relatively easy to graph.

Example 4   Using the same set of axes, graph $f(x) = x^2$ and $g(x) = x^3$.

Solution   We set up a table of values, plot the points, and then draw the graphs.

x	x^2	x^3
-2	4	-8
-1	1	-1
$-\frac{1}{2}$	$\frac{1}{4}$	$-\frac{1}{8}$
0	0	0
$\frac{1}{2}$	$\frac{1}{4}$	$\frac{1}{8}$
1	1	1
2	4	8

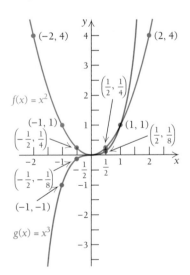

TECHNOLOGY   CONNECTION

Solving Polynomial Equations

The INTERSECT Feature

Consider solving the equation

$$x^3 = 3x + 1.$$

Doing so amounts to finding the x-coordinates of the point(s) of intersection of the graphs of the two functions

$$f(x) = x^3 \quad \text{and} \quad g(x) = 3x + 1.$$

We enter the functions as

$$y_1 = x^3 \quad \text{and} \quad y_2 = 3x + 1$$

and then graph. We use a $[-3, 3, -5, 8]$ window to see the curvature and possible points of intersection.

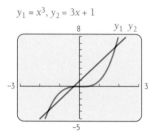

$y_1 = x^3, y_2 = 3x + 1$

There appear to be at least three points of intersection. Using the INTERSECT feature in the CALC menu, we see that the point of intersection on the left is about $(-1.53, -3.60)$.

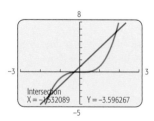

In a similar manner, we find the other points of intersection to be about $(-0.35, -0.04)$ and $(1.88, 6.64)$. The solutions of $x^3 = 3x + 1$ are the x-coordinates of these points:

$$-1.53, \ -0.35, \quad \text{and} \quad 1.88.$$

The ZERO Feature

A **ZERO**, or **ROOT**, feature can be used to solve an equation. The word "zero" in this context refers to an input, or x-value, for which the output of a function is 0. That is, c is a **zero** of the function f if $f(c) = 0$.

To use such a feature, we must first get a 0 on one side of the equation. Thus, to solve $x^3 = 3x + 1$, we consider $x^3 - 3x - 1 = 0$ by subtracting $3x$ and then 1. Graphing $y = x^3 - 3x - 1$ and using the **ZERO** feature to find the zero on the left, we obtain a screen like the following.

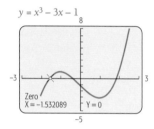

$y = x^3 - 3x - 1$

We see that $x^3 - 3x - 1 = 0$ when $x \approx -1.53$, so -1.53 is an approximate solution of the equation $x^3 = 3x + 1$. Proceeding in a similar manner, we can approximate the other solutions as -0.35 and 1.88.

Exercises

Using the **INTERSECT** feature, solve the equation.

1. $x^2 = 10 - 3x$
2. $2x + 24 = x^2$
3. $x^3 = 3x - 2$
4. $x^4 - 2x^2 = 0$

Using the **ZERO** feature, solve the equation.

5. $0.4x^2 = 280x$
   (*Hint:* Use $[-200, 800, -100{,}000, 200{,}000]$.)
6. $\frac{1}{3}x^3 - \frac{1}{2}x^2 = 2x - 1$
7. $x^2 = 0.1x^4 + 0.4$
8. $0 = 2x^4 - 4x^2 + 2$
9. $x^4 + x^3 = 4x^2 + 2x - 4$
10. $11x^2 - 9x - 18 = x^4 - x^3$

Find the zeros of the function.

11. $f(x) = 3x^2 - 4x - 2$
12. $f(x) = -x^3 + 6x^2 + 5$

Rational Functions

> **Definition**
> Functions given by the quotient, or ratio, of two polynomials are called **rational functions.**

The following are examples of rational functions:

$$f(x) = \frac{x^2 - 9}{x - 3},$$

$$g(x) = \frac{x^2 - 16}{x + 4},$$

$$h(x) = \frac{x - 3}{x^2 - x - 2}.$$

The domain of a rational function is restricted to those input values that do not result in division by zero. Thus for f above, the domain consists of all real numbers except 3. To determine the domain of h, we set the denominator equal to 0 and solve:

$$x^2 - x - 2 = 0$$
$$(x + 1)(x - 2) = 0$$
$$x = -1 \quad or \quad x = 2.$$

Therefore, -1 and 2 are not in the domain. The domain of h consists of all real numbers except -1 and 2. We will refer to -1 and 2 as **split numbers** because, in the context of rational functions, they "split," or "separate," the intervals in the domain of the function.

The graphing of most rational functions is rather complicated and is best dealt with using the tools of calculus that we will develop in Chapters 2 and 3. At this point, we will consider graphs that are fairly easy and leave the more complicated graphs for Chapter 3 or for a grapher.

Example 5   Graph: $f(x) = \dfrac{x^2 - 9}{x - 3}$.

Solution   This particular function can be simplified before we graph it. We do so by factoring the numerator and removing a factor of 1 as follows:

$$f(x) = \frac{x^2 - 9}{x - 3}$$

$$= \frac{(x - 3)(x + 3)}{x - 3}$$

$$= \frac{x - 3}{x - 3} \cdot \frac{x + 3}{1}$$

$$= x + 3, \quad x \neq 3.$$

This simplification assumes that x is not 3. The number 3 is not in the domain because it would result in division by zero. Thus we can express the function as follows:

$$y = f(x) = x + 3, \quad x \neq 3.$$

To find function values, we substitute any value for x other than 3. We make calculations as in the following table and draw the graph. The open circle at the point $(3, 6)$ indicates that it is not part of the graph.

x	1	0	2	−3	4	−1	−2
y	4	3	5	0	7	2	1

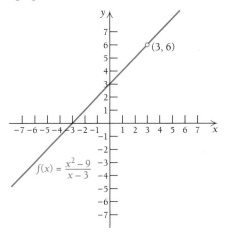

$$f(x) = \frac{x^2 - 9}{x - 3}$$

One important class of rational functions is given by $f(x) = k/x$, where k is constant.

Example 6   Graph: $f(x) = 1/x$.

Solution   We make a table of values, plot the points, and then draw the graph.

x	f(x)
−3	$-\frac{1}{3}$
−2	$-\frac{1}{2}$
−1	−1
$-\frac{1}{2}$	−2
$-\frac{1}{4}$	−4
$\frac{1}{4}$	4
$\frac{1}{2}$	2
1	1
2	$\frac{1}{2}$
3	$\frac{1}{3}$

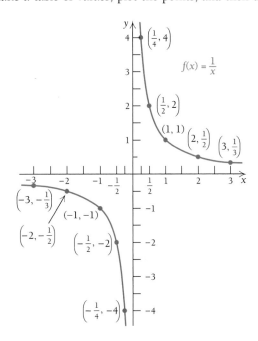

$$f(x) = \frac{1}{x}$$

TECHNOLOGY   CONNECTION

Graphs of Rational Functions

Consider the rational function given by

$$f(x) = \frac{2x + 1}{x - 3}.$$

Let's consider its graph, two versions of which are shown below.

CONNECTED Mode

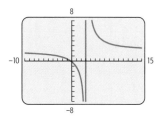

DOT Mode

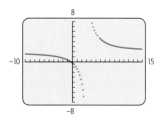

Here we see a disadvantage of the grapher. **CONNECTED** mode can lead to an *incorrect* graph. Because, in **CONNECTED** mode, a grapher connects plotted points with line segments, it connects

branches of the graph, making it "appear" as though the vertical line $x = 3$ is part of the graph.

On the other hand, in **DOT** mode, a grapher simply plots dots representing coordinates of points.

If you have a choice in plotting functions given by rational expressions, use **DOT** mode.

Exercises

Graph each of the following functions using **DOT** mode.

1. $f(x) = \dfrac{4}{x - 2}$ 2. $f(x) = \dfrac{x}{x + 2}$

3. $f(x) = \dfrac{x^2 - 1}{x^2 + x - 6}$ 4. $f(x) = \dfrac{x^2 - 4}{x - 1}$

5. $f(x) = \dfrac{10}{x^2 + 4}$ 6. $f(x) = \dfrac{8}{x^2 - 4}$

7. $f(x) = \dfrac{2x + 3}{3x^2 + 7x - 6}$ 8. $f(x) = \dfrac{2x^3}{x^2 + 1}$

9. Graph

$$f(x) = \frac{x^2 - 9}{x - 3}.$$

Compare your result with Example 5. Try using both **DOT** and **CONNECTED** modes, and different viewing windows and the **TRACE** feature as well. Does the "hole" ever become evident?

In Example 6, note that 0 is not in the domain of the function because it would yield a denominator of zero. The function is decreasing over the intervals $(-\infty, 0)$ and $(0, \infty)$. The function $f(x) = 1/x$ is an example of **inverse variation.**

Definition

y **varies inversely** as *x* if there is some positive number k such that $y = k/x$. We also say that *y* is **inversely proportional** to *x*.

Example 7 *Business: Stocks and Gold.* Certain economists theorize that stock prices are inversely proportional to the price of gold. That is, when the price of gold goes up, the prices of stocks go down; and when the price of gold goes down, the prices of stocks go up. Let's assume that the Dow Jones Industrial Average *D*, an index of the overall price of stock, is inversely proportional to the price of gold *G*, in dollars per ounce. One day the Dow Jones was 9177 and the price of gold was $364 per ounce. What will the Dow Jones be if the price of gold drops to $300?

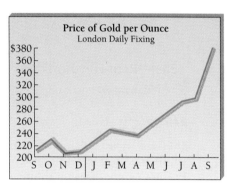

Source: Dow Jones Industrial Average

Solution We know that $D = k/G$, so $9177 = k/364$ and $k = 3,340,428$. Thus,

$$D = \frac{3,340,428}{G}.$$

We substitute 300 for *G* and compute *D*:

$$D = \frac{3,340,428}{300} \approx 11,134.76.$$

Warning! Do not put too much "stock" in the equation of this example. It is meant only to give us an idea of economic relationships. An equation to predict the stock market accurately has not been found!   ◄

Absolute-Value Functions

The following is an example of an absolute-value function and its graph. The absolute value of a number is its distance from 0 on the number line. We denote the absolute value of a number *x* as $|x|$.

Example 8 Graph: $f(x) = |x|$.

Solution We make a table of values, plot the points, and then draw the graph.

x	-3	-2	-1	0	1	2	3
$f(x)$	3	2	1	0	1	2	3

TECHNOLOGY
CONNECTION

Exercises
Graph.
1. $f(x) = |x|$
2. $f(x) = |x^2 - 4|$

We can think of this function as being defined piecewise by considering the definition of absolute value:

$$f(x) = |x| = \begin{cases} x, & \text{if } x \geq 0, \\ -x, & \text{if } x < 0. \end{cases}$$

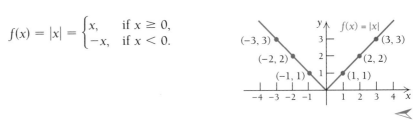

Square-Root Functions

The following is an example of a square-root function and its graph.

Example 9   Graph: $f(x) = -\sqrt{x}$.

Solution   The domain of this function is just the nonnegative numbers—the interval $[0, \infty)$. You can find approximate values of square roots on your calculator.
We set up a table of values, plot the points, and then draw the graph.

x	0	1	2	3	4	5
$f(x)$, or $-\sqrt{x}$	0	-1	-1.4	-1.7	-2	-2.2

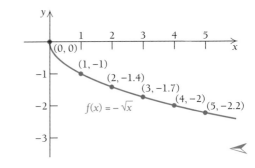

Power Functions with Rational Exponents

We are motivated to define rational exponents so that the laws of exponents still hold (see the appendix). For example, if the laws of exponents are to hold, we would have

$$a^{1/2} \cdot a^{1/2} = a^{1/2+1/2} = a^1 = a.$$

Thus we are led to define $a^{1/2}$ as $\sqrt{a}$. Similarly, we are led to define $a^{1/3}$ as the cube root of a, $\sqrt[3]{a}$. In general,

$$a^{1/n} = \sqrt[n]{a}, \quad \text{provided } \sqrt[n]{a} \text{ is defined.}$$

Again, if the laws of exponents are to hold, we would have

$$\sqrt[n]{a^m} = (a^m)^{1/n} = (a^{1/n})^m = a^{m/n}.$$

An expression $a^{-m/n}$ is defined by

$$a^{-m/n} = \frac{1}{a^{m/n}} = \frac{1}{\sqrt[n]{a^m}}.$$

Example 10   Convert to rational exponents.

a) $\sqrt[3]{x^2} = x^{2/3}$

b) $\sqrt[4]{y} = y^{1/4}$

c) $\dfrac{1}{\sqrt[3]{b^5}} = \dfrac{1}{b^{5/3}} = b^{-5/3}$

d) $\dfrac{1}{\sqrt{x}} = \dfrac{1}{x^{1/2}} = x^{-1/2}$

e) $\sqrt{x^8} = x^{8/2}$, or x^4

Example 11   Convert to radical notation.

a) $x^{1/3} = \sqrt[3]{x}$

b) $t^{6/7} = \sqrt[7]{t^6}$

c) $x^{-2/3} = \dfrac{1}{x^{2/3}} = \dfrac{1}{\sqrt[3]{x^2}}$

d) $e^{-1/4} = \dfrac{1}{e^{1/4}} = \dfrac{1}{\sqrt[4]{e}}$

Example 12   Simplify.

a) $8^{5/3} = (8^{1/3})^5 = (\sqrt[3]{8})^5 = 2^5 = 32$

b) $81^{3/4} = (81^{1/4})^3 = (\sqrt[4]{81})^3 = 3^3 = 27$

TECHNOLOGY   CONNECTION

Graphing Radical Functions

Graphing functions defined by radical expressions involves approximating roots. Since the square root of a negative number is not a real number, y-values may not exist for some x-values. For example, y-values for the graph of $f(x) = \sqrt{x} - 1$ do not exist for x-values that are less than 1 because square roots of negative numbers would result.

We must enter $y = \sqrt{x} - 1$ using parentheses around the radicand as $y_1 = \sqrt{(x - 1)}$. Some graphers supply the left parenthesis automatically.

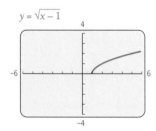

$y = \sqrt{x - 1}$

Similarly, y-values for the graph of $f(x) = \sqrt{2} - x$ do not exist for x-values that are greater than 2.

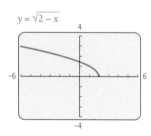

$y = \sqrt{2 - x}$

Exercises

Graph each of the following functions. Then use the **TABLE** and **TRACE** features to determine the domain and the range of each function. The **MATH** menu contains $\boxed{\sqrt[3]{}}$ and $\boxed{\sqrt[x]{}}$ keys to enter kth roots.

1. $f(x) = \sqrt{x}$
2. $g(x) = \sqrt{x} + 2$
3. $f(x) = \sqrt[3]{x}$
4. $f(x) = \sqrt[3]{x} - 2$
5. $f(x) = \sqrt[4]{x - 1}$
6. $F(x) = \sqrt[5]{6 - x}$
7. $g(x) = 5 - \sqrt{x} + 3$
8. $f(x) = 4 - \sqrt[3]{x}$
9. $f(x) = x^{2/3}$
10. $g(x) = x^{1.41}$

Use the **GRAPH** and **TABLE** features to determine whether each of the following is correct.

11. $\sqrt{x + 4} = \sqrt{x} + 2$
12. $\sqrt{25x} = 5\sqrt{x}$

Earlier when we graphed $f(x) = \sqrt{x}$, we were also graphing $f(x) = x^{1/2}$, or $f(x) = x^{0.5}$. The power functions

$$f(x) = ax^k, \quad k \text{ fractional},$$

do occur in applications.

Caribou in their territorial area.

Example 13 *Life Science: Home Range.* The *home range* of an animal is defined as the region to which the animal confines its movements. It has been hypothesized in statistical studies* that the area H of that region can be approximated by the function

$$H = W^{1.41},$$

where W is the weight of the animal. Graph the function.

Solution We can approximate function values using a power key $\boxed{y^x}$ on a calculator.

W	0	10	20	30	40	50
H	0	26	68	121	182	249

We see that

$$H = W^{1.41} = W^{141/100} = \sqrt[100]{W^{141}}.$$

The graph is shown below. Note that the function values increase from left to right. As body weight increases, the area over which the animal moves increases.

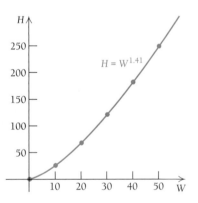

Supply and Demand Functions

Supply and demand in economics are modeled by increasing and decreasing functions.

---

*Source: Emlen, J. M., *Ecology: An Evolutionary Approach*, p. 200 (Reading, MA: Addison-Wesley, 1973).

DEMAND FUNCTIONS

Look at the following table and graph. They show the relationship between the price p per bag of sugar and the quantity x of 5-lb bags that consumers will buy at that price.

DEMAND SCHEDULE

Price (p) per 5-lb Bag	Quantity (x) of 5-lb Bags (in millions)
$5	4
4	5
3	7
2	10
1	15

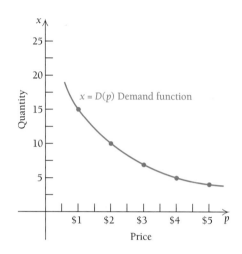

Note that as the price per bag decreases, the quantity demanded by consumers increases; and as the price per bag increases, the quantity demanded by consumers decreases. Thus it is natural to think of x as a function of p,

$$x = D(p),$$

and to graph the function with the price p on the horizontal axis and the quantity x on the vertical axis. Thus, for a **demand function** D, $D(p)$ is the quantity x of units demanded by consumers when the price is p.

In some situations, it may be appropriate to consider the demand function as $D(x) = p$, where the price is a function of the quantity demanded.

SUPPLY FUNCTIONS

Look at the table and graph shown on the following page. They show the relationship between the price p per bag of sugar and the quantity x of 5-lb bags that sellers are willing to supply, or sell, at that price. Note that as the price per bag increases, the more the sellers are willing to supply; and as the price per bag decreases, the less sellers are willing to supply. Thus it is natural to think of x as a function of p,

$$x = S(p),$$

and to graph the function with the price p on the horizontal axis and the quantity x on the vertical axis. Thus, for a **supply function** S, $S(p)$ is the quantity x of items that the sellers are willing to supply, or sell, at price p.

In some situations, it may be appropriate to consider the supply function as $S(x) = p$, where the price is a function of the quantity supplied.

SUPPLY SCHEDULE

Price (p) per 5-lb Bag	Quantity (x) of 5-lb Bags (in millions)
$1	0
2	10
3	16
4	20
5	22

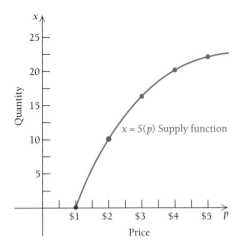

Let's now look at these curves together. As price increases, supply increases and demand decreases; and as price decreases, supply decreases and demand increases. The point of intersection of the two curves (p_E, x_E) is called the **equilibrium point.** The equilibrium price p_E (in this case, $2 per bag) is the point at which the amount x_E (in this case, 10 million bags) that sellers willingly supply is the same as the amount that consumers willingly demand. The situation is analogous to a buyer and seller haggling over the sale of an item. The equilibrium point, or selling price, is what they finally agree on.

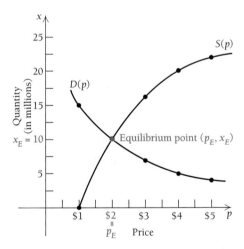

Example 14 *Economics: Equilibrium Point.* Find the equilibrium point for the demand and supply functions

$$D(p) = 300 - 49p$$

and

$$S(p) = 100 + p.$$

TECHNOLOGY
CONNECTION

Exercise

1. Use the **INTERSECT** feature to find the equilibrium point for the demand and supply functions

$$D(p) = 1123.6 - 61.4p$$

and

$$S(p) = 201.8 + 4.6p.$$

Solution   To find the equilibrium point, we set $D(p) = S(p)$ and solve:

$$300 - 49p = 100 + p$$
$$-50p = -200$$
$$p = \frac{-200}{-50} = 4.$$

Thus, $p_E = \$4$. To find x_E, we substitute p_E into either $D(p)$ or $S(p)$. We choose $S(p)$:

$$x_E = S(p_E) = S(4) = 100 + 4 = 104.$$

Thus the equilibrium quantity is 104 units and the equilibrium point is ($4, 104). ◄

EXERCISE SET 1.5

Using the same set of axes, graph the pair of equations.

1. $y = \frac{1}{2}x^2$ and $y = -\frac{1}{2}x^2$

2. $y = \frac{1}{4}x^2$ and $y = -\frac{1}{4}x^2$

3. $y = x^2$ and $y = (x - 1)^2$

4. $y = x^2$ and $y = (x - 3)^2$

5. $y = x^2$ and $y = (x + 1)^2$

6. $y = x^2$ and $y = (x + 3)^2$

7. $y = |x|$ and $y = |x + 3|$

8. $y = |x|$ and $y = |x + 1|$

9. $y = x^3$ and $y = x^3 + 1$

10. $y = x^3$ and $y = x^3 - 1$

11. $y = \sqrt{x}$ and $y = \sqrt{x + 1}$

12. $y = \sqrt{x}$ and $y = \sqrt{x - 2}$

Graph.

13. $y = x^2 - 4x + 3$

14. $y = x^2 - 6x + 5$

15. $y = -x^2 + 2x - 1$

16. $y = -x^2 - x + 6$

17. $y = \frac{2}{x}$

18. $y = \frac{3}{x}$

19. $y = \frac{-2}{x}$

20. $y = \frac{-3}{x}$

21. $y = \frac{1}{x^2}$

22. $y = \frac{1}{x - 1}$

23. $y = \sqrt[3]{x}$

24. $y = \frac{1}{|x|}$

25. $f(x) = \frac{x^2 - 9}{x + 3}$

26. $g(x) = \frac{x^2 - 4}{x - 2}$

27. $f(x) = \frac{x^2 - 1}{x - 1}$

28. $g(x) = \frac{x^2 - 25}{x + 5}$

Solve.

29. $x^2 - 2x = 2$

30. $x^2 - 2x + 1 = 5$

31. $x^2 + 6x = 1$

32. $x^2 + 4x = 3$

33. $4x^2 = 4x + 1$

34. $-4x^2 = 4x - 1$

35. $3y^2 + 8y + 2 = 0$

36. $2p^2 - 5p = 1$

Convert to expressions with rational exponents.

37. $\sqrt{x^3}$

38. $\sqrt{x^5}$

39. $\sqrt[5]{a^3}$

40. $\sqrt[4]{b^2}$

41. $\sqrt[7]{t}$

42. $\sqrt[8]{c}$

43. $\frac{1}{\sqrt[3]{t^4}}$

44. $\frac{1}{\sqrt[5]{b^6}}$

45. $\frac{1}{\sqrt{t}}$

46. $\frac{1}{\sqrt{m}}$

47. $\frac{1}{\sqrt{x^2 + 7}}$

48. $\sqrt{x^3 + 4}$

Convert to radical notation.

49. $x^{1/5}$

50. $t^{1/7}$

51. $y^{2/3}$

52. $t^{2/5}$

53. $t^{-2/5}$

54. $y^{-2/3}$

55. $b^{-1/3}$

56. $b^{-1/5}$

57. $e^{-17/6}$

58. $m^{-19/6}$

59. $(x^2 - 3)^{-1/2}$

60. $(y^2 + 7)^{-1/4}$

Simplify.

61. $9^{3/2}$ **62.** $16^{5/2}$ **63.** $64^{2/3}$

64. $8^{2/3}$ **65.** $16^{3/4}$ **66.** $25^{5/2}$

Determine the domain of the function.

67. $f(x) = \dfrac{x^2 - 25}{x - 5}$ **68.** $f(x) = \dfrac{x^2 - 4}{x + 2}$

69. $f(x) = \dfrac{x^3}{x^2 - 5x + 6}$ **70.** $f(x) = \dfrac{x^4 + 7}{x^2 + 6x + 5}$

71. $f(x) = \sqrt{5x + 4}$ **72.** $f(x) = \sqrt{2x - 6}$

Applications ....

BUSINESS AND ECONOMICS

Find the equilibrium point for the given demand and supply functions.

73. $D(p) = 1000 - 10p, \quad S(p) = 250 + 5p$

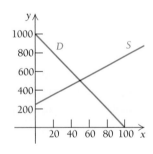

74. $D(p) = 8800 - 30p, \quad S(p) = 7000 + 15p$

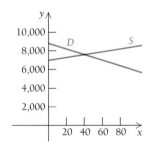

75. $D(p) = \dfrac{5}{p}, \quad S(p) = \dfrac{p}{5}$

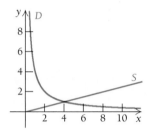

76. $D(p) = \dfrac{4}{p}, \quad S(p) = \dfrac{p}{4}$

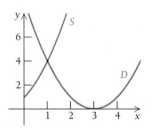

77. $D(p) = (p - 3)^2, \quad S(p) = p^2 + 2p + 1$

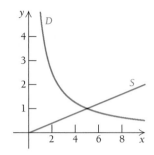

78. $D(p) = (p - 4)^2, \quad S(p) = p^2 + 2p + 6$

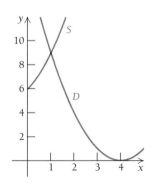

79. $D(p) = 5 - p, \quad S(p) = \sqrt{p + 7}$

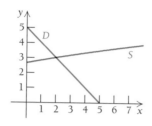

80. $D(p) = 7 - p, \quad S(p) = 2\sqrt{p + 1}$

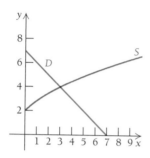

81. *Stock Prices and Prime Rate.* It is theorized that the price per share of a stock is inversely proportional to the prime (interest) rate. Recently, the price per share S of Silicon Graphics, Inc., was $\$13\frac{7}{8}$ and the prime rate R was $8\frac{1}{4}\%$. The prime rate rose to $9\frac{1}{2}\%$. What would the price per share be if the assumption of inverse proportionality is correct?

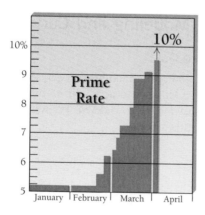

82. *Demand.* The quantity sold x of a certain kind of fax machine is inversely proportional to the price p. It was found that 240,000 fax machines will be

sold if the price is $320 each. How many will be sold if the price is $480 each?

LIFE AND PHYSICAL SCIENCES

83. *Fruit Stacking.* The number of oranges in a pile of the type shown below is approximated by the function

$$f(x) = \frac{1}{6}x^3 + \frac{1}{2}x^2 + \frac{1}{3}x,$$

where $f(x)$ is the number of oranges and x is the number of layers. Each layer is an equilateral triangle. Find the number of oranges when the number of layers is 7, 10, and 12.

84. *Territorial Area.* Refer to Example 13. The *territorial area* of an animal is defined to be its defended region, or exclusive region. For example, a lion has a certain region over which it is ruler. The area T of that region can be approximated by the power function

$$T = W^{1.31},$$

where W is the weight of the animal. Complete the table of approximate function values and graph the function.

W	0	10	20	30	40	50	100	150
T	0	20						

Synthesis ....

85. *Life Science: Pollution Control.* Pollution control has become a very important concern in all countries. If controls are not put in place, it has been predicted that the function

$$P = 1000t^{5/4} + 14{,}000$$

will describe the average pollution, in particles of pollution per cubic centimeter, in most cities at time t, in years, where $t = 0$ corresponds to 1970 and $t = 32$ corresponds to 2002.

  a) Predict the pollution in 2002; 2005; 2010.
  b) Graph the function over the interval $[0, 40]$.

tw 86. At most, how many y-intercepts can a function have? Explain.

tw 87. Explain the difference between a rational function and a polynomial function. Is every polynomial function a rational function?

 TECHNOLOGY CONNECTION

Use the **ZERO** feature or the **INTERSECT** feature to approximate the zeros of the function to three decimal places.

88. $f(x) = x^3 - x$
  (Also, use algebra to find the zeros of the function.)

89. $f(x) = x^3 - 2x^2 - 2$

90. $f(x) = 2x^3 - x^2 - 14x - 10$

91. $f(x) = \frac{1}{2}(|x - 4| + |x - 7|) - 4$

92. $f(x) = x^4 + 4x^3 - 36x^2 - 160x + 300$

93. $f(x) = \sqrt{7 - x^2}$

94. $f(x) = |x + 1| + |x - 2| - 5$

95. $f(x) = |x + 1| + |x - 2|$

96. $f(x) = |x + 1| + |x - 2| - 3$

97. $f(x) = x^8 + 8x^7 - 28x^6 - 56x^5 + 70x^4$
      $+ 56x^3 - 28x^2 - 8x + 1$

98. Find the equilibrium point for the demand and supply functions

$$D(p) = 83 - p$$

and

$$S(p) = 24\sqrt{p + 1.9}.$$

1.6

OBJECTIVES

➤ Use curve fitting to find a mathematical model for a set of data and use the model to make predictions.

Mathematical Modeling and Curve Fitting

Fitting Functions to Data

We have developed a library of functions that can serve as models for many applications. Although others will be introduced later, let's look at those that we have considered. We will not consider rational functions in this section.

Linear function:
$f(x) = mx + b$

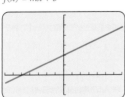

Quadratic function:
$f(x) = ax^2 + bx + c, a > 0$

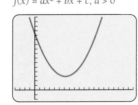

Quadratic function:
$f(x) = ax^2 + bx + c, a < 0$

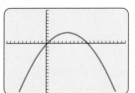

Absolute-value function:
$f(x) = |x|$

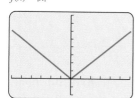

Cubic function:
$f(x) = ax^3 + bx^2 + cx + d, a > 0$

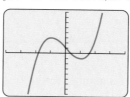

Quartic function:
$f(x) = ax^4 + bx^3 + cx^2 + dx + e, a > 0$

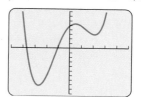

Now let's consider some real-world data. How can we decide which, if any, type of function might fit the data? One simple way is to examine a graph of the data called a **scatterplot.** Then we look for a pattern resembling one of the graphs above. For example, data might be modeled by a linear function if the graph resembles a straight line. The data might be modeled by a quadratic function if the graph rises and then falls, or falls and then rises, in a curved manner resembling a parabola. We will not consider the cubic, quartic, or polynomial functions in detail (that will be left to Chapter 3), but we show them for reference.

Let's now use our library of functions to see which, if any, might fit certain data situations.

Examples *Choosing Models.* For the scatterplots and graphs below, determine which, if any, of the following functions might be used as a model for the data.

Linear, $f(x) = mx + b$;

Quadratic, $f(x) = ax^2 + bx + c, a > 0$;

Quadratic, $f(x) = ax^2 + bx + c, a < 0$;

Polynomial, neither quadratic nor linear.

1.

The data rise and then fall in a curved manner fitting a quadratic function

$$f(x) = ax^2 + bx + c, a < 0.$$

2.

The data seem to fit a linear function

$$f(x) = mx + b.$$

3.

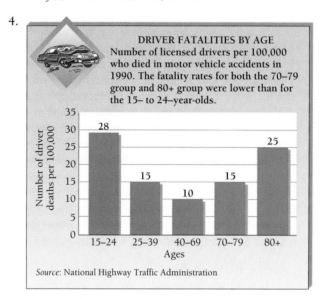

The data rise in a manner fitting the right side of a quadratic function

$$f(x) = ax^2 + bx + c, a > 0.$$

4.

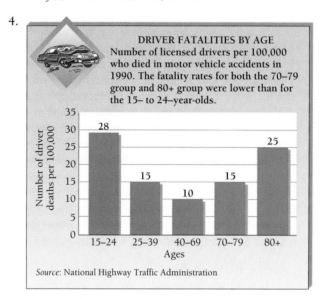

DRIVER FATALITIES BY AGE
Number of licensed drivers per 100,000 who died in motor vehicle accidents in 1990. The fatality rates for both the 70–79 group and 80+ group were lower than for the 15– to 24–year-olds.

Source: National Highway Traffic Administration

The data fall and then rise in a curved manner fitting a quadratic function

$$f(x) = ax^2 + bx + c, a > 0.$$

5.

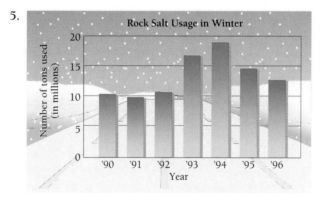

Rock Salt Usage in Winter

Source: Salt Institute

The data fall, then rise, then fall again, so they do not fit a linear or quadratic function but might fit a polynomial function that is neither quadratic nor linear.

The following is a procedure that sometimes works for finding mathematical models.

Curve Fitting

Given a set of data:

1. Graph the data (in the form of a scatterplot).
2. Look at the scatterplot to determine whether a known function seems to fit.
3. Find a function that fits the data by using data points to derive the constants.

Example 6 *Business: Admission to Disney World.* The following table lists data showing the price P of a one-day adult admission to Disney World for years since 1993.

Years, x (since 1993)	Price P of a One-Day Adult Admission to Disney World	Scatterplot
0. 1993	$34.00	
1. 1994	36.00	
2. 1995	37.00	**Disney World**
3. 1996	40.81	It appears that the data points can be represented or fitted by a straight line.
4. 1997	42.14	The graph is linear.
5. 1998	44.52	

Source: Amusement Business

a) Make a scatterplot of the data and determine whether the data seem to fit a linear function.
b) Find a linear function that (approximately) fits the data.
c) Use the model to predict the price of a one-day adult admission in 2000 ($x = 7$).

Solution

a) The scatterplot is shown above. The data tend to follow a straight line, although a "perfect" straight line cannot be drawn through the data.

b) We consider the linear function

$$P(x) = mx + b, \tag{1}$$

where P is the price of a one-day adult admission to Disney World x years after

1993. To derive the constants (or parameters) m and b, we choose two data points. Although this procedure is somewhat arbitrary, we try to choose two points that follow the general linear pattern. In this case, we pick (1, 36) and (5, 44.52). Since the points are to be solutions of equation (1), it follows that

$$44.52 = m \cdot 5 + b, \quad \text{or} \quad 44.52 = 5m + b, \tag{2}$$

$$36.00 = m \cdot 1 + b, \quad \text{or} \quad 36.00 = m + b. \tag{3}$$

We now have a system of equations. We solve by subtracting each side of equation (3) from equation (2) to eliminate b:

$$8.52 = 4m.$$

Then we have

$$m = \frac{8.52}{4}$$

$$= 2.13.$$

Substituting 2.13 for m in equation (3), we can solve for b:

$$36 = 2.13 + b$$

$$33.87 = b.$$

Substituting these values of m and b into equation (1), we get the function (model) given by

$$P(x) = 2.13x + 33.87. \tag{4}$$

We can see the representative data points and the linear function in the graph below.

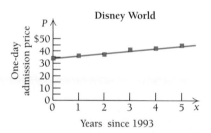

c) The price of a one-day adult admission in 2000 is found by letting $x = 7$ in equation (4):

$$P(x) = 2.13(7) + 33.87$$

$$= 48.78.$$

Thus we predict the price to be $48.78 in 2000.

TECHNOLOGY   CONNECTION

Linear Regression: Fitting a Linear Function to Data

We now consider **linear regression,** a procedure that can be used to fit a linear function to a set of data. Although the complete basis for this method belongs in Section 7.5, we consider it here because we can carry out the procedure easily using technology. One advantage of linear regression is that it uses *all* data points rather than just two. The grapher gives us the powerful capability to find linear models and make predictions.

Example   *Business: Admission to Disney World.* Consider the data in Example 6.

a) Fit a regression line to the data using the **REGRESSION** feature on a grapher.
b) Graph the regression line with the scatterplot.
c) Use the model to predict the price of a one-day adult admission in 2000 ($x = 7$).

Solution

a) We can fit a linear function to the data using linear regression. We enter the data in **STAT** lists, use the **LINEAR REGRESSION** feature from the **STAT CALC** menu, and copy the regression equation to the Y = screen. Consult the manual for your particular grapher or the GCM for the keystrokes.

b) We set up a **STAT PLOT** to draw a scatterplot and select a window that shows all the data.

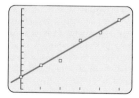

If the scatterplot had not been shown on p. 77, we would begin this example by creating a scatterplot to determine whether it appears as though a linear function might fit the data points.

c) To predict the price of admission in 2000, we enter $Y_1(7)$ on the home screen.

Thus we predict the price of admission in 2000 to be about $48.70. Note that this disagrees with the estimate found in Example 6 by only $0.08. Such a slight discrepancy may not always be the case.

Exercise

1. *Study Time and Test Scores.* The data in the following table relate study time and test scores.

Study Time (in hours)	Test Grade (in percent)
19	83
20	85
21	88
22	91
23	?

a) Fit a regression line to the data using the **REGRESSION** feature on a grapher.
b) Make a scatterplot of the data. Then graph the regression line with the scatterplot.
c) Use the linear model to predict the test score received when one has studied for 23 hr.
tw d) Discuss the appropriateness of a linear model of this data.

Example 7 *Life Science: Hours of Sleep and Death Rate.* In a study by Dr. Harold J. Morowitz of Yale University, data were gathered that showed the relationship between the death rate of men and the average number of hours per day that the men slept. These data are listed in the following table.

Average Number of Hours of Sleep, x	Death Rate per 100,000 Males, y
5	1121
6	805
7	626
8	813
9	967

Source: Morowitz, Harold J., "Hiding in the Hammond Report," *Hospital Practice.*

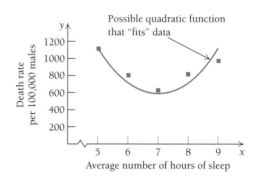

a) Make a scatterplot of the data and determine whether the data seem to fit a quadratic function.

b) Find a quadratic function that fits the data.

c) Use the model to find the death rate for males who sleep 2 hr, 8 hr, and 10 hr.

Solution

a) The scatterplot is shown above. Note that the rate drops and then rises, which suggests that a quadratic function might fit the data.

b) We consider the quadratic model

$$y = ax^2 + bx + c. \tag{1}$$

To derive the constants (or parameters) a, b, and c, we use the three data points (5, 1121), (7, 626), and (9, 967). Since these points are to be solutions of equation (1), it follows that

$$1121 = a \cdot 5^2 + b \cdot 5 + c, \quad \text{or} \quad 1121 = 25a + 5b + c,$$
$$626 = a \cdot 7^2 + b \cdot 7 + c, \quad \text{or} \quad 626 = 49a + 7b + c,$$
$$967 = a \cdot 9^2 + b \cdot 9 + c, \quad \text{or} \quad 967 = 81a + 9b + c.$$

We solve this system of three equations in three variables using procedures of algebra and get

$$a = 104.5, \quad b = -1501.5, \quad \text{and} \quad c = 6016.$$

Substituting these values of a, b, and c into equation (1), we get the function (model) given by

$$y = 104.5x^2 - 1501.5x + 6016.$$

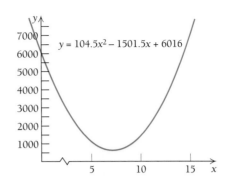

c) The death rate for 2 hr is given by

$$y = 104.5(2)^2 - 1501.5(2) + 6016 = 3431.$$

The death rate for 8 hr is given by

$$y = 104.5(8)^2 - 1501.5(8) + 6016 = 692.$$

The death rate for 10 hr is given by

$$y = 104.5(10)^2 - 1501.5(10) + 6016 = 1451.$$

TECHNOLOGY   CONNECTION

Mathematical Modeling Using Regression: Fitting Quadratic and Other Polynomial Functions to Data

Regression can be extended to quadratic, cubic, and quartic polynomial functions. The grapher gives us the powerful capability to find these polynomial models, select which seems best, and make predictions.

Example   *Life Science: Live Births to Women of Age x.* The following chart relates the number of live births to women of a particular age.

a) Fit a quadratic function to the data using the REGRESSION feature on a grapher.
b) Make a scatterplot of the data. Then graph the quadratic function with the scatterplot.
c) Fit a cubic function to the data using the REGRESSION feature on a grapher.
d) Make a scatterplot of the data. Then graph the cubic function with the scatterplot.
 e) Decide which function seems to fit the data better.

Age, x	Average Number of Live Births per 1000 Women
16	34
18.5	86.5
22	111.1
27	113.9
32	84.5
37	35.4
42	6.8

Source: Centers for Disease Control and Prevention

f) Use the function from part (e) to estimate the average number of live births by women of ages 20 and 30.

Solution   We proceed as follows.

a) We fit a quadratic function to the data using the REGRESSION feature. The procedure is

(*continued*)

similar to what is outlined in the preceding Technology Connection on linear regression. We just choose QUADREG instead of LINREG($ax + b$). (Consult your manual for further details.) We obtain the following function:

$$y_1 = f(x) = -0.49x^2 + 25.95x - 238.49.$$

b) The quadratic function is graphed with the scatterplot above.

c) We fit a cubic function to the data using the REGRESSION feature. We choose cubicReg and obtain the following function:

$$y_2 = f(x)$$
$$= 0.03x^3 - 3.22x^2 + 101.18x - 886.93.$$

d) The cubic function is graphed with the scatterplot above.

e) The graph of the cubic function seems to fit closer to the data points. Thus we choose it as a model.

f) Using the grapher to do the calculation, we get Y2(20) ≈ 99.6 and Y2(30) ≈ 97.4 as shown.

Thus the average number of live births is 99.6 per 1000 by women age 20 and 97.4 per 1000 by women age 30.

Exercises

1. *Life Science: Live Births.*

   a) Use the REGRESSION feature to fit a quartic equation to the live-birth data. Make a scatterplot of the data. Then graph the quartic function with the scatterplot. Decide whether the quartic function gives a better fit than either the quadratic or the cubic function.

   b) Explain why the domain of the cubic live-birth function should probably be restricted to women whose ages are in the interval [15, 45].

2. *Business: Median Household Income by Age.*

Age, x	Median Income in 1996
19.5	$21,438
29.5	35,888
39.5	44,420
49.5	50,472
59.5	39,815
65	19,448

   Source: U.S. Bureau of the Census; The Conference Board: Simmons Bureau of Labor Statistics

   a) Fit a quadratic function to the data using the REGRESSION feature.

   b) Make a scatterplot of the data. Then graph the quadratic function with the scatterplot.

   c) Fit a cubic function to the data using the REGRESSION feature.

   d) Make a scatterplot of the data. Then graph the cubic function with the scatterplot.

   e) Fit a quartic function to the data using the REGRESSION feature.

   f) Make a scatterplot of the data. Then graph the quartic function with the scatterplot.

   g) Decide which of the quadratic, cubic, or quartic functions seems to best fit the data.

   h) Use the function from part (e) to estimate the median household income of people ages 25 and 45.

3. *Life Science: Hours of Sleep and Death Rate.* Repeat Example 7 using quadratic regression to fit a function to the data.

EXERCISE SET 1.6

Choosing Models. For the scatterplots and graphs in Exercises 1–6, determine which, if any, of the following functions might be used as a model for the data:

Linear, $f(x) = mx + b$;

quadratic, $f(x) = ax^2 + bx + c$, $a > 0$;

quadratic, $f(x) = ax^2 + bx + c$, $a < 0$;

polynomial, neither quadratic nor linear.

1.
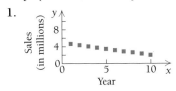

2.
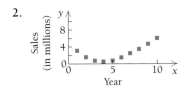

3.
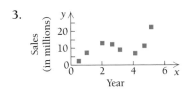

4.
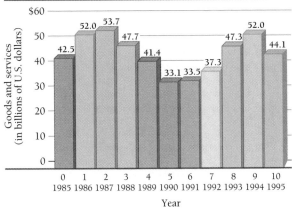

5.
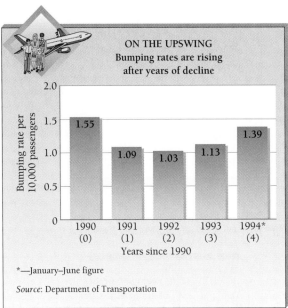

6.
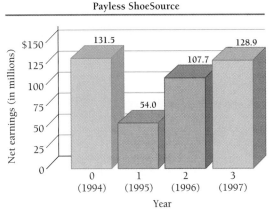

Source: Payless 1997 Annual Report

For each set of data in Exercises 7 and 8:

a) Make a scatterplot of the data and determine whether the data could be fit (or modeled) by a linear function.

b) Find a linear function that fits the data using the years 1993 and 1997.

c) Predict the number of home runs in 2000 and in 2005.

7. *Home Runs per Game in the National League.*

Years, x (since 1992)	Average Number of Home Runs per Game, H
0. 1992	1.30
1. 1993	1.72
2. 1994	1.91
3. 1995	1.90
4. 1996	2.17
5. 1997	1.83

8. *Home Runs per Game in the American League.*

Years, x (since 1992)	Average Number of Home Runs per Game, H
0. 1992	1.57
1. 1993	1.83
2. 1994	2.23
3. 1995	2.14
4. 1996	2.48
5. 1997	2.09

9. *Net Sales of The Gap.*

Years, x (since 1992)	Net Sales, S (in billions)
0. 1992	$2.5
1. 1993	3.0
2. 1994	3.3
3. 1995	3.7
4. 1996	4.4
5. 1997	5.3
6. 1998	6.5

Source: The Gap, Inc.

a) Choose two points from the data and find a linear function that fits the data.

b) Graph the scatterplot and the function on the same set of axes.

c) Use the function to predict net sales of The Gap in 2000 and in 2005.

10. *Net Sales of Toys R Us.*

Years, x (since 1992)	Net Sales, S (in billions)
0. 1992	$ 6.1
1. 1993	7.2
2. 1994	8.7
3. 1995	9.4
4. 1996	9.9
5. 1997	11.0

Source: Toys R Us

a) Choose two points from the data and find a linear function that fits the data.

b) Graph the scatterplot and the function on the same set of axes.

c) Use the function to predict net sales of Toys R Us in 2000 and in 2010.

11. *Payless ShoeSource Net Earnings.* Use the data from Exercise 6.

a) Find a quadratic function that fits the data using the data points (0, 131.5), (1, 54.0), and (2, 107.7).

b) Use the function to predict net earnings in 2002 and in 2010.

tw c) How well does this quadratic function predict the earnings for 1997?

12. *Bumping Rates.* Use the data from Exercise 5.

a) Find a quadratic function that fits the data using the data points (0, 1.55), (2, 1.03), and (4, 1.39).

b) Use the function to predict the bumping rates in 2000 and in 2005.

tw c) How well does this quadratic function predict the bumping rate for 1993?

13. *Nighttime Accidents.*

a) Find a quadratic function that fits the following data.

Travel Speed (in kilometers per hour)	Number of Nighttime Accidents (for every 200 million kilometers driven)
60	400
80	250
100	250

b) Use the function to estimate the number of nighttime accidents that occur at 50 km/h.

14. *Daytime Accidents.*

a) Find a quadratic function that fits the following data.

Travel Speed (in kilometers per hour)	Number of Daytime Accidents (for every 200 million kilometers driven)
60	100
80	130
100	200

b) Use the function to estimate the number of daytime accidents that occur at 50 km/h.

Synthesis ....

tw 15. Explain the restrictions that should be placed on the domains of the quadratic functions found in

Exercises 5 and 8 and why such restrictions are needed.

tw 16. Explain the restrictions that should be placed on the domains of the quadratic functions found in Exercises 9 and 10 and why such restrictions are needed.

   TECHNOLOGY   CONNECTION

17. *Home Runs per Game in the National League.*

a) Use the **REGRESSION** feature to fit a linear function to the data in Exercise 7.

b) Use the function to predict the number of home runs per game in 2000 and in 2005.

tw c) Compare your answers to those found in Exercise 7.

18. *Net Sales of Toys R Us.*

a) Use the **REGRESSION** feature to fit a linear function to the data in Exercise 10.

b) Use the function to predict net sales in 2000 and in 2010.

tw c) Compare your answers to those found in Exercise 10.

19. *Payless ShoeSource Net Earnings.*

a) Use the **REGRESSION** feature to fit a quadratic function to the data in Exercise 6.

b) Use the function to predict net earnings in 2002 and in 2010.

tw c) Compare your answers to those found in Exercise 11.

20. *Japanese Trade Deficit.*

a) Use the **REGRESSION** feature to fit a quadratic function to the data in Exercise 4.

b) Use the function to predict the trade deficit in 2000 and in 2004.

tw c) What restrictions, if any, should be placed on the domain? Explain.

CHAPTER 1 SUMMARY AND REVIEW

Terms to Know

Ordered pair, p. 4
First coordinate, p. 4
Second coordinate, p. 4
Solution, p. 4
Graph, p. 4
Mathematical model, p. 9
Compound interest, p. 10
Domain, p. 15
Range, p. 15
Function, p. 15
Relation, p. 16
Input, p. 16
Output, p. 16
Vertical-line test, p. 22
Functions defined piecewise, p. 22
Independent variable, p. 25
Dependent variable, p. 25
Set, p. 29

Real numbers, p. 29
Roster method, p. 29
Set-builder notation, p. 29
Interval notation, p. 30
Endpoints, p. 30
Slope, p. 42
Direct variation, p. 43
Variation constant, p. 43
Linear function, p. 44
Constant function, p. 44
Slope–intercept equation, p. 45
Point–slope equation, p. 46
Average rate of change, p. 48
Fixed costs, p. 49
Variable costs, p. 49
Total cost, p. 49
Break-even value, p. 50
Quadratic function, p. 56

Parabola, p. 56
Vertex, p. 56
Quadratic formula, p. 58
x-intercepts, p. 59
Polynomial function, p. 60
Coefficients of a polynomial, p. 60
Power function, p. 60
Rational function, p. 62
Split numbers, p. 62
Inverse variation, p. 64
Absolute-value function, p. 65
Square-root function, p. 66
Demand function, p. 69
Supply function, p. 69
Equilibrium point, p. 70
Scatterplot, p. 75
Curve fitting, p. 77
Linear regression, p. 79

Review Exercises

These review exercises are for test preparation. They can also be used as a lengthened practice test. Answers are at the back of the book. The answers also contain bracketed section references, which tell you where to restudy if your answer is incorrect.

1. *Life Science: Live Births to Women of Age x.* The following graph relates the number of live births B per 1000 women of age x.

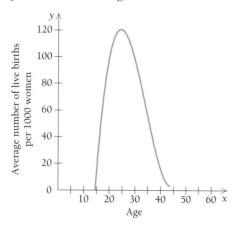

Age

Use the graph to answer the following.

a) What is the incidence of live births to women of age 20?

b) For what ages is the incidence of live births approximately 100 per 1000 women?

tw c) Make an estimate of the domain of the function and explain why it should be so.

2. *Business: Compound Interest.* Suppose that $1100 is invested at 10%, compounded semiannually. How much is in the account at the end of 1 yr?

3. *Business: Compound Interest.* Suppose that $4000 is invested at 12%, compounded annually. How much is in the account at the end of 2 yr?

Write interval notation.

4. $\{x \mid -6 < x \le 1\}$

5. The set of all positive numbers

A function f is given by $f(x) = 2x^2 - x + 3$. Find each of the following.

6. $f(-2)$ 7. $f(1 + h)$ 8. $f(0)$

A function f is given by $f(x) = (1 - x)^2$. Find each of the following.

9. $f(-5)$ 10. $f(2 - h)$ 11. $f(4)$

Graph.

12. $y = |x + 1|$

13. $f(x) = (x - 2)^2$

14. $f(x) = \dfrac{x^2 - 16}{x + 4}$

Use the vertical-line test to determine whether each of the following is the graph of a function.

15.   16.

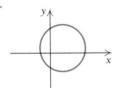   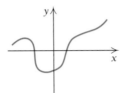

17.

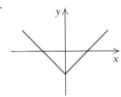

18. For the following graph of function f, determine
(a) $f(2)$; (b) the domain; (c) all x-values such that $f(x) = 2$; and (d) the range.

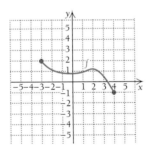

Graph.

19. $x = -2$

20. $y = 4 - 2x$

21. Write an equation of the line containing the points $(4, -2)$ and $(-7, 5)$.

22. Write an equation of the line with slope 8, containing the point $\left(\frac{1}{2}, 11\right)$.

23. For the linear equation $y = 3 - \frac{1}{6}x$, find the slope and the y-intercept.

Find the average rate of change.

24.

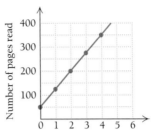

Number of days spent reading

25.

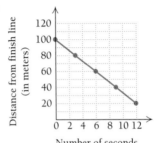

Number of seconds
spent running

26. *Life Science: Muscle Weight and Body Weight.* The weight M of muscles in a human is directly proportional to body weight W. It is known that a person who weighs 150 lb has 60 lb of muscles. What is the muscle weight of a person who weighs 180 lb?

27. *Business: Profit-and-Loss Analysis.* A furniture manufacturer has fixed costs of $80,000 for producing a certain type of classroom chair for universities. The variable cost for producing x chairs is $16x$ dollars. The chairs will be sold at a price of $28 each.

a) Formulate a function $R(x)$ for the total revenue from the sale of x chairs.

b) Formulate a function $C(x)$ for the total cost of producing x chairs.

c) Formulate a function $P(x)$ for the total profit from the production and sale of x chairs.

d) What profit or loss will the company realize from expected sales of 8000 chairs?

e) How many chairs must the company sell in order to break even?

28. Convert to radical notation: $y^{1/6}$.

29. Convert to rational exponents: $\sqrt[20]{x^3}$.

30. Simplify: $25^{3/2}$.

Find the domain of the function.

31. $f(x) = \sqrt{18 - 6x}$

32. $g(x) = \dfrac{1}{x^2 - 1}$

33. Graph:

$$f(x) = \begin{cases} x - 5, & \text{for } x > 2, \\ x + 3, & \text{for } x \le 2. \end{cases}$$

34. *Social Sciences: Study Time and Grades.* A math instructor asked her students to keep track of how much time each spent studying the chapter on percent notation in her basic mathematics course. She collected the information together with test scores from that chapter's test. The data are listed in the following table.

Study Time, x (in hours)	Test Grade, G (in percent)
9	74
11	94
13	81
15	86
16	87
17	81
21	87
23	92

a) Using the data points (9, 74) and (23, 92), find a linear function that fits the data.

b) Use the function to predict the test scores of a student who studies for 18 hr and for 25 hr.

35. *Business: Credit Cards.* More people are paying off their credit-card debts on a monthly basis. The percentage who do so has increased, according to the data in the following table.

a) Using the data points (0, 29), (3, 32), and (7, 41), find a quadratic function that fits the data.

b) Use the function to predict the percentage who will pay off their debts in 2000 and in 2008.

tw c) Make an estimate of the domain of this function. Explain its restrictions.

Years, x (since 1990)	Percentage, P Who Pay Off Their Debts
0. 1990	29
1. 1991	30
2. 1992	31
3. 1993	32
4. 1994	33
5. 1995	34
6. 1996	36
7. 1997	41

Source: RAM Research

Synthesis

36. Simplify: $(64^{5/3})^{-1/2}$.

   TECHNOLOGY   CONNECTION

Graph the function and find the zeros and the domain and the range.

37. $f(x) = x^3 - 4x$

38. $f(x) = \sqrt[3]{|9 - x^2|} - 1$

39. Approximate the points of intersection of the graphs of the two functions in Exercises 37 and 38.

40. *Social Sciences: Study Time and Grades.* Use the data in Exercise 34.

a) Use the **REGRESSION** feature to fit a linear function to the data.

b) Use the function to predict the test scores of a student who studies for 18 hr and for 25 hr.

tw c) Compare your answers to those found in Exercise 34.

41. *Business: Credit Cards.* Use the data in Exercise 35.

a) Use the **REGRESSION** feature to fit a quadratic function to the data.

b) Use the function to predict the percentage who will pay off their debts in 2000 and in 2008.

tw c) Make an estimate of the domain of this function. Explain its restrictions.

CHAPTER 1 TEST

1. *Social Sciences: Time Spent on Home Computer.* The following graph relates the average number of minutes spent per month A on a home computer to a person's age x.

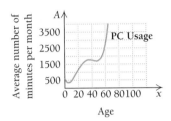

Source: Media Matrix; The PC Meter Company

Use the graph to answer the following.

a) What is the average use, in minutes per month, of persons of age 20?

b) For what ages is the average use approximately 3000 minutes per month?

tw c) Make an estimate of the domain of the function and explain why it should be so.

2. *Business: Compound Interest.* A person makes an investment at 13%, compounded annually. It has grown to $1039.60 at the end of 1 yr. How much was originally invested?

3. A function is given by $f(x) = x^2 - 4$. Find
   (a) $f(-3)$ and (b) $f(x + h)$.

4. What are the slope and the y-intercept of $y = -3x + 2$?

5. Find an equation of the line with slope $\frac{1}{4}$, containing the point $(8, -5)$.

6. Find the slope of the line containing the points $(-2, 3)$ and $(-4, -9)$.

Find the average rate of change.

7.
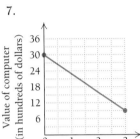

8.
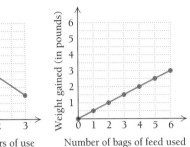

9. *Life Science: Body Fluids.* The weight F of fluids in a human is directly proportional to body weight W. It is known that a person who weighs 180 lb has 120 lb of fluids. Find an equation of variation expressing F as a function of W.

10. *Business: Profit-and-Loss Analysis.* A music company has fixed costs of $10,000 for producing a CD master. Thereafter, the variable costs are $0.50 per CD for duplicating from the CD master. The revenue from each CD is expected to be $7.50.

    a) Formulate a function $C(x)$ for the total cost of producing x CDs.
    b) Formulate a function $R(x)$ for the total revenue from the sale of x CDs.
    c) Formulate a function $P(x)$ for the total profit from the production and sale of x CDs.
    d) How many CDs must the company sell in order to break even?

11. *Economics: Equilibrium Point.* Find the equilibrium point for the demand and supply functions

$$D(p) = (p - 7)^2, \quad 0 \le p \le 7,$$

and

$$S(p) = p^2 + p + 4.$$

Use the vertical-line test to determine whether each of the following is the graph of a function.

12.

13.

14. For the following graph of function f, determine (a) $f(1)$; (b) the domain; (c) all x-values such that $f(x) = 4$; and (d) the range.

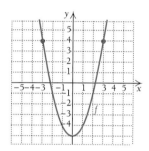

15. Graph: $f(x) = 4/x$.

16. Convert to rational exponents: $1/\sqrt{t}$.

17. Convert to radical notation: $t^{-3/5}$.

18. Graph: $f(x) = \dfrac{x^2 - 1}{x + 1}$.

Determine the domain of the function.

19. $f(x) = \dfrac{x^2 + 20}{(x - 2)(x + 7)}$

20. $f(x) = \sqrt{5x + 10}$

21. Write interval notation for the following graph.

22. Graph:

$$f(x) = \begin{cases} x^2 + 2, & \text{for } x \geq 0, \\ x^2 - 2, & \text{for } x < 0. \end{cases}$$

23. *Life Science: Maximum Heart Rate.* A person exercising should not exceed a maximum heart rate, which depends on his or her gender, age, and resting heart rate. The following table shows data relating resting heart rate and maximum heart rate for a 20-yr-old woman.

Resting Heart Rate, r (in beats per minute)	Maximum Heart Rate, M (in beats per minute)
50	170
60	172
70	174
80	176

Source: American Heart Association

a) Using the data points (50, 170) and (80, 176), find a linear function that fits the data.
b) Use the linear function to predict the maximum heart rate of a woman whose resting heart rate is 62 and one whose resting heart rate is 75.

24. *Business: Ticket Profits.* Valley Community College is running a play. Data relating its profits P, in dollars, after x days are given in the following table.

Days, x	Profit, P
0	−100
90	560
180	872
270	870
360	548
450	−100

a) Make a scatterplot of the data.
b) Decide whether the data seem to fit a quadratic function.
c) Using the data points (0, −100), (180, 872), and (360, 548), find a quadratic function that fits the data.
d) Use the function to estimate the profits after 225 days.
tw e) Make an estimate of the domain of this function. Explain its restrictions.

Synthesis ....

25. *Economics: Demand.* The demand function for a product is given by

$$x = D(p) = 800 - p^3, \quad 0 \leq p \leq 9.28.$$

a) Find the number of units sold when the price per unit is $6.50.

b) Find the price per unit when 720 units are sold.

 TECHNOLOGY CONNECTION

Graph the function and find the zeros and the domain and the range.

26. $f(x) = x^3 - 9x^2 + 27x + 50$

27. $f(x) = \sqrt[3]{|4 - x^2|} + 1$

28. Approximate the points of intersection of the graphs of the two functions in Exercises 26 and 27.

29. *Life Science: Maximum Heart Rate.* Use the data in Exercise 23.

a) Use the **REGRESSION** feature to fit a linear function to the data.

b) Use the linear function to predict the maximum heart rate of a woman whose resting heart rate is 62 and one whose resting heart rate is 75.

tw c) Compare your answers to those found in Exercise 23.

30. *Business: Ticket Profits.* Use the data in Exercise 24.

a) Use the **REGRESSION** feature to fit a quadratic function to the data.

b) Use the function to estimate the profits after 225 days.

tw c) Make an estimate of the domain of this function. Explain its restrictions.

31. *Social Sciences: Time Spent on Home Computer.* The following data relate the average number of minutes spent per month A on a home computer to a person's age x.

Age (in years)	Average Use (in minutes per month)
6.5	363
14.5	645
21	1377
29.5	1727
39.5	1696
49.5	2052
55	2299

Source: Media Matrix; The PC Meter Company

a) Use the **REGRESSION** feature to fit linear, quadratic, cubic, and quartic functions to the data.

b) Make a scatterplot of the data and graph each function on the scatterplot.

tw c) Decide which function best fits the data and explain your result.

tw d) Compare the result of part (a) with the answer found in Exercise 1.

 EXTENDED TECHNOLOGY APPLICATION

The Ecological Effect of Global Warming

Extended Technology Applications occur at the end of each chapter. They are designed to consider certain applications in greater depth, make use of grapher skills, and allow for possible group or collaborative learning.

Ecologists continue to be concerned about global warming, that is, the trend of average global temperatures to rise over recent years. One possible effect of such a warming is the melting of the icecaps. Concern is so great that in 1992, over 150 nations signed a treaty designed to reduce the emission of gases believed to compound the effects of global warming.

(continued)

Greenhouse Effect

One cause of global warming may be the so-called *greenhouse effect*. Smog and pollution in the atmosphere create a cover over the surface of the earth. The heat of the sun hits the surface and bounces off, but tends not to leave the surface because of the cover of pollution. Some sources of pollution are the exhaust of automobiles and deforestation.

Note the data in the table at right regarding average global temperature. Let's examine the data, make a graph, find some models, and make predictions.

To find a mathematical model of global warming, let's first graph the temperature data and look for a pattern.

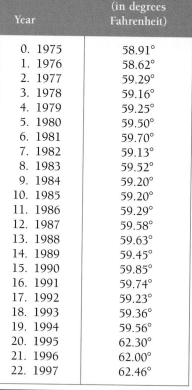

Year		Average Global Temperature (in degrees Fahrenheit)
0.	1975	58.91°
1.	1976	58.62°
2.	1977	59.29°
3.	1978	59.16°
4.	1979	59.25°
5.	1980	59.50°
6.	1981	59.70°
7.	1982	59.13°
8.	1983	59.52°
9.	1984	59.20°
10.	1985	59.20°
11.	1986	59.29°
12.	1987	59.58°
13.	1988	59.63°
14.	1989	59.45°
15.	1990	59.85°
16.	1991	59.74°
17.	1992	59.23°
18.	1993	59.36°
19.	1994	59.56°
20.	1995	62.30°
21.	1996	62.00°
22.	1997	62.46°

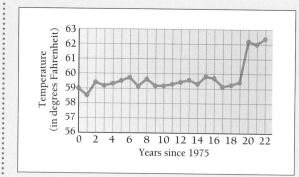

If we look at the graph above and try to determine a model, we see erratic behavior, but there is an overall trend toward an increase. Indeed, if we draw a line through the data as shown to the right, we see that the line does seem to have a positive slope and to increase.

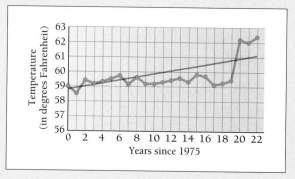

Exercises

1. **a)** Use a grapher that performs linear regression. Consider all the data to find a linear function that fits the data.
   b) Graph the linear function.
   c) Use the function to predict the average global temperature in 1999, 2000, and 2050.
   d) Use the function to predict when the average global temperature will reach 70° and 80°.

2. **a)** Consider all the data to find a cubic polynomial function

   $$y = ax^3 + bx^2 + cx + d$$

   that fits the data.
   b) Graph the cubic polynomial function.
   c) Use the function to predict the average global temperature in 1999, 2000, and 2050.
   d) Use the function to predict when the average global temperature will reach 70° and 80°.

3. **a)** Consider all the data to find a quartic polynomial function

   $$y = ax^4 + bx^3 + cx^2 + dx + e$$

   that fits the data.
   b) Graph the quartic polynomial function.
   c) Use the function to predict the average global temperature in 1999, 2000, and 2050.
   d) Use the function to predict when the average global temperature will reach 70° and 80°.

4. Discuss the merits of each type of function for predicting global warming.

5. There is great controversy on the issue of global warming. Read the following article and use the preceding analysis to make your own decision on this issue:

   Robinson, Arthur B., and Zachary W. Robinson, "Science Has Spoken: Global Warming Is a Myth," *The Wall Street Journal*, 12/4/97.

2

Differentiation

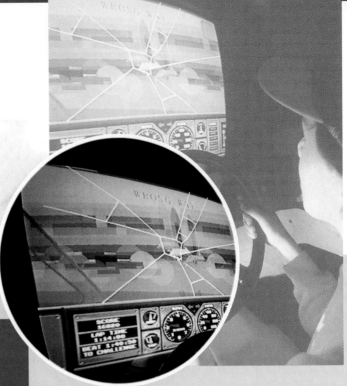

INTRODUCTION

With this chapter, we begin our study of calculus. The first concepts we consider are those of limits and continuity. Then we apply those concepts to establishing the first of the two main building blocks of calculus: differentiation.

Differentiation is a process that takes a formula for a function and derives a formula for another function, called a *derivative,* that allows us to find the slope of the tangent line to a curve at a point. We also find that a derivative can represent an instantaneous rate of change. Throughout the chapter, we will learn various techniques for finding derivatives.

AN APPLICATION

When a consumer receives x units of a product, a certain amount of pleasure, or utility, U, is derived. Suppose that for a new video game, the utility related to the number of cartridges x obtained is

$$U(x) = 80\sqrt{\frac{2x + 1}{3x + 4}}.$$

Find the marginal utility.

The marginal utility is the instantaneous rate of change of U with respect to x and is given by

$$U'(x) = \frac{200}{(3x + 4)^2} \cdot \sqrt{\frac{3x + 4}{2x + 1}}.$$

This function is a *derivative.*

This problem appears as Exercise 68 in Exercise Set 2.8.

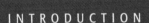

2.1

O B J E C T I V E S

➢ Find limits of functions, if they exist, using numerical or graphical methods.
➢ Determine the continuity of a function from its graph.
➢ Determine continuity of a function at a point.

Limits and Continuity: Numerically and Graphically

In this section, we give an intuitive (meaning "based on prior and present experience") treatment of two important concepts: *limits* and *continuity*.

Limits

Suppose that a defensive football team is on its own 5-yd line and is given a penalty. Such a penalty is half the distance to the goal. Suppose that it keeps getting such a penalty. Then the offense moves from the 5-yd line to the $2\frac{1}{2}$-yd line to the $1\frac{1}{4}$-yd line, and so on. Note that it can never score under this premise, but its distance to the goal continues to get closer and closer to 0. We say that the *limit* is 0.

One important aspect of the study of calculus is the analysis of how function values, or outputs, change when the inputs change. Basic to this study is the notion of limit. Suppose that the inputs get closer and closer to some number. If the corresponding outputs get closer and closer to a number, then that number is called a *limit*.

Consider the function f given by

$$f(x) = 2x + 3.$$

Suppose we select input numbers x closer and closer to the number 4, and look at the output numbers $2x + 3$. Study the input–output table and the graph that follow.

In the table and the graph, we see that as input numbers approach 4, but do not equal 4, from the left, output numbers approach 11.

Limit Numerically

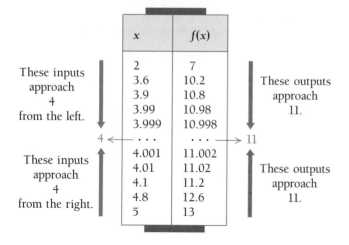

x	$f(x)$
2	7
3.6	10.2
3.9	10.8
3.99	10.98
3.999	10.998
...	...
4.001	11.002
4.01	11.02
4.1	11.2
4.8	12.6
5	13

These inputs approach 4 from the left.

These inputs approach 4 from the right.

These outputs approach 11.

These outputs approach 11.

Limit Graphically

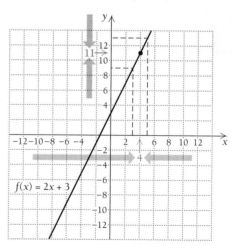

$f(x) = 2x + 3$

As input numbers approach 4, but do not equal 4, from the right, output numbers approach 11. Thus we say:

> As *x approaches* 4 from either side, $2x + 3$ *approaches* 11.

An arrow, →, is often used for the wording "approaches from either side." Thus the statement above can be written:

> As $x \to 4$, $2x + 3 \to 11$.

The number 11 is said to be the *limit* of $2x + 3$ as *x* approaches 4 from either side, but does not equal 4. We can abbreviate this statement as follows:

$$\lim_{x \to 4} (2x + 3) = 11.$$

This is read, "The limit, as *x* approaches 4 from either side, of $2x + 3$ is 11."

Definition

A function *f* has the **limit** *L* as *x* approaches *a* from either side, written

$$\lim_{x \to a} f(x) = L,$$

if *all* values $f(x)$ for *f* are close to *L* for values of *x* that are arbitrarily close, but not equal, to *a*.

The phrase "from either side" used in the preceding definition is very important. We use the notation

$$\lim_{x \to a^+} f(x) \quad \text{to indicate the limit from the right}$$

and

$$\lim_{x \to a^-} f(x) \quad \text{to indicate the limit from the left.}$$

Then in order for a limit to exist, both of the limits above must exist and be the same. We can rephrase the definition as follows.

Definition

A function *f* has the **limit** *L* as *x* approaches *a* if the limit from the left exists and the limit from the right exists and both limits are *L*—that is,

$$\lim_{x \to a^+} f(x) = \lim_{x \to a^-} f(x) = L = \lim_{x \to a} f(x).$$

Finding Limits Using the TABLE and TRACE Features

E x p l o r a t o r y . Consider the function $f(x) = 3x - 1$. Let's use the TABLE feature to complete the following table. Note that the inputs do not have the same increment from one to the next, but do approach 6 from either the left or the right. We set up a table with Indpnt in ASK mode. Then we enter the inputs shown and use the corresponding outputs to complete the table.

x	f(x)
5	14
5.8	16 4
5.9	16.7
5.99	16.97
5.999	16.997
6.001	17.003
6.01	17.03
6.1	17.3
6.4	18.2
7	20

6 ← ——————— → ?

Now, we set the table in AUTO mode and starting (TblStart) with a number near 6, we make tables for some increments (ΔTbl) like 0.1, 0.01, and so on, and like −0.1, −0.01, and so on, to determine $\lim_{x \to 6} f(x)$.

```
TABLE SETUP
  TblStart = 5.97
  ΔTbl = .01 ▮
Indpnt: Auto Ask
Depend: Auto Ask
```

X	Y₁	
5.97	16.91	
5.98	16.94	
5.99	16.97	
6	17	
6.01	17.03	
6.02	17.06	
6.03	17.09	

Now, using the TRACE feature with the graph, we move the cursor from left to right so that the x-coordinate approaches 6 from the left. We may need to make several window changes to see what

happens. For example, let's try [5.3, 6.4, 14, 18]. Next, we move the cursor from right to left so that the x-coordinate approaches 6 from the right. In general, the TRACE feature is not an efficient way to find limits, but it will help you to visualize the limit process in this early stage of your learning.

With what we have observed using the TABLE and TRACE features, let's complete the following:

$$\lim_{x \to 6^+} f(x) = \underline{\quad 17 \quad} \quad \text{and} \quad \lim_{x \to 6^-} f(x) = \underline{\quad 17 \quad}.$$

Thus,

$$\lim_{x \to 6} f(x) = \underline{\quad 17 \quad}.$$

Exercises

Consider $f(x) = 3x - 1$. Use the TABLE and TRACE features, making up your own tables, to find each of the following.

1. $\lim_{x \to 2} f(x)$ 2. $\lim_{x \to -1} f(x)$

Consider $g(x) = x^3 - 2x - 2$ for Exercises 3–5.

3. Complete the following table.

x	g(x)
7	
7.8	
7.9	
7.99	
7.999	
8.001	
8.01	
8.1	
8.4	
8.9	

8 ← ——————— → ?

Use the TABLE feature to find each of the following.

4. $\lim_{x \to 8} g(x)$ 5. $\lim_{x \to -1} g(x)$

Example 1   Consider the function H defined as follows:

$$H(x) = \begin{cases} 2x + 2, & \text{for } x < 1, \\ 2x - 2, & \text{for } x \geq 1. \end{cases}$$

Graph the function and find each of the following limits, if they exist.

a) $\lim\limits_{x \to 1} H(x)$
b) $\lim\limits_{x \to -3} H(x)$

Solution   We check the limits from the left and from the right both numerically, with an input–output table, and graphically.

a) *Limit Numerically*   *Limit Graphically*

$x \to 1^-, (x < 1)$	$H(x)$
0	2
0.5	3
0.8	3.6
0.9	3.8
0.99	3.98
0.999	3.998

$x \to 1^+, (x > 1)$	$H(x)$
2	2
1.8	1.6
1.1	0.2
1.01	0.02
1.001	0.002
1.0001	0.0002

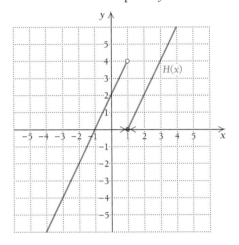

As inputs x approach 1 from the left, outputs $H(x)$ approach 4. Thus the limit from the left is 4. That is,

$$\lim_{x \to 1^-} H(x) = 4.$$

But as inputs x approach 1 from the right, outputs $H(x)$ approach 0. Thus the limit from the right is 0. That is,

$$\lim_{x \to 1^+} H(x) = 0.$$

Since the limit from the left, 4, is not the same as the limit from the right, 0, we say that

$$\lim_{x \to 1} H(x) \quad \text{does not exist.}$$

Note that $H(1) = 0$, but the limit is not 0. The function value exists, but the limit does not exist.

b) *Limit Numerically*   *Limit Graphically*

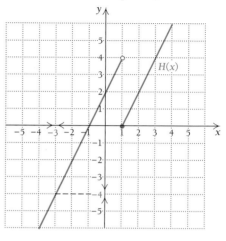

$x \to -3^{-}, (x < -3)$	$H(x)$
-4	-6
-3.5	-5
-3.1	-4.2
-3.01	-4.02
-3.001	-4.002

$x \to -3^{+}, (x > -3)$	$H(x)$
-2	-2
-2.5	-3
-2.9	-3.8
-2.99	-3.98
-2.999	-3.998

As inputs x approach -3 from the left, outputs $H(x)$ approach -4, so the limit from the left is -4. That is,

$$\lim_{x \to -3^{-}} H(x) = -4.$$

As inputs x approach -3 from the right, outputs $H(x)$ approach -4, so the limit from the right is -4. That is,

$$\lim_{x \to -3^{+}} H(x) = -4.$$

Since the limits from the left and from the right exist and are the same, we have

$$\lim_{x \to -3} H(x) = -4.$$

THE "WALL" METHOD

Let's look at the limits found in Example 1 in another way that might make the concept more meaningful. We draw a "wall" at $x = 1$, shown in blue on the graphs at the top of the following page. Then we follow the curve from the left with a pencil until we hit the wall and mark the location with an $\times$, assuming it can be determined. Then we follow the curve from the right until we hit the wall and mark that location with an $\times$. If the locations are the same, we have a limit. Thus,

$$\lim_{x \to 1} H(x) \text{ does not exist}$$

and

$$\lim_{x \to -3} H(x) = -4.$$

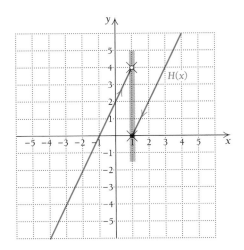
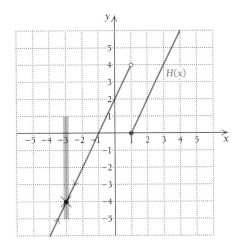

The following is also important in finding limits.

The limit at a number *a does not depend* on the function value at *a* even if that function value, $f(a)$, exists. That is, whether or not a limit exists at *a* has *nothing* to do with the function value $f(a)$.

Example 2   Consider the function G defined as follows:

$$G(x) = \begin{cases} 5, & \text{for } x = 1, \\ x + 1, & \text{for } x \neq 1. \end{cases}$$

Graph the function and find each of the following limits, if they exist.

a) $\lim\limits_{x \to 1} G(x)$ b) $\lim\limits_{x \to 3} G(x)$

Solution   The graph follows.

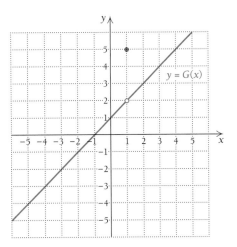

a) As inputs x approach 1 from the left, outputs $G(x)$ approach 2, so the limit from the left is 2. As inputs x approach 1 from the right, outputs $G(x)$ approach 2, so the limit from the right is 2. Since the limit from the left, 2, is the same as the limit from the right, 2, we have

$$\lim_{x \to 1} G(x) = 2.$$

Note that the limit, 2, is not the same as the function value at 1, which is $G(1) = 5$.

b) We have

$$\lim_{x \to 3} G(x) = 4.$$

We also know that

$$G(3) = 4.$$

In this case, the function value and the limit are the same.

After working through Example 2, we might ask, "When can we substitute to find a limit?" The answer lies in the following development of the concept of continuity.

Continuity

When the limit of a function is the same as its function value, it satisfies a condition called **continuity at a point.**

The following are the graphs of functions that are *continuous* over the whole real line $(-\infty, \infty)$.

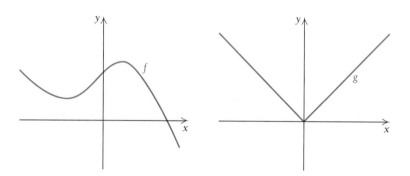

Note that there are no "jumps" or holes in the graphs. For now we will use a somewhat intuitive definition of continuity, which we will refine. We say that a function is **continuous over, or on, some interval** of the real line if its graph can be traced without lifting the pencil from the paper. If a function has one point at which it fails to be continuous, then we say that it is *not continuous* over the whole real line. The graphs of functions F, G, and H, which follow, show that these functions are *not* continuous over the whole real line. For functions G and H, the open circle indicates that the point is not part of the graph.

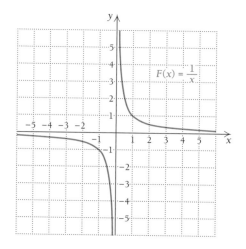

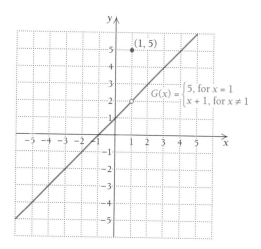

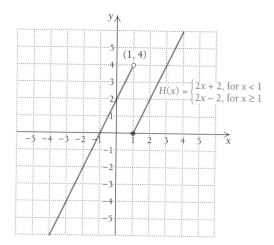

A continuous curve.

In each case, the graph *cannot* be traced without lifting the pencil from the pa-per. However, each case represents a different situation. Let's discuss why each case fails to be continuous over the whole real line.

The function F fails to be continuous over the whole real line $(-\infty, \infty)$. Since F is not defined at $x = 0$, the point $x = 0$ is not part of the domain, so $F(0)$ does not exist and there is no point $(0, F(0))$ on the graph. Thus there is no point to trace at $x = 0$. However, F is continuous over the intervals $(-\infty, 0)$ and $(0, \infty)$.

The function G is not continuous over the whole real line since it is not continu-ous at $x = 1$. Let's trace the graph of G starting to the left of $x = 1$. As x approaches 1, $G(x)$ seems to approach 2. However, at $x = 1$, $G(x)$ *jumps* up to 5, whereas to the right of $x = 1$, $G(x)$ *drops* back to some value close to 2. Thus G is discontinuous at $x = 1$.

The function H is not continuous over the whole real line since it is not continu-lous at $x = 1$. Let's trace the graph of H starting to the left of $x = 1$. As x approaches 1, $H(x)$ approaches 4. However, at $x = 1$, $H(x)$ *drops* down to 0; and just to the right of $x = 1$, $H(x)$ is close to 0. Since the limit from the left is 4 and the limit from the right is 0, we see that $\lim_{x \to 1} f(x)$ does not exist. Thus $H(x)$ is discontinuous at $x = 1$.

Limits and Continuity

Now we can answer the question, "When can we substitute to find a limit?" The answer lies in the more formal definition of continuity.

> **Definition**
>
> A function f is **continuous** at $x = a$ if:
>
> a) $f(a)$ exists,   (The output $f(a)$ exists.)
> b) $\lim_{x \to a} f(x)$ exists, and   (The limit as $x \to a$ exists.)
> c) $\lim_{x \to a} f(x) = f(a)$.   (The limit is the same as the output.)
>
> A function is **continuous over an interval** I if it is continuous at each point in I.

Note that we will also use the notation "$\lim_{x \to a} f(x)$" for a limit.

Example 3   Determine whether the function given by

$$f(x) = 2x + 3$$

is continuous at $x = 4$.

Solution   This function is continuous at 4 because:

a) $f(4)$ exists, $(f(4) = 11)$
b) $\lim_{x \to 4} f(x)$ exists, and $[\lim_{x \to 4} f(x) = 11$ (as shown earlier)$]$,
c) $\lim_{x \to 4} f(x) = 11 = f(4)$.

In fact, $f(x) = 2x + 3$ is continuous at any point on the real line. ◄

Example 4   Determine whether the function F previously considered on p. 103 is continuous at $x = 0$.

Solution   The function is *not* continuous at $x = 0$ because $f(0)$ does not exist. ◄

Example 5   Determine whether the function H previously considered on p. 103 is continuous at $x = 1$.

Solution   The function is *not* continuous at $x = 1$ because $\lim_{x \to 1} H(x)$ does not exist. ◄

Example 6   Determine whether the function G previously considered on p. 103 is continuous at $x = 1$.

Solution The function is *not* continuous at $x = 1$ because $G(1) = 5$, but $\lim_{x \to 1} G(x) = 2$. ◄

Continuity Principles

The following continuity principles, which we will not prove, allow us to determine whether a function is continuous.

C1. Any constant function is continuous (such a function never varies).

C2. For any positive integer n and any continuous function f, $[f(x)]^n$ and $\sqrt[n]{f(x)}$ are continuous. When n is even, the inputs of f in $\sqrt[n]{f(x)}$ are restricted to inputs x for which $f(x) \geq 0$.

C3. If $f(x)$ and $g(x)$ are continuous, then so are $f(x) + g(x)$, $f(x) - g(x)$, and $f(x) \cdot g(x)$.

C4. If $f(x)$ and $g(x)$ are continuous, so is $g(x)/f(x)$, so long as the inputs x do not yield outputs $f(x) = 0$.

Example 7 Provide an argument to show that

$$f(x) = x^2 - 3x + 2$$

is continuous.

Solution First we note that x is a continuous function. We determine this by the definition of continuity. We know that x^2 is continuous by C2. The constant function 3 is continuous by C1 and the function x is continuous, so the product $3x$ is continuous by C3. Thus, $x^2 - 3x$ is continuous by C3, and since the constant 2 is continuous, we can apply C3 again to show that $x^2 - 3x + 2$ is continuous. ◄

In similar fashion, we can show that any linear function $f(x) = mx + b$, and indeed any polynomial function, such as

$$f(x) = x^4 - 5x^3 + x^2 - 7,$$

is continuous.

A rational function is a quotient of two polynomials

$$r(x) = \frac{g(x)}{f(x)}.$$

Thus by C4, a rational function is continuous so long as the inputs x are not such that $f(x) = 0$. Thus a function such as

$$r(x) = \frac{x^4 - 5x^3 + x^2 - 7}{x - 3}$$

is continuous for all values of x except $x = 3$.

EXERCISE SET 2.1

Determine whether each of the following is continuous.

1.

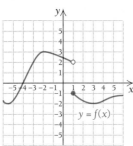

2.

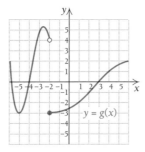

3.

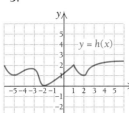

4.

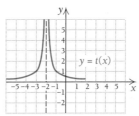

Use the graphs and functions in Exercises 1–4 to answer each of the following.

5. a) Find $\lim\limits_{x \to 1^+} f(x)$, $\lim\limits_{x \to 1^-} f(x)$, and $\lim\limits_{x \to 1} f(x)$.

  b) Find $f(1)$.

  c) Is f continuous at $x = 1$?

  d) Find $\lim\limits_{x \to -2} f(x)$.

  e) Find $f(-2)$.

  f) Is f continuous at $x = -2$?

6. a) Find $\lim\limits_{x \to 1^+} g(x)$, $\lim\limits_{x \to 1^-} g(x)$, and $\lim\limits_{x \to 1} g(x)$.

  b) Find $g(1)$.

  c) Is g continuous at $x = 1$?

  d) Find $\lim\limits_{x \to -2} g(x)$.

  e) Find $g(-2)$.

  f) Is g continuous at $x = -2$?

7. a) Find $\lim\limits_{x \to 1} h(x)$.

  b) Find $h(1)$.

  c) Is h continuous at $x = 1$?

  d) Find $\lim\limits_{x \to -2} h(x)$.

  e) Find $h(-2)$.

  f) Is h continuous at $x = -2$?

8. a) Find $\lim\limits_{x \to 1} t(x)$.

  b) Find $t(1)$.

  c) Is t continuous at $x = 1$?

  d) Find $\lim\limits_{x \to -2} t(x)$.

  e) Find $t(-2)$.

  f) Is t continuous at $x = -2$?

In Exercises 9–12, use the graphs to find the limits and answer the related questions.

9. Consider $f(x) = \begin{cases} 4 - x, & \text{for } x \neq 1, \\ 2, & \text{for } x = 1. \end{cases}$

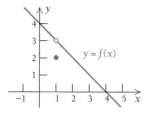

  a) Find $\lim\limits_{x \to 1^+} f(x)$.

  b) Find $\lim\limits_{x \to 1^-} f(x)$.

  c) Find $\lim\limits_{x \to 1} f(x)$.

  d) Find $f(1)$.

  e) Is f continuous at $x = 1$?

  f) Is f continuous at $x = 2$?

10. Consider $f(x) = \begin{cases} 4 - x^2, & \text{for } x \neq -2, \\ 3, & \text{for } x = -2. \end{cases}$

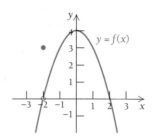

  a) Find $\lim\limits_{x \to -2^+} f(x)$.

  b) Find $\lim\limits_{x \to -2^-} f(x)$.

  c) Find $\lim\limits_{x \to -2} f(x)$.

  d) Find $f(-2)$.

  e) Is f continuous at $x = -2$?

  f) Is f continuous at $x = 1$?

11. Refer to the graph of f below to determine whether each statement is true or false.

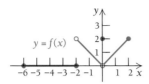

a) $\lim\limits_{x \to -2^+} f(x) = 1$

b) $\lim\limits_{x \to -2^-} f(x) = 0$

c) $\lim\limits_{x \to -2^-} f(x) = \lim\limits_{x \to -2^+} f(x)$

d) $\lim\limits_{x \to -2} f(x)$ exists.

e) $\lim\limits_{x \to -2} f(x) = 2$

f) $\lim\limits_{x \to 0} f(x) = 0$

g) $f(0) = 2$

h) f is continuous at $x = -2$.

i) f is continuous at $x = 0$.

j) f is continuous at $x = -1$.

12. Refer to the graph of f below to determine whether each statement is true or false.

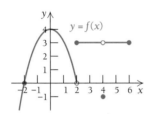

a) $\lim\limits_{x \to 2^-} f(x) = 3$

b) $\lim\limits_{x \to 2^+} f(x) = 0$

c) $\lim\limits_{x \to 2^-} f(x) = \lim\limits_{x \to 2^+} f(x)$

d) $\lim\limits_{x \to 2} f(x)$ exists.

e) $\lim\limits_{x \to 4} f(x)$ exists.

f) $\lim\limits_{x \to 4} f(x) = f(4)$

g) f is continuous at $x = 4$.

h) f is continuous at $x = 0$.

i) $\lim\limits_{x \to 3} f(x) = \lim\limits_{x \to 5} f(x)$

j) f is continuous at $x = 2$.

Applications ....

BUSINESS AND ECONOMICS

The Postage Function. Postal rates are 33¢ for the first ounce and 22¢ for each additional ounce or fraction thereof. Formally speaking, if x is the weight of a letter in ounces, then $p(x)$ is the cost of mailing the letter, where

$$p(x) = 33¢, \quad \text{if } 0 < x \le 1,$$
$$p(x) = 55¢, \quad \text{if } 1 < x \le 2,$$
$$p(x) = 77¢, \quad \text{if } 2 < x \le 3,$$

and so on, up to 13 ounces (at which point postal cost also depends on distance). The graph of p is shown here.

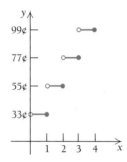

13. Is p continuous at 1? at $1\frac{1}{2}$? at 2? at 2.53?

14. Is p continuous at 3? at $3\frac{1}{4}$? at 4? at 3.98?

Using the graph of the postage function, find each of the following limits, if it exists.

15. $\lim\limits_{x \to 1^-} p(x),\ \lim\limits_{x \to 1^+} p(x),\ \lim\limits_{x \to 1} p(x)$

16. $\lim\limits_{x \to 2^-} p(x),\ \lim\limits_{x \to 2^+} p(x),\ \lim\limits_{x \to 2} p(x)$

17. $\lim\limits_{x \to 2.3} p(x)$

18. $\lim\limits_{x \to 1/2} p(x)$

Total Cost. The total cost $C(x)$ of producing x units of a product is shown in the graph below. Note that the curve increases gradually, but jumps suddenly at a certain value. This is probably the point at which new machines must be purchased and extra employees must be employed.

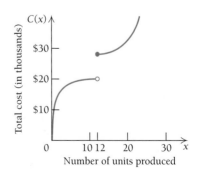

Number of units produced

Using the graph, find each of the following limits, if it exists.

19. $\lim\limits_{x \to 12^-} C(x)$, $\lim\limits_{x \to 12^+} C(x)$, $\lim\limits_{x \to 12} C(x)$

20. $\lim\limits_{x \to 20^-} C(x)$, $\lim\limits_{x \to 20^+} C(x)$, $\lim\limits_{x \to 20} C(x)$

21. Is C continuous at 12? at 20?

22. Is C continuous at 6? at 18?

Social Sciences

A Learning Curve. In psychology one often takes a certain amount of time t to learn a task. Suppose that the goal is to do a task perfectly and that you are practicing the ability to master it. After a certain time period, what is known to psychologists as an "I've got it!" experience occurs, and you are able to perform the task perfectly.

tw 23. At what point do you think the "I've got it!" experience happens on the learning curve below?

tw 24. Why do you think the curve below is constant for inputs $t \geq 20$?

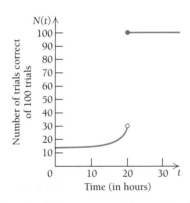

Using the graph above, find each of the following limits, if it exists.

25. $\lim\limits_{t \to 20^+} N(t)$, $\lim\limits_{t \to 20^-} N(t)$, $\lim\limits_{t \to 20} N(t)$

26. $\lim\limits_{t \to 30^-} N(t)$, $\lim\limits_{t \to 30^+} N(t)$, $\lim\limits_{t \to 30} N(t)$

27. Is N continuous at 20? at 30?

28. Is N continuous at 10? at 26?

Synthesis

tw 29. Discuss three ways in which a function may not be continuous at a point a. Draw graphs to illustrate your discussion.

 TECHNOLOGY CONNECTION

In Exercises 30–34, use a grapher that can graph functions defined piecewise.

30. Graph the function f given by

$$f(x) = \begin{cases} -3, & \text{for } x = -2, \\ x^2, & \text{for } x \neq -2. \end{cases}$$

Use the graph and the TRACE feature to find each of the following limits, if it exists.

a) $\lim\limits_{x \to -2^+} f(x)$

b) $\lim\limits_{x \to -2^-} f(x)$

c) $\lim\limits_{x \to -2} f(x)$

d) $\lim\limits_{x \to 2^+} f(x)$

e) $\lim\limits_{x \to 2^-} f(x)$

f) Does $\lim\limits_{x \to -2} f(x) = f(-2)$?

g) Does $\lim\limits_{x \to 2} f(x) = f(2)$?

31. Graph the function f given by

$$f(x) = \begin{cases} 2 - x^2, & \text{for } x \geq 0, \\ x^2 - 2, & \text{for } x < 0. \end{cases}$$

Use the graph and the TRACE feature to find each of the following limits, if it exists.

a) $\lim\limits_{x \to 0} f(x)$

b) $\lim\limits_{x \to -2} f(x)$

tw 32. Graph the function f given by

$$f(x) = \begin{cases} -3, & \text{for } x = -2, \\ x^2, & \text{for } x \neq -2. \end{cases}$$

Using the TRACE feature, determine whether this function is continuous at $x = -2$. Explain.

tw 33. Graph the function f given by

$$f(x) = \begin{cases} 2 - x^2, & \text{for } x \geq 0, \\ x^2 - 2, & \text{for } x < 0. \end{cases}$$

Using the TRACE feature, determine whether this function is continuous at $x = 0$. Explain.

2.2

Limits: Algebraically

Using Limit Principles

If a function is continuous at a, we can substitute to find the limit.

Example 1   Find $\lim_{x\to2} (x^4 - 5x^3 + x^2 - 7)$.

Solution   It follows from the Continuity Principles that $x^4 - 5x^3 + x^2 - 7$ is continuous. Thus we can find the limit by substitution:

$$\lim_{x\to2} (x^4 - 5x^3 + x^2 - 7) = 2^4 - 5\cdot 2^3 + 2^2 - 7$$
$$= 16 - 40 + 4 - 7$$
$$= -27.$$

Example 2   Find $\lim_{x\to0} \sqrt{x^2 - 3x + 2}$.

Solution   Using the Continuity Principles, we have shown (in Example 7 of Section 2.1) that $x^2 - 3x + 2$ is continuous for all values of x. When we restrict x to values for which $x^2 - 3x + 2$ is nonnegative, it follows from Principle C2 that $\sqrt{x^2 - 3x + 2}$ is continuous. Since $x^2 - 3x + 2$ is nonnegative when $x = 0$, we can substitute to find the limit:

$$\lim_{x\to0} \sqrt{x^2 - 3x + 2} = \sqrt{0^2 - 3\cdot 0 + 2}$$
$$= \sqrt{2}.$$

There are Limit Principles that correspond to the Continuity Principles. We can use them to find limits when we are uncertain of the continuity of a function at a given point.

Limit Principles

If $\lim_{x\to a} f(x) = L$ and $\lim_{x\to a} g(x) = M$, then we have the following.

L1. $\lim_{x\to a} c = c$.

(The limit of a constant is the constant.)

L2. $\lim_{x\to a} [f(x)]^n = \left[\lim_{x\to a} f(x)\right]^n = L^n$,

$\lim_{x\to a} \sqrt[n]{f(x)} = \sqrt[n]{\lim_{x\to a} f(x)} = \sqrt[n]{L}$, for any positive integer.

When n is even, the inputs x in $\sqrt[n]{f(x)}$ must be restricted to those inputs x for which $f(x) \ge 0$.

(The limit of a power is the power of the limit, and the limit of a root is the root of the limit.)

(continued)

L3. $\lim\limits_{x\to a} [f(x) \pm g(x)] = \lim\limits_{x\to a} f(x) \pm \lim\limits_{x\to a} g(x) = L \pm M.$

(The limit of a sum or a difference is the sum or the difference of the limits.)

$$\lim\limits_{x\to a} [f(x) \cdot g(x)] = \left[\lim\limits_{x\to a} f(x)\right] \cdot \left[\lim\limits_{x\to a} g(x)\right] = L \cdot M.$$

(The limit of a product is the product of the limits.)

L4. $\lim\limits_{x\to a} \dfrac{g(x)}{f(x)} = \dfrac{\lim\limits_{x\to a} g(x)}{\lim\limits_{x\to a} f(x)} = \dfrac{M}{L}$, provided $L \neq 0.$

(The limit of a quotient is the quotient of the limits.)

L5. $\lim\limits_{x\to a} cf(x) = c \cdot \lim\limits_{x\to a} f(x) = c \cdot L.$

(The limit of a constant times a function is the constant times the limit.)

Principle L5 can actually be proved using Principles L1 and L3, but we state it for emphasis.

Example 3   Find

$$\lim\limits_{x\to -3} \frac{x^2 - 9}{x + 3}.$$

Solution   Consider the graph of the rational function $r(x) = (x^2 - 9)/(x + 3)$, shown here. We can see that the limit is -6. But let's also examine the limit using the Limit Principles.

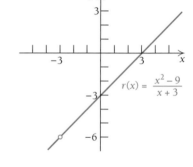

$r(x) = \dfrac{x^2 - 9}{x + 3}$

The function $(x^2 - 9)/(x + 3)$ is not continuous at $x = -3$. We use some algebraic simplification and then some Limit Principles:

$$\lim\limits_{x\to -3} \frac{x^2 - 9}{x + 3} = \lim\limits_{x\to -3} \frac{(x + 3)(x - 3)}{x + 3}$$

$$= \lim\limits_{x\to -3} (x - 3) \qquad \text{Simplifying, assuming } x \neq -3$$

$$= \lim\limits_{x\to -3} x - \lim\limits_{x\to -3} 3 \qquad \text{By L3}$$

$$= -3 - 3 = -6.$$

It is important to keep in mind in Example 3 that the simplification

$$\frac{x^2 - 9}{x + 3} = x - 3, \quad x \neq -3,$$

Exercises

Find each limit, if it exists, using the
TABLE feature.

1. $\lim\limits_{x\to-2} (x^4 - 5x^3 + x^2 - 7)$

2. $\lim\limits_{x\to1} \sqrt{x^2 + 3x + 4}$

3. $\lim\limits_{x\to3} \dfrac{x - 3}{x^2 - 9}$

can be done only for *x*-values *not* equal to -3. When finding the
limit, we need consider only *x*-values *close* to -3, but different
from -3, so the simplification and reasoning are valid.

In Section 2.3, we will encounter expressions with two variables, *x* and *h*. Our interest is in those limits where *x* is fixed as a
constant and *h* approaches zero.

Example 4   Find $\lim_{h\to0} (3x^2 + 3xh + h^2)$.

Solution   We treat *x* as a constant since we are interested only in
the way in which the expression varies when *h* approaches 0. We
use the Limit Principles to find that

$$\lim_{h\to0} (3x^2 + 3xh + h^2) = 3x^2 + 3x(0) + 0^2$$
$$= 3x^2.$$

The student can check any limit about which there is uncertainty by using an input–
output table. The following is a table for this limit.

h	$3x^2 + 3xh + h^2$	
1	$3x^2 + 3x \cdot 1 + 1^2,$	or $3x^2 + 3x + 1$
0.8	$3x^2 + 3x(0.8) + (0.8)^2,$	or $3x^2 + 2.4x + 0.64$
0.5	$3x^2 + 3x(0.5) + (0.5)^2,$	or $3x^2 + 1.5x + 0.25$
0.1	$3x^2 + 3x(0.1) + (0.1)^2,$	or $3x^2 + 0.3x + 0.01$
0.01	$3x^2 + 3x(0.01) + (0.01)^2,$	or $3x^2 + 0.03x + 0.0001$
0.001	$3x^2 + 3x(0.001) + (0.001)^2,$	or $3x^2 + 0.003x + 0.000001$

From the pattern in the table, it appears that

$$\lim_{h\to0} (3x^2 + 3xh + h^2) = 3x^2.$$

Summary

The following is a summary of the three methods we can use to determine a limit.
Consider

$$\lim_{x\to2} f(x),$$

where

$$f(x) = \frac{x^2 - 4}{x - 2}.$$

Limit Numerically

$x \to 2^-$	$f(x)$
1	3
1.5	3.5
1.8	3.8
1.9	3.9
1.99	3.99
1.999	3.999

$x \to 2^+$	$f(x)$
2.8	4.8
2.2	4.2
2.1	4.1
2.01	4.01
2.001	4.001
2.0001	4.0001

Limit Graphically

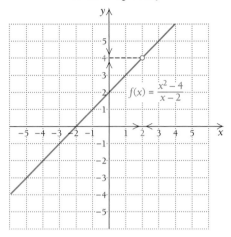

$f(x) = \dfrac{x^2 - 4}{x - 2}$

We see from either the input–output tables on the left or the graph on the right that

$$\lim_{x \to 2^-} f(x) = 4 \quad \text{and} \quad \lim_{x \to 2^+} f(x) = 4,$$

so

$$\lim_{x \to 2} f(x) = 4.$$

Limit Algebraically

The function is not continuous at $x = 2$. We use some algebraic simplification and then some Limit Principles:

$$\lim_{x \to 2} \frac{x^2 - 4}{x - 2} = \lim_{x \to 2} \frac{(x + 2)(x - 2)}{x - 2}$$

$$= \lim_{x \to 2} (x + 2) \qquad \text{Simplifying, assuming } x \neq 2$$

$$= \lim_{x \to 2} x + \lim_{x \to 2} 2 \qquad \text{By L3}$$

$$= 2 + 2$$

$$= 4.$$

EXERCISE SET 2.2

Find the limit. Use any of the three methods: numerical, graphical, and algebraic.

1. $\lim_{x \to 1} (x^2 - 3)$

2. $\lim_{x \to 1} (x^2 + 4)$

3. $\lim_{x \to 0} \dfrac{3}{x}$

4. $\lim_{x \to 0} \dfrac{-4}{x}$

5. $\lim_{x \to 3} (2x + 5)$

6. $\lim_{x \to 4} (5 - 3x)$

7. $\lim_{x \to -5} \dfrac{x^2 - 25}{x + 5}$

8. $\lim_{x \to -4} \dfrac{x^2 - 16}{x + 4}$

9. $\lim_{x \to -2} \dfrac{5}{x}$

10. $\lim_{x \to -5} \dfrac{-2}{x}$

11. $\lim_{x \to 2} \dfrac{x^2 + x - 6}{x - 2}$

12. $\lim_{x \to -4} \dfrac{x^2 - x - 20}{x + 4}$

Find the limit. Use any method.

13. $\lim_{x \to 5} \sqrt[3]{x^2 - 17}$

14. $\lim_{x \to 2} \sqrt{x^2 + 5}$

15. $\lim_{x \to 1} (x^4 - x^3 + x^2 + x + 1)$

16. $\lim_{x \to 2} (2x^5 - 3x^4 + x^3 - 2x^2 + x + 1)$

17. $\lim_{x \to 2} \dfrac{1}{x - 2}$

18. $\lim_{x \to 1} \dfrac{1}{(x - 1)^2}$

19. $\lim_{x \to 2} \dfrac{3x^2 - 4x + 2}{7x^2 - 5x + 3}$

20. $\lim_{x \to -1} \dfrac{4x^2 + 5x - 7}{3x^2 - 2x + 1}$

21. $\lim_{x \to 2} \dfrac{x^2 + x - 6}{x^2 - 4}$

22. $\lim_{x \to 4} \dfrac{x^2 - 16}{x^2 - x - 12}$

23. $\lim_{h \to 0} (6x^2 + 6xh + 2h^2)$

24. $\lim_{h \to 0} (10x + 5h)$

25. $\lim_{h \to 0} \dfrac{-2x - h}{x^2(x + h)^2}$

26. $\lim_{h \to 0} \dfrac{-5}{x(x + h)}$

27. Consider
$$f(x) = \begin{cases} 1, & \text{for } x \neq 2, \\ -1, & \text{for } x = 2. \end{cases}$$

Find each of the following.

a) $\lim_{x \to 0} f(x)$

b) $\lim_{x \to 2^-} f(x)$

c) $\lim_{x \to 2^+} f(x)$

d) $\lim_{x \to 2} f(x)$

e) Is f continuous at 0? at 2?

28. Consider
$$g(x) = \begin{cases} -4, & \text{for } x = 3, \\ 2x + 5, & \text{for } x \neq 3. \end{cases}$$

Find each of the following.

a) $\lim_{x \to 3^-} g(x)$

b) $\lim_{x \to 3^+} g(x)$

c) $\lim_{x \to 3} g(x)$

d) $\lim_{x \to 2} g(x)$

e) Is g continuous at 3? at 2?

Synthesis

Find the limit, if it exists.

29. $\lim_{x \to 0} \dfrac{|x|}{x}$

30. $\lim_{x \to -2} \dfrac{x^3 + 8}{x^2 - 4}$

TECHNOLOGY CONNECTION

Further Use of the TABLE Feature. In Section 2.1, we discussed how to use the **TABLE** feature to find limits. Consider
$$\lim_{x \to 0} \dfrac{\sqrt{1 + x} - 1}{x}.$$

Input–output tables for this function are shown below. The table on the left uses TblStart $= -1$ and ΔTbl $= 0.5$. By using smaller and smaller step values and beginning closer to 0, we can refine the table and obtain a better estimate of the limit. On the right is an input–output table with TblStart $= -0.03$ and ΔTbl $= 0.01$.

x	y
−1	1
−0.5	0.585786
0	ERROR
0.5	0.449490
1	0.414214
1.5	0.387426
2	0.366025

x	y
−0.03	0.503807
−0.02	0.502525
−0.01	0.501256
0	ERROR
0.01	0.498756
0.02	0.497525
0.03	0.496305

It appears that the limit is 0.5. We can verify this by graphing

$$y = \frac{\sqrt{1 + x} - 1}{x}$$

and tracing the curve near $x = 0$, zooming in on that portion of the curve.

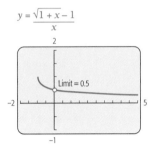

We see that

$$\lim_{x \to 0} \frac{\sqrt{1 + x} - 1}{x} = 0.5.$$

This can be verified algebraically. (*Hint*: Multiply by 1, using

$$\frac{\sqrt{1 + x} + 1}{\sqrt{1 + x} + 1}.$$

In Exercises 31–38, find the limit. Use the **TABLE** feature and start with ΔTbl = 0.1. Then move to 0.01, 0.001, and 0.0001. When you think you know the limit, graph and use the **TRACE** feature to verify your assertion. Then try to verify algebraically.

31. $\lim\limits_{a \to -2} \dfrac{a^2 - 4}{\sqrt{a^2 + 5} - 3}$

32. $\lim\limits_{x \to 1} \dfrac{\sqrt{x} - 1}{x - 1}$

33. $\lim\limits_{x \to 0} \dfrac{\sqrt{3 - x} - \sqrt{3}}{x}$

34. $\lim\limits_{x \to 0} \dfrac{\sqrt{4 + x} - \sqrt{4 - x}}{x}$

35. $\lim\limits_{x \to 1} \dfrac{x - \sqrt[4]{x}}{x - 1}$

36. $\lim\limits_{x \to 0} \dfrac{\sqrt{7 + 2x} - \sqrt{7}}{x}$

37. $\lim\limits_{x \to 4} \dfrac{2 - \sqrt{x}}{4 - x}$

38. $\lim\limits_{x \to 0} \dfrac{7 - \sqrt{49 - x^2}}{x}$

2.3

Average Rates of Change

Let's say that a car travels 110 mi in 2 hr. Its *average rate of change* (*speed*) is 110 mi/2 hr, or 55 mi/hr (55 mph). Suppose that you are on the freeway and you begin accelerating. Glancing at the speedometer, you see that at that *instant* your *instantaneous rate of change* is 55 mph. These are two quite different concepts. The first you are probably familiar with. The second involves ideas of limits and calculus. To understand *instantaneous rate of change,* we first use this section to develop a solid understanding of *average rate of change.*

The following graph shows the total production of suits by Raggs, Ltd., during one morning of work. Industrial psychologists have found curves like this typical of the production of factory workers.

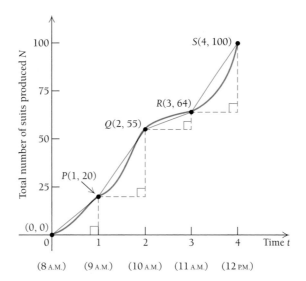

Example 1 *Business: Production.* What was the number of suits produced from 9 A.M. to 10 A.M.?

Solution   At 9 A.M., 20 suits had been produced. At 10 A.M., 55 suits had been produced. In the hour from 9 A.M. to 10 A.M., the number of suits produced was

   55 suits − 20 suits,   or   35 suits.

Note that 35 is the slope of the line from P to Q.   ◁

Example 2 *Business: Average Rate of Change.* What was the average number of suits produced per hour from 9 A.M. to 11 A.M.?

Solution   We have

$$\frac{64 \text{ suits} - 20 \text{ suits}}{11 \text{ A.M.} - 9 \text{ A.M.}} = \frac{44 \text{ suits}}{2 \text{ hr}}$$

$$= 22 \, \frac{\text{suits}}{\text{hr}} \text{ (suits per hour)}.$$

Note that 22 is the slope of the line from P to R. The line from P to R is not shown in the graph.   ◁

   Let's consider a function $y = f(x)$ and two inputs x_1 and x_2. The *change in input*, or the *change in x*, is

   $x_2 - x_1$.

The *change in output*, or the *change in y*, is

   $y_2 - y_1$,

where $y_1 = f(x_1)$ and $y_2 = f(x_2)$.

> *Definition*
>
> The **average rate of change of y with respect to x,** as x changes from x_1 to x_2, is the ratio of the change in output to the change in input:
>
> $$\frac{y_2 - y_1}{x_2 - x_1}, \quad \text{where } x_2 \neq x_1.$$

If we look at a graph of the function, we see that

$$\frac{y_2 - y_1}{x_2 - x_1} = \frac{f(x_2) - f(x_1)}{x_2 - x_1}$$

and that this is the slope of the line from $P(x_1, y_1)$ to $Q(x_2, y_2)$. The line $\overleftrightarrow{PQ}$ is called a **secant line.**

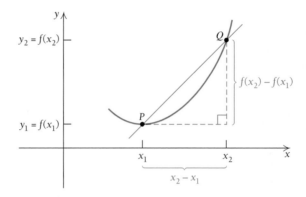

Example 3   For $y = f(x) = x^2$, find the average rate of change as:

a) x changes from 1 to 3.

b) x changes from 1 to 2.

c) x changes from 2 to 3.

Solution   The following graph is not necessary to the computations, but gives us a look at two of the secant lines whose slopes are being computed.

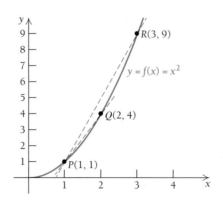

a) When $x_1 = 1$,
$$y_1 = f(x_1) = f(1) = 1^2 = 1;$$
and when $x_2 = 3$,
$$y_2 = f(x_2) = f(3) = 3^2 = 9.$$
The average rate of change is
$$\frac{y_2 - y_1}{x_2 - x_1} = \frac{f(x_2) - f(x_1)}{x_2 - x_1}$$
$$= \frac{9 - 1}{3 - 1}$$
$$= \frac{8}{2} = 4.$$

b) When $x_1 = 1$,
$$y_1 = f(x_1) = f(1) = 1^2 = 1;$$
and when $x_2 = 2$,
$$y_2 = f(x_2) = f(2) = 2^2 = 4.$$
The average rate of change is
$$\frac{4 - 1}{2 - 1} = \frac{3}{1} = 3.$$

c) When $x_1 = 2$,
$$y_1 = f(x_1) = f(2) = 2^2 = 4;$$
and when $x_2 = 3$,
$$y_2 = f(x_2) = f(3) = 3^2 = 9.$$
The average rate of change is
$$\frac{9 - 4}{3 - 2} = \frac{5}{1} = 5.$$

For a linear function, the average rates of change are the same for any choice of x_1 and x_2; that is, they are equal to the slope m of the line. As we saw in Example 3, a function that is not linear has average rates of change that vary with the choice of x_1 and x_2.

TECHNOLOGY
CONNECTION

Exercises

Use the **TABLE** feature to show that $f(x + h) \neq f(x) + h$ for each of the following functions.

1. $f(x) = x^2$; let $x = 6$ and $h = 2$.
2. $f(x) = x^3 - 2x^2 + 4$; let $x = 6$ and $h = 0.1$.

Difference Quotients as Average Rates of Change

We now use a different notation for average rates of change by eliminating the subscripts. Instead of x_1, we will write simply x.

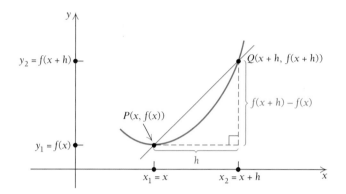

To get from x_1, or x, to x_2, we move a distance h. Thus, $x_2 = x + h$. Then the average rate of change, also called a **difference quotient,** is given by

$$\frac{y_2 - y_1}{x_2 - x_1} = \frac{f(x_2) - f(x_1)}{x_2 - x_1} = \frac{f(x + h) - f(x)}{(x + h) - x} = \frac{f(x + h) - f(x)}{h}.$$

> ### Definition
>
> The average rate of change of f with respect to x is also called the **difference quotient.** It is given by
>
> $$\frac{f(x + h) - f(x)}{h}, \quad \text{where } h \neq 0.$$
>
> The difference quotient is equal to the slope of the line from a point $P(x, f(x))$ to a point $Q(x + h, f(x + h))$.

Keep in mind that, in general, $f(x + h) \neq f(x) + h$. You should check this using a function like $f(x) = x^2$.

Example 4   For $f(x) = x^2$, find the difference quotient when:

a) $x = 5$ and $h = 3$.

b) $x = 5$ and $h = 0.1$.

Solution

a) We substitute $x = 5$ and $h = 3$ into the formula:

$$\frac{f(x + h) - f(x)}{h} = \frac{f(5 + 3) - f(5)}{3} = \frac{f(8) - f(5)}{3}.$$

Now $f(8) = 8^2 = 64$ and $f(5) = 5^2 = 25$, and we have

$$\frac{f(8) - f(5)}{3} = \frac{64 - 25}{3} = \frac{39}{3} = 13.$$

The difference quotient is 13. It is also the slope of the line from $(5, 25)$ to $(8, 64)$.

b) We substitute $x = 5$ and $h = 0.1$ into the formula:

$$\frac{f(x + h) - f(x)}{h} = \frac{f(5 + 0.1) - f(5)}{0.1} = \frac{f(5.1) - f(5)}{0.1}.$$

Now $f(5.1) = (5.1)^2 = 26.01$ and $f(5) = 25$, and we have

$$\frac{f(5.1) - f(5)}{0.1} = \frac{26.01 - 25}{0.1} = \frac{1.01}{0.1} = 10.1. \qquad \blacktriangleleft$$

For the function in Example 4, let's find a general form of the difference quotient. This will allow more efficient computations.

Example 5   For $f(x) = x^2$, find a **simplified** form of the **difference quotient.** Then find the value of the difference quotient when $x = 5$ and $h = 0.1$.

Solution   We have

$$f(x) = x^2,$$

so

$$f(x + h) = (x + h)^2 = x^2 + 2xh + h^2.$$

Then

$$f(x + h) - f(x) = (x^2 + 2xh + h^2) - x^2 = 2xh + h^2.$$

Thus,

$$\frac{f(x + h) - f(x)}{h} = \frac{2xh + h^2}{h} = \frac{h(2x + h)}{h} = 2x + h, \quad h \neq 0.$$

It is important to note that a difference quotient is defined only when $h \neq 0$. The simplification above is valid only for nonzero values of h.

When $x = 5$ and $h = 0.1$,

$$\frac{f(x + h) - f(x)}{h} = 2x + h = 2 \cdot 5 + 0.1 = 10 + 0.1 = 10.1. \qquad \blacktriangleleft$$

Example 6   For $f(x) = x^3$, find a simplified form of the difference quotient.

Solution   Now $f(x) = x^3$, so

$$f(x + h) = (x + h)^3$$
$$= x^3 + 3x^2h + 3xh^2 + h^3.$$

(This is shown in the appendix.) Then

$$f(x + h) - f(x) = (x^3 + 3x^2h + 3xh^2 + h^3) - x^3$$
$$= 3x^2h + 3xh^2 + h^3.$$

Thus,

$$\frac{f(x + h) - f(x)}{h} = \frac{3x^2h + 3xh^2 + h^3}{h}$$
$$= \frac{h(3x^2 + 3xh + h^2)}{h}$$
$$= 3x^2 + 3xh + h^2, \quad h \neq 0.$$

Again, this is true *only* for $h \neq 0$.

Example 7 For $f(x) = 3/x$, find a simplified form of the difference quotient.

Solution Now

$$f(x) = \frac{3}{x},$$

so

$$f(x + h) = \frac{3}{x + h}.$$

Then

$$f(x + h) - f(x) = \frac{3}{x + h} - \frac{3}{x}$$
$$= \frac{3}{x + h} \cdot \frac{x}{x} - \frac{3}{x} \cdot \frac{x + h}{x + h}$$

Here we are multiplying by 1 to get a common denominator.

$$= \frac{3x - 3(x + h)}{x(x + h)}$$
$$= \frac{3x - 3x - 3h}{x(x + h)}$$
$$= \frac{-3h}{x(x + h)}.$$

Thus,

$$\frac{f(x + h) - f(x)}{h} = \frac{\dfrac{-3h}{x(x + h)}}{h}$$
$$= \frac{-3h}{x(x + h)} \cdot \frac{1}{h} = \frac{-3}{x(x + h)}, \quad h \neq 0.$$

This is true *only* for $h \neq 0$.

EXERCISE SET 2.3

For the functions in each of Exercises 1–12, (a) find a simplified form of the difference quotient and (b) complete the following table.

x	h	$\dfrac{f(x + h) - f(x)}{h}$
4	2	
4	1	
4	0.1	
4	0.01	

1. $f(x) = 7x^2$
2. $f(x) = 5x^2$
3. $f(x) = -7x^2$
4. $f(x) = -5x^2$
5. $f(x) = 7x^3$
6. $f(x) = 5x^3$
7. $f(x) = \dfrac{5}{x}$
8. $f(x) = \dfrac{4}{x}$
9. $f(x) = -2x + 5$
10. $f(x) = 2x + 3$
11. $f(x) = x^2 - x$
12. $f(x) = x^2 + x$

Applications ....

BUSINESS AND ECONOMICS

13. *Utility.* Utility is a type of function that occurs in economics. When a consumer receives x units of a certain product, a certain amount of pleasure, or utility U, is derived. The following is a graph of a typical utility function.

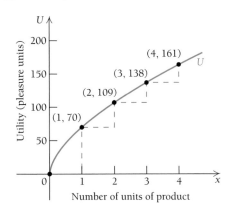

Number of units of product

a) Find the average rate of change of U as x changes from 0 to 1; from 1 to 2; from 2 to 3; from 3 to 4.

tw b) Why do you think the average rates of change are decreasing?

14. *Advertising Results.* The following graph shows a typical response to advertising. After an amount a is spent on advertising, the company sells $N(a)$ units of a product.

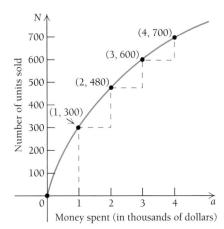

Money spent (in thousands of dollars)

a) Find the average rate of change of N as a changes from 0 to 1; from 1 to 2; from 2 to 3; from 3 to 4.

tw b) Why do you think the average rates of change are decreasing?

15. *Total Revenue.* A firm determines that the total revenue from the sale of x units of a certain product is given by

$$R(x) = -0.01x^2 + 1000x,$$

where $R(x)$ is in dollars.

a) Find $R(301)$.
b) Find $R(300)$.
c) Find $R(301) - R(300)$.
d) Find $\dfrac{R(301) - R(300)}{301 - 300}$.

tw e) What does the ratio in part (d) mean to the company?

16. *Total Cost.* A firm determines that the total cost C of producing x units of a certain product is

given by

$$C(x) = -0.05x^2 + 50x,$$

where $C(x)$ is in dollars.

a) Find $C(301)$.
b) Find $C(300)$.
c) Find $C(301) - C(300)$.
d) Find $\dfrac{C(301) - C(300)}{301 - 300}$.

tw e) What does the ratio in part (d) mean to the company?

17. *Total Revenue.* The Gap, Inc., is a specialty retailer that operates stores selling casual apparel under private-label brand names. The company has experienced great sales growth during the past few years, as shown in the following bar graph.

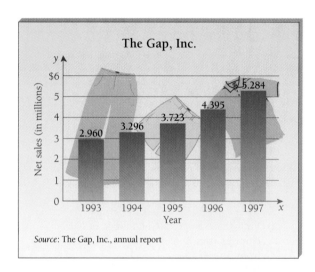

The Gap, Inc.

Source: The Gap, Inc., annual report

a) In 1993, the total revenue was $2.960 million. In 1997, it was $5.284 million. Find the average rate of change.
b) Find the average rate of change from 1995 to 1997.
tw c) Find the average rates of change for each of the successive years from 1993 to 1997. Have the rates been constant? How have they changed over the years?

LIFE AND PHYSICAL SCIENCES

18. *Temperature During an Illness.* The temperature T, in degrees Fahrenheit, of a patient during an illness is shown in the following graph, where t is the time, in days.

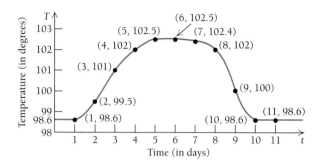

a) Find the average rate of change of T as t changes from 1 to 10. Using this rate of change, would you know that the person was sick?
b) Find the rate of change of T with respect to t, as t changes from 1 to 2; from 2 to 3; from 3 to 4; from 4 to 5; from 5 to 6; from 6 to 7; from 7 to 8; from 8 to 9; from 9 to 10; from 10 to 11.
c) When do you think the temperature began to rise? reached its peak? began to subside? was back to normal?
tw d) Explain your answers to part (c).

19. *Memory.* The total number of words $M(t)$ that a person can memorize in time t, in minutes, is shown in the following graph.

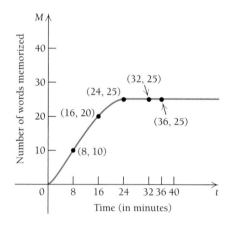

a) Find the average rate of change of M as t changes from 0 to 8; from 8 to 16; from 16 to 24; from 24 to 32; from 32 to 36.
tw b) Why do the average rates of change become 0 after 24 min?

20. *Average Velocity.* A car is at a distance s, in miles, from its starting point in t hours, given by

$$s(t) = 10t^2.$$

a) Find $s(2)$ and $s(5)$.
b) Find $s(5) - s(2)$. What does this represent?
c) Find the average rate of change of distance with respect to time as t changes from $t_1 = 2$ to $t_2 = 5$. This is known as **average velocity,** or **speed.**

21. *Average Velocity.* An object is dropped from a certain height. It is known that it will fall a distance s, in feet, in t seconds, given by

$$s(t) = 16t^2.$$

a) How far will the object fall in 3 sec?
b) How far will the object fall in 5 sec?
c) What is the average rate of change of distance with respect to time during the period from 3 to 5 sec? This is also *average velocity.*

22. *Gas Mileage.* At the beginning of a trip, the odometer on a car reads 30,680 and the car has a full tank of gas. At the end of the trip, the odometer reads 30,970. It takes 20 gal of gas to refill the tank.

a) What is the average rate of change of the number of miles with respect to the number of gallons?
b) What is the average rate of consumption (that is, the rate of change of the number of miles with respect to the number of gallons)?

SOCIAL SCIENCES

23. *Population Growth.* The two curves shown in the following figure describe the number of people in each of two countries at time t, in years.

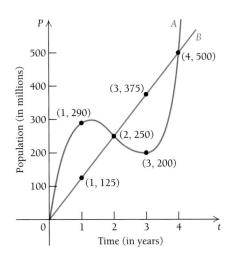

a) Find the average rate of change of each population (the number of people in the population) with respect to time t as t changes from 0 to 4. This is often called an **average growth rate.**
tw b) If the calculation in part (a) were the only one made, would we detect the fact that the populations were growing differently? Explain.
c) Find the average rates of change of each population as t changes from 0 to 1; from 1 to 2; from 2 to 3; from 3 to 4.
tw d) For which population does the statement "the population grew by 125 million each year" convey the least information about what really took place? Explain.

Synthesis

tw 24. *Business: Comparing Rates of Change.* The following two graphs show the number of federally insured banks and the Nasdaq Composite Stock Index over a six-month period.

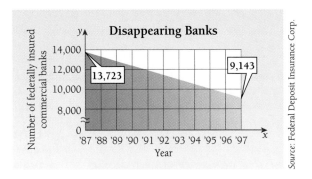

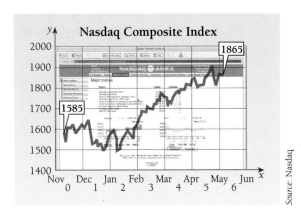

Explain the differences between these graphs in as many ways as you can. Be sure to consider average rates of change.

tw 25. *Business: 401(k) Retirement Plans.* The following graph shows the unit sales, in millions, of two types of retirement plans, pension and 401(k).

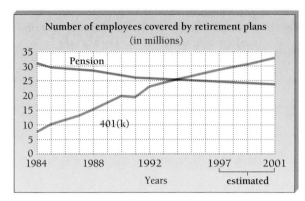

Source: Cerulli Associates

a) Estimate average rates of change for the pension graph (blue). Were these changes positive or negative? Relate this to slope and to the shape of the graph.

b) Estimate average rates of change for the 401(k) graph (red). Were these changes positive or negative? Relate this to slope and to the shape of the graph.

Find the simplified difference quotient.

26. $f(x) = mx + b$

27. $f(x) = ax^2 + bx + c$

28. $f(x) = ax^3 + bx^2$

29. $f(x) = \sqrt{x}$

$\left(\textit{Hint:} \text{ Multiply by 1 using } \dfrac{\sqrt{x + h} + \sqrt{x}}{\sqrt{x + h} + \sqrt{x}}.\right)$

30. $f(x) = x^4$

31. $f(x) = \dfrac{1}{x^2}$

32. $f(x) = \dfrac{1}{1 - x}$

33. $f(x) = \dfrac{x}{1 + x}$

34. $f(x) = \sqrt{3 - 2x}$

2.4

Differentiation Using Limits of Difference Quotients

We will see in Sections 2.4–2.6 that an instantaneous rate of change is given by the slope of a tangent line to the graph of a function. Here we see how to take the limit of a simplified difference quotient as h approaches 0 to define the slope of a tangent line.

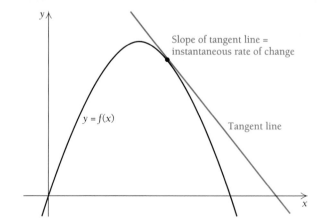

Tangent Lines

A line tangent to a circle is a line that touches the circle exactly once.

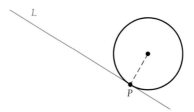

The path of the loose ski follows a tangent line to the ski chute at the point where it breaks loose.

This definition becomes unworkable with other curves. For example, consider the curve shown in Fig. 1. Line L touches the curve at point P but meets the curve at other places as well. It is considered a tangent line, but "touching at one point" cannot be its definition.

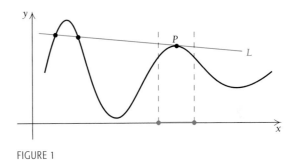

FIGURE 1

Note in Fig. 1 that over a small interval containing P, line L does touch the curve exactly once. This is still not a suitable definition of a *tangent line* because it allows a line like M in Fig. 2 to be a tangent line, which we will not accept.

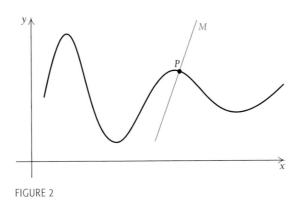

FIGURE 2

Later we will give a definition of a tangent line, but for now we will rely on intuition. In Fig. 3, lines L_1 and L_2 are not tangent lines. All the others are.

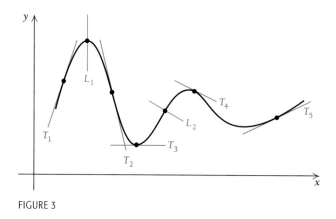

FIGURE 3

TECHNOLOGY
CONNECTION

Exploratory.
Graph
$y_1 = 3x^5 - 20x^3$ with
the viewing window
$[-3, 3, -80, 80]$, with
$Xscl = 1$ and
$Yscl = 10$. Then also
graph the lines
$y_2 = -7x - 10$,
$y_3 = -30x + 13$, and
$y_4 = -45x + 28$.
Which appears to be
tangent to the graph
of y_1 at $(1, -17)$? It
might be helpful to
zoom in near the
point $(1, -17)$. Use
smaller viewing
windows to try to
refine your guess.

Why Do We Study Tangent Lines?

One reason for studying tangent lines is so that we can use their slopes to define instantaneous rates of change. Another reason will become apparent in Chapter 3. For now, look at the following graph of a total-profit function. Note that the largest (or maximum) value of the function occurs at the point where the graph has a horizontal tangent, that is, where the tangent line has slope 0.

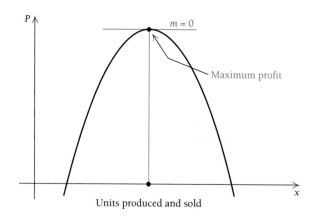

Units produced and sold

Differentiation Using Limits

We will define *tangent line* in such a way that it makes sense for *any* curve. To do this, we use the notion of limit.

We obtain the line tangent to the curve at point P by considering secant lines through P and neighboring points Q_1, Q_2, and so on. As the points Q approach P, the secant lines approach the tangent line. Each secant line has a slope. The slopes m_1, m_2, m_3, and so on, of the secant lines approach the slope m of the tangent line. In fact, we *define* the **tangent line** as the line that contains the point P and has slope m, where m is the limit of the slopes of the secant lines as the points Q approach P.

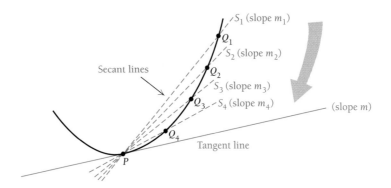

How might we calculate the limit m? Suppose that in Fig. 4 P has coordinates $(x, f(x))$. Then the first coordinate of Q is x plus some number h, or $x + h$. The coordinates of Q are $(x + h, f(x + h))$.

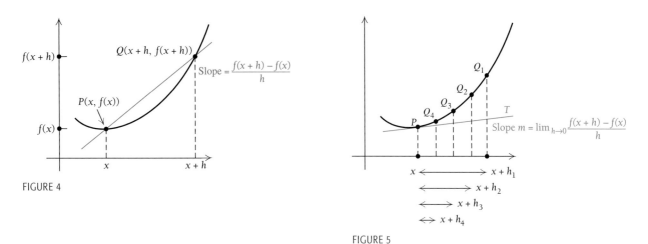

FIGURE 4

FIGURE 5

From Section 2.3, we know that the slope of the secant line $\overleftrightarrow{PQ}$ is given by the difference quotient

$$\frac{f(x + h) - f(x)}{h}.$$

Now, as we see in Fig. 5, as the points Q approach P, $x + h$ approaches x. That is, h approaches 0. Thus we have the following.

$$\text{The slope of the tangent line} = m = \lim_{h \to 0} \frac{f(x + h) - f(x)}{h}.$$

The formal definition of the *derivative of a function f* can now be given. We will designate the derivative at x as $f'(x)$, rather than $m(x)$.

> ### Definition
>
> For a function $y = f(x)$, its **derivative** at x is the function f' defined by
>
> $$f'(x) = \lim_{h \to 0} \frac{f(x + h) - f(x)}{h},$$
>
> provided the limit exists. If $f'(x)$ exists, then we say that f is **differentiable** at x. We sometimes call f' the **derived function.**

This is the basic definition of *differential calculus.*

Let's now calculate some formulas for derivatives. That is, given a formula for a function f, we will be trying to find a formula for f'.

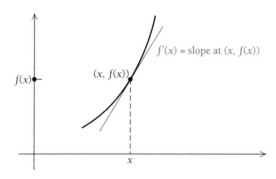

There are three steps in calculating a derivative.

1. Write down the difference quotient $[f(x + h) - f(x)]/h$.
2. Simplify the difference quotient.
3. Find the limit as h approaches 0.

Example 1   For $f(x) = 3x - 4$, find $f'(x)$.

Solution   We follow the steps above. Thus:

1. $\dfrac{f(x + h) - f(x)}{h} = \dfrac{[3(x + h) - 4] - (3x - 4)}{h}$;

2. $\dfrac{f(x + h) - f(x)}{h} = \dfrac{3x + 3h - 4 - 3x + 4}{h}$

$$= \frac{3h}{h} = 3, \quad h \neq 0;$$

3. $\lim\limits_{h \to 0} \dfrac{f(x + h) - f(x)}{h} = \lim\limits_{h \to 0} 3 = 3,$

   since 3 is a constant.

Thus if $f(x) = 3x - 4$, then $f'(x) = 3$.

A general formula for the derivative of a linear function

$$f(x) = mx + b$$

is $f'(x) = m.$

The formula could be verified in a manner similar to that used in Example 1.

Consider the graph of $f(x) = x^2$ that follows. Tangent lines are drawn at various points on the graph. Let $m(x) =$ the slope at the point $(x, f(x))$. Estimate the slope of each line and complete the table. Can you guess a formula for $m(x)$?

Lines	x	$m(x)$
L_1	-1	
L_2	$-\frac{1}{2}$	
L_3	0	
L_4	$\frac{1}{2}$	
L_5	1	

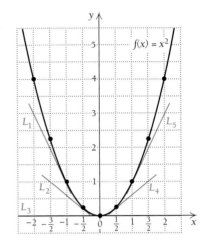

Now let's see if we can find a formula for the derivative of

$$f(x) = x^2.$$

We first find the slope of the tangent line at $x = 4$. That would be $f'(4)$. Then we find the general formula for $f'(x)$.

Example 2   For $f(x) = x^2$, find $f'(4)$.

Solution   We have

1. $\dfrac{f(4 + h) - f(4)}{h} = \dfrac{(4 + h)^2 - 4^2}{h};$

2. $\dfrac{f(4 + h) - f(4)}{h} = \dfrac{16 + 8h + h^2 - 16}{h}$

   $= \dfrac{8h + h^2}{h}$

   $= \dfrac{h(8 + h)}{h}$

   $= 8 + h, \quad h \neq 0;$

3. $\lim\limits_{h \to 0} \dfrac{f(4 + h) - f(4)}{h} = \lim\limits_{h \to 0} (8 + h) = 8.$

Thus, $f'(4) = 8.$

Example 3   For $f(x) = x^2$, find (the general formula) $f'(x)$.

Solution

1. We have

$$\frac{f(x + h) - f(x)}{h} = \frac{(x + h)^2 - x^2}{h}.$$

2. In Example 5 of Section 2.3, we showed how this difference quotient can be simplified to

$$\frac{f(x + h) - f(x)}{h} = 2x + h.$$

3. We want to find

$$\lim_{h \to 0} \frac{f(x + h) - f(x)}{h} = \lim_{h \to 0} (2x + h).$$

As $h \to 0$, we see that $2x + h \to 2x$. Thus,

$$\lim_{h \to 0} (2x + h) = 2x,$$

and we have

$$f'(x) = 2x.$$

We can check the result of Example 3 in Example 2. We know that a general formula is $f'(x) = 2x$. Thus, $f'(4) = 2(4) = 8$, as we found in Example 2. This formula also tells us, for example, that at $x = -3$, the curve has a tangent line whose slope is

$$f'(-3) = 2(-3) = -6.$$

We can also say:

- The tangent line to the curve at the point $(-3, 9)$ has slope -6.
- The tangent line to the curve at the point $(4, 16)$ has slope 8.

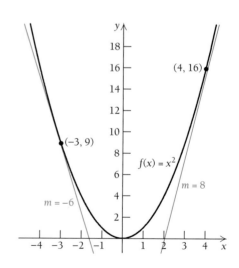

Example 4   For $f(x) = x^3$, find $f'(x)$. Then find $f'(-1)$ and $f'(1.5)$.

Solution

1. We have

$$\frac{f(x + h) - f(x)}{h} = \frac{(x + h)^3 - x^3}{h}.$$

2. In Example 6 of Section 2.3, we showed how this difference quotient can be simplified to

$$\frac{f(x + h) - f(x)}{h} = 3x^2 + 3xh + h^2.$$

3. We then have

$$\lim_{h \to 0} \frac{f(x + h) - f(x)}{h} = \lim_{h \to 0} (3x^2 + 3xh + h^2) = 3x^2.$$

(An input–output table for this is shown in Example 4 of Section 2.2.) Thus, for $f(x) = x^3$, we have $f'(x) = 3x^2$. Then

$$f'(-1) = 3(-1)^2 = 3 \quad \text{and} \quad f'(1.5) = 3(1.5)^2 = 6.75.$$

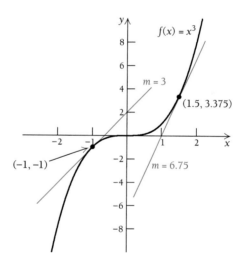

You should know that Examples 1–4 and Example 5, which follows, involve a somewhat lengthy process, but in Section 2.5 we will begin to develop some faster techniques. It is very important in this section, however, to fully understand the concept of a derivative.

Example 5   For $f(x) = 3/x$:

a) Find $f'(x)$.

b) Find $f'(1)$ and $f'(2)$.

c) Find an equation of the tangent line to the curve at the point $(1, 3)$.

d) Find an equation of the tangent line to the curve at the point $\left(2, \frac{3}{2}\right)$.

Solution

a) 1. We have

$$\frac{f(x + h) - f(x)}{h} = \frac{[3/(x + h)] - (3/x)}{h}.$$

2. In Example 7 of Section 2.3, we showed that this difference quotient can be simplified to

$$\frac{f(x + h) - f(x)}{h} = \frac{-3}{x(x + h)}.$$

3. We want to find

$$\lim_{h \to 0} \frac{f(x + h) - f(x)}{h} = \lim_{h \to 0} \frac{-3}{x(x + h)}.$$

As $h \to 0$, $x + h \to x$, so we have

$$f'(x) = \lim_{h \to 0} \frac{-3}{x(x + h)}$$

$$= \frac{-3}{x^2}.$$

b) Then

$$f'(1) = \frac{-3}{1^2} = -3$$

and

$$f'(2) = \frac{-3}{2^2} = -\frac{3}{4}.$$

c) We know that the point $(1, 3)$ is on the graph of the function because $f(1) = 3$. To find an equation of the tangent line to the curve at the point $(1, 3)$, we use the fact that the slope at $x = 1$ is -3, as we found in the preceding work. Now we have

Point: $(1, 3)$,

Slope: -3.

We substitute into the point–slope equation (see Section 1.4):

$$y - y_1 = m(x - x_1)$$
$$y - 3 = -3(x - 1)$$
$$y = -3x + 3 + 3$$
$$= -3x + 6.$$

The equation of the tangent line to the curve at $x = 1$ is $y = -3x + 6$ (see the following figure).

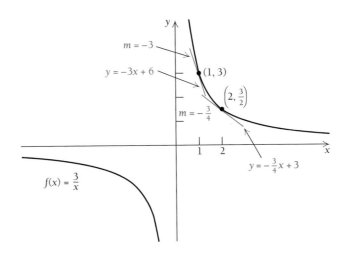

$f(x) = \frac{3}{x}$

d) To find an equation of the tangent line to the curve at $x = 2$, we use the fact that the slope at $x = 2$ is $-\frac{3}{4}$, as we found in part (b). Now we have

$$\text{Point: } \left(2, \tfrac{3}{2}\right),$$
$$\text{Slope: } -\tfrac{3}{4}.$$

We substitute into the point–slope equation (see Section 1.4):

$$y - y_1 = m(x - x_1)$$
$$y - \tfrac{3}{2} = -\tfrac{3}{4}(x - 2)$$
$$y = -\tfrac{3}{4}x + \tfrac{3}{2} + \tfrac{3}{2}$$
$$= -\tfrac{3}{4}x + 3.$$

The equation of the tangent line to the curve at $x = 2$ is

$$y = -\tfrac{3}{4}x + 3.$$

Because $f(0)$ does not exist, we cannot evaluate the difference quotient

$$\frac{f(0 + h) - f(0)}{h}.$$

Thus, $f'(0)$ does not exist. We say that "f is not differentiable at 0."

TECHNOLOGY
CONNECTION

Curves and Tangent Lines

Exercises

1. For $f(x) = 3/x$, find $f'(x)$, $f'(-2)$, and $f'\left(-\tfrac{1}{2}\right)$.

2. Find an equation of the tangent line to the curve $f(x) = 3/x$ at $\left(-2, -\tfrac{3}{2}\right)$ and an equation of the tangent line to the curve at $\left(-\tfrac{1}{2}, -6\right)$. Then graph the curve $f(x) = 3/x$ and both tangent lines. Use different viewing windows near the points of tangency.

When a function is not defined at a point, it is not differentiable at that point. Also, if a function is discontinuous at a point, it is not differentiable at that point.

It can happen that a function f is defined and continuous at a point but that its derivative f' is not defined. The function f given by

$$f(x) = |x|$$

is an example. Note that

$$f(0) = |0| = 0,$$

so the function is defined at 0.

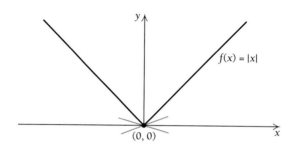

Suppose we try to draw a tangent line at $(0, 0)$. A function like this with a corner (not smooth) would seem to have many tangent lines at $(0, 0)$, and thus many slopes. The derivative at such a point would not be unique. Let's try to calculate the derivative at 0.

Since

$$f'(x) = \lim_{h \to 0} \frac{|x + h| - |x|}{h},$$

at $x = 0$, we have

$$f'(0) = \lim_{h \to 0} \frac{|0 + h| - |0|}{h} = \lim_{h \to 0} \frac{|h|}{h}.$$

$h \to 0^-$	$\dfrac{\|h\|}{h}$
-2	$\dfrac{\|-2\|}{-2}$, or $\dfrac{2}{-2}$, or -1
-1	-1
-0.1	-1
-0.01	-1
-0.001	-1

$h \to 0^+$	$\dfrac{\|h\|}{h}$
2	$\dfrac{\|2\|}{2}$, or $\dfrac{2}{2}$, or 1
1	1
0.1	1
0.01	1
0.001	1

Look at the input–output tables. Note that as h approaches 0 from the left, $|h|/h$ approaches -1, but as h approaches 0 from the right, $|h|/h$ approaches 1. Thus,

$$\lim_{h \to 0} \frac{|h|}{h} \quad \text{does not exist,}$$

so

$f'(0)$ does not exist.

If a function has a "sharp point" or "corner," it will not have a derivative at that point.

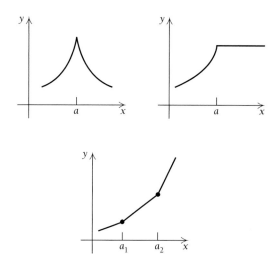

A function may also fail to be differentiable at a point by having a vertical tangent at that point. For example, the function shown below has a vertical tangent at point a. Recall that since the slope of a vertical line is undefined, there is no derivative at such a point.

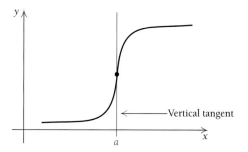

The function $f(x) = |x|$ illustrates the fact that although a function may be continuous at each point in an interval I, it may not be differentiable at each point in I. That is, continuity does not imply differentiability. On the other hand, if we know that a function is differentiable at each point in an interval I, then it is continuous over I. That is, if $f'(a)$ exists, then f is continuous at a. The function $f(x) = x^2$ is an example of a function that is differentiable over the interval $(-\infty, \infty)$ and is thus continuous everywhere. Also, if a function is discontinuous at some point a, then it is not differentiable at a. Thus when we know that a function is differentiable over an interval, it is *smooth* in the sense that there are no "sharp points," "corners," or "breaks" in the graph.

TECHNOLOGY   CONNECTION

Numerical Differentiation and Drawing Tangent Lines

Graphers have the capability of finding slopes of tangent lines, that is, of taking the derivative at a specific x-value. Let's consider the function

$$f(x) = x(100 - x)$$

and find the value of the derivative at $x = 70$.

Calculating the Derivative

Select nDeriv (numerical derivative) from the **MATH** menu and enter the function, the variable, and the value at which the derivative is to be evaluated. When we enter nDeriv($x(100 - x)$, x, 70), the grapher returns -40.

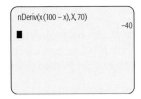

We see that the derivative of the function at $x = 70$ is -40. That is, the slope of the tangent line to the curve at $x = 70$ is -40.

Drawing the Tangent Line

We first graph the function using the viewing window $[-10, 100, -10, 3000]$, with Xscl = 10 and Yscl = 1000.

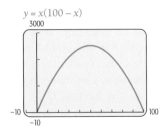

To draw a tangent line at $x = 70$, we go to the home screen and select the **TANGENT** feature from the **DRAW** menu. Then we enter Tangent(Y_1, 70). We see the graph of $f(x)$ and the tangent line at $x = 70$.

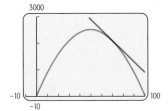

Exercises

For each of the following functions, evaluate the derivative at the given point. Then draw the graph and the tangent line.

1. $f(x) = x(100 - x)$;
   $x = 20$, $x = 37$, $x = 50$, $x = 90$

2. $f(x) = -\frac{1}{3}x^3 + 6x^2 - 11x - 50$;
   $x = -5$, $x = 0$, $x = 7$, $x = 12$, $x = 15$

3. $f(x) = 6x^2 - x^3$;
   $x = -2$, $x = 0$, $x = 2$, $x = 4$, $x = 6.3$

4. $f(x) = x\sqrt{4 - x^2}$;
   $x = -2$, $x = -1.3$, $x = -0.5$, $x = 0$, $x = 1$,
   $x = 2$

tw 5. For the function in Exercise 4, try to draw a tangent line at $x = 3$ and estimate the derivative. What goes wrong? Explain.

EXERCISE SET 2.4

In Exercises 1–14:

a) Graph the function.

b) Draw tangent lines to the graph at points whose x-coordinates are -2, 0, and 1.

c) Find $f'(x)$ by determining $\lim\limits_{h \to 0} \dfrac{f(x + h) - f(x)}{h}$.

d) Find $f'(-2)$, $f'(0)$, and $f'(1)$. How do these slopes compare with those of the lines you drew in part (b)?

1. $f(x) = 5x^2$

2. $f(x) = 7x^2$

3. $f(x) = -5x^2$

4. $f(x) = -7x^2$

5. $f(x) = x^3$

6. $f(x) = -x^3$

7. $f(x) = 2x + 3$

8. $f(x) = -2x + 5$

9. $f(x) = -4x$

10. $f(x) = \frac{1}{2}x$

11. $f(x) = x^2 + x$

12. $f(x) = x^2 - x$

13. $f(x) = \dfrac{1}{x}$

14. $f(x) = \dfrac{5}{x}$

15. Find $f'(x)$ for $f(x) = mx$.

16. Find $f'(x)$ for $f(x) = ax^2 + bx + c$.

17. Find an equation of the tangent line to the graph of $f(x) = x^2$ at the point $(3, 9)$, at $(-1, 1)$, and at $(10, 100)$. See Example 3.

18. Find an equation of the tangent line to the graph of $f(x) = x^3$ at the point $(-2, -8)$, at $(0, 0)$, and at $(4, 64)$. See Example 4.

19. Find an equation of the tangent line to the graph of $f(x) = 5/x$ at the point $(1, 5)$, at $(-1, -5)$, and at $(100, 0.05)$. See Exercise 14.

20. Find an equation of the tangent line to the graph of $f(x) = 2/x$ at the point $(-1, -2)$, at $(2, 1)$, and at $\left(10, \frac{1}{5}\right)$.

21. Find an equation of the tangent line to the graph of $f(x) = 4 - x^2$ at the point $(-1, 3)$, at $(0, 4)$, and at $(5, -21)$.

22. Find an equation of the tangent line to the graph of $f(x) = x^2 - 2x$ at the point $(-2, 8)$, at $(1, -1)$, and at $(4, 8)$.

List the points in the graph at which each function is not differentiable.

23.

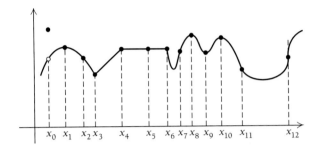

24.

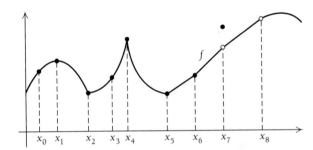

Applications
BUSINESS AND ECONOMICS

25. *The Postage Function.* Consider the postage function defined in Exercise Set 2.1. At what values is the function not differentiable?

Synthesis

tW 26. Which of the following appear to be tangent lines? Try to explain why or why not.

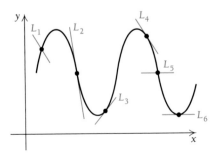

tw **27.** In the following figure, use a blue colored pencil and draw each secant line from point P to the points Q. Then use a red colored pencil and draw a tangent line to the curve at P. Describe what happens.

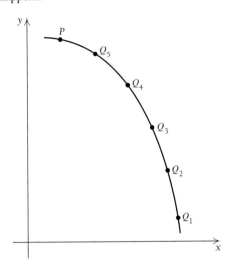

Find $f'(x)$.

28. $f(x) = x^4$

29. $f(x) = \dfrac{1}{x^2}$

30. $f(x) = \dfrac{1}{1 - x}$

31. $f(x) = \dfrac{x}{1 + x}$

32. $f(x) = \sqrt{x}$

$\left(\text{Multiply by 1, using } \dfrac{\sqrt{x + h} + \sqrt{x}}{\sqrt{x + h} + \sqrt{x}}.\right)$

33. Consider the function f given by

$$f(x) = \frac{x^2 - 9}{x + 3}.$$

For what values is this function not differentiable?

  TECHNOLOGY   CONNECTION

34.–39. Use a grapher to do the numerical differentiation and draw the tangent lines in each of Exercises 17–22.

40. *Business: Growth of an Investment.* A company determines that the value of an investment is V, in millions of dollars, after time t, in years, where V is given by

$$V(t) = 5t^3 - 30t^2 + 45t + 5\sqrt{t}.$$

Note: Graphers usually graph functions using only the variables y and x, so you may need to change the variables when entering this function.

a) Graph V over the interval $[0, 5]$.
b) Find the equation of the secant line passing through the points $(1, V(1))$ and $(5, V(5))$. Then sketch this secant line using the same axes as in step (a).
c) Find the average rate of change of the investment between year 1 and year 5.
d) Repeat steps (b) and (c) for the pairs of points $(1, V(1))$ and $(4, V(4))$; $(1, V(1))$ and $(3, V(3))$; $(1, V(1))$ and $(1.5, V(1.5))$.
e) What appears to be the slope of the tangent line to the graph at the point $(1, V(1))$?

2.5

O B J E C T I V E S

➤ Differentiate using the Power or Sum–Difference Rule.
➤ Differentiate a constant or a constant times a function.

Differentiation Techniques: The Power and Sum–Difference Rules

Leibniz's Notation

When y is a function of x, we will also designate the derivative, $f'(x)$, as*

$$\frac{dy}{dx},$$

---

*The notation $D_x y$ is also used.

Historical Note: The German mathematician and philosopher Gottfried Wilhelm von Leibniz (1646−1716) and the English mathematician, philosopher, and physicist Sir Isaac Newton (1642−1727) are both credited with the invention of the calculus, though each made the invention independently of the other. Newton used the dot notation $\dot{y}$ for dy/dt, where y is a function of time, and this notation is still used, though it is not as common as Leibniz's notation.

which is read "the derivative of y with respect to x." This notation was invented by the German mathematician Leibniz. It does *not* mean dy divided by dx nor does it mean y/x, with the d's having been removed. Think of dy/dx as a single entity made up of many parts. That is, we cannot interpret dy/dx as a quotient until meanings are given to dy and dx, which we will not do here. For example, if $y = x^2$, then

$$\frac{dy}{dx} = 2x.$$

We can also write

$$\frac{d}{dx} f(x)$$

to denote the derivative of f with respect to x. For example,

$$\frac{d}{dx} x^2 = 2x.$$

The value of dy/dx when $x = 5$ can be denoted by

$$\left. \frac{dy}{dx} \right|_{x=5}.$$

Thus for $dy/dx = 2x$,

$$\left. \frac{dy}{dx} \right|_{x=5} = 2 \cdot 5, \quad \text{or} \quad 10.$$

In general, for $y = f(x)$,

$$\left. \frac{dy}{dx} \right|_{x=a} = f'(a).$$

The Power Rule

In the remainder of this section, we will develop rules and techniques for efficient differentiation.

   Look for a pattern in the following table, which contains functions and derivatives that we have found in previous work.

Function	Derivative
x^2	$2x^1$
x^3	$3x^2$
x^4	$4x^3$
$\dfrac{1}{x} = x^{-1}$	$-1 \cdot x^{-2} = \dfrac{-1}{x^2}$
$\dfrac{1}{x^2} = x^{-2}$	$-2 \cdot x^{-3} = \dfrac{-2}{x^3}$

Perhaps you have discovered the following theorem.

Theorem 1

The Power Rule

For any real number k,

$$\frac{d}{dx}x^k = k \cdot x^{k-1}.$$

We proved this theorem for the cases $k = 2$, 3, and -1 in Examples 3 and 4 and Exercise 13, respectively, of Section 2.4. We will not prove the other cases in this text. Note that this rule holds no matter what the exponent. That is, to differentiate x^k, we write the exponent k as the coefficient, followed by x with an exponent 1 less than k.

① Write the exponent as the coefficient.

$$\overset{①}{\underset{k \cdot x^{k-1}}{x^k}} \ ②$$

② Subtract 1 from the exponent.

Example 1 $\dfrac{d}{dx}x^5 = 5x^4$ ◄

Example 2 $\dfrac{d}{dx}x = 1 \cdot x^{1-1} = 1 \cdot x^0 = 1$ ◄

Example 3 $\dfrac{d}{dx}x^{-4} = -4 \cdot x^{-4-1} = -4x^{-5}$, or $-4 \cdot \dfrac{1}{x^5}$, or $-\dfrac{4}{x^5}$ ◄

The Power Rule also allows us to differentiate $\sqrt{x}$. To do so, it helps to first convert to an expression with a rational exponent.

Example 4 $\dfrac{d}{dx}\sqrt{x} = \dfrac{d}{dx}x^{1/2} = \dfrac{1}{2} \cdot x^{(1/2)-1}$

$$= \dfrac{1}{2}x^{-1/2}, \quad \text{or } \dfrac{1}{2} \cdot \dfrac{1}{x^{1/2}}, \quad \text{or } \dfrac{1}{2} \cdot \dfrac{1}{\sqrt{x}}, \quad \text{or } \dfrac{1}{2\sqrt{x}}$$ ◄

Example 5 $\dfrac{d}{dx}x^{-2/3} = -\dfrac{2}{3}x^{(-2/3)-1}$

$$= -\dfrac{2}{3}x^{-5/3}, \quad \text{or } -\dfrac{2}{3}\dfrac{1}{x^{5/3}}, \quad \text{or } -\dfrac{2}{3\sqrt[3]{x^5}}$$ ◄

TECHNOLOGY CONNECTION

More on Numerical Differentiation and Tangent Lines

Consider $f(x) = x\sqrt{4 - x^2}$, graphed below.

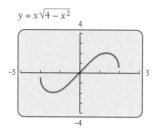

To find the value of dy/dx at a point, we select dy/dx from the **CALC** menu.

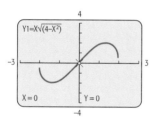

We see the graph and the cursor at the point $(0, 0)$. We key in the desired x-value or use the arrow keys to move the cursor to the desired point. We then press $\boxed{\text{ENTER}}$ to obtain the value of the derivative at the given x-value.

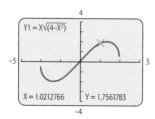

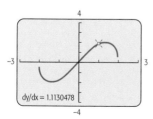

If the arrow keys were used to position the cursor when the derivative was found, we can use the **TANGENT** feature from the **DRAW** menu to draw the tangent line at the point where the derivative was found. This can be done directly from the Graph screen. The equation of the tangent line will also be displayed.

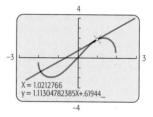

If an x-value was keyed in when the derivative was found, we can also use the **TANGENT** feature to draw the tangent lines. In this case, we must access the **TANGENT** feature from the home screen and supply the name of the function and the x-coordinate of the point of tangency. The equation of the tangent line will not be displayed when this is done.

Exercises

For each of the following functions, find the derivative and draw the tangent line at the given point. (*Hint:* When selecting the viewing window, choose x-dimensions that include the given points.)

1. $f(x) = x(200 - x)$;
   $x = 24, x = 138, x = 150, x = 190$
2. $f(x) = -\frac{1}{3}x^3 + 6x^2 - 11x - 50$;
   $x = -5, x = 0, x = 7, x = 12, x = 15$
3. $f(x) = 6x^2 - x^3$;
   $x = -2, x = 0, x = 2, x = 4, x = 6.3$
4. $f(x) = x\sqrt{4 - x^2}$;
   $x = -2, x = -1.3, x = -0.5, x = 0, x = 1,$
   $x = 2$

Exploratory.
Graph the constant
function $y = -3$.
Then find the
derivative of this
function at $x = -6$,
$x = 0$, and $x = 8$.
What do you conclude
about the derivative of
a constant function?

The Derivative of a Constant Function

Look at the graph of the constant function $F(x) = c$ shown below. What is the slope at each point on the graph?

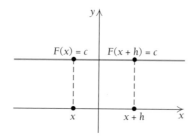

We now have the following.

Theorem 2

The derivative of a constant function is 0. That is, $\dfrac{d}{dx}c = 0$.

Proof: Let F be the function given by $F(x) = c$. Then

$$\frac{F(x + h) - F(x)}{h} = \frac{c - c}{h}$$

$$= \frac{0}{h} = 0.$$

The difference quotient is always 0. Thus, as h approaches 0, the limit of the difference quotient approaches 0, so $F'(x) = 0$.

The Derivative of a Constant Times a Function

Now let's consider differentiating functions such as

$$f(x) = 5x^2 \quad \text{and} \quad g(x) = -7x^4.$$

Note that we already know how to differentiate x^2 and x^4. Let's again look for a pattern in the results of Exercise Set 2.3.

Function	Derivative
$5x^2$	$10x$
$4x^{-1}$	$-4x^{-2}$
$-7x^2$	$-14x$
$5x^3$	$15x^2$

Perhaps you have discovered the following.

> **Theorem 3**
> The derivative of a constant times a function is the constant times the derivative of the function. Using derivative notation, we can write this as
> $$\frac{d}{dx}[c \cdot f(x)] = c \cdot \frac{d}{dx} f(x).$$

Proof: Let F be the function given by $F(x) = cf(x)$. Then

$$\frac{F(x+h) - F(x)}{h} = \frac{cf(x+h) - cf(x)}{h}$$

$$= c\left[\frac{f(x+h) - f(x)}{h}\right].$$

As h approaches 0, the limit of the preceding expression is the same as c times $f'(x)$. Thus, $F'(x) = cf'(x)$.

Combining this rule with the Power Rule allows us to find many derivatives.

Example 6 $\dfrac{d}{dx} 5x^4 = 5\dfrac{d}{dx} x^4 = 5 \cdot 4 \cdot x^{4-1} = 20x^3$

Example 7 $\dfrac{d}{dx}(-9x) = -9\dfrac{d}{dx} x = -9 \cdot 1 = -9$

With practice you will be able to differentiate many such functions in one step.

Example 8 $\dfrac{d}{dx} \dfrac{-4}{(x^2)} = \dfrac{d}{dx}(-4x^{-2}) = -4 \cdot \dfrac{d}{dx} x^{-2}$

$$= -4(-2)x^{-2-1}$$

$$= 8x^{-3}, \quad \text{or} \quad \frac{8}{x^3}$$

Example 9 $\dfrac{d}{dx}(-x^{0.7}) = -1 \cdot \dfrac{d}{dx} x^{0.7}$

$$= -1 \cdot 0.7 \cdot x^{0.7-1}$$

$$= -0.7x^{-0.3}$$

E x p l o r a t o r y .
Enter the following
functions into a
grapher:

$$f(x) = x(100 - x),$$
$$g(x) = x\sqrt{100 - x^2}.$$

Then enter a third
function as the sum of
the above two:
$Y_3 = Y_1 + Y_2$. Find
the derivative of each
of the three functions
at $x = 50$ using the
grapher's numerical
differentiation feature.
(See the Technology
Connection on p. 141.)
    Compare your
answers. How do you
think you can find
the derivative of a
sum?

The Derivative of a Sum or a Difference

In Exercise 11 of Exercise Set 2.4, you found that the derivative of

$$f(x) = x^2 + x$$

is

$$f'(x) = 2x + 1.$$

Note that the derivative of x^2 is $2x$, the derivative of x is 1, and the sum of these derivatives is $f'(x)$. This illustrates the following.

Theorem 4

The Sum–Difference Rule

Sum. The derivative of a sum is the sum of the derivatives:

$$\frac{d}{dx}[f(x) + g(x)] = \frac{d}{dx}f(x) + \frac{d}{dx}g(x).$$

Difference. The derivative of a difference is the difference of the derivatives:

$$\frac{d}{dx}[f(x) - g(x)] = \frac{d}{dx}f(x) - \frac{d}{dx}g(x).$$

Proof: For the Sum Rule, the proof is based on the fact that the limit of a sum is the sum of the limits. Let F be the function defined by $F(x) = f(x) + g(x)$. Then

$$\frac{F(x + h) - F(x)}{h} = \frac{[f(x + h) + g(x + h)] - [f(x) + g(x)]}{h}$$

$$= \frac{f(x + h) - f(x)}{h} + \frac{g(x + h) - g(x)}{h}.$$

As h approaches 0, the two terms on the right approach $f'(x)$ and $g'(x)$, respectively, so their sum approaches $f'(x) + g'(x)$. Thus, $F'(x) = f'(x) + g'(x)$.
    The proof of the Difference Rule is similar.
    Any function that is a sum or a difference of several terms can be differentiated term by term.

Example 10 $\dfrac{d}{dx}(3x + 7) = \dfrac{d}{dx}(3x) + \dfrac{d}{dx}(7)$

$$= 3\frac{d}{dx}(x) + \frac{d}{dx}(7)$$

$$= 3 \cdot 1 + 0$$

$$= 3$$

Example 11
$$\frac{d}{dx}(5x^3 - 3x^2) = \frac{d}{dx}(5x^3) - \frac{d}{dx}(3x^2)$$
$$= 5\frac{d}{dx}x^3 - 3\frac{d}{dx}x^2$$
$$= 5 \cdot 3x^2 - 3 \cdot 2x$$
$$= 15x^2 - 6x$$

Example 12
$$\frac{d}{dx}\left(24x - \sqrt{x} + \frac{2}{x}\right) = \frac{d}{dx}(24x) - \frac{d}{dx}(\sqrt{x}) + \frac{d}{dx}\left(\frac{2}{x}\right)$$
$$= 24 \cdot \frac{d}{dx}x - \frac{d}{dx}x^{1/2} + 2 \cdot \frac{d}{dx}x^{-1}$$
$$= 24 \cdot 1 - \frac{1}{2}x^{(1/2)-1} + 2(-1)x^{-1-1}$$
$$= 24 - \frac{1}{2}x^{-1/2} - 2x^{-2}$$
$$= 24 - \frac{1}{2\sqrt{x}} - \frac{2}{x^2}$$

A *word of caution!* The derivative of

$$f(x) + c,$$

a function plus a constant, is just the derivative of the function,

$$f'(x).$$

The derivative of

$$c \cdot f(x),$$

a function times a constant, is the constant times the derivative

$$c \cdot f'(x).$$

That is, for a product the constant is retained, but for a sum it is not.

Slopes of Tangent Lines

It is important to be able to determine points at which the tangent line to a curve has a certain slope, that is, points at which the derivative attains a certain value.

Example 13 Find the points on the graph of $y = -x^3 + 6x^2$ at which the tangent line is horizontal.

Horizontal Tangents

The following graph of

$$f(x) = -\tfrac{1}{3}x^3 + 6x^2 - 11x - 50$$

shows a tangent line at $x = 12$.

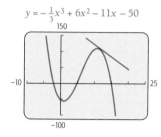

Exercises

Graph each of the following functions and draw tangent lines at various points. Then estimate those x-coordinates at which tangent lines are horizontal. In Examples 13 and 14, you will learn to use calculus techniques to obtain exact values.

1. $f(x) = -\tfrac{1}{3}x^3 + 6x^2 - 11x - 50$
2. $f(x) = -x^3 + 6x^2$
3. $f(x) = \tfrac{1}{3}x^3 - 2x^2 + 4x$
4. $f(x) = x\sqrt{4 - x^2}$

Solution   A horizontal tangent line has slope 0. Thus we seek the values of x for which $dy/dx = 0$. That is, we want to find x such that

$$-3x^2 + 12x = 0.$$

We factor and solve:

$$-3x(x - 4) = 0$$
$$-3x = 0 \quad or \quad x - 4 = 0$$
$$x = 0 \quad or \qquad x = 4.$$

We are to find the points *on the graph*, so we must determine the second coordinates from the original equation, $y = -x^3 + 6x^2$.

$$\text{For } x = 0, \; y = -0^3 + 6 \cdot 0^2 = 0.$$
$$\text{For } x = 4, \; y = -4^3 + 6 \cdot 4^2 = -64 + 96 = 32.$$

Thus the points we are seeking are $(0, 0)$ and $(4, 32)$. This is shown on the graph.

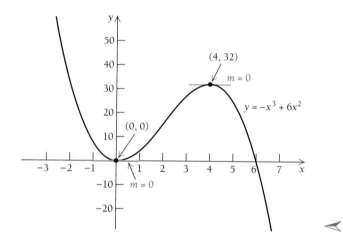

Example 14   Find the points on the graph of $y = -x^3 + 6x^2$ at which the tangent line has slope 6.

Solution   We want to find values of x for which $dy/dx = 6$. That is, we want to find x such that

$$-3x^2 + 12x = 6.$$

To solve, we add -6 on both sides and get

$$-3x^2 + 12x - 6 = 0.$$

We can simplify this equation by multiplying by $-\tfrac{1}{3}$ since each term has a common factor of -3. This gives us

$$x^2 - 4x + 2 = 0.$$

This is a quadratic equation, not readily factorable, so we use the quadratic formula, where $a = 1$, $b = -4$, and $c = 2$:

$$x = \frac{-b \pm \sqrt{b^2 - 4ac}}{2a}$$

$$= \frac{-(-4) \pm \sqrt{(-4)^2 - 4 \cdot 1 \cdot 2}}{2 \cdot 1}$$

$$= \frac{4 \pm \sqrt{8}}{2}$$

$$= \frac{2 \cdot 2 \pm 2\sqrt{2}}{2 \cdot 1}$$

$$= \frac{2}{2} \cdot \frac{2 \pm \sqrt{2}}{1}$$

$$= 2 \pm \sqrt{2}.$$

The solutions are $2 + \sqrt{2}$ and $2 - \sqrt{2}$.

We determine the second coordinates from the original equation. For $x = 2 + \sqrt{2}$,

$$y = -(2 + \sqrt{2})^3 + 6(2 + \sqrt{2})^2$$

$$= -[(2 + \sqrt{2})^2(2 + \sqrt{2})] + 6(4 + 4\sqrt{2} + 2)$$

$$= -[(6 + 4\sqrt{2})(2 + \sqrt{2})] + 6(6 + 4\sqrt{2})$$

$$= -[12 + 6\sqrt{2} + 8\sqrt{2} + 8] + 36 + 24\sqrt{2}$$

$$= -[20 + 14\sqrt{2}] + 36 + 24\sqrt{2}$$

$$= -20 - 14\sqrt{2} + 36 + 24\sqrt{2}$$

$$= 16 + 10\sqrt{2}.$$

Similarly, for $x = 2 - \sqrt{2}$,

$$y = 16 - 10\sqrt{2}.$$

Thus the points that we are seeking are $(2 + \sqrt{2}, 16 + 10\sqrt{2})$ and $(2 - \sqrt{2}, 16 - 10\sqrt{2})$. This is shown on the following graph.

TECHNOLOGY CONNECTION

Exercises

1. Graph $y = \frac{1}{3}x^3 - 2x^2 + 4x$ and draw tangent lines at various points. Estimate points at which the tangent line is horizontal. Then use the calculus (analytic techniques) of Examples 13 and 14 to find the exact results.

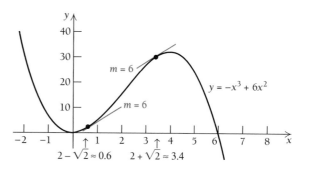

E X E R C I S E   S E T   2 . 5

Find $\dfrac{dy}{dx}$.

1. $y = x^7$

2. $y = x^8$

3. $y = 15$

4. $y = 78$

5. $y = 4x^{150}$

6. $y = 7x^{200}$

7. $y = x^3 + 3x^2$

8. $y = x^4 - 7x$

9. $y = 8\sqrt{x}$

10. $y = 4\sqrt{x}$

11. $y = x^{0.07}$

12. $y = x^{0.78}$

13. $y = \frac{1}{2}x^{4/5}$

14. $y = -4.8x^{1/3}$

15. $y = x^{-3}$

16. $y = x^{-4}$

17. $y = 3x^2 - 8x + 7$

18. $y = 4x^2 - 7x + 5$

19. $y = \sqrt[4]{x} - \dfrac{1}{x}$

20. $y = \sqrt[5]{x} - \dfrac{2}{x}$

Find $f'(x)$.

21. $f(x) = 0.64x^{2.5}$

22. $f(x) = 0.32x^{12.5}$

23. $f(x) = \dfrac{5}{x} - x$

24. $f(x) = \dfrac{4}{x} - x$

25. $f(x) = 4x - 7$

26. $f(x) = 7x + 11$

27. $f(x) = 4x + 9$

28. $f(x) = 7x - 14$

29. $f(x) = \dfrac{x^4}{4}$

30. $f(x) = \dfrac{x^3}{3}$

31. $f(x) = -0.01x^2 - 0.5x + 70$

32. $f(x) = -0.01x^2 + 0.4x + 50$

33. $f(x) = 3x^{-2/3} + x^{3/4} + x^{6/5} + \dfrac{8}{x^3}$

34. $f(x) = x^{-3/4} - 3x^{2/3} + x^{5/4} + \dfrac{2}{x^4}$

35. $f(x) = \dfrac{2}{x} - \dfrac{x}{2}$

36. $f(x) = \dfrac{x}{5} + \dfrac{5}{x}$

37. $f(x) = \dfrac{16}{x} - \dfrac{8}{x^3} + \dfrac{1}{x^4}$

38. $f(x) = \dfrac{20}{x^5} + \dfrac{1}{x^3} - \dfrac{2}{x}$

39. $f(x) = \sqrt{x} + \sqrt[3]{x} - \sqrt[4]{x} + \sqrt[5]{x}$

40. $f(x) = \dfrac{x^5 - 3x^4 + 2x^3 - 5x^2 - 8x + 4}{x^2}$

41. Find an equation of the tangent line to the graph of $f(x) = x^3 - 2x + 1$ at the point $(2, 5)$, at $(-1, 2)$, and at $(0, 1)$.

42. Find an equation of the tangent line to the graph of $f(x) = x^2 - \sqrt{x}$ at the point $(1, 0)$, at $(4, 14)$, and at $(9, 78)$.

For each function, find the points on the graph at which the tangent line is horizontal.

43. $y = x^2$

44. $y = -x^2$

45. $y = -x^3$

46. $y = x^3$

47. $y = 3x^2 - 5x + 4$

48. $y = 5x^2 - 3x + 8$

49. $y = -0.01x^2 - 0.5x + 70$

50. $y = -0.01x^2 + 0.4x + 50$

51. $y = 2x + 4$

52. $y = -2x + 5$

53. $y = 4$

54. $y = -3$

55. $y = -x^3 + x^2 + 5x - 1$

56. $y = -\frac{1}{3}x^3 + 6x^2 - 11x - 50$

57. $y = \frac{1}{3}x^3 - 3x + 2$

58. $y = x^3 - 6x + 1$

For each function, find the points on the graph at which the tangent line has slope 1.

59. $y = 20x - x^2$

60. $y = 6x - x^2$

61. $y = -0.025x^2 + 4x$

62. $y = -0.01x^2 + 2x$

63. $y = \frac{1}{3}x^3 + 2x^2 + 2x$

64. $y = \frac{1}{3}x^3 - x^2 - 4x + 1$

Synthesis ....

65. Find the points on the graph of
$$y = x^4 - \tfrac{4}{3}x^2 - 4$$
at which the tangent line is horizontal.

66. Find the points on the graph of
$$y = 2x^6 - x^4 - 2$$
at which the tangent line is horizontal.

Find dy/dx. Each of the following can be differentiated using the rules developed in this section, but some algebra may be required beforehand.

67. $y = x(x - 1)$

68. $y = (x - 1)(x + 1)$

69. $y = (x - 2)(x + 3)$

70. $y = \dfrac{5x^2 - 8x + 3}{8}$

71. $y = \dfrac{x^5 + x}{x^2}$

72. $y = (5x)^2$

73. $y = (-4x)^3$

74. $y = \sqrt{7x}$

75. $y = \sqrt[3]{8x}$

76. $y = (x - 3)^2$

77. $y = (x + 1)^3$

78. $y = (x - 2)^3(x + 1)$

79. Prove Theorem 4 (the Difference Rule).

tw 80. Write a paragraph comparing the Power Rule, the Sum–Difference Rule, and the rules for differentiating a constant or a constant times a function.

tw 81. Write a short biographical paper on the lives of Leibniz and/or Newton. Emphasize the contributions they made to many areas of science and society.

 TECHNOLOGY   CONNECTION

Graph each of the following. Draw tangent lines at various points. Estimate those values at which tangent lines are horizontal.

82. $f(x) = x^4 - 3x^2 + 1$

83. $f(x) = 1.6x^3 - 2.3x - 3.7$

84. $f(x) = 10.2x^4 - 6.9x^3$

85. $f(x) = \dfrac{5x^2 + 8x - 3}{3x^2 + 2}$

Some graphers can graph both a function f and its derivative f' using the same viewing window. For each of the following functions, graph f and f'.

86. $f(x) = 20x^3 - 3x^5$

87. $f(x) = x^4 - 3x^2 + 1$

88. $f(x) = x^3 - 2x - 2$

89. $f(x) = x^4 - x^3$

90. $f(x) = \dfrac{4x}{x^2 + 1}$

91. $f(x) = \dfrac{5x^2 + 8x - 3}{3x^2 + 2}$

2.6

OBJECTIVES

➤ Given a formula for distance, find velocity and acceleration.

➤ Find instantaneous rates of change.

Instantaneous Rates of Change; Business Applications

Now we are ready to move from average rates of change to instantaneous rates of change. In Section 2.3, we found that for a function f:

> **Average rate of change = Simplified difference quotient**
> $$= \frac{f(x + h) - f(x)}{h}.$$

If we let h approach 0, we find the instantaneous rate of change.

> **Definition**
> For any function f,
> $$\textbf{Instantaneous rate of change} = f'(x) = \lim_{h \to 0} \frac{f(x + h) - f(x)}{h}.$$

FUEL ECONOMY

INSTANTANEOUS	AVERAGE
20 MPG	**22.5** MPG

RANGE
300 MILES

TRIP COMPUTER	TRIP DATA

Some of the newer automobiles describe fuel economy by giving average miles per gallon and instantaneous miles per gallon. Next time you see such a display, think of the instantaneous mpg part of the display as a derivative.

In Sections 2.4 and 2.5, we focused on this limit as the slope of the tangent line to a curve, finding some fast ways to compute. Now we focus on this limit as an instantaneous rate of change.

Let's say that a car travels 108 mi in 2 hr. Its *average speed* (or *average velocity*) is 108 mi/2 hr, or 54 mi/hr. This is the *average rate of change* of distance with respect to time. At various times during the trip, however, the speedometer did not read 54. Thus we say that 54 is the *average*. A snapshot of the speedometer taken at any instant would indicate *instantaneous speed,* or **instantaneous rate of change.**

Average rates of change are given by difference quotients. If distance s is a function of time t and h is the duration of the trip, then the average rate of change of distance with respect to time, called **average velocity,** is given by

$$\text{Average velocity} = \frac{\text{Difference in distance}}{\text{Difference in time}} = \frac{s(t + h) - s(t)}{h}.$$

Instantaneous rates of change are found by letting h approach 0. Thus,

$$\text{Instantaneous velocity} = \lim_{h \to 0} \frac{s(t + h) - s(t)}{h} = s'(t).$$

Example 1 *Physical Science: Velocity.* An object travels in such a way that distance s (in miles) from the starting point is a function of time t (in hours) as follows:

$$s(t) = 10t^2.$$

a) Find the average velocity between the times $t = 2$ and $t = 5$.

b) Find the (instantaneous) velocity when $t = 4$.

Solution

a) From $t = 2$ to $t = 5$, $h = 3$, so

$$\frac{\text{Difference in miles}}{\text{Difference in hours}} = \frac{s(t + h) - s(t)}{h} = \frac{s(2 + 3) - s(2)}{3}$$

$$= \frac{s(5) - s(2)}{3} = \frac{10 \cdot 5^2 - 10 \cdot 2^2}{3} = 70 \frac{\text{mi}}{\text{hr}}.$$

b) The instantaneous velocity is given by

$$\lim_{h \to 0} \frac{s(t + h) - s(t)}{h} = s'(t).$$

An instantaneous velocity.

We know how to find this limit quickly from the special techniques learned in Section 2.5. Thus, $s'(t) = 20t$ and

$$s'(4) = 20 \cdot 4 = 80 \frac{\text{mi}}{\text{hr}}.$$

We generally use the letter v for velocity. We have the following definition.

Definition

Velocity $= v(t) = \lim_{h \to 0} \dfrac{s(t + h) - s(t)}{h} = s'(t)$

The rate of change of velocity is called **acceleration.** We generally use the letter a for acceleration. Thus the following definition applies.

Definition

Acceleration $= a(t) = v'(t)$

Example 2   *Physical Science: Distance, Velocity, and Acceleration.* For $s(t) = 10t^2$, find $v(t)$ and $a(t)$, where s is the distance from the starting point, in miles, and t is in hours. Then find the distance, velocity, and acceleration when $t = 4$ hr.

Solution   We have

$$v(t) = s'(t) = 20t,$$
$$a(t) = v'(t) = 20.$$

Then $s(4) = 10(4)^2 = 160$ miles,

$$v(4) = 20(4) = 80 \text{ mi/hr}, \quad \text{and}$$
$$a(4) = 20 \text{ mi/hr}^2.$$

If this distance function applies to a vehicle, then at time $t = 4$ hr, the distance is 160 mi, the velocity, or instantaneous speed, is 80 mi/hr, and the acceleration is 20 miles per hour per hour, which we abbreviate as 20 mi/hr^2.

In general, derivatives give instantaneous rates of change.

Definition

If y is a function of x, $y = f(x)$, then the (*instantaneous*) **rate of change of y with respect to x** is given by the derivative

$$\frac{dy}{dx} = f'(x) = \lim_{h \to 0} \frac{f(x + h) - f(x)}{h}.$$

Example 3 *Life Science: Volume of a Cancer Tumor.* The spherical volume V of a cancer tumor is given by

$$V(r) = \tfrac{4}{3}\pi r^3,$$

where r is the radius of the tumor, in centimeters.

a) Find the rate of change of the volume with respect to the radius.

b) Find the rate of change of the volume at $r = 1.2$ cm.

Solution

a) $\dfrac{dV}{dr} = V'(r) = 3 \cdot \dfrac{4}{3} \cdot \pi r^2 = 4\pi r^2$

(This turns out to be the surface area.)

b) $V'(1.2) = 4\pi(1.2)^2 = 5.76\pi \approx 18 \dfrac{\text{cm}^3}{\text{cm}} = 18 \text{ cm}^2$

Example 4 *Life Science: Population Growth.* The initial population in a bacteria colony is 10,000. After t hours, the colony has grown to a number $P(t)$ given by

$$P(t) = 10{,}000(1 + 0.86t + t^2).$$

a) Find the rate of change of the population P with respect to time t. This is also known as the **growth rate.**

b) Find the number of bacteria present after 5 hr. Also, find the growth rate when $t = 5$.

Solution

a) Note that $P(t) = 10{,}000 + 8600t + 10{,}000t^2$. Then

$$P'(t) = 8600 + 20{,}000t.$$

b) The number of bacteria present when $t = 5$ is given by

$$P(5) = 10{,}000 + 8600 \cdot 5 + 10{,}000 \cdot 5^2 = 303{,}000.$$

The growth rate when $t = 5$ is given by

$$P'(5) = 8600 + 20{,}000 \cdot 5 = 108{,}600 \; \frac{\text{bacteria}}{\text{hr}}.$$

Thus at $t = 5$, there are 303,000 bacteria present, and the colony is growing at the rate of 108,600 bacteria per hour.

Rates of Change in Business and Economics

In the study of business and economics, we are frequently interested in how such quantities as cost, revenue, and profit change with an increase in product quantity. In particular, we are interested in what is called **marginal*** cost or profit (or what-

---

*The term "marginal" comes from the Marginalist School of Economic Thought, which originated in Austria for the purpose of applying mathematics and statistics to the study of economics.

ever). This term is used to signify the *rate of change with respect to quantity*. Thus, if

$$C(x) = \text{the } total \ cost \text{ of producing } x \text{ units of a product}$$
$$\text{(usually considered in some time period)},$$

then

$$C'(x) = \text{the } \textbf{marginal cost}$$
$$= \text{the rate of change of the total cost with respect to}$$
$$\text{the number of units, } x, \text{ produced.}$$

Let's think about these interpretations. The total cost of producing 5 units of a product is $C(5)$. The rate of change $C'(5)$ is the cost per unit at that stage in the production process. That this cost per unit does not include fixed costs is seen in this example:

$$C(x) = \underbrace{(x^2 + 4x)}_{\text{Variable costs}} + \underbrace{\$10{,}000.}_{\text{Fixed costs (constant)}}$$

Because the derivative of a constant is 0,

$$C'(x) = 2x + 4.$$

This illustrates an economic principle stating that the fixed costs of a company have no effect on marginal cost.

Following are some other marginal functions. Recall that

$$R(x) = \text{the } total \ revenue \text{ from the sale of } x \text{ units.}$$

Then

$$R'(x) = \text{the } \textbf{marginal revenue}$$
$$= \text{the rate of change of the total revenue with respect}$$
$$\text{to the number of units, } x, \text{ sold.}$$

Also,

$$P(x) = \text{the } total \ profit \text{ from the production and sale}$$
$$\text{of } x \text{ units of a product}$$
$$= R(x) - C(x).$$

Then

$$P'(x) = \text{the } \textbf{marginal profit}$$
$$= \text{the rate of change of the total profit with respect}$$
$$\text{to the number of units, } x, \text{ produced and sold}$$
$$= R'(x) - C'(x).$$

Example 5   *Business: Marginal Revenue, Cost, and Profit.* Given

$$R(x) = 50x, \quad \text{and}$$
$$C(x) = 2x^3 - 12x^2 + 40x + 10,$$

find each of the following.

a) Total profit $P(x)$

TECHNOLOGY
CONNECTION

Business: Marginal Revenue, Cost, and Profit

Exercise

1. Using the viewing window [0, 100, 0, 2000], graph these total-revenue and total-cost functions:

$$R(x) = 50x - 0.5x^2 \quad \text{and}$$
$$C(x) = 10x + 3.$$

Then find $P(x)$ and graph it using the same viewing window. Find $R'(x)$, $C'(x)$, and $P'(x)$ and graph them using [0, 60, 0, 60]. Then find $R(40)$, $C(40)$, $P(40)$, $R'(40)$, $C'(40)$, and $P'(40)$. Which marginal function is constant?

b) Total revenue $R(2)$, cost $C(2)$, and profit $P(2)$ from the production and sale of 2 units of the product

c) Marginal revenue $R'(x)$, cost $C'(x)$, and profit $P'(x)$

d) Marginal revenue $R'(2)$, cost $C'(2)$, and profit $P'(2)$ at that point in the process where 2 units have been produced and sold

Solution

a) The total profit $P(x) = R(x) - C(x)$
$$= 50x - (2x^3 - 12x^2 + 40x + 10)$$
$$= -2x^3 + 12x^2 + 10x - 10$$

b) $R(2) = 50 \cdot 2 = \$100$ (the total revenue from the sale of the first 2 units);

$C(2) = 2 \cdot 2^3 - 12 \cdot 2^2 + 40 \cdot 2 + 10 = \$58$
(the total cost of producing the first 2 units);

$P(2) = R(2) - C(2) = \$100 - \$58 = \$42$ (the total profit from the production and sale of the first 2 units)

c) The marginal revenue $R'(x) = 50$;

The marginal cost $C'(x) = 6x^2 - 24x + 40$;

The marginal profit $P'(x) = R'(x) - C'(x)$
$$= 50 - (6x^2 - 24x + 40)$$
$$= -6x^2 + 24x + 10$$

d) $R'(2) = \$50$ per unit;

$C'(2) = 6 \cdot 2^2 - 24 \cdot 2 + 40 = \$16$ per unit;

$P'(2) = \$50 - \$16 = \$34$ per unit

Note that the marginal revenue in this example is constant. No matter how much is produced and sold, the revenue per unit stays the same. This may not always be the case. Also note that $C'(2)$, or \$16 per unit, is not the average cost per unit, which is given by

$$\frac{\text{Total cost of producing 2 units}}{2 \text{ units}} = \frac{\$58}{2}$$
$$= \$29 \text{ per unit.}$$

In general, we have the following.

Definition

$A(x) =$ the **average cost** of producing x units $= \dfrac{C(x)}{x}$

Let's look at a typical marginal-cost function C' and its associated total-cost function C.

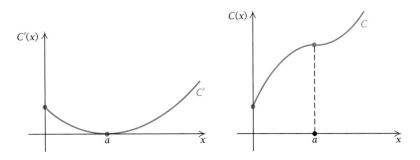

Marginal cost generally decreases as more units are produced until it reaches some minimum value at a, and then it increases. (This is probably due to factors such as paying overtime or buying more machinery.) Since $C'(x)$ represents the slope of $C(x)$ and is positive and decreasing up to a, the graph of $C(x)$ turns down as x goes from 0 to a. Then past a, it turns up.

EXERCISE SET 2.6

1. Given
$$s(t) = t^3 + t,$$
where s is in feet and t is in seconds, find each of the following.
   a) $v(t)$
   b) $a(t)$
   c) The velocity and acceleration when $t = 4$ sec

2. Given
$$s(t) = 3t + 10,$$
where s is in miles and t is in hours, find each of the following.
   a) $v(t)$
   b) $a(t)$
   c) The velocity and acceleration when $t = 2$ hr. When the distance function is given by a linear function, we have *uniform motion*.
   tw d) What does uniform motion mean in terms of velocity and acceleration?

Applications ....
BUSINESS AND ECONOMICS

3. *Marginal Revenue, Cost, and Profit.* Given
$$R(x) = 5x \quad \text{and}$$
$$C(x) = 0.001x^2 + 1.2x + 60,$$
find each of the following.
   a) $P(x)$
   b) $R(100)$, $C(100)$, and $P(100)$
   c) $R'(x)$, $C'(x)$, and $P'(x)$
   d) $R'(100)$, $C'(100)$, and $P'(100)$
   tw e) Describe in words the meaning of each quantity in parts (b) and (d).

4. *Marginal Revenue, Cost, and Profit.* Given
$$R(x) = 50x - 0.5x^2 \quad \text{and} \quad C(x) = 4x + 10,$$
find each of the following.
   a) $P(x)$
   b) $R(20)$, $C(20)$, and $P(20)$
   c) $R'(x)$, $C'(x)$, and $P'(x)$
   d) $R'(20)$, $C'(20)$, and $P'(20)$

5. *Advertising.* A firm estimates that it will sell N units of a product after spending a dollars on advertising, where

$$N(a) = -a^2 + 300a + 6$$

and a is in thousands of dollars.

a) What is the rate of change of the number of units sold with respect to the amount spent on advertising?
b) How many units will be sold after spending $10,000 on advertising?
c) What is the rate of change at $a = 10$?
tw d) Explain the meaning of your answers to parts (a) and (c).

6. *Sales.* A company determines that monthly sales S, in thousands of dollars, after t months of marketing a product is given by

$$S(t) = 2t^3 - 40t^2 + 220t + 160.$$

a) Find the monthly sales after 1 month; 4 months; 6 months; 9 months; 20 months.
b) Find the rate of change $S'(t)$.
c) Find the rate of change at $t = 1$; $t = 4$; $t = 6$; $t = 9$; $t = 20$.
tw d) Explain why the CEO of this company should not be upset by decreasing sales in month 6.

7. *Marginal Productivity.* An employee's monthly productivity M, in numbers of units produced, is found to be a function of the number of years of service t. For a certain product, the productivity function is given by

$$M(t) = -2t^2 + 100t + 180.$$

a) Find the productivity of an employee after 5 yr, 10 yr, 25 yr, and 45 yr of service.
b) Find the marginal productivity.

c) Find the marginal productivity at $t = 5$; $t = 10$; $t = 25$; $t = 45$.
tw d) Explain how the employee's marginal productivity might be related to experience and to age.

8. *Supply.* A supply function for a certain product is given by

$$S(p) = 0.08p^3 + 2p^2 + 10p + 11,$$

where S is the number of items sold at price p, in dollars (see Section 1.5).

a) Find the rate of change of supply with respect to price, dS/dp.
b) How many units will the seller allow to be sold when the price is $3 per unit?
c) What is the rate of change at $p = 3$?
tw d) Would you expect dS/dp to be positive or negative for values of p? Why?

9. *Demand.* A demand function for a certain product is given by

$$D(p) = 100 - \sqrt{p},$$

where D is the number of items purchased at price p, in dollars (see Section 1.5).

a) Find the rate of change of quantity with respect to price, dD/dp.
b) How many units will the consumer want to buy when the price is $25 per unit?
c) What is the rate of change at $p = 25$?
tw d) Would you expect dD/dp to be positive or negative for values of p? Why?

LIFE AND PHYSICAL SCIENCES

10. *Stopping Distance on Glare Ice.* The stopping distance on glare ice (at some fixed speed) of regular tires is given by a linear function of the air temperature F,

$$D(F) = 2F + 115,$$

where $D(F)$ is the stopping distance, in feet, when the air temperature is F, in degrees Fahrenheit.

a) Find the rate of change of the stopping distance D with respect to the air temperature F.
tw b) Explain the meaning of your answer to part (a).

11. *Healing Wound.* The circumference C, in centimeters, of a healing wound is given by

$$C(r) = 2\pi r,$$

where r is the radius, in centimeters.

a) Find the rate of change of the circumference with respect to the radius.

tw b) Explain the meaning of your answer to part (a).

12. *Healing Wound.* The circular area A, in square centimeters, of a healing wound is given by

$$A(r) = \pi r^2,$$

where r is the radius, in centimeters.

a) Find the rate of change of the area with respect to the radius.

tw b) Explain the meaning of your answer to part (a).

13. *Temperature During an Illness.* The temperature T of a person during an illness is given by

$$T(t) = -0.1t^2 + 1.2t + 98.6,$$

where T is the temperature, in degrees Fahrenheit, at time t, in days.

a) Find the rate of change of the temperature with respect to time.

b) Find the temperature at $t = 1.5$ days.

c) Find the rate of change at $t = 1.5$ days.

tw d) Why would the sign of $T'(x)$ be significant to a doctor?

14. *Blood Pressure.* For a certain dosage of x cubic centimeters (cc) of a drug, the resulting blood pressure B is approximated by

$$B(x) = 0.05x^2 - 0.3x^3.$$

a) Find the rate of change of the blood pressure with respect to the dosage. Such a rate of change is often called the *sensitivity.*

tw b) Explain the meaning of your answer to part (a).

15. *Territorial Area.* The territorial area T of an animal is defined as its defended, or exclusive,

These bears may be determining territorial area.

region. The area T of that region can be approximated using the animal's body weight W by

$$T = W^{1.31}$$

(see Section 1.5).

a) Find dT/dW.

tw b) Explain the meaning of your answer to part (a).

16. *Home Range.* The home range H of an animal is defined as the region to which the animal confines its movements. The area of that region can be approximated using the animal's body weight W by

$$H = W^{1.41}$$

(see Section 1.5).

a) Find dH/dW.

tw b) Explain the meaning of your answer to part (a).

17. *Sensitivity.* The reaction R of the body to a dose Q of medication is often represented by the general function

$$R(Q) = Q^2\left(\frac{k}{2} - \frac{Q}{3}\right),$$

where k is a constant and R is in millimeters, if the reaction is a change in blood pressure, or in degrees of temperature, if the reaction is a change in temperature. The rate of change dR/dQ is defined to be the *sensitivity.*

a) Find a formula for the sensitivity.

tw b) Explain the meaning of your answer to part (a).

SOCIAL SCIENCES

18. *Population Growth Rate.* The population of a city grows from an initial size of 100,000 to an amount P given by

$$P(t) = 100{,}000 + 2000t^2,$$

where t is in years.

a) Find the growth rate.

b) Find the number of people in the city after 10 yr (at $t = 10$).

c) Find the growth rate at $t = 10$.

tw d) Explain the meaning of your answers to parts (a) and (c).

19. *Median Age of Women at First Marriage.* The median age of women at first marriage can be approximated by the linear function

$$A(t) = 0.08t + 19.7,$$

where $A(t)$ is the median age of women at first marriage the tth year after 1950.

a) Find the rate of change of the median age A with respect to time t.

tw b) Explain the meaning of your answer to part (a).

GENERAL INTEREST

20. *View to the Horizon.* The view V, or distance, in miles that one can see to the horizon from a height h, in feet, is given by

$$V = 1.22\sqrt{h}.$$

a) Find the rate of change of V with respect to h.
b) How far can one see to the horizon from an airplane window from a height of 40,000 ft?

c) Find the rate of change at $h = 40,000$.

tw d) Explain the meaning of your answers to parts (a) and (c).

Synthesis

tw **21.** Explain and compare average rate of change and instantaneous rate of change.

tw **22.** Discuss as many interpretations as you can of the derivative of a function at a point x.

 TECHNOLOGY   CONNECTION

In Exercises 23–26, graph the total-revenue and total-cost functions. Then graph the marginal-revenue and marginal-cost functions.

23. $R(x) = 50x - 0.5x^2$, $C(x) = 4x + 10$

24. $R(x) = 5x$, $C(x) = 0.001x^2 + 1.2x + 60$

25. $R(x) = 9x$, $C(x) = x^3 - 6x^2 + 15x$

26. $R(x) = 50x - 0.5x^2$, $C(x) = 10x + 3$

2.7

OBJECTIVES

➤ Differentiate using the Product and Quotient Rules.

Differentiation Techniques: The Product and Quotient Rules

The Product Rule

The derivative of a sum is the sum of the derivatives, but the derivative of a product is *not* the product of the derivatives. To see this, consider x^2 and x^5. The product is x^7, and the derivative of this product is $7x^6$. The individual derivatives are $2x$ and $5x^4$, and the product of these derivatives is $10x^5$, which is not $7x^6$.

The following is the rule for finding the derivative of a product.

Theorem 5

The Product Rule

Suppose that $F(x) = f(x) \cdot g(x)$, where $f(x)$ is the "first" factor and $g(x)$ is the "second" factor. Then

$$F'(x) = \frac{d}{dx}[f(x) \cdot g(x)] = f(x) \cdot \left[\frac{d}{dx}g(x)\right] + \left[\frac{d}{dx}f(x)\right] \cdot g(x).$$

The derivative of a product is the first factor times the derivative of the second factor, plus the derivative of the first factor times the second factor.

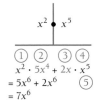

$x^2 \bullet x^5$

① ② ③ ④
$x^2 \cdot 5x^4 + 2x \cdot x^5$
$= 5x^6 + 2x^6$ ⑤
$= 7x^6$

Let's check this for $x^2 \cdot x^5$. There are five steps.

1. Write down the first factor.
2. Multiply it by the derivative of the second factor.
3. Write down the derivative of the first factor.
4. Multiply it by the second factor.
5. Add the result of steps (1) and (2) to the result of steps (3) and (4).

Example 1   Find $\dfrac{d}{dx}[(x^4 - 2x^3 - 7)(3x^2 - 5x)]$.

Solution   We have

$$\frac{d}{dx}(x^4 - 2x^3 - 7)(3x^2 - 5x) = (x^4 - 2x^3 - 7)(6x - 5)$$
$$+ (4x^3 - 6x^2)(3x^2 - 5x).$$

Note that we could have multiplied the polynomials and then differentiated, avoiding the use of the Product Rule, but this would have been more work.   ◄

Example 2   For $F(x) = (x^2 + 4x - 11)(7x^3 - \sqrt{x})$, find $F'(x)$.

Solution   We rewrite this as

$$F(x) = (x^2 + 4x - 11)(7x^3 - x^{1/2}).$$

Then using the Product Rule, we have

$$F'(x) = (x^2 + 4x - 11)\left(21x^2 - \tfrac{1}{2}x^{-1/2}\right) + (2x + 4)(7x^3 - x^{1/2}).\quad ◄$$

The Quotient Rule

The derivative of a quotient is *not* the quotient of the derivatives. To see why, consider x^5 and x^2. The quotient x^5/x^2 is x^3, and the derivative of this quotient is $3x^2$. The individual derivatives are $5x^4$ and $2x$, and the quotient of these derivatives $5x^4/2x$ is $(5/2)x^3$, which is not $3x^2$.

The rule for differentiating quotients is as follows.

Theorem 6

The Quotient Rule

If

$$Q(x) = \frac{N(x)}{D(x)},$$

then

$$Q'(x) = \frac{D(x) \cdot N'(x) - D'(x) \cdot N(x)}{[D(x)]^2}.$$

(continued)

The derivative of a quotient is the denominator times the derivative of the numerator, minus the derivative of the denominator times the numerator, all divided by the square of the denominator.

(If we think of the function in the numerator as the first function and the function in the denominator as the second function, then we can reword the Quotient Rule as "the second function times the derivative of the first function minus the derivative of the second function times the first function, all divided by the square of the second function.")

The Quotient Rule is illustrated below.

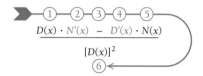

$$\frac{\overset{①}{D(x)} \cdot \overset{②}{N'(x)} \quad \overset{③}{-} \quad \overset{④}{D'(x)} \cdot \overset{⑤}{N(x)}}{\underset{⑥}{[D(x)]^2}}$$

1. Write down the denominator.
2. Multiply the denominator by the derivative of the numerator.
3. Write a minus sign.
4. Write down the derivative of the denominator.
5. Multiply it by the numerator.
6. Divide by the square of the denominator.

Example 3   For $Q(x) = x^5/x^3$, find $Q'(x)$.

Solution

$$Q'(x) = \frac{x^3 \cdot 5x^4 - 3x^2 \cdot x^5}{(x^3)^2}$$

$$= \frac{5x^7 - 3x^7}{x^6} = \frac{2x^7}{x^6} = 2x$$

Example 4   Differentiate: $f(x) = \dfrac{1 + x^2}{x^3}$.

Solution

$$f'(x) = \frac{x^3 \cdot 2x - 3x^2(1 + x^2)}{(x^3)^2}$$

$$= \frac{2x^4 - 3x^2 - 3x^4}{x^6} = \frac{-x^4 - 3x^2}{x^6}$$

$$= \frac{-x^2 \cdot x^2 - 3x^2}{x^6} = \frac{x^2(-x^2 - 3)}{x^2 \cdot x^4}$$

$$= \frac{-x^2 - 3}{x^4}$$

TECHNOLOGY CONNECTION

Checking Derivatives Graphically

Consider the derivative in Example 5. To check the answer, first we enter the function:

$$y_1 = \frac{x^2 - 3x}{x - 1}.$$

Then we enter the possible derivative:

$$y_2 = \frac{x^2 - 2x + 3}{(x - 1)^2}.$$

For the third function, we enter

$$y_3 = \text{nDERIV}(Y_1, x, x).$$

Next, we deselect y_1 and graph y_2 and y_3. We use different graph styles so that we see each graph as it appears on the screen.

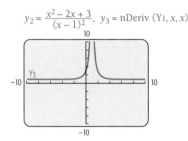

Since the graphs appear to coincide, it appears that $y_2 = y_3$ and we have a check of the result. This is considered a partial check, however, because the graphs might not coincide at a point that is not in the viewing window.

We can also use a table to check that $y_2 = y_3$.

X	Y2	Y3
5.97	1.081	1.081
5.98	1.0806	1.0806
5.99	1.0803	1.0803
6	1.08	1.08
6.01	1.0797	1.0797
6.02	1.0794	1.0794
6.03	1.079	1.079
X = 5.97		

Suppose that we had incorrectly calculated the derivative to be $y_2 = (x^2 - 2x - 8)/(x - 1)^2$. We see in the following that the graphs do not agree:

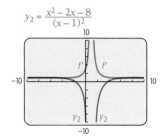

Exercises

1. For the function

$$f(x) = \frac{x^2 - 4x}{x + 2},$$

use graphs and tables to determine which of the following seems to be the correct derivative.

a) $f'(x) = \dfrac{-x^2 - 4x - 8}{(x + 2)^2}$

b) $f'(x) = \dfrac{x^2 - 4x + 8}{(x^2 - 4x)^2}$

c) $f'(x) = \dfrac{x^2 + 4x - 8}{(x + 2)^2}$

2.–5. Check the results of Examples 1–4 using a grapher.

Example 5   Differentiate: $f(x) = \dfrac{x^2 - 3x}{x - 1}$.

Solution   We have

$$f'(x) = \frac{(x-1)(2x-3) - 1(x^2 - 3x)}{(x-1)^2}$$

$$= \frac{2x^2 - 5x + 3 - x^2 + 3x}{(x-1)^2}$$

$$= \frac{x^2 - 2x + 3}{(x-1)^2}.$$

It is not necessary to multiply out $(x-1)^2$ to get $x^2 - 2x + 1$.

TECHNOLOGY CONNECTION

Business: Demand and Revenue

Exercises

1. A company determines that the demand function for a certain product is given by $D(p) = 200 - p$. Find an expression for the total revenue $R(p)$. Then find the marginal revenue $R'(p)$. Graph all three functions.

An Application

We discussed earlier that it is typical for a total-revenue function to vary depending on the number of units x sold. Let's see what can determine this. Recall the consumer's demand function, or price function, $x = D(p)$, discussed in Section 1.5. It is the quantity x of units of a product that a consumer will purchase when the price is p dollars per unit. This is typically a decreasing function.

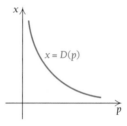

The total revenue when the price is p dollars per unit is then

$$R(p) = (\text{Number of units sold}) \cdot (\text{Price charged per unit}),$$

or $R(p) = x \cdot p = D(p) \cdot p = pD(p).$

The graph of a typical revenue function follows.

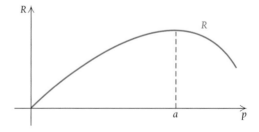

As the price increases, $D(p)$ decreases. Because we have a product $p \cdot D(p)$, the revenue typically rises for a while as p increases, but tapers off as $D(p)$ gets smaller and smaller.

Using the Product Rule, we can obtain an expression for the marginal revenue $R'(p)$ in terms of p and $D'(p)$. If

$$R(p) = pD(p),$$

then

$$R'(p) = p \cdot D'(p) + 1 \cdot D(p) = pD'(p) + D(p).$$

You need not memorize this. You can merely repeat the Product Rule where necessary.

EXERCISE SET 2.7

Differentiate.

1. $y = x^3 \cdot x^8$, two ways
2. $y = x^4 \cdot x^9$, two ways
3. $y = \dfrac{-1}{x}$, two ways   4. $y = \dfrac{1}{x}$, two ways
5. $y = \dfrac{x^8}{x^5}$, two ways   6. $y = \dfrac{x^9}{x^5}$, two ways
7. $y = (8x^5 - 3x^2 + 20)(8x^4 - 3\sqrt{x})$
8. $f(x) = (7x^6 + 4x^3 - 50)(9x^{10} - 7\sqrt{x})$
9. $f(x) = x(300 - x)$, two ways
10. $f(x) = x(400 - x)$, two ways
11. $f(x) = (4\sqrt{x} - 6)(x^3 - 2x + 4)$
12. $f(x) = (\sqrt[3]{x} - 5x^2 + 4)(4x^2 + 11x - 5)$
13. $f(x) = (x + 3)^2$
    [*Hint*: $(x + 3)^2 = (x + 3)(x + 3)$.]
14. $f(x) = (5x - 4)^2$
15. $f(x) = (x^3 - 4x)^2$
16. $f(x) = (3x^2 - 4x + 5)^2$
17. $f(x) = 5x^{-3}(x^4 - 5x^3 + 10x - 2)$
18. $f(x) = 6x^{-4}(6x^3 + 10x^2 - 8x + 3)$
19. $f(x) = \left(x + \dfrac{2}{x}\right)(x^2 - 3)$
20. $f(x) = (4x^3 - x^2)\left(x - \dfrac{5}{x}\right)$
21. $f(x) = \dfrac{x}{300 - x}$ 22. $f(x) = \dfrac{x}{400 - x}$
23. $f(x) = \dfrac{3x - 1}{2x + 5}$ 24. $f(x) = \dfrac{2x + 3}{x - 5}$

25. $y = \dfrac{x^2 + 1}{x^3 - 1}$ 26. $y = \dfrac{x^3 - 1}{x^2 + 1}$
27. $y = \dfrac{x}{1 - x}$ 28. $y = \dfrac{x}{3 - x}$
29. $y = \dfrac{x - 1}{x + 1}$ 30. $y = \dfrac{x + 2}{x - 2}$
31. $f(x) = \dfrac{1}{x - 3}$ 32. $f(x) = \dfrac{1}{x + 2}$
33. $f(x) = \dfrac{3x^2 + 2x}{x^2 + 1}$ 34. $f(x) = \dfrac{3x^2 - 5x}{x^2 - 1}$
35. $f(x) = \dfrac{3x^2 - 5x}{x^8}$, two ways
36. $f(x) = \dfrac{3x^2 + 2x}{x^5}$, two ways
37. $g(x) = \dfrac{4x + 3}{\sqrt{x}}$
38. $g(x) = \dfrac{6x^2 - 3x}{3\sqrt{x}}$

39.–76. Use a grapher to check the results of Exercises 1–38.

77. Find an equation of the tangent line to the graph of $y = 8/(x^2 + 4)$ at the point $(0, 2)$ and at $(-2, 1)$.

78. Find an equation of the tangent line to the graph of $y = 4x/(1 + x^2)$ at the point $(0, 0)$ and at $(-1, -2)$.

Applications
BUSINESS AND ECONOMICS

79. *Marginal Demand.* The demand function for a certain product is given by

$$D(p) = \frac{2p + 300}{10p + 11}.$$

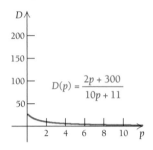

a) Find the marginal demand $D'(p)$.
b) Find $D'(4)$.

80. *Marginal Revenue.* A company sells fix-it-yourself books. It finds that its total revenue from the sale of x books, where x is in thousands, is given by

$$R(x) = \frac{120,000\sqrt{x}}{4 + x^{3/2}}.$$

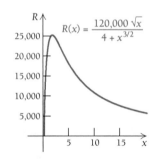

a) Find the marginal revenue $R'(x)$.
b) Find $R'(4)$.

Marginal Revenue. In each of Exercises 81–84, a demand function $x = D(p)$ is given. Find **(a)** the total revenue $R(p)$ and **(b)** the marginal revenue $R'(p)$.

81. $D(p) = 400 - p$
82. $D(p) = 500 - p$
83. $D(p) = \dfrac{4000}{p} + 3$
84. $D(p) = \dfrac{3000}{p} + 5$

85. *Marginal Average Cost.* In Section 2.6, we defined the average cost of producing x units of a product in terms of the total cost $C(x)$ by

$$A(x) = \frac{C(x)}{x}.$$

Use the Quotient Rule to find a general expression for *marginal average cost $A'(x)$*.

86. *Marginal Demand.* In this section, we determined that

$$R(p) = p\,D(p).$$

Then

$$D(p) = \frac{R(p)}{p}$$

= the number of units sold when the price is p dollars per unit

= the demand function.

Use the Quotient Rule to find a general expression for *marginal demand $D'(p)$*.

LIFE AND PHYSICAL SCIENCES

87. *Temperature During an Illness.* The temperature T of a person during an illness is given by

$$T(t) = \frac{4t}{t^2 + 1} + 98.6,$$

where T is the temperature, in degrees Fahrenheit, at time t, in hours.

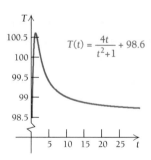

a) Find the rate of change of the temperature with respect to time.
b) Find the temperature at $t = 2$ hr.
c) Find the rate of change at $t = 2$ hr.

SOCIAL SCIENCES

88. *Population Growth.* The population P, in thousands, of a small city is given by

$$P(t) = \frac{500t}{2t^2 + 9},$$

where t is the time, in months.

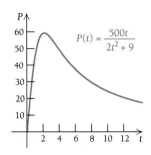

$$P(t) = \frac{500t}{2t^2 + 9}$$

a) Find the growth rate.
b) Find the population after 12 months.
c) Find the growth rate at $t = 12$ months.

Synthesis

Differentiate.

89. $f(x) = \dfrac{x^3}{\sqrt{x} - 5}$

90. $g(t) = \dfrac{1 + \sqrt{t}}{t^5 + 3}$

91. $f(v) = \dfrac{3}{1 + v + v^2}$

92. $g(z) = \dfrac{1 + z + z^2}{1 - z + z^2}$

93. $p(t) = \dfrac{t}{1 - t + t^2 - t^3}$

94. $f(x) = \dfrac{\dfrac{2}{3x} - 1}{\dfrac{3}{x^2} + 5}$

95. $h(x) = \dfrac{x^3 + 5x^2 - 2}{\sqrt{x}}$

96. $y(t) = 5t(t - 1)(2t + 3)$

97. $f(x) = x(3x^3 + 6x - 2)(3x^4 + 7)$

98. $g(x) = (x^3 - 8) \cdot \dfrac{x^2 + 1}{x^2 - 1}$

99. $f(t) = (t^5 + 3) \cdot \dfrac{t^3 - 1}{t^3 + 1}$

100. $f(x) = \dfrac{(x^2 + 3x)(x^5 - 7x^2 - 3)}{x^4 - 3x^3 - 5}$

101. $f(x) = \dfrac{(2x^2 + 3)(4x^3 - 7x + 2)}{x^7 - 2x^6 + 9}$

102. $s(t) = \dfrac{5t^8 - 2t^3}{(t^5 - 3)(t^4 + 7)}$

 103. Try to develop a rule for finding the derivative of the product of three functions. Describe the rule in words.

TECHNOLOGY   CONNECTION

For the function in each of Exercises 104–109, graph f and f'. Then estimate points at which the tangent line is horizontal.

104. $f(x) = x^2(x - 2)(x + 2)$

105. $f(x) = \left(x + \dfrac{2}{x}\right)(x^2 - 3)$

106. $f(x) = \dfrac{x^3 - 1}{x^2 + 1}$

107. $f(x) = \dfrac{0.3x}{0.04 + x^2}$

108. $f(x) = \dfrac{0.01x^2}{x^4 + 0.0256}$

109. $f(x) = \dfrac{4x}{x^2 + 1}$

110. Decide graphically which of the following seems to be the correct derivative of the function of Exercise 109.

$$y_1 = \frac{2}{x},$$

$$y_2 = \frac{4 - 4x}{x^2 + 1},$$

$$y_3 = \frac{4 - 4x^2}{(x^2 + 1)^2},$$

$$y_4 = \frac{4x^2 - 4}{(x^2 + 1)^2}$$

2.8

The Chain Rule

The Extended Power Rule

How can we differentiate more complicated functions such as

$$y = (1 + x^2)^3,$$
$$y = (1 + x^2)^{89},$$

or

$$y = (1 + x^2)^{1/3}?$$

For $(1 + x^2)^3$, we can expand and then differentiate. Although this could be done for $(1 + x^2)^{89}$, it would certainly be time-consuming, and such an expansion would not work for $(1 + x^2)^{1/3}$. Not knowing a rule, we might conjecture that the derivative of the function $y = (1 + x^2)^3$ is

$$3(1 + x^2)^2. \tag{1}$$

To check this, we expand $(1 + x^2)^3$ and then differentiate. From algebra, we recall that $(a + h)^3 = a^3 + 3a^2h + 3ah^2 + h^3$, so

$$(1 + x^2)^3 = 1^3 + 3 \cdot 1^2 \cdot (x^2)^1 + 3 \cdot 1 \cdot (x^2)^2 + (x^2)^3$$
$$= 1 + 3x^2 + 3x^4 + x^6.$$

(We could also have done this by finding $(1 + x^2)^2$ and then multiplying again by $1 + x^2$.) It follows that

$$\frac{dy}{dx} = 6x + 12x^3 + 6x^5$$
$$= (1 + 2x^2 + x^4)6x$$
$$= 3(1 + x^2)^2 \cdot 2x. \tag{2}$$

Comparing this with equation (1), we see that the Power Rule is not sufficient for such a differentiation. Note that the factor $2x$ in the actual derivative, equation (2), is the derivative of the "inside" function, $1 + x^2$. This is consistent with the following new rule.

Theorem 7

The Extended Power Rule

Suppose that $g(x)$ is a function of x. Then for any real number k,

$$\frac{d}{dx}[g(x)]^k = k[g(x)]^{k-1} \cdot \frac{d}{dx}g(x).$$

Let's differentiate $(1 + x^3)^5$. There are three steps to carry out.

$(1 + x^3)^5$ 1. Mentally block out the "inside" function, $1 + x^3$.

$5(1 + x^3)^4$ 2. Differentiate the "outside" function, $(1 + x^3)^5$.

$5(1 + x^3)^4 \cdot 3x^2$ 3. Multiply by the derivative of the "inside" function.

$= 15x^2(1 + x^3)^4$

Step (3) is most commonly overlooked. Try not to forget it!

**TECHNOLOGY
CONNECTION**

For Example 1, graph f and f'. Check the results of Example 1 graphically.

Example 1   Differentiate: $f(x) = (1 + x^3)^{1/2}$.

Solution

$$\frac{d}{dx}(1 + x^3)^{1/2} = \frac{1}{2}(1 + x^3)^{1/2-1} \cdot 3x^2$$

$$= \frac{1}{2}(1 + x^3)^{-1/2} \cdot 3x^2$$

$$= \frac{3x^2}{2\sqrt{1 + x^3}}$$

Example 2   Differentiate: $y = (1 - x^2)^3 - (1 - x^2)^2$.

Solution   Here we combine the Difference Rule and the Extended Power Rule:

$$\frac{dy}{dx} = 3(1 - x^2)^2(-2x) - 2(1 - x^2)(-2x).$$ We differentiate each term using the Extended Power Rule.

Thus,

$$\frac{dy}{dx} = -6x(1 - x^2)^2 + 4x(1 - x^2)$$

$$= x(1 - x^2)[-6(1 - x^2) + 4]$$ Here we factor out $x(1 - x^2)$.

$$= x(1 - x^2)[-6 + 6x^2 + 4]$$

$$= x(1 - x^2)(6x^2 - 2)$$

$$= 2x(1 - x^2)(3x^2 - 1).$$

Example 3   Differentiate: $f(x) = (x - 5)^4(7 - x)^{10}$.

Solution   Here we combine the Product Rule and the Extended Power Rule:

$$f'(x) = (x - 5)^4 \cdot 10(7 - x)^9(-1) + 4(x - 5)^3(1)(7 - x)^{10}$$

$$= -10(x - 5)^4(7 - x)^9 + 4(x - 5)^3(7 - x)^{10}$$

$$= 2(x - 5)^3(7 - x)^9[-5(x - 5) + 2(7 - x)]$$ We factor out $2(x - 5)^3(7 - x)^9$.

$$= 2(x - 5)^3(7 - x)^9[-5x + 25 + 14 - 2x]$$

$$= 2(x - 5)^3(7 - x)^9(39 - 7x).$$

Example 4   Differentiate: $f(x) = \sqrt[4]{\dfrac{x + 3}{x - 1}}$.

Solution   We can use the Quotient Rule to differentiate the inside function, $(x + 3)/(x - 1)$:

$$\frac{d}{dx} \sqrt[4]{\frac{x + 3}{x - 1}} = \frac{d}{dx}\left(\frac{x + 3}{x - 1}\right)^{1/4} = \frac{1}{4}\left(\frac{x + 3}{x - 1}\right)^{1/4-1}\left[\frac{(x - 1)1 - 1(x + 3)}{(x - 1)^2}\right]$$

$$= \frac{1}{4}\left(\frac{x + 3}{x - 1}\right)^{-3/4}\left[\frac{x - 1 - x - 3}{(x - 1)^2}\right]$$

$$= \frac{1}{4}\left(\frac{x + 3}{x - 1}\right)^{-3/4} \cdot \frac{-4}{(x - 1)^2}$$

$$= \left(\frac{x + 3}{x - 1}\right)^{-3/4} \cdot \frac{-1}{(x - 1)^2}.$$

Composition of Functions and the Chain Rule

Jordan YMCA, Marv Bittinger

The Extended Power Rule is a special case of a more general rule called the *Chain Rule*. Before discussing it, we define the *composition* of functions.

The author of this text exercises three times a week at a local YMCA. When he recently bought a pair of athletic shoes, he found a label like the following on the box. This pair of shoes is labeled size $11\frac{1}{2}$ in the United States, but differently in other countries. The numbers at the bottom of the label indicate equivalent shoe sizes in five countries.

This label suggests that there are functions that convert shoe sizes in one country to those in another. There is, indeed, a function g that gives a correspondence between shoe sizes in the United States and those in France:

$$g(x) = \frac{4x + 92}{3},$$

where x is the U.S. size and $g(x)$ is the French size. Thus a U.S. size $11\frac{1}{2}$ corresponds to a French size

$$g\left(11\tfrac{1}{2}\right) = \frac{4 \cdot 11\frac{1}{2} + 92}{3}, \quad \text{or } 46.$$

There is also a function f that gives a correspondence between shoe sizes in France and those in Japan. The function is given by

$$f(x) = \frac{15x - 100}{2},$$

where x is the French size and $f(x)$ is the corresponding Japanese size. Thus a French size 46 corresponds to a Japanese size

$$f(46) = \frac{15 \cdot 46 - 100}{2}, \quad \text{or } 295.$$

It seems reasonable to conclude that a shoe size of $11\frac{1}{2}$ in the United States corresponds to a size of 295 in Japan and that some function h describes this correspondence. Can we find a formula for h?

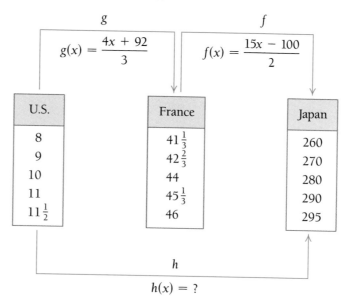

A shoe size x in the United States corresponds to a shoe size $g(x)$ in France, where

$$g(x) = \frac{4x + 92}{3}.$$

Now $(4x + 92)/3$ is a shoe size in France. If we replace x in $f(x)$ with $(4x + 92)/3$, we can find the corresponding shoe size in Japan:

$$f(g(x)) = \frac{15\left(\dfrac{4x + 92}{3}\right) - 100}{2}$$

$$= \frac{5(4x + 92) - 100}{2} = \frac{20x + 460 - 100}{2}$$

$$= \frac{20x + 360}{2} = 10x + 180.$$

This gives a formula for h: $h(x) = 10x + 180$. Thus a shoe size of $11\frac{1}{2}$ in the United States corresponds to a shoe size of $h\left(11\frac{1}{2}\right) = 10\left(11\frac{1}{2}\right) + 180 = 295$ in Japan. The function h is the **composition** of f and g, symbolized by $f \circ g$.

Definition

The **composed** function $f \circ g$, the **composition** of f and g, is defined as

$$f \circ g(x) = f(g(x)), \quad \text{or} \quad (f \circ g)(x) = f[g(x)].$$

We can visualize the composition of functions as shown below.

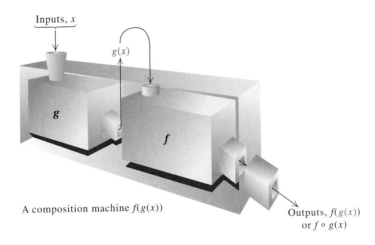

A composition machine $f(g(x))$

To find $f \circ g(x)$, we substitute $g(x)$ for x in $f(x)$.

Example 5   For $f(x) = x^3$ and $g(x) = 1 + x^2$, find $f \circ g(x)$ and $g \circ f(x)$.

Solution   Consider each function separately:

$$f(x) = x^3 \qquad\qquad \text{This function cubes each input.}$$

and

$$g(x) = 1 + x^2. \qquad \text{This function adds 1 to the square of each input.}$$

a) $f \circ g$ first does what g does (adds 1 to the square) and then does what f does (cubes). We find $f(g(x))$ by substituting $g(x)$ for x:

$$
\begin{aligned}
f \circ g(x) = f(g(x)) &= f(1 + x^2) \qquad \text{Substituting } 1 + x^2 \text{ for } x \\
&= (1 + x^2)^3 \\
&= 1 + 3x^2 + 3x^4 + x^6.
\end{aligned}
$$

b) $g \circ f$ first does what f does (cubes) and then does what g does (adds 1 to the square). We find $g(f(x))$ by substituting $f(x)$ for x:

$$
\begin{aligned}
g \circ f(x) = g(f(x)) &= g(x^3) \qquad \text{Substituting } x^3 \text{ for } x \\
&= 1 + (x^3)^2 = 1 + x^6.
\end{aligned}
$$

Example 6   For $f(x) = \sqrt{x}$ and $g(x) = x - 1$, find $f \circ g(x)$ and $g \circ f(x)$.

Solution

$$f \circ g(x) = f(g(x)) = f(x - 1) = \sqrt{x - 1},$$
$$g \circ f(x) = g(f(x)) = g(\sqrt{x}) = \sqrt{x} - 1$$

How do we differentiate the composition of functions? The following theorem tells us.

Theorem 8

The Chain Rule

The derivative of the composition $f \circ g$ is given by

$$\frac{d}{dx}[f \circ g(x)] = \frac{d}{dx}[f(g(x))] = f'(g(x)) \cdot \frac{d}{dx}g(x).$$

Note that the Extended Power Rule is a special case of the Chain Rule. Consider the function $f(x) = x^k$. Then for any other function $g(x)$, $f \circ g(x) = [g(x)]^k$ and the derivative of the composition is

$$\frac{d}{dx}[g(x)]^k = k[g(x)]^{k-1} \cdot \frac{d}{dx}g(x).$$

The Chain Rule often appears in another form. Suppose that $y = f(u)$ and $u = g(x)$. Then

$$\frac{dy}{dx} = \frac{dy}{du} \cdot \frac{du}{dx}.$$

Example 7   For $y = 2 + \sqrt{u}$ and $u = x^3 + 1$, find dy/du, du/dx, and dy/dx.

Solution   First we find dy/du and du/dx:

$$\frac{dy}{du} = \frac{1}{2}u^{-1/2} \quad \text{and} \quad \frac{du}{dx} = 3x^2.$$

Then

$$\frac{dy}{dx} = \frac{dy}{du} \cdot \frac{du}{dx}$$

$$= \frac{1}{2\sqrt{u}} \cdot 3x^2$$

$$= \frac{3x^2}{2\sqrt{x^3 + 1}}. \qquad \text{Substituting } x^3 + 1 \text{ for } u$$

It is important to be able to recognize how a function can be expressed as a composition.

Example 8 Find $f(x)$ and $g(x)$ such that $h(x) = f \circ g(x)$:

$$h(x) = (4x^3 - 7)^6.$$

Solution This is $4x^3 - 7$ to the 6th power. Two functions that can be used for the composition are $f(x) = x^6$ and $g(x) = 4x^3 - 7$. We can check by forming the composition:

$$h(x) = f \circ g(x) = f(g(x)) = f(4x^3 - 7) = (4x^3 - 7)^6.$$

This is the most "obvious" answer to the question. There can be other less obvious answers—for example, $f(x) = x^3$ and $g(x) = (4x^3 - 7)^2$.   ◄

EXERCISE SET 2.8

Differentiate.

1. $y = (1 - x)^{55}$

2. $y = (1 - x)^{100}$

3. $y = \sqrt{1 + 8x}$

4. $y = \sqrt{1 - x}$

5. $y = \sqrt{3x^2 - 4}$

6. $y = \sqrt{4x^2 + 1}$

7. $y = (3x^2 - 6)^{-40}$

8. $y = (4x^2 + 1)^{-50}$

9. $y = x\sqrt{2x + 3}$

10. $y = x\sqrt{4x - 7}$

11. $y = x^2\sqrt{x - 1}$

12. $y = x^3\sqrt{x + 1}$

13. $y = \dfrac{1}{(3x + 8)^2}$

14. $y = \dfrac{1}{(4x + 5)^2}$

15. $f(x) = (1 + x^3)^3 - (1 + x^3)^4$

16. $f(x) = (1 + x^3)^5 - (1 + x^3)^4$

17. $f(x) = x^2 + (200 - x)^2$

18. $f(x) = x^2 + (100 - x)^2$

19. $f(x) = (x + 6)^{10}(x - 5)^4$

20. $f(x) = (x - 4)^8(x + 3)^9$

21. $f(x) = (x - 4)^8(3 - x)^4$

22. $f(x) = (x + 6)^{10}(5 - x)^9$

23. $f(x) = -4x(2x - 3)^3$

24. $f(x) = -5x(3x + 5)^6$

25. $f(x) = \sqrt{\dfrac{1 - x}{1 + x}}$

26. $f(x) = \sqrt{\dfrac{3 + x}{2 - x}}$

27. $f(x) = \left(\dfrac{3x - 1}{5x + 2}\right)^4$

28. $f(x) = \left(\dfrac{x}{x^2 + 1}\right)^3$

29. $f(x) = \sqrt[3]{x^4 + 3x^2}$

30. $f(x) = \sqrt{x^2 + 5x}$

31. $f(x) = (2x^3 - 3x^2 + 4x + 1)^{100}$

32. $f(x) = (7x^4 + 6x^3 - x)^{204}$

33. $g(x) = \left(\dfrac{2x + 3}{5x - 1}\right)^{-4}$

34. $h(x) = \left(\dfrac{1 - 3x}{2 - 7x}\right)^{-5}$

35. $f(x) = \sqrt{\dfrac{x^2 + 1}{x^2 - 1}}$

36. $f(x) = \sqrt[3]{\dfrac{4 - x^3}{1 - x^2}}$

37. $f(x) = \dfrac{(2x + 3)^4}{(3x - 2)^5}$

38. $f(x) = \dfrac{(5x - 4)^7}{(6x + 1)^3}$

39. $f(x) = 12(2x + 1)^{2/3}(3x - 4)^{5/4}$

40. $y = 6\sqrt[3]{x^2 + x}(x^4 - 6x)^3$

Find $\dfrac{dy}{du}, \dfrac{du}{dx}$, and $\dfrac{dy}{dx}$.

41. $y = \sqrt{u}$ and $u = x^2 - 1$

42. $y = \dfrac{15}{u^3}$ and $u = 2x + 1$

43. $y = u^{50}$ and $u = 4x^3 - 2x^2$

44. $y = \dfrac{u + 1}{u - 1}$ and $u = 1 + \sqrt{x}$

45. $y = u(u + 1)$ and $u = x^3 - 2x$

46. $y = (u + 1)(u - 1)$ and $u = x^3 + 1$

47. Find an equation for the tangent line to the graph of $y = \sqrt{x^2 + 3x}$ at the point $(1, 2)$.

48. Find an equation for the tangent line to the graph of $y = (x^3 - 4x)^{10}$ at the point $(2, 0)$.

49. Consider

$$f(x) = \dfrac{x^2}{(1 + x)^5}.$$

a) Find $f'(x)$ using the Quotient Rule and the Extended Power Rule.
b) Note that $f(x) = x^2(1 + x)^{-5}$. Find $f'(x)$ using the Product Rule and the Extended Power Rule.
c) Compare your answers to parts (a) and (b).

50. Consider

$$g(x) = (x^3 + 5x)^2.$$

a) Find $g'(x)$ using the Extended Power Rule.
b) Note that $g(x) = x^6 + 10x^4 + 25x^2$. Find $g'(x)$.
c) Compare your answers to parts (a) and (b).

Find $f \circ g(x)$ and $g \circ f(x)$.

51. $f(x) = 3x^2 + 2, \quad g(x) = 2x - 1$

52. $f(x) = 4x + 3, \quad g(x) = 2x^2 - 5$

53. $f(x) = 4x^2 - 1, \quad g(x) = \dfrac{2}{x}$

54. $f(x) = \dfrac{3}{x}, \quad g(x) = 2x^2 + 3$

55. $f(x) = x^2 + 1, \quad g(x) = x^2 - 1$

56. $f(x) = \dfrac{1}{x^2}, \quad g(x) = x + 2$

Find $f(x)$ and $g(x)$ such that $h(x) = f \circ g(x)$. Answers may vary.

57. $h(x) = (3x^2 - 7)^5$

58. $h(x) = \dfrac{1}{\sqrt{7x + 2}}$

59. $h(x) = \dfrac{x^3 + 1}{x^3 - 1}$

60. $h(x) = (\sqrt{x} + 5)^4$

Do Exercises 61–64 in two ways. First, use the Chain Rule to find the answer. Next, check your answer by finding $f(g(x))$, taking the derivative, and substituting.

61. $f(u) = u^3, \quad g(x) = u = 2x^4 + 1$
Find $(f \circ g)'(-1)$.

62. $f(u) = \dfrac{u + 1}{u - 1}, \quad g(x) = u = \sqrt{x}$
Find $(f \circ g)'(4)$.

63. $f(u) = \sqrt[3]{u}, \quad g(x) = u = 1 - 3x^2$
Find $(f \circ g)'(2)$.

64. $f(u) = 2u^5, \quad g(x) = u = \dfrac{3 - x}{4 + x}$
Find $(f \circ g)'(-10)$.

Applications
BUSINESS AND ECONOMICS

65. *Marginal Cost.* A total-cost function is given by
$$C(x) = 1000\sqrt{x^3 + 2}.$$
a) Find the marginal costs $C'(x)$ and $C'(10)$.
tw b) Interpret the meaning of $C'(x)$.

66. *Marginal Revenue.* A total-revenue function is given by
$$R(x) = 2000\sqrt{x^2 + 3}.$$
a) Find the marginal revenues $R'(x)$ and $R'(20)$.
tw b) Interpret the meaning of $R'(x)$.

67. *Marginal Profit.*
a) Use the total-cost and total-revenue functions in Exercises 65 and 66 to find the marginal profit, $P'(x)$.
tw b) Interpret the meaning of marginal profit.

68. *Utility.* Utility is a type of function that occurs in economics. When a consumer receives x units of a product, a certain amount of pleasure, or utility, U, is derived. Suppose for a new video game, the utility related to the number of cartridges x obtained is
$$U(x) = 80\sqrt{\dfrac{2x + 1}{3x + 4}}.$$
a) Find the marginal utility.
tw b) Interpret the meaning of marginal utility.

69. *Compound Interest.* If $1000 is invested at interest rate i, compounded annually, it will grow in 3 yr to an amount A given by
$$A = \$1000(1 + i)^3$$
(see Section 1.1).
a) Find the rate of change, dA/di.
tw b) Interpret the meaning of dA/di.

70. *Compound Interest.* If $1000 is invested at interest rate i, compounded quarterly, it will grow in 5 yr to an amount A given by
$$A = \$1000\left(1 + \dfrac{i}{4}\right)^{20}.$$
a) Find the rate of change, dA/di.
tw b) Interpret the meaning of dA/di.

71. *Marginal Demand.* Suppose that the demand function for a product is given by

$$D(p) = \frac{80{,}000}{p}$$

and that price p is a function of time given by $p = 1.6t + 9$, where t is in days.

a) Find the demand as a function of time t.
b) Find the marginal demand as a function of time.
c) Find the rate of change of the quantity demanded when $t = 100$ days.

72. *Marginal Profit.* A company is selling microcomputers. It determines that its total profit is given by

$$P(x) = 0.08x^2 + 80x + 260,$$

where x is the number of units produced and sold. Suppose that x is a function of time, in months, where $x = 5t + 1$.

a) Find the total profit as a function of time t.
b) Find the marginal profit as a function of time.
c) Find the rate of change of total profit when $t = 48$ months.

Synthesis ....

Differentiate.

73. $y = \sqrt[3]{x^3 - 6x + 1}$ **74.** $s = \sqrt[4]{t^4 + 3t^2 + 8}$

75. $y = \dfrac{x}{\sqrt{x - 1}}$ **76.** $y = \dfrac{(x + 1)^2}{(x^2 + 1)^3}$

77. $u = \dfrac{(1 + 2v)^4}{v^4}$ **78.** $y = x\sqrt{1 + x^2}$

79. $y = \dfrac{\sqrt{1 - x^2}}{1 - x}$ **80.** $w = \dfrac{u}{\sqrt{1 + u^2}}$

81. $y = \left(\dfrac{x^2 - x - 1}{x^2 + 1}\right)^3$ **82.** $y = \sqrt{1 + \sqrt{x}}$

83. $s = \dfrac{\sqrt{t} - 1}{\sqrt{t} + 1}$ **84.** $y = x^{2/3} \cdot \sqrt[3]{1 + x^2}$

tw 85. The following is the beginning of an alternative proof of the Quotient Rule for finding the derivative of a quotient that uses the Product Rule and the Power Rule. Complete the proof, giving reasons for each step.

Proof: Let

$$Q(x) = \frac{N(x)}{D(x)}.$$

Then

$$Q(x) = N(x) \cdot [D(x)]^{-1}.$$

Therefore, ...

tw 86. Describe composition of functions in as many ways as possible.

TECHNOLOGY CONNECTION

For the function in each of Exercises 87 and 88, graph f and f' over the given interval. Then estimate points at which the tangent line is horizontal.

87. $f(x) = 1.68x\sqrt{9.2 - x^2}; [-3, 3]$

88. $f(x) = \sqrt{6x^3 - 3x^2 - 48x + 45}; [-5, 5]$

Find the derivative of each of the following functions analytically. Then use a grapher to check the results.

89. $f(x) = x\sqrt{4 - x^2}$

90. $f(x) = \dfrac{4x}{\sqrt{x - 10}}$

2.9

OBJECTIVES

➤ Find derivatives of higher order.

Higher-Order Derivatives

Consider the function given by

$$y = f(x) = x^5 - 3x^4 + x.$$

Its derivative f' is given by

$$y' = f'(x) = 5x^4 - 12x^3 + 1.$$

The derivative function f' can also be differentiated. We can think of its derivative as the rate of change of the slope of the tangent lines of f. We use the notation f'' for the derivative $(f')'$. That is,

$$f''(x) = \frac{d}{dx} f'(x).$$

We call f'' the *second derivative* of f. It is given by

$$y'' = f''(x) = 20x^3 - 36x^2.$$

Continuing in this manner, we have

$$f'''(x) = 60x^2 - 72x, \qquad \text{The third derivative of } f$$
$$f''''(x) = 120x - 72, \qquad \text{The fourth derivative of } f$$
$$f'''''(x) = 120. \qquad \text{The fifth derivative of } f$$

When notation like $f'''(x)$ gets lengthy, we abbreviate it using a symbol in parentheses. Thus, $f^{(n)}(x)$ is the nth derivative. For the function above,

$$f^{(4)}(x) = 120x - 72,$$
$$f^{(5)}(x) = 120,$$
$$f^{(6)}(x) = 0, \quad \text{and}$$
$$f^{(n)}(x) = 0, \quad \text{for any integer } n \geq 6.$$

Leibniz's notation for the second derivative of a function given by $y = f(x)$ is

$$\frac{d^2y}{dx^2}, \quad \text{or} \quad \frac{d}{dx}\left(\frac{dy}{dx}\right),$$

read "the second derivative of y with respect to x." The 2's in this notation are *not* exponents. If $y = x^5 - 3x^4 + x$, then

$$\frac{d^2y}{dx^2} = 20x^3 - 36x^2.$$

Leibniz's notation for the third derivative is d^3y/dx^3; for the fourth derivative, d^4y/dx^4; and so on:

$$\frac{d^3y}{dx^3} = 60x^2 - 72x,$$

$$\frac{d^4y}{dx^4} = 120x - 72,$$

$$\frac{d^5y}{dx^5} = 120.$$

Example 1   For $y = 1/x$, find d^2y/dx^2.

Solution   We have $y = x^{-1}$, so

$$\frac{dy}{dx} = -1 \cdot x^{-1-1} = -x^{-2}, \quad \text{or} \quad -\frac{1}{x^2}.$$

Then

$$\frac{d^2y}{dx^2} = (-2)(-1)x^{-2-1} = 2x^{-3}, \quad \text{or} \quad \frac{2}{x^3}.$$

Example 2   For $y = (x^2 + 10x)^{20}$, find y' and y''.

Solution   To find y', we use the Extended Power Rule:

$$y' = 20(x^2 + 10x)^{19}(2x + 10)$$
$$= 20(x^2 + 10x)^{19} \cdot 2(x + 5)$$
$$= 40(x^2 + 10x)^{19}(x + 5).$$

To find y'', we use the Product Rule and the Extended Power Rule:

$$y'' = 40(x^2 + 10x)^{19}(1) + 19 \cdot 40(x^2 + 10x)^{18}(2x + 10)(x + 5)$$
$$= 40(x^2 + 10x)^{19} + 760(x^2 + 10x)^{18} \cdot 2(x + 5)(x + 5)$$
$$= 40(x^2 + 10x)^{19} + 1520(x^2 + 10x)^{18}(x + 5)^2$$
$$= 40(x^2 + 10x)^{18}[(x^2 + 10x) + 38(x + 5)^2]$$
$$= 40(x^2 + 10x)^{18}[x^2 + 10x + 38(x^2 + 10x + 25)]$$
$$= 40(x^2 + 10x)^{18}[39x^2 + 390x + 950].$$

We encountered an application of second derivatives when we considered acceleration in Section 2.6. Acceleration can be regarded as a second derivative. As an object moves, its distance from a fixed point after time t is some function of the time, say, $s(t)$. Then

$$v(t) = s'(t) = \text{the velocity at time } t$$

and

$$a(t) = v'(t) = s''(t) = \text{the acceleration at time } t.$$

Whenever a quantity is a function of time, the first derivative gives the rate of change with respect to time and the second derivative gives the acceleration. For example, if $y = P(t)$ gives the number of people in a population at time t, then $P'(t)$ represents how fast the size of the population is changing and $P''(t)$ gives the acceleration in the size of the population.

EXERCISE SET 2.9

Find d^2y/dx^2.

1. $y = 3x + 5$

2. $y = -4x + 7$

3. $y = -\dfrac{1}{x}$

4. $y = -\dfrac{3}{x}$

5. $y = x^{1/4}$

6. $y = \sqrt{x}$

7. $y = x^4 + \dfrac{4}{x}$

8. $y = x^3 - \dfrac{3}{x}$

9. $y = x^{-3}$

10. $y = x^{-4}$

11. $y = x^n$

12. $y = x^{-n}$

13. $y = x^4 - x^2$

14. $y = x^4 + x^3$

15. $y = \sqrt{x - 1}$

16. $y = \sqrt{x + 1}$

17. $y = ax^2 + bx + c$

18. $y = \sqrt{a - bx}$

19. $y = (x^2 - 8x)^{43}$

20. $y = (x^3 + 15x)^{20}$

21. $y = (x^4 - 4x^2)^{50}$

22. $y = (x^2 - 3x + 1)^{12}$

23. $y = x^{2/3} + 4x$

24. $y = x^{5/4} - x^{4/5}$

25. $y = (x - 8)^{3/4}$

26. $y = \dfrac{x^5}{40} - \dfrac{x^3}{36}$

27. $y = \dfrac{1}{x^2} + \dfrac{2}{x^3}$

28. $y = \dfrac{4}{x^5} - \dfrac{7}{3x^4}$

29. For $y = x^4$, find d^4y/dx^4.

30. For $y = x^5$, find d^4y/dx^4.

31. For $y = x^6 - x^3 + 2x$, find d^5y/dx^5.

32. For $y = x^7 - 8x^2 + 2$, find d^6y/dx^6.

33. For $y = (x^2 - 5)^{10}$, find d^2y/dx^2.

34. For $y = x^k$, find d^5y/dx^5.

35. If s is a distance given by $s(t) = t^3 + t^2 + 2t$, find the acceleration.

36. If s is a distance given by $s(t) = t^4 + t^2 + 3t$, find the acceleration.

Applications
LIFE AND PHYSICAL SCIENCES

37. *Population Growth.* A population grows from an initial size of 100,000 to an amount $P(t)$, given by

$$P(t) = 100{,}000(1 + 0.6t + t^2).$$

What is the acceleration in the size of the population?

38. *Population Growth.* A population grows from an initial size of 100,000 to an amount $P(t)$, given by

$$P(t) = 100{,}000(1 + 0.4t + t^2).$$

What is the acceleration in the size of the population?

Synthesis

Find y', y'', and y'''.

39. $y = x^{-1} + x^{-2}$

40. $y = \dfrac{1}{1 - x}$

41. $y = x\sqrt{1 + x^2}$

42. $y = 3x^5 + 8\sqrt{x}$

43. $y = \dfrac{3x - 1}{2x + 3}$

44. $y = \dfrac{1}{\sqrt{x - 1}}$

45. $y = \dfrac{x}{\sqrt{x - 1}}$

46. $y = \dfrac{\sqrt{x} - 1}{\sqrt{x} + 1}$

Find $f''(x)$.

47. $f(x) = \dfrac{x}{x - 1}$

48. $f(x) = \dfrac{1}{1 + x^2}$

Find the first through the fifth derivatives. Be sure to simplify at each stage before continuing.

49. $f(x) = \dfrac{x - 1}{x + 2}$

50. $f(x) = \dfrac{x + 3}{x - 2}$

TECHNOLOGY CONNECTION

For the function in each of Exercises 51–54, graph f, f', and f'' over the given interval. Analyze and compare the behavior of these functions.

51. $f(x) = 0.1x^4 - x^2 + 0.4;\ [-5, 5]$

52. $f(x) = -x^3 + 3x;\ [-3, 3]$

53. $f(x) = x^4 + x^3 - 4x^2 - 2x + 4;\ [-3, 3]$

54. $f(x) = x^3 - 3x^2 + 2;\ [-3, 5]$

CHAPTER 2 SUMMARY AND REVIEW

Terms to Know

Limit, p. 97
Continuity at a point, pp. 102, 104
Continuous over an interval, pp. 102, 104
Continuous function, p. 104
Average rate of change, pp. 116, 149
Secant line, p. 116
Difference quotient, p. 118
Simplified difference quotient, p. 119
Tangent line, p. 126

Derivative, p. 128
Differentiation, p. 128
Leibniz notation, p. 138
Power Rule, p. 140
Sum–Difference Rule, p. 144
Instantaneous rate of change, p. 149
Velocity, pp. 150, 151
Speed, p. 150
Acceleration, p. 151
Growth rate, p. 152

Marginal cost, p. 153
Marginal revenue, p. 153
Marginal profit, p. 153
Average cost, p. 154
Product Rule, p. 158
Quotient Rule, p. 159
Extended Power Rule, p. 166
Composition of functions, p. 170
Chain Rule, p. 171
Higher-order derivative, p. 174

Review Exercises

These review exercises are for test preparation. They can also be used as a lengthened practice test. Answers are at the back of the book. The answers also contain bracketed section references, which tell you where to restudy if your answer is incorrect.

Consider

$$\lim_{x \to -7} f(x), \quad \text{where } f(x) = \frac{x^2 + 4x - 21}{x + 7},$$

for Exercises 1–3.

1. *Limit Numerically.*

   a) Complete the following input–output tables.

$x \to -7^-$	$f(x)$
-8	
-7.5	
-7.1	
-7.01	
-7.001	
-7.0001	

$x \to -7^+$	$f(x)$
-6	
-6.5	
-6.9	
-6.99	
-6.999	
-6.9999	

   b) Find $\lim_{x \to -7^-} f(x)$, $\lim_{x \to -7^+} f(x)$, and $\lim_{x \to -7} f(x)$, if each exists.

2. *Limit Graphically.* Graph the function and use the graph to find the limit.

3. *Limit Algebraically.* Find the limit algebraically. Show your work.

Find the limit, if it exists.

4. $\lim\limits_{x \to -2} \dfrac{8}{x}$

5. $\lim\limits_{x \to 1} (4x^3 - x^2 + 7x)$

6. $\lim\limits_{x \to -7} \dfrac{x^2 + 4x - 21}{x + 7}$

7. $\lim\limits_{x \to 4} \sqrt{x^2 + 9}$

Determine whether the function is continuous.

8.

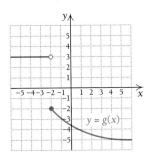

9.

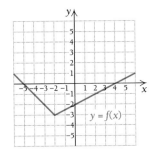

For the function in Exercise 8, answer the following.

10. Find $\lim_{x \to 1} g(x)$.

11. Find $g(1)$.

12. Is g continuous at 1?

13. Find $\lim_{x \to -2} g(x)$.

14. Find $g(-2)$.

15. Is g continuous at -2?

16. For $f(x) = x^3 - 2x$, find the average rate of change as x changes from -2 to 1.

17. Find a simplified difference quotient for $g(x) = -3x + 5$.

18. Find a simplified difference quotient for $f(x) = 2x^2 - 3$.

19. Find an equation of the tangent line to the graph of $y = x^2 + 3x$ at the point $(-1, -2)$.

20. Find the points on the graph of $y = -x^2 + 8x - 11$ at which the tangent line is horizontal.

21. Find the points on the graph of $y = 5x^2 - 49x + 12$ at which the tangent line has slope 1.

Find dy/dx.

22. $y = 4x^5$

23. $y = 3\sqrt[3]{x}$

24. $y = \dfrac{-8}{x^8}$

25. $y = 15x^{2/5}$

26. $y = 0.1x^7 - 3x^4 - x^3 + 6$

Differentiate.

27. $f(x) = \dfrac{1}{6}x^6 + 8x^4 - 5x$

28. $y = \dfrac{x^3 + x}{x}$

29. $y = \dfrac{x^2 + 8}{8 - x}$

30. $g(x) = (5 - x)^2(2x - 1)^5$

31. $f(x) = (x^5 - 2)^7$

32. $f(x) = x^2(4x + 3)^{3/4}$

33. For $y = x^3 - \dfrac{2}{x}$, find $\dfrac{d^5y}{dx^5}$.

34. For $y = x^7 + 3x^2$, find $\dfrac{d^4y}{dx^4}$.

35. For $s(t) = t + t^4$, find each of the following.
   a) $v(t)$
   b) $a(t)$
   c) The velocity and the acceleration when $t = 2$ sec

36. *Business: Marginal Revenue, Cost, and Profit.* Given $R(x) = 40x$ and $C(x) = 8x^2 - 7x - 10$, find each of the following.
   a) $P(x)$
   b) $R(20)$, $C(20)$, and $P(20)$
   c) $R'(x)$, $C'(x)$, and $P'(x)$
   d) $R'(20)$, $C'(20)$, and $P'(20)$

37. *Life Science: Growth Rate.* The population of a city grows from an initial size of 10,000 to an amount P, given by $P = 10,000 + 50t^2$, where t is in years.
   a) Find the growth rate.
   b) Find the number of people in the city after 20 yr (at $t = 20$).
   c) Find the growth rate at $t = 20$.

38. Find $f \circ g(x)$ and $g \circ f(x)$, given that $f(x) = x^2 + 5$ and $g(x) = 1 - 2x$.

Synthesis ....

39. Differentiate $y = \dfrac{x\sqrt{1 + 3x}}{1 + x^3}$.

 TECHNOLOGY CONNECTION

Use a grapher that creates input–output tables. Find each of the following limits. Start with Step = 0.1 and then go to 0.01, 0.001, and 0.0001. When you think you know the limit, graph, and use the TRACE feature to further verify your assertion.

40. $\lim_{x \to 1} \dfrac{2 - \sqrt{x + 3}}{x - 1}$

41. $\lim_{x \to 11} \dfrac{\sqrt{x - 2} - 3}{x - 11}$

42. Graph f and f' over the given interval. Then estimate points at which the tangent line to f is horizontal.

$$f(x) = 3.8x^5 - 18.6x^3; \quad [-3, 3]$$

CHAPTER 2 TEST

Consider

$$\lim_{x \to 6} f(x), \quad \text{where } f(x) = \frac{x^2 - 36}{x - 6}$$

for Exercises 1–3.

1. *Limit Numerically.*

   a) Complete the following input–output tables.

$x \to 6^-$	$f(x)$
5	
5.7	
5.9	
5.99	
5.999	
5.9999	

$x \to 6^+$	$f(x)$
7	
6.5	
6.1	
6.01	
6.001	
6.0001	

   b) Find $\lim\limits_{x \to 6^-} f(x)$, $\lim\limits_{x \to 6^+} f(x)$, and $\lim\limits_{x \to 6} f(x)$, if each exists.

2. *Limit Graphically.* Graph the function and use the graph to find the limit.

3. *Limit Algebraically.* Find the limit algebraically. Show your work.

Limits Graphically. Consider the following graph of function f for Questions 4–11.

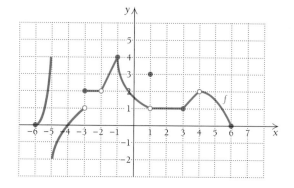

Find the limit, if it exists.

4. $\lim\limits_{x \to -5} f(x)$

5. $\lim\limits_{x \to -4} f(x)$

6. $\lim\limits_{x \to -3} f(x)$

7. $\lim\limits_{x \to -2} f(x)$

8. $\lim\limits_{x \to -1} f(x)$

9. $\lim\limits_{x \to 1} f(x)$

10. $\lim\limits_{x \to 2} f(x)$

11. $\lim\limits_{x \to 3} f(x)$

Determine whether the function is continuous.

12.

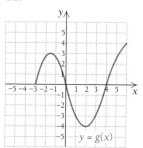

13.

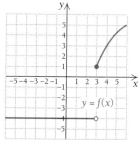

For the function in Question 13, answer the following.

14. Find $\lim\limits_{x \to 3} f(x)$.

15. Find $f(3)$.

16. Is f continuous at 3?

17. Find $\lim\limits_{x \to 4} f(x)$.

18. Find $f(4)$.

19. Is f continuous at 4?

Find the limit, if it exists.

20. $\lim\limits_{x \to 1} (3x^4 - 2x^2 + 5)$

21. $\lim\limits_{x \to 1} \dfrac{x - 1}{x^2 - 1}$

22. $\lim\limits_{x \to 0} \dfrac{7}{x}$

23. Find a simplified difference quotient for

$$f(x) = 3x^2 + 1.$$

24. Find an equation of the tangent line to the graph of $y = x + (4/x)$ at the point $(4, 5)$.

25. Find the points on the graph of $y = x^3 - 3x^2$ at which the tangent line is horizontal.

Find dy/dx.

26. $y = x^{84}$

27. $y = 10\sqrt{x}$

28. $y = \dfrac{-10}{x}$

29. $y = x^{5/4}$

30. $y = -0.5x^2 + 0.61x + 90$

Differentiate.

31. $y = \dfrac{1}{3}x^3 - x^2 + 2x + 4$

32. $y = \dfrac{2x - 5}{x^4}$ **33.** $f(x) = \dfrac{x}{5 - x}$

34. $f(x) = (x + 3)^4(7 - x)^5$

35. $y = (x^5 - 4x^3 + x)^{-5}$

36. $f(x) = x\sqrt{x^2 + 5}$

37. For $y = x^4 - 3x^2$, find $\dfrac{d^3y}{dx^3}$.

38. *Business: Marginal Revenue, Cost, and Profit.* Given $R(x) = 50x$ and $C(x) = 0.001x^2 + 1.2x + 60$, find each of the following.

   a) $P(x)$
   b) $R(10)$, $C(10)$, and $P(10)$
   c) $R'(x)$, $C'(x)$, and $P'(x)$
   d) $R'(10)$, $C'(10)$, and $P'(10)$

39. *Social Sciences: Memory.* In a certain memory experiment, a person is able to memorize M words after t minutes, where $M = -0.001t^3 + 0.1t^2$.

   a) Find the rate of change of the number of words memorized with respect to time.
   b) How many words are memorized during the first 10 min (at $t = 10$)?
   c) What is the memory rate at $t = 10$ min?

40. Find $f \circ g(x)$ and $g \circ f(x)$, given that $f(x) = x + x^2$ and $g(x) = x^3$.

Synthesis

41. Differentiate $y = (1 - 3x)^{2/3}(1 + 3x)^{1/3}$.

42. Find $\lim\limits_{x \to 2} \dfrac{x^3 - 8}{x - 2}$.

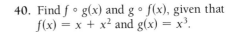 TECHNOLOGY   CONNECTION

43. Graph f and f' over the given interval. Then estimate points at which the tangent line to f is horizontal.

$$f(x) = 5x^3 - 30x^2 + 45x + 5\sqrt{x};\quad [0, 5]$$

44. Find the following limit using tables on a grapher:

$$\lim_{x \to 0} \frac{\sqrt{5x + 25} - 5}{x}.$$

Start with Step $= 0.1$ and then go to 0.01, 0.001, and 0.0001. When you think you know the limit, graph

$$y = \frac{\sqrt{5x + 25} - 5}{x},$$

and use the **TRACE** feature to verify your assertion.

 EXTENDED   TECHNOLOGY   APPLICATION

Path of a Baseball: The Tale of the Tape

Have you ever been to a baseball game and seen a home run ball strike some obstacle after it has cleared the fence? Suppose the ball hits the obstacle at a location that is 60 ft above the ground at a distance of 400 ft from home plate. The scoreboard would display a message reading, "According to the tale of the tape, the ball would have traveled 442 ft." Did you ever wonder how the calculation is made? The answer is related to the curve formed by the path of a baseball.

Whatever the path of a well-hit baseball is, it is *not* the graph of a parabola,

$$f(x) = ax^2 + bx + c.$$

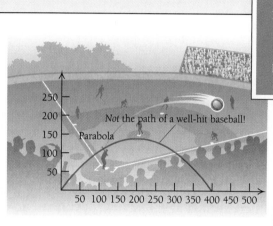

A well-hit baseball follows a path of a "skewed" parabola, as shown below.

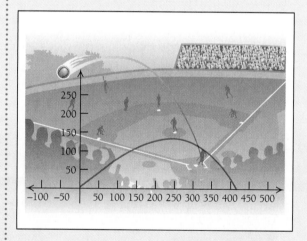

One reason that the ball does not follow the graph of a true parabola is that a well-hit ball has backspin. This fact, combined with the frictional effect of the stitches on the ball with the air, tends to make the path of the ball skewed toward the direction in which it is hit.

Let's see if we can model the path of a baseball. Consider the following data. Assume that $(0, 4.5)$ is the point at home plate at which the ball is hit, roughly 4.5 ft off the ground. Also, assume

that the ball has hit an obstacle 60 ft above the ground and 400 ft from home plate.

Horizontal Distance, x (in feet)	0	50	100	200	285	300	360	400
Vertical Distance, y (in feet)	4.5	43	82	130	142	134	100	60

Exercises

You will need a grapher that performs curve fitting using regression and finds zeros or roots of functions.

1. Plot the points and connect them with line segments. This can be done on some graphers using a **STAT PLOTS** command. Does it seem to fit the preceding graph?

2. a) Use all the data from the table and your grapher to find a cubic polynomial function $y = ax^3 + bx^2 + cx + d$ that fits the data.
   b) Graph the function over the interval $[0, 500]$.
   c) Analyze the validity of using the curve to fit the data in the table.
   d) Predict the horizontal distance from home plate at which the ball would have hit the ground had it not hit the obstacle.
   e) Find the rate of change of the vertical height with respect to the horizontal distance.
   f) Find the points at which the graph has a horizontal tangent line.

3. a) Use all the data to find a quartic poly-nomial function $y = ax^4 + bx^3 + cx^2 + dx + e$ that fits the data.
   b) Graph the function over the interval $[0, 500]$.
   c) Analyze the validity of using the curve to fit the data in the table.
   d) Predict the horizontal distance from home plate at which the ball would have hit the ground had it not hit the obstacle.
   e) Find the rate of change of the vertical height with respect to the horizontal distance.

f) Find the points at which the graph has a horizontal tangent line.

4. a) Although your grapher probably cannot fit such a function to the data, assume that the equation

$$y = 0.0015x\sqrt{202{,}500 - x^2}$$

has been found using some type of curve-fitting technique. Graph the function over the interval $[0, 500]$.

b) Predict the horizontal distance from home plate at which the ball would have hit the ground had it not hit the obstacle.

c) Find the rate of change of the vertical height with respect to the horizontal distance.

d) Find the points at which the graph has a horizontal tangent line.

5. Compare the answers in Exercises 2(d), 3(d), and 4(b). Discuss the relative merits of using the quartic model in Exercise 3 with the model in Exercise 4 to make the prediction.

Tale of the Tape. Actually, scoreboard operators in the major leagues use different models to predict the distance that a home run will travel. The models are linear and are related to the trajectory of the ball, that is, how high the ball is hit. See the following graph.

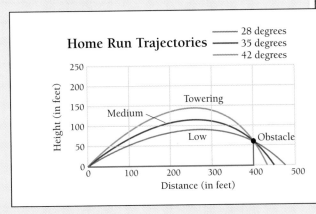

Home Run Trajectories

Suppose that a ball hits an obstacle d feet horizontally from home plate at a vertical height of H feet. Then the estimated horizontal distance that the ball would have traveled is given by

$$D = kH + d,$$

where D is the estimated distance, H is the vertical height of the ball when obstructed, and d is the horizontal distance of the ball when obstructed. Then if we consider the trajectory of the ball, the formula becomes as follows:

Low trajectory: $D = 1.1H + d,$

Medium trajectory: $D = 0.7H + d,$

Towering trajectory: $D = 0.5H + d.$

6. For a ball striking an obstacle at $d = 400$ ft and $H = 60$ ft, estimate how far the ball would have traveled if it were low trajectory, medium trajectory, and towering trajectory.

7. In 1953, Hall-of-Famer Mickey Mantle hit a towering home run in old Griffith Stadium in Washington, D.C., that hit a sign 60 ft tall and 460 ft from home plate. The press asserted at the time that the ball would have traveled 565 ft. Is this estimate valid?

*Many thanks to Robert K. Adair, professor of physics at Yale University, for many of the ideas presented in this section.

3

Applications of Differentiation

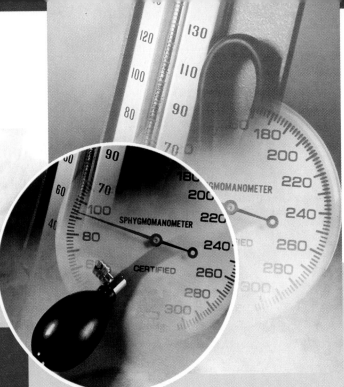

INTRODUCTION

In this chapter, we discover many applications of differentiation. We learn to find maximum and minimum values of functions, and that skill allows us to solve many kinds of problems in which we need to find the largest and/or smallest value of a function. We also apply our differentiation skills to graphing functions.

AN APPLICATION

For a dosage of x cubic centimeters of a certain drug, the resulting blood pressure B is approximated by

$$B(x) = 0.05x^2 - 0.3x^3, \quad 0 \le x \le 0.16.$$

Find the maximum blood pressure and the dosage at which it occurs.

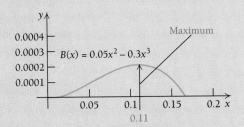

If we find the first derivative, set it equal to 0, and solve

$$B'(x) = 0.1x - 0.9x^2 = 0,$$

the resulting value of x will be the dosage at which the maximum blood pressure occurs.

This problem appears as Exercise 100 in Exercise Set 3.4.

3.1

O B J E C T I V E S

➤ Find relative extrema of a continuous function using the First-Derivative Test.
➤ Sketch graphs of continuous functions.

Using First Derivatives to Find Maximum and Minimum Values and Sketch Graphs

The graph below illustrates a typical life cycle of a retail product. Note that the number of items sold varies with respect to time. Sales begin at a small level and increase to a point of maximum sales after which they decline and taper off to a low level, probably due to the effect of new competitive products. Then the company rejuvenates the product by making product improvements. Think about products such as black and white TVs, color TVs, records, compact discs, and CD-ROM computers. Where might each be in a typical life cycle and is that curve appropriate?

The graph of a typical life cycle illustrates many of the ideas that we will consider in this chapter.

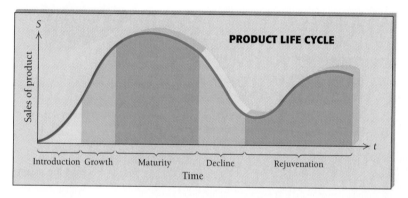

Finding the largest and smallest values of a function—that is, the maximum and minimum values—has extensive application. The first and second derivatives of a function are tools of calculus that give us information about the shape of a graph that may be helpful in finding maximum and minimum values of functions and in graphing functions. Throughout this section, we will assume that the functions f are continuous, but this does not necessarily imply that f' and f'' are continuous.

Increasing and Decreasing Functions

If the graph of a function rises from left to right over an interval I, it is said to be **increasing** on I.

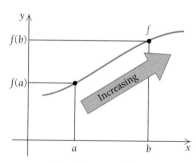

f is an increasing function.
If $a < b$, then $f(a) < f(b)$.

TECHNOLOGY CONNECTION

E x p l o r a t o r y .   Graph the function

$$f(x) = -\frac{1}{3}x^3 + 6x^2 - 11x - 50$$

and its derivative

$$f'(x) = -x^2 + 12x - 11$$

using a viewing window of $[-10, 25, -100, 150]$, with Xscl $= 5$ and Yscl $= 25$. Then use the TRACE feature, moving from left to right along each graph. As you move the cursor from left to right, note that the x-coordinate always increases. If a function is increasing over an interval, the y-coordinate will be increasing. If a function is decreasing over an interval, the y-coordinate will be decreasing.

Over what intervals is the function increasing?

Over what intervals is the function decreasing?

Over what intervals is the derivative positive?

Over what intervals is the derivative negative?

What can you conjecture?

If the graph drops from left to right, it is said to be **decreasing** over *I*. We can describe this mathematically as follows.

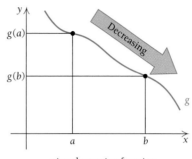

g is a decreasing function.
If $a < b$, then $g(a) > g(b)$.

Definition

A function f is **increasing** over I if, for every a and b in I,

if $a < b$,   then $f(a) < f(b)$.

(If the input a is less than the input b, then the output for a is less than the output for b.)

A function f is **decreasing** over I if, for every a and b in I,

if $a < b$,   then $f(a) > f(b)$.

(If the input a is less than the input b, then the output for a is greater than the output for b.)

Note that the directions of the inequalities stay the same for an increasing function, but they differ for a decreasing function.

In Chapter 1, we saw how the slope of a linear function determines whether that function is increasing or decreasing (or neither). For a general function, the derivative yields similar information.

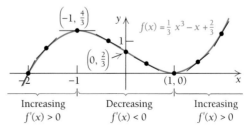

f is increasing over the intervals $(-\infty, -1)$ and $(1, \infty)$; slopes of tangent lines are positive.
f is decreasing over the interval $(-1, 1)$; slopes of tangent lines are negative.

The following theorem shows how we can use derivatives to determine whether a function is increasing or decreasing.

> ### Theorem 1
>
> If $f'(x) > 0$ for all x in an interval I, then f is increasing over I.
> If $f'(x) < 0$ for all x in an interval I, then f is decreasing over I.

Critical Points

Consider this graph of a continuous function.

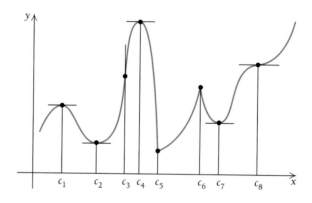

Note the following:

1. $f'(c) = 0$ at points c_1, c_2, c_4, c_7, and c_8. That is, the tangent line to the graph is horizontal at these points.

2. $f'(c)$ does not exist at points c_3, c_5, and c_6. The tangent line is vertical at c_3 and there is a corner point, or sharp point, at both c_5 and c_6. (See also the discussion at the end of Section 2.4.)

> ### Definition
>
> A **critical point** of a function is an interior point c of its domain at which the tangent line to the graph at $(c, f(c))$ is horizontal or at which the derivative does not exist. That is, c is a critical point if
>
> $$f'(c) = 0 \quad \text{or} \quad f'(c) \text{ does not exist.}$$

Thus, in the preceding graph:

1. c_1, c_2, c_4, c_7, and c_8 are critical points because $f'(c) = 0$ for each point.
2. c_3, c_5, and c_6 are critical points because $f'(c)$ does not exist at each point.

Note too that a function can change from increasing to decreasing or from decreasing to increasing *only* at a critical point. In the graph above, c_1, c_2, c_4, c_5, c_6, and c_7 separate the intervals over which the function changes from increasing to decreasing or from decreasing to increasing. The points c_3 and c_8 are critical points but do not separate intervals over which the function changes from increasing to decreasing or from decreasing to increasing.

Finding Relative Maximum and Minimum Values

Consider the graph below. Note the "peaks" and "valleys" at the interior points c_1, c_2, and c_3.

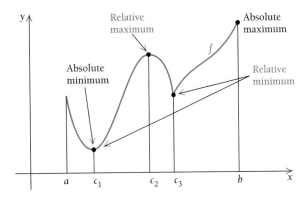

The function value $f(c_2)$ is called a **relative maximum.** Each of the function values $f(c_1)$ and $f(c_3)$ is called a **relative minimum.**

Definition

Suppose that f is a function whose value $f(c)$ exists at input c in the domain of f. Then:

$f(c)$ is a **relative minimum** if there exists an open interval I_1 containing c in the domain such that $f(c) \le f(x)$, for all x in I_1;

and

$f(c)$ is a **relative maximum** if there exists an open interval I_2 containing c in the domain such that $f(c) \ge f(x)$, for all x in I_2.

A relative maximum can be thought of as a high point that may or may not be the highest point, or *absolute maximum*, on an interval I. Similarly, a relative minimum can be thought of as a low point that may or may not be the lowest point, or *absolute minimum*, on I. For now we will consider how to find relative maximum or minimum values, stated simply as **relative extrema.**

Look again at the preceding graph. The points at which a continuous function has relative extrema are points where the derivative is 0 or where the derivative does not exist—the critical points.

Theorem 2

If a function f has a relative extreme value $f(c)$, then c is a critical point, so

$$f'(c) = 0 \quad \text{or} \quad f'(c) \text{ does not exist.}$$

Theorem 2 is very useful, but it is important to understand it precisely. What it says is that when we are looking for points that are relative extrema, the only points we need consider are those where the derivative is 0 or where the derivative does not exist. We can think of a critical point as a *candidate* for a relative maximum or minimum, and the candidate might or might not provide a relative extremum. That is, Theorem 2 does not say that if a point is a critical point, its function value will necessarily be a relative maximum or minimum. The existence of a critical point does *not* guarantee that a function has a relative maximum or minimum. A useful counterexample is the graph of

$$f(x) = (x - 1)^3 + 2,$$

shown below. Note that

$$f'(x) = 3(x - 1)^2,$$

and

$$f'(1) = 3(1 - 1)^2 = 0.$$

The function has a critical point at $c = 1$, but has no relative maximum or minimum at $c = 1$.

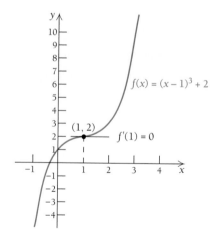

Now how can we tell when the existence of a critical point leads us to a relative extremum? The following graph leads us to a test.

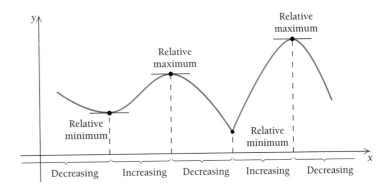

At a critical point for which there is a relative extremum, the function is increasing on one side of the critical point and decreasing on the other side.

$f(c)$	Sign of $f'(x)$ for x in (a, c)	Sign of $f'(x)$ for x in (c, b)	Graph over the interval (a, b)
Relative minimum	−	+	
Relative maximum	+	−	
No relative maxima or minima	−	−	
No relative maxima or minima	+	+	

Derivatives tell us when a function is increasing or decreasing. This leads us to the First-Derivative Test.

Theorem 3

The First-Derivative Test for Relative Extrema

For any continuous function f that has exactly one critical point c in an open interval (a, b):

F1. f has a relative minimum at c if $f'(x) < 0$ on (a, c) and $f'(x) > 0$ on (c, b). That is, f is decreasing to the left of c and increasing to the right of c.

F2. f has a relative maximum at c if $f'(x) > 0$ on (a, c) and $f'(x) < 0$ on (c, b). That is, f is increasing to the left of c and decreasing to the right of c.

F3. f has neither a relative maximum nor a relative minimum at c if $f'(x)$ has the same sign on (a, c) as on (c, b).

Now let's see how we can use the First-Derivative Test to create and understand graphs and to find relative extrema.

Example 1   Graph the function f given by

$$f(x) = 2x^3 - 3x^2 - 12x + 12$$

and find the relative extrema.

Solution   Suppose that we are trying to graph this function, but don't know any calculus. What can we do? We could plot several points to determine in which direction the graph seems to be turning. Let's guess some x-values and see what happens.

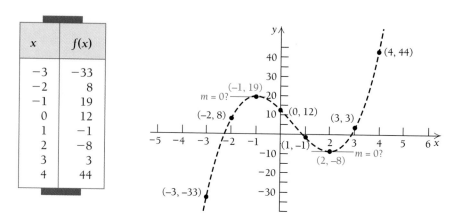

x	$f(x)$
-3	-33
-2	8
-1	19
0	12
1	-1
2	-8
3	3
4	44

We plot the points and use them to sketch a "best guess" of the graph, shown as the dashed line in the figure above. According to this rough sketch, it would seem that the graph has a tangent line with slope 0 somewhere around $x = -1$ and

$x = 2$. But how do we know for sure? We can begin by finding a general expression for the derivative:

$$f'(x) = 6x^2 - 6x - 12.$$

We then determine where $f'(x)$ does not exist or where $f'(x) = 0$. We can replace x in $f'(x) = 6x^2 - 6x - 12$ with any real number. Thus, $f'(x)$ exists for all real numbers. So the only possibilities for critical points are where $f'(x) = 0$, at which there are horizontal tangents. To find such points, we solve $f'(x) = 0$:

$$
\begin{aligned}
6x^2 - 6x - 12 &= 0 \\
x^2 - x - 2 &= 0 & &\text{Dividing by 6 on both sides} \\
(x + 1)(x - 2) &= 0 & &\text{Factoring} \\
x + 1 = 0 \quad &\text{or} \quad x - 2 = 0 & &\text{Using the Principle of Zero Products} \\
x = -1 \quad &\text{or} \qquad\quad x = 2.
\end{aligned}
$$

The critical points are -1 and 2. Since it is at these points that a relative maximum or minimum will exist, if there is one, we examine the intervals on each side of the critical points. We use the critical points to divide the real-number line into three intervals: $A(-\infty, -1)$, $B(-1, 2)$, and $C(2, \infty)$, as shown below.

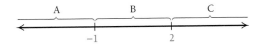

Then we analyze the sign of the derivative on each interval. If $f'(x)$ is positive for one value in the interval, then it will be positive for all numbers in the interval. Similarly, if it is negative for one value, it will be negative for all values in the interval. This is because in order for the derivative to change signs, it must become 0 or be undefined at some point. Such a point will be a critical point. Thus we merely choose a test value in each interval and make a substitution. The test values we choose are -2, 0, and 4.

A: Test -2, $f'(-2) = 6(-2)^2 - 6(-2) - 12$
$$= 24 + 12 - 12 = 24 > 0;$$

B: Test 0, $f'(0) = 6(0)^2 - 6(0) - 12 = -12 < 0;$

C: Test 4, $f'(4) = 6(4)^2 - 6(4) - 12$
$$= 96 - 24 - 12 = 60 > 0.$$

Interval	$(-\infty, -1)$	$(-1, 2)$	$(2, \infty)$
Test Value	$x = -2$	$x = 0$	$x = 4$
Sign of $f'(x)$	$f'(-2) > 0$	$f'(0) < 0$	$f'(4) > 0$
Result	f is increasing	f is decreasing	f is increasing

Change
indicates a
relative
maximum.

Change
indicates a
relative
minimum.

Therefore, by the First-Derivative Test,

f has a relative maximum at $x = -1$ given by

$$f(-1) = 2(-1)^3 - 3(-1)^2 - 12(-1) + 12 \qquad \text{Substituting into the original function}$$

$$= 19$$

and f has a relative minimum at $x = 2$ given by

$$f(2) = 2(2)^3 - 3(2)^2 - 12(2) + 12 = -8.$$

Thus there is a relative maximum at $(-1, 19)$ and a relative minimum at $(2, -8)$, as we suspected from the sketch of the graph.

The information we have obtained can be very useful in sketching a graph of the function. We know that this polynomial is continuous, and we know where the function is increasing, where it is decreasing, and where it has relative extrema. We complete the graph by using a calculator to generate some additional function values. Some calculators can actually be programmed to generate function values by first entering a formula and then generating many outputs. The graph of the function, shown below in red, has been scaled to clearly show its curving nature.

E x p l o r a t o r y .
Consider the function f given by

$$f(x) = x^3 - 3x + 2.$$

Graph both f and f' using the same set of axes. Examine the graphs using the TABLE and TRACE features. Where do you think the relative extrema of $f(x)$ occur? Where is the derivative 0? Where does $f(x)$ have critical points?

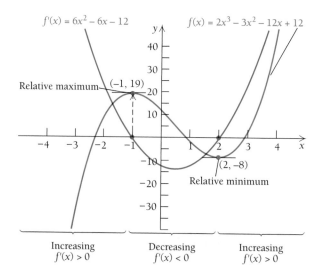

For reference, the graph of the derivative is shown in blue. Note how the critical points reveal themselves. The derivative is 0 where the function has relative extrema.

Keep the following in mind as you find relative extrema.

The *derivative f'* is used to find the critical points of f. The test values, in the intervals defined by the critical points, are substituted into the *derivative f'*, and the function values are found using the *original* function f. Use the derivative f' to find information about the shape of the graph of f.

TECHNOLOGY   CONNECTION

Finding Relative Extrema

There are several methods for approximating relative extrema on a grapher. As an example, consider finding the relative extrema of

$$f(x) = -0.4x^3 + 6.2x^2 - 11.3x - 54.8.$$

We first graph the function, using a viewing window that reveals the curvature.

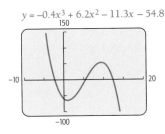

Method 1: TRACE

Beginning with the window shown above, we press **TRACE** and move the cursor along the curve, noting where relative extrema might occur.

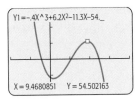

A relative maximum seems to be about Y = 54.5 at X = 9.47. We can refine the approximation by zooming in to obtain the following window. We press **TRACE** and move the cursor along the curve, again noting where the y-value is largest. The approximation seems to be about Y = 54.61 at X = 9.34.

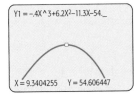
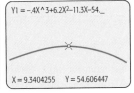

We can continue in this manner until the desired accuracy is achieved.

Method 2: TABLE

We can also use the **TABLE** feature, adjusting starting points and step values to improve accuracy.

X	Y1	
9.3	54.605	
9.31	54.607	
9.32	54.608	
9.33	54.608	
9.34	54.607	
9.35	54.604	
9.36	54.601	
X = 9.32		

Now the approximation seems to be about Y = 54.61 at an x-value between 9.32 and 9.33. We could set up a new table showing function values between $f(9.32)$ and $f(9.33)$ to refine the approximation.

Method 3: MAXIMUM, MINIMUM

Using the **MAXIMUM** feature from the **CALC** menu, we find that a relative maximum of about 54.61 occurs at $x \approx 9.32$.

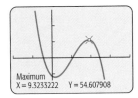

Method 4: fMax or fMin

This feature calculates a relative maximum or minimum value over any specified closed interval. We see from the initial graph that a relative maximum occurs in the interval $[-10, 20]$. Using the fMax feature from the **MATH** menu, we see that a relative maximum occurs on $[-10, 20]$ when $x \approx 9.32$.

(continued)

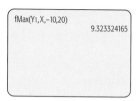

To obtain the maximum value, we evaluate the function at the given *x*-value, obtaining the following.

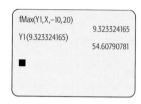

The approximation is about Y = 54.61 at X = 9.32.

Using any of these methods, we find the relative minimum to be about Y = −60.30 at X = 1.01.

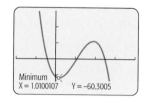

Exercise

1. Approximate the relative extrema of the function in Example 1.

Example 2   Find the relative extrema of the function *f* given by

$$f(x) = 2x^3 - x^4.$$

Then sketch the graph.

Solution   First, we must determine the critical points. To do so, we find $f'(x)$:

$$f'(x) = 6x^2 - 4x^3.$$

Next, we find where $f'(x)$ does not exist or where $f'(x) = 0$. We can replace *x* in $f'(x) = 6x^2 - 4x^3$ with any real number. Thus, $f'(x)$ exists for all real numbers. So the only possibilities for critical points are where $f'(x) = 0$, that is, where there are horizontal tangent lines. To find such points, we solve $f'(x) = 0$:

$$6x^2 - 4x^3 = 0$$
$$2x^2(3 - 2x) = 0 \qquad \text{Factoring}$$
$$2x^2 = 0 \quad or \quad 3 - 2x = 0$$
$$x^2 = 0 \quad or \qquad 3 = 2x$$
$$x = 0 \quad or \qquad x = \tfrac{3}{2}.$$

The critical points are 0 and $\frac{3}{2}$. We use these points to divide the real-number line into three intervals: $A(-\infty, 0)$, $B(0, \frac{3}{2})$, and $C(\frac{3}{2}, \infty)$, as shown below.

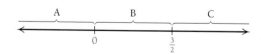

We now analyze the sign of the derivative on each interval. We begin by choosing a test value in each interval and making a substitution. We generally choose as

Exercises

Graph each of the following functions and approximate the relative extrema using a grapher.

1. $f(x) = x^4 - 8x^3 + 18x^2$
2. $f(x) = 2.01x^3 - 0.9x^4$
3. $f(x) = \frac{1}{3}x^3 - \frac{1}{2}x^2 - 2x + 1$
4. $f(x) = 0.21x^4 - 4.3x^2 + 22$

test values numbers for which it is easy to compute outputs of the derivative—in this case, -1, 1, and 2.

A: Test -1, $f'(-1) = 6(-1)^2 - 4(-1)^3$
$$= 6 + 4 = 10 > 0;$$

B: Test 1, $f'(1) = 6(1)^2 - 4(1)^3$
$$= 6 - 4 = 2 > 0;$$

C: Test 2, $f'(2) = 6(2)^2 - 4(2)^3$
$$= 24 - 32 = -8 < 0.$$

Interval	$(-\infty, 0)$	$\left(0, \frac{3}{2}\right)$	$\left(\frac{3}{2}, \infty\right)$
Test Value	$x = -1$	$x = 1$	$x = 2$
Sign of $f'(x)$	$f'(-1) > 0$	$f'(1) > 0$	$f'(2) < 0$
Result	f is increasing	f is increasing	f is decreasing

└── No change ──┘ └── Change ──┘
indicates a
relative
maximum.

Therefore, by the First-Derivative Test, f has neither a relative maximum nor a relative minimum at $x = 0$ since the function is increasing on both sides of 0, and f has a relative maximum at $x = \frac{3}{2}$ given by

$$f\left(\tfrac{3}{2}\right) = 2\left(\tfrac{3}{2}\right)^3 - \left(\tfrac{3}{2}\right)^4 = \tfrac{27}{16}.$$ Remember to substitute into the original function.

Thus there is a relative maximum at $\left(\frac{3}{2}, \frac{27}{16}\right)$.

We use the information obtained to sketch the graph. Other function values are listed in the table at left. (More can be generated by the student.) The graph follows.

x	$f(x)$, approximately
-1	-3
-0.75	-1.16
-0.5	-0.31
-0.25	-0.04
0	0
0.25	0.03
0.5	0.19
0.75	0.53
1	1
1.25	1.46
1.5	1.69
1.75	1.34
2	0

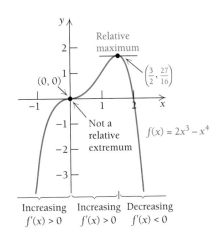

Example 3   Find the relative extrema of the function f given by

$$f(x) = (x - 2)^{2/3} + 1.$$

Then sketch the graph.

Solution   First, we determine the critical points. To do so, we find $f'(x)$:

$$f'(x) = \frac{2}{3}(x - 2)^{-1/3}$$

$$= \frac{2}{3\sqrt[3]{x - 2}}.$$

Next, we find where $f'(x)$ does not exist or where $f'(x) = 0$. Note that the derivative $f'(x)$ does not exist at 2, although $f(x)$ does. Thus the number 2 is a critical point. The equation $f'(x) = 0$ has no solution, so the only critical point is 2. We use 2 to divide the real-number line into two intervals: $A(-\infty, 2)$ and $B(2, \infty)$, as shown below.

We analyze the derivative on each interval. We begin by choosing a test value in each interval and making a substitution. We choose test points 0 and 3. It is not necessary to find an exact value of the derivative; we need only determine the sign. Sometimes we can do this by just examining the formula for the derivative:

A: Test 0, $f'(0) = \dfrac{2}{3\sqrt[3]{0 - 2}} < 0$;

B: Test 3, $f'(3) = \dfrac{2}{3\sqrt[3]{3 - 2}} > 0.$

Interval	$(-\infty, 2)$	$(2, \infty)$
Test Value	$x = 0$	$x = 3$
Sign of $f'(x)$	$f'(0) < 0$	$f'(3) > 0$
Result	f is decreasing	f is increasing

Change
indicates a
relative minimum.

Since we have a change from decreasing to increasing, we conclude from the First-Derivative Test that

f has a relative minimum at $x = 2$ given by

$$f(2) = (2 - 2)^{2/3} + 1 = 1.$$

Thus there is a relative minimum at $(2, 1)$. (The graph has *no* tangent line at $(2, 1)$.)

Exercises

In Exercises 1–3, consider the function f given by

$$f(x) = 2 - (x - 1)^{2/3}.$$

1. Graph the function using the viewing window $[-4, 6, -2, 4]$.
2. Graph the first derivative. What happens to the graph of the derivative at the critical points?
3. Approximate the relative extrema.

We use the information obtained to sketch the graph. Other function values are listed in the table below. (More can be generated by the student.) The graph follows.

x	$f(x)$, approximately
-1	3.08
-0.5	2.84
0	2.59
0.5	2.31
1	2
1.5	1.63
2	1
2.5	1.63
3	2
3.5	2.31
4	2.59

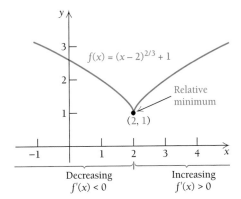

EXERCISE SET 3.1

Find the relative extrema of the function, if they exist. List your answers in terms of ordered pairs. Then sketch a graph of the function.

1. $f(x) = x^2 - 4x + 5$
2. $f(x) = x^2 - 6x - 3$
3. $f(x) = 5 + x - x^2$
4. $f(x) = 2 - 3x - 2x^2$
5. $f(x) = 1 + 6x + 3x^2$
6. $f(x) = 0.5x^2 - 2x - 11$
7. $f(x) = x^3 - x^2 - x + 2$
8. $f(x) = x^3 + \frac{1}{2}x^2 - 2x + 5$
9. $f(x) = x^3 - 3x + 6$
10. $f(x) = x^3 - 3x^2$
11. $f(x) = 3x^2 - 2x^3$
12. $f(x) = x^3 - 3x$

13. $f(x) = 2x^3$

14. $f(x) = 1 - x^3$

15. $f(x) = x^3 - 6x^2 + 10$

16. $f(x) = 12 + 9x - 3x^2 - x^3$

17. $f(x) = x^3 - x^4$

18. $f(x) = x^4 - 2x^3$

19. $f(x) = x^4 - 8x^2 + 3$

20. $f(x) = x^4 - 2x^2 + 5$

21. $f(x) = 1 - x^{2/3}$

22. $f(x) = (x + 3)^{2/3} - 5$

23. $f(x) = \dfrac{-8}{x^2 + 1}$

24. $f(x) = \dfrac{5}{x^2 + 1}$

25. $f(x) = \dfrac{4x}{x^2 + 1}$

26. $f(x) = \dfrac{x^2}{x^2 + 1}$

27. $f(x) = \sqrt[3]{x}$

28. $f(x) = (x + 1)^{1/3}$

 29.–56. Check the results of each of Exercises 1–28 using a grapher.

Applications

BUSINESS AND ECONOMICS

57. *Advertising.* A firm estimates that it will sell N units of a product after spending a dollars on advertising, where

$$N(a) = -a^2 + 300a + 6, \quad 0 \le a \le 300,$$

and a is in thousands of dollars. Find the relative extrema and sketch a graph of the function.

LIFE AND PHYSICAL SCIENCES

58. *Temperature During an Illness.* The temperature of a person during an intestinal illness is given by

$$T(t) = -0.1t^2 + 1.2t + 98.6, \quad 0 \le t \le 12,$$

where T is the temperature (°F) at time t, in days. Find the relative extrema and sketch a graph of the function.

GENERAL INTEREST

59. *Path of the Olympic Arrow.* The Olympic flame at the 1992 summer Olympics was lit by a flaming arrow. As the arrow moved d feet horizontally from the archer, its height h, in feet, was approximated by the function

$$h = -0.002d^2 + 0.8d + 6.6.$$

Find the relative maximum and sketch a graph of the function.

60. Use the graph in Exercise 59 to describe the words "the path of the Olympic arrow." Include the height from which it was launched, its maximum height, and the horizontal distance from the archer at which it hit the ground.

61. Consider this graph.

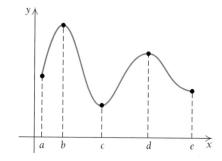

Using the graph and the intervals noted, explain how to relate the concept of the function being increasing or decreasing to the first derivative.

TECHNOLOGY CONNECTION

tw 62. Consider this graph.

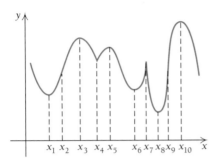

Explain the idea of a critical point. Then determine which points on the graph are critical points and why.

Graph the function. Then estimate any relative extrema.

63. $f(x) = x^4 + 4x^3 - 36x^2 - 160x + 400$
64. $f(x) = -x^6 - 4x^5 + 54x^4 + 160x^3 - 641x^2 - 828x + 1200$
65. $f(x) = x\sqrt{9 - x^2}$
66. $f(x) = \sqrt[3]{|4 - x^2|} + 1$

Business and Life Science: Caloric Intake and Life Expectancy. The following data relate for various countries daily caloric intake in 1992, projected life expectancy in 2000, and infant mortality. Use the table for Exercises 67 and 68.

Country	Daily Caloric Intake (1992)	Life Expectancy (2000)	Infant Mortality (in deaths per 1000 births)
Argentina	2880	72.3	26.1
Bolivia	2100	62.0	60.2
Canada	3482	80.0	5.5
Dominican Republic	2359	70.4	40.8
Germany	3443	76.7	22.2
Haiti	1707	50.2	98.4
Mexico	3181	75.0	20.7
United States	3671	76.3	6.2
Venezuela	2622	73.3	25.5
Australia	3216	80.4	5.0

Source: The Universal Almanac

tw 67. *Infant Mortality and Daily Caloric Intake.*
   a) Use the regression procedures of Section 1.6 to fit a cubic function $y = f(x)$ to the data, where x is daily caloric intake and y is infant mortality. Then fit a quartic function and decide which fits best. Explain.
   b) What is the domain of the function?
   c) Does the function have any relative extrema? Explain.

tw 68. *Life Expectancy and Daily Caloric Intake.*
   a) Use the regression procedures of Section 1.6 to fit a cubic function $y = f(x)$ to the data, where x is daily caloric intake and y is life expectancy. Then fit a quartic function and decide which fits best. Explain.
   b) What is the domain of the function?
   c) Does the function have any relative extrema? Explain.

3.2

OBJECTIVES

➤ Find the relative extrema of a function using the Second-Derivative Test.
➤ Sketch the graph of a continuous function.

Using Second Derivatives to Find Maximum and Minimum Values and Sketch Graphs

Concavity: Increasing and Decreasing Derivatives

The graphs of two functions are shown in Fig. 1. The graph in Fig. 1(a) is turning up and the graph in Fig. 1(b) is turning down. Let's see if we can relate this to their derivatives.

Consider the graph of f (Fig. 1a). Take a ruler, or straightedge, and draw tangent lines as you move along the curve from left to right. What happens to the slopes of the tangent lines? Do the same for the graph of g (Fig. 1b). Look for a pattern.

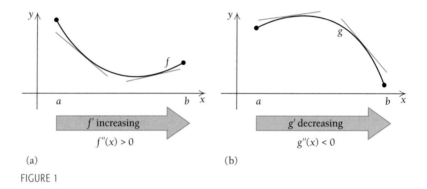

(a)   (b)

FIGURE 1

In Fig. 1(a), the slopes are increasing. That is, f' is increasing over the interval. We know that f' is increasing if we know that f'' is positive, since the relationship between f' and f'' is like the relationship between f and f'. Note also that all the tangent lines are below the graph. In Fig. 1(b), the slopes are decreasing. That is, we know that g' is decreasing if we know that g'' is negative. All the tangent lines are above the graph.

Definition

Suppose that f is a function whose derivative f' exists at every point in an open interval I. Then:

1. f is **concave up** on the interval I if f' is increasing over I.
2. f is **concave down** on the interval I if f' is decreasing over I.

For example, the graph in Fig. 1(a) is concave up and the graph in Fig. 1(b) is concave down. We then have the following theorem, which allows us to use second derivatives to determine concavity.

E x p l o r a t o r y .   Graph the function
$$f(x) = -\tfrac{1}{3}x^3 + 6x^2 - 11x - 50$$
and its second derivative
$$f''(x) = -2x + 12$$
using the viewing window
$[-10, 25, -100, 150]$, with Xscl = 5 and Yscl = 25.

Over what intervals is the graph of f concave up?

Over what intervals is the graph of f concave down?

Over what intervals is the graph of f'' positive?

Over what intervals is the graph of f'' negative?

What can you conjecture?

Now graph the first derivative
$$f'(x) = -x^2 + 12x - 11$$
and the second derivative
$$f''(x) = -2x + 12$$
using the viewing window
$[-10, 25, -200, 50]$, with Xscl = 5 and Yscl = 25.

Over what intervals is the first derivative f' increasing?

Over what intervals is the first derivative f' decreasing?

Over what intervals is the graph of f'' positive?

Over what intervals is the graph of f'' negative?

What can you conjecture?

Theorem 4

A Test for Concavity

1. If $f''(x) > 0$ on an interval I, then the graph of f is turning up. (f' is increasing, so f is concave up on I.)
2. If $f''(x) < 0$ on an interval I, then the graph of f is turning down. (f' is decreasing, so f is concave down on I.)

A helpful memory device follows.

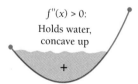

$f''(x) > 0$:
Holds water,
concave up

$f''(x) < 0$:
Loses water,
concave down

Finding Relative Extrema Using Second Derivatives

In the following discussion, we see how we can use second derivatives to determine whether a function has a relative extremum on an open interval.

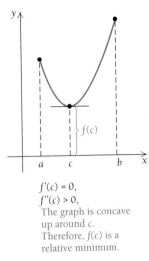

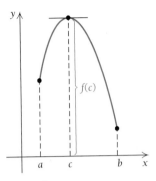

$f'(c) = 0,$
$f''(c) > 0,$
The graph is concave
up around c.
Therefore, $f(c)$ is a
relative minimum.

$f'(c) = 0,$
$f''(c) < 0,$
The graph is concave
down around c.
Therefore, $f(c)$ is a
relative maximum.

Theorem 5

The Second-Derivative Test for Relative Extrema

Suppose that f is a function for which $f'(x)$ exists for every x in an open interval (a, b) contained in its domain, and that there is a critical point c in (a, b) for which $f'(c) = 0$. Then:

1. $f(c)$ is a relative minimum if $f''(c) > 0$. positive
2. $f(c)$ is a relative maximum if $f''(c) < 0$. negative

The test fails if $f''(c) = 0$. The First-Derivative Test would then have to be used.

Note that $f''(c) = 0$ does not tell us that there is no relative extremum. It just tells us that we do not know at this point and that we must use some other means, such as the First-Derivative Test, to determine whether there is a relative extremum.

Consider the following graphs. In each one, f' and f'' are both 0 at $c = 2$, but the first function has an extremum and the second function has *no* extremum. Also note that if $f'(c)$ does not exist, then $f''(c)$ does not exist and the Second-Derivative Test cannot be used.

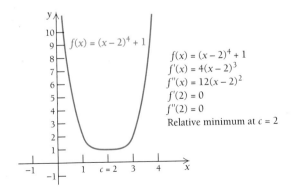

$f(x) = (x - 2)^4 + 1$
$f'(x) = 4(x - 2)^3$
$f''(x) = 12(x - 2)^2$
$f'(2) = 0$
$f''(2) = 0$
Relative minimum at $c = 2$

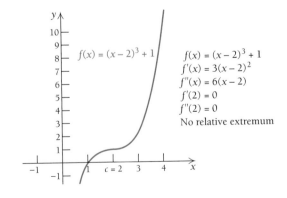

$f(x) = (x - 2)^3 + 1$
$f'(x) = 3(x - 2)^2$
$f''(x) = 6(x - 2)$
$f'(2) = 0$
$f''(2) = 0$
No relative extremum

Example 1   Graph the function f given by

$$f(x) = x^3 + 3x^2 - 9x - 13$$

and find the relative extrema.

Solution   We begin by plotting a few points and using them and our knowledge of calculus to create the graph.

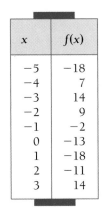

x	$f(x)$
-5	-18
-4	7
-3	14
-2	9
-1	-2
0	-13
1	-18
2	-11
3	14

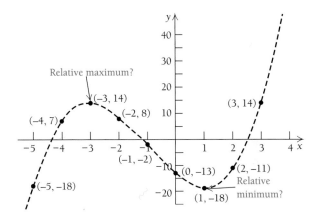

TECHNOLOGY CONNECTION

Exploratory.
Consider the function f given by

$$f(x) = x^3 - 3x^2 - 9x - 1.$$

Use a grapher to estimate the relative extrema. Then find the first and second derivatives. Graph both in the same window. Use the **ZERO** feature to determine where the first derivative is zero. Verify that relative extrema occur at those x-values by checking the sign of the second derivative. Then check your work using the analytic method of Example 1.

We plot these points and begin thinking about the shape of the graph. The dashed graph shown above can be considered a first guess. Are the points $(-3, 14)$ and $(1, -18)$ relative extrema? It could be that we did not even choose them to plot at the outset. To refine our thoughts, we use calculus. We find both the first and second derivatives, $f'(x)$ and $f''(x)$:

$$f'(x) = 3x^2 + 6x - 9,$$
$$f''(x) = 6x + 6.$$

Then we solve $f'(x) = 0$:

$$3x^2 + 6x - 9 = 0$$
$$x^2 + 2x - 3 = 0 \qquad \text{Dividing by 3 on both sides}$$
$$(x + 3)(x - 1) = 0 \qquad \text{Factoring}$$
$$x + 3 = 0 \quad or \quad x - 1 = 0 \qquad \text{Using the Principle of Zero Products}$$
$$x = -3 \quad or \qquad x = 1.$$

We then find second coordinates by substituting in the original function:

$$f(-3) = (-3)^3 + 3(-3)^2 - 9(-3) - 13 = 14;$$
$$f(1) = (1)^3 + 3(1)^2 - 9(1) - 13 = -18.$$

Are the points $(-3, 14)$ and $(1, -18)$ relative extrema? Let's look at the second derivative. We use the Second-Derivative Test with the numbers -3 and 1:

$$f''(-3) = 6(-3) + 6 = -12 < 0; \longrightarrow \quad \text{Relative maximum}$$
$$f''(1) = 6(1) + 6 = 12 > 0. \longrightarrow \quad \text{Relative minimum}$$

Thus there is a relative maximum at $(-3, 14)$ and a relative minimum at $(1, -18)$.

The following figures illustrate the information concerning the function f that can be found from the first and second derivatives of f. The relative extrema are shown in Figs. 1 and 2. In Fig. 2, we see that the x-coordinates of the x-intercepts of f' are the critical points of f. We also see the intervals over which f is increasing and decreasing from the intervals over which f' is positive and negative.

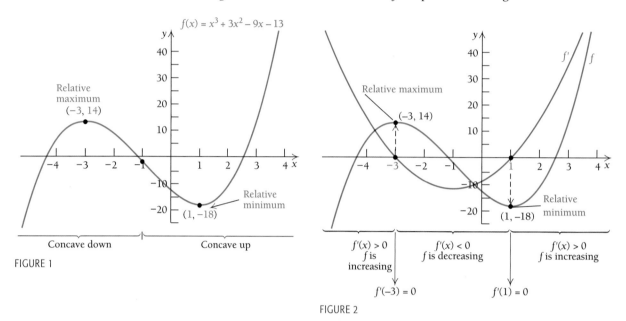

FIGURE 1

FIGURE 2

In Fig. 3, the intervals over which f' is increasing and decreasing are seen from the intervals over which f'' is positive and negative. And finally in Fig. 4, we note that when $f''(x) < 0$, f is concave down, and when $f''(x) > 0$, f is concave up.

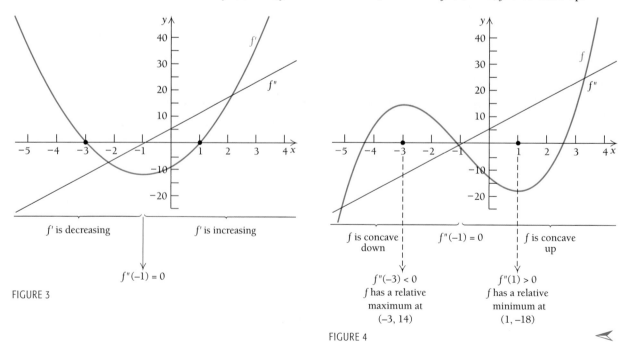

FIGURE 3

FIGURE 4

Example 2   Find the relative extrema of the function f given by

$$f(x) = 3x^5 - 20x^3$$

and sketch the graph.

Solution   We find both the first and second derivatives, $f'(x)$ and $f''(x)$:

$$f'(x) = 15x^4 - 60x^2,$$
$$f''(x) = 60x^3 - 120x.$$

Then we solve for $f'(x) = 0$:

$$15x^4 - 60x^2 = 0$$
$$15x^2(x^2 - 4) = 0$$
$$15x^2(x + 2)(x - 2) = 0 \qquad \text{Factoring}$$
$$15x^2 = 0 \quad or \quad x + 2 = 0 \quad or \quad x - 2 = 0 \qquad \text{Using the Principle of Zero Products}$$
$$x = 0 \quad or \qquad x = -2 \quad or \qquad x = 2.$$

We then find second coordinates by substituting in the original function:

$$f(-2) = 3(-2)^5 - 20(-2)^3 = 64;$$
$$f(2) = 3(2)^5 - 20(2)^3 = -64;$$
$$f(0) = 3(0)^5 - 20(0)^3 = 0.$$

The points $(-2, 64)$, $(2, -64)$, and $(0, 0)$ are candidates for relative extrema. We now use the Second-Derivative Test with the numbers -2, 2, and 0:

$$f''(-2) = 60(-2)^3 - 120(-2) = -240 < 0; \longrightarrow \qquad \text{Relative maximum}$$
$$f''(2) = 60(2)^3 - 120(2) = 240 > 0; \longrightarrow \qquad \text{Relative minimum}$$
$$f''(0) = 60(0)^3 - 120(0) = 0. \longrightarrow \qquad \text{The Second-Derivative Test fails. Use the First-Derivative Test.}$$

Thus there is a relative maximum at $(-2, 64)$ and a relative minimum at $(2, -64)$. Checking the first derivative, we know that the function decreases to the left and to the right of $x = 0$. Thus we know by the First-Derivative Test that it has no relative extremum at the point $(0, 0)$. We complete the graph, plotting other points as needed. The extrema are shown in the graph at left.

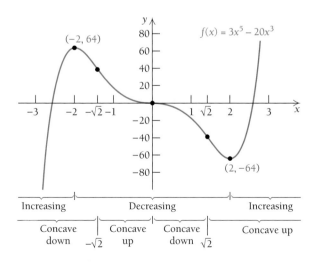

Points of Inflection

A **point of inflection,** or an **inflection point,** is a point across which the direction of concavity changes. For example, in Figs. 5–7, point P is an inflection point.

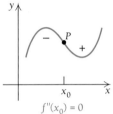

FIGURE 5

FIGURE 6

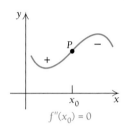

FIGURE 7

There are points of inflection in other examples that we have considered in this section. In Example 1, the graph has a point of inflection at $(-1, -2)$, and in Example 2, the graph has points of inflection at $(-\sqrt{2}, 28\sqrt{2})$, $(0, 0)$, and $(\sqrt{2}, -28\sqrt{2})$.

As we move to the right along the curve in Fig. 5, the concavity changes from concave down, $f''(x) < 0$, on the left of P to concave up, $f''(x) > 0$, on the right of P. Since, as we move through P, $f''(x)$ changes sign from $-$ to $+$, the value of $f''(x_0)$ at P must be 0, as in Fig 5; or $f''(x_0)$ does not exist, as in Fig. 6. A similar change in concavity occurs at P in Fig. 7.

Theorem 6

Finding Points of Inflection

If a function f has a point of inflection, it occurs at a point x_0, where

$$f''(x_0) = 0 \quad \text{or} \quad f''(x_0) \text{ does not exist.}$$

Thus we find candidates for points of inflection by looking for numbers x_0 for which $f''(x_0) = 0$ or for which $f''(x_0)$ does not exist. Then if $f''(x)$ changes sign as x moves through x_0 (see Figs. 5–7), we have a point of inflection at $x = x_0$. If $f''(x) = k$, where $k \neq 0$, then x is not a candidate for a point of inflection.

Curve Sketching

What we have learned thus far in this chapter will greatly enhance our ability to sketch curves. We use the following strategy, writing it first in abbreviated form.

Strategy for Sketching Graphs

a) Derivatives
b) Critical points of f
c) Increasing and/or decreasing relative extrema
d) Inflection points
e) Concavity
f) Sketch

Below we expand on the strategy.

Strategy for Sketching Graphs

a) *Derivatives.* Find $f'(x)$ and $f''(x)$.
b) *Critical points of f.* Find the critical points of f by solving $f'(x) = 0$ and finding where $f'(x)$ does not exist. These numbers yield candidates for relative maxima or minima. Find the function values at these points.
c) *Increasing and/or decreasing relative extrema.* Use the critical points of f from step (b) to define intervals. Determine whether f is increasing or decreasing over the intervals. Do this by selecting test values and substituting into $f'(x)$. Use this information and/or the second derivative to determine the relative maxima and minima.
d) *Inflection points.* Determine candidates for inflection points by finding where $f''(x) = 0$ or where $f''(x)$ does not exist. Find the function values at these points.
e) *Concavity.* Use the candidates for inflection points from step (d) to define intervals. Determine the concavity by checking to see where f' is increasing—that is, where $f''(x) > 0$—and where f' is decreasing—that is, where $f''(x) < 0$. Do this by selecting test values and substituting into $f''(x)$.
f) *Sketch the graph.* Sketch the graph using the information from steps (a) through (e), plotting extra points (computing them with your calculator) if the need arises.

Example 3   Find the relative maxima and minima of the function f given by

$$f(x) = x^3 - 3x + 2$$

and sketch the graph.

Solution

a) *Derivatives.* Find $f'(x)$ and $f''(x)$:

$$f'(x) = 3x^2 - 3,$$
$$f''(x) = 6x.$$

b) *Critical points of f.* Find the critical points of f by finding where $f'(x)$ does not exist and by solving $f'(x) = 0$. We know that $f'(x) = 3x^2 - 3$ exists for all values of x, so the only critical points are where

$$3x^2 - 3 = 0$$
$$3x^2 = 3$$
$$x^2 = 1$$
$$x = \pm 1.$$

Now $f(-1) = 4$ and $f(1) = 0$, which gives the points $(-1, 4)$ and $(1, 0)$ on the graph.

c) *Increasing and/or decreasing relative extrema.* Find the intervals over which f is increasing and the intervals over which f is decreasing. The critical points are -1 and 1. We use these points to divide the real-number line into three intervals: $A(-\infty, -1)$, $B(-1, 1)$, and $C(1, \infty)$. We choose a test point in each interval and make a substitution. The test values we select are -2, 0, and 3:

A: Test -2, $f'(-2) = 3(-2)^2 - 3 = 9 > 0$;
B: Test 0, $f'(0) = 3(0)^2 - 3 = -3 < 0$;
C: Test 3, $f'(3) = 3(3)^2 - 3 = 24 > 0$.

Interval	$(-\infty, -1)$	$(-1, 1)$	$(1, \infty)$
Test Value	$x = -2$	$x = 0$	$x = 3$
Sign of $f'(x)$	$f'(-2) > 0$	$f'(0) < 0$	$f'(3) > 0$
Result	f is increasing	f is decreasing	f is increasing

Change indicates a relative maximum; $f''(-1) < 0$.

Change indicates a relative minimum; $f''(1) > 0$.

Therefore, by the First-Derivative Test, there is a relative maximum at $(-1, 4)$ and a relative minimum at $(1, 0)$. That these are relative extrema can also be verified by the Second-Derivative Test: $f''(-1) < 0$ tells us that $(-1, 4)$ is a relative maximum, and $f''(1) > 0$ tells us that $(1, 0)$ is a relative minimum.

d) *Inflection points.* Find possible inflection points by finding where $f''(x)$ does not exist and by solving $f''(x) = 0$. We know that $f''(x) = 6x$ exists for all values of x, so we try to solve $f''(x) = 0$:

$$6x = 0$$
$$x = 0.$$

Now $f(0) = 2$, which gives us another point, $(0, 2)$, that lies on the graph.

e) *Concavity.* Find the intervals on which f is concave up and concave down. We do this by determining where f' is increasing and decreasing using the numbers found in step (d). There is only one such number, 0. We use this point to divide the real-number line into two intervals: $A(-\infty, 0)$ and $B(0, \infty)$. We choose a test point in each interval and make a substitution into f''. The test points we select are -4 and 4:

A: Test -4, $f''(-4) = 6(-4) = -24 < 0$;

B: Test 4, $f''(4) = 6(4) = 24 > 0$.

Interval	$(-\infty, 0)$	$(0, \infty)$
Test Value	$x = -4$	$x = 4$
Sign of $f''(x)$	$f''(-4) < 0$	$f''(4) > 0$
Result	f' is decreasing; f is concave down.	f' is increasing; f is concave up.

Change indicates a point of inflection.

Since f' is decreasing over the interval $(-\infty, 0)$ and f' is increasing over the interval $(0, \infty)$, f is concave down over $(-\infty, 0)$ and concave up over $(0, \infty)$. The graph changes concavity across $(0, 2)$, so it is a point of inflection.

f) *Sketch the graph.* Sketch the graph using the information in the following table. Calculate some extra function values if desired. The graph follows.

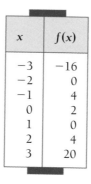

x	$f(x)$
-3	-16
-2	0
-1	4
0	2
1	0
2	4
3	20

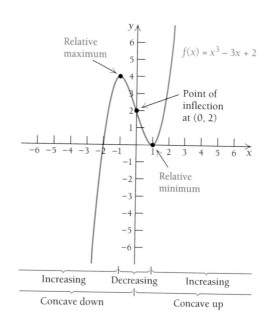

Increasing   Decreasing   Increasing

Concave down   Concave up

Example 4   Find the relative maxima and minima of the function f given by

$$f(x) = x^4 - 2x^2$$

and sketch the graph.

Solution

a) *Derivatives.* Find $f'(x)$ and $f''(x)$:

$$f'(x) = 4x^3 - 4x,$$
$$f''(x) = 12x^2 - 4.$$

b) *Critical points of f.* Since $f'(x) = 4x^3 - 4x$ exists for all values of x, the only critical points of f are where

$$4x^3 - 4x = 0$$
$$4x(x^2 - 1) = 0$$
$$4x = 0 \quad \text{or} \quad x^2 - 1 = 0$$
$$x = 0 \quad \text{or} \quad x^2 = 1$$
$$x = \pm 1.$$

Now $f(0) = 0$, $f(-1) = -1$, and $f(1) = -1$, which gives the points $(0, 0)$, $(-1, -1)$, and $(1, -1)$ on the graph.

c) *Increasing and/or decreasing relative extrema.* Use the critical points of f— namely, -1, 0, and 1—to divide the real-number line into four intervals: $A(-\infty, -1)$, $B(-1, 0)$, $C(0, 1)$, and $D(1, \infty)$. We choose a test value in each interval and make a substitution. The test values we select are -2, $-\frac{1}{2}$, $\frac{1}{2}$, and 2:

A: Test -2, $f'(-2) = 4(-2)^3 - 4(-2) = -24 < 0$;

B: Test $-\frac{1}{2}$, $f'\left(-\frac{1}{2}\right) = 4\left(-\frac{1}{2}\right)^3 - 4\left(-\frac{1}{2}\right) = \frac{3}{2} > 0$;

C: Test $\frac{1}{2}$, $f'\left(\frac{1}{2}\right) = 4\left(\frac{1}{2}\right)^3 - 4\left(\frac{1}{2}\right) = -\frac{3}{2} < 0$;

D: Test 2, $f'(2) = 4(2)^3 - 4(2) = 24 > 0$.

Interval	$(-\infty, -1)$	$(-1, 0)$	$(0, 1)$	$(1, \infty)$
Test Value	$x = -2$	$x = -\frac{1}{2}$	$x = \frac{1}{2}$	$x = 2$
Sign of $f'(x)$	$f'(-2) < 0$	$f'\left(-\frac{1}{2}\right) > 0$	$f'\left(\frac{1}{2}\right) < 0$	$f'(2) > 0$
Result	f is decreasing	f is increasing	f is decreasing	f is increasing

Change indicates a relative minimum; $f''(-1) > 0$.

Change indicates a relative maximum; $f''(0) < 0$.

Change indicates a relative minimum; $f''(1) > 0$.

Thus, by the First-Derivative Test, there is a relative maximum at $(0, 0)$ and two relative minima at $(-1, -1)$ and $(1, -1)$. That these are relative extrema can also be verified by the Second-Derivative Test: $f''(0) < 0, f''(-1) > 0,$ and $f''(1) > 0$.

d) *Inflection points.* Find where $f''(x)$ does not exist and where $f''(x) = 0$. Since $f''(x)$ exists for all real numbers, we just solve $f''(x) = 0$:

$$12x^2 - 4 = 0$$
$$4(3x^2 - 1) = 0$$
$$3x^2 - 1 = 0$$
$$3x^2 = 1$$
$$x^2 = \frac{1}{3}$$
$$x = \pm\sqrt{\frac{1}{3}}$$
$$= \pm\frac{1}{\sqrt{3}}.$$

Now

$$f\left(\frac{1}{\sqrt{3}}\right) = \left(\frac{1}{\sqrt{3}}\right)^4 - 2\left(\frac{1}{\sqrt{3}}\right)^2$$
$$= \frac{1}{9} - \frac{2}{3} = -\frac{5}{9}$$

and

$$f\left(-\frac{1}{\sqrt{3}}\right) = -\frac{5}{9}.$$

This gives the points

$$\left(-\frac{1}{\sqrt{3}}, -\frac{5}{9}\right) \quad \text{and} \quad \left(\frac{1}{\sqrt{3}}, -\frac{5}{9}\right)$$

on the graph—$(-0.6, -0.6)$ and $(0.6, -0.6)$, approximately.

e) *Concavity.* Find the intervals over which f is concave up and concave down. We do this by determining where f' is increasing and decreasing using the numbers found in step (d). Those numbers divide the real-number line into three intervals:

$$A\left(-\infty, -\frac{1}{\sqrt{3}}\right), \quad B\left(-\frac{1}{\sqrt{3}}, \frac{1}{\sqrt{3}}\right), \quad \text{and} \quad C\left(\frac{1}{\sqrt{3}}, \infty\right).$$

Next, we choose a test value in each interval and make a substitution into f''. The test points we select are $-1, 0,$ and 1:

A: Test -1, $f''(-1) = 12(-1)^2 - 4 = 8 > 0$;

B: Test 0, $f''(0) = 12(0)^2 - 4 = -4 < 0$;

C: Test 1, $f''(1) = 12(1)^2 - 4 = 8 > 0$.

TECHNOLOGY
CONNECTION

Check the results of Example 4 using a grapher.

Interval	$(-\infty, -1/\sqrt{3})$	$(-1/\sqrt{3}, 1/\sqrt{3})$	$(1/\sqrt{3}, \infty)$
Test Value	$x = -1$	$x = 0$	$x = 1$
Sign of $f''(x)$	$f''(-1) > 0$	$f''(0) < 0$	$f''(1) > 0$
Result	f' is increasing; f is concave up.	f' is decreasing; f is concave down.	f' is increasing; f is concave up.

Change indicates a point of inflection.   Change indicates a point of inflection.

Exercise

1. Consider $f(x) = x^3(x - 2)^3$. How many relative extrema do you anticipate finding? Where do you think they will be?

    Graph f, f', and f'' using $[-1, 3, -2, 6]$ as a viewing window. Estimate the relative extrema and the inflection points of f. Then check your work using the analytic methods of Examples 3 and 4.

Therefore, f' is increasing over the interval $(-\infty, -1/\sqrt{3})$, decreasing over the interval $(-1/\sqrt{3}, 1/\sqrt{3})$, and increasing over the interval $(1/\sqrt{3}, \infty)$. The function f is concave up over $(-\infty, -1/\sqrt{3})$ and $(1/\sqrt{3}, \infty)$ and concave down over $(-1/\sqrt{3}, 1/\sqrt{3})$. The graph changes concavity across $(-1/\sqrt{3}, -5/9)$ and $(1/\sqrt{3}, -5/9)$, so these are points of inflection.

f) *Sketch the graph.* Sketch the graph using the information in the following table. By solving $x^4 - 2x^2 = 0$, we can find the x-intercepts easily. They are $(-\sqrt{2}, 0)$, $(0, 0)$, and $(\sqrt{2}, 0)$. This also aids the graphing. Extra function values can be calculated if desired. The graph is shown below.

x	$f(x)$, approximately
-2	8
-1.5	0.56
-1	-1
-0.5	-0.44
0	0
0.5	-0.44
1	-1
1.5	0.56
2	8

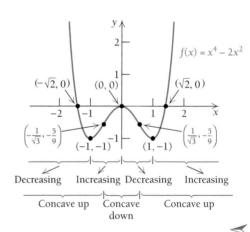

Find the relative extrema of the function. List your answers in terms of ordered pairs. Use the Second-Derivative Test, where possible. Then sketch the graph.

1. $f(x) = 2 - x^2$
2. $f(x) = 3 - x^2$
3. $f(x) = x^2 + x - 1$

4. $f(x) = x^2 - x$

5. $f(x) = -4x^2 + 3x - 1$

6. $f(x) = 7 - 8x + 5x^2$

7. $f(x) = 2x^3 - 3x^2 - 36x + 28$

8. $f(x) = 3x^3 - 36x - 3$

9. $f(x) = \frac{8}{3}x^3 - 2x + \frac{1}{3}$

10. $f(x) = 80 - 9x^2 - x^3$

11. $f(x) = -x^3 + 3x^2 - 4$

12. $f(x) = -x^3 + 3x - 2$

13. $f(x) = 3x^4 - 16x^3 + 18x^2$

14. $f(x) = 3x^4 + 4x^3 - 12x^2 + 5$

15. $f(x) = (x + 1)^{2/3}$

16. $f(x) = (x - 1)^{2/3}$

17. $f(x) = x^4 - 6x^2$

18. $f(x) = 2x^2 - x^4$

19. $f(x) = x^3 - 2x^2 - 4x + 3$

20. $f(x) = x^3 - 6x^2 + 9x + 1$

21. $f(x) = 3x^4 + 4x^3$

22. $f(x) = x^4 - 2x^3$

23. $f(x) = x^3 - 6x^2 - 135x$

24. $f(x) = x^3 - 3x^2 - 144x - 140$

25. $f(x) = \dfrac{x}{x^2 + 1}$ 26. $f(x) = \dfrac{8x}{x^2 + 1}$

27. $f(x) = \dfrac{3}{x^2 + 1}$ 28. $f(x) = \dfrac{-4}{x^2 + 1}$

29. $f(x) = (x - 1)^3$ 30. $f(x) = (x + 2)^3$

31. $f(x) = x^2(1 - x)^2$ 32. $f(x) = x^2(3 - x)^2$

33. $f(x) = 20x^3 - 3x^5$ 34. $f(x) = 5x^3 - 3x^5$

35. $f(x) = x\sqrt{4 - x^2}$ 36. $f(x) = -x\sqrt{1 - x^2}$

37. $f(x) = (x - 1)^{1/3} - 1$ 38. $f(x) = 2 - x^{1/3}$

Find all points of inflection, if they exist.

39. $f(x) = x^3 + 3x + 1$

40. $f(x) = x^3 - 6x^2 + 12x - 6$

41. $f(x) = \frac{4}{3}x^3 - 2x^2 + x$

42. $f(x) = x^4 - 4x^3 + 10$

 43.–84. Check the results of each of Exercises 1–42 using a grapher.

Applications
BUSINESS AND ECONOMICS

Total Revenue, Cost, and Profit. Using the same set of axes, sketch the graphs of the total-revenue, total-cost,

and total-profit functions.

85. $R(x) = 50x - 0.5x^2$, $C(x) = 4x + 10$

86. $R(x) = 50x - 0.5x^2$, $C(x) = 10x + 3$

LIFE AND PHYSICAL SCIENCES

87. *Coughing Velocity.* A person coughs when a foreign object is in the windpipe. The velocity of the cough depends on the size of the object. Suppose a person has a windpipe with a 20-mm radius. If a foreign object has a radius r, in millimeters, then the velocity V, in millimeters/second, needed to remove the object by a cough is given by

$$V(r) = k(20r^2 - r^3), \quad 0 \leq r \leq 20,$$

where k is some positive constant. For what size object is the maximum velocity needed to remove the object?

88. *Temperature in January.* Suppose that the temperature T, in degrees Fahrenheit, during a 24-hr day in January is given by

$$T(x) = 0.0027(x^3 - 34x + 240), \quad 0 \leq x \leq 24,$$

where x is the number of hours since midnight. Estimate the relative minimum temperature and when it occurs.

Synthesis

tW In each of Exercises 89 and 90, determine which graph is the derivative of the other and explain why.

89.   90.

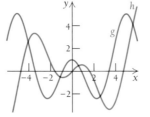 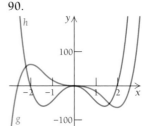

Social Sciences: Three Aspects of Love. Researchers at Yale University have suggested that the following graphs may represent three different aspects of love.

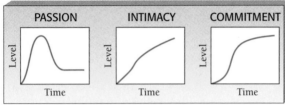

Source: From "A Triangular Theory of Love," by R. J. Sternberg, 1986, *Psychological Review,* **93**(2), 119–135. Copyright 1986 by the American Psychological Association, Inc. Reprinted by permission.

tw 91. Analyze each of these graphs in terms of the concepts you have learned in Sections 3.1 and 3.2: relative extrema, concavity, increasing, decreasing, and so on.

tw 92. Do you agree with the researchers regarding the shape of these graphs? Explain your reasons.

 TECHNOLOGY CONNECTION

Graph the function. Then estimate any relative extrema.

93. $f(x) = 3x^{2/3} - 2x$

94. $f(x) = 4x - 6x^{2/3}$

95. $f(x) = x^2(x - 2)^3$

96. $f(x) = x^2(1 - x)^3$

97. $f(x) = x - \sqrt{x}$

98. $f(x) = (x - 1)^{2/3} - (x + 1)^{2/3}$

tw 99. *Business: Time Spent on Home Computer.* The following data relate the average number of minutes spent per month on a home computer to a person's age.

Age (in years)	Average Use (in minutes per month)
6.5	363
14.5	645
21	1377
29.5	1727
39.5	1696
49.5	2052
55 and up	2299

Source: Media Matrix; The PC Meter Company

a) Use the regression procedures of Section 1.5 to fit linear, cubic, and quartic functions $y = f(x)$ to the data, where x is age and y is average number of minutes per month. Decide which function best fits the data. Explain.

b) What is the domain of the function?

c) Does the function have any relative extrema? Explain.

3.3

OBJECTIVES

➤ Find limits involving infinity.
➤ Graph rational functions.

Graph Sketching: Asymptotes and Rational Functions

Rational Functions

Thus far we have considered a strategy for graphing a continuous function using the tools of calculus. We now want to consider some discontinuous functions, most of which are rational functions. Our graphing skills will now have to take into account the discontinuities of the graph and certain lines called *asymptotes*.

Let's reconsider the definition of a rational function.

> ### Definition
>
> A **rational function** is a function f that can be described by
> $$f(x) = \frac{P(x)}{Q(x)},$$
> where $P(x)$ and $Q(x)$ are polynomials and with $Q(x)$ not the zero polynomial. The domain of f consists of all inputs x for which $Q(x) \neq 0$.

Polynomials are themselves a special kind of rational function, since $Q(x)$ can be the polynomial 1. Here we are considering graphs of rational functions in which the denominator is not a constant. Before we do so, however, we need to reconsider limits.

Limits and Infinity

Let's look again at the graph of the rational function $F(x) = 1/x$. We see that

$$\lim_{x \to 0} \frac{1}{x} \quad \text{does not exist.}$$

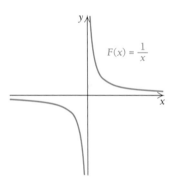

Looking at the graph above, we note that as x approaches 0 from the right, the outputs increase without bound. These numbers do not approach any real number, though it might be said that the limit from the right is infinity (∞). That is,

$$\lim_{x \to 0^+} \frac{1}{x} = \infty.$$

As x approaches 0 from the left, the outputs become more and more negative without bound. These numbers do not approach any real number, though it might be said that the limit from the left is negative infinity ($-\infty$). That is,

$$\lim_{x \to 0^-} \frac{1}{x} = -\infty.$$

Keep in mind that ∞ and $-\infty$ are not real numbers. We associate ∞ with numbers increasing without bound in a positive direction, as in the interval notation $(2, \infty)$. Thus we associate $-\infty$ with numbers decreasing without bound in a negative direction, as in the interval notation $(-\infty, 8)$.

Occasionally we need to determine limits when the inputs get larger and larger without bound, that is, when they approach infinity. In such cases, we are finding *limits at infinity*. Such a limit is expressed as

$$\lim_{x \to \infty} f(x).$$

Example 1 Find

$$\lim_{x \to \infty} \frac{3x - 1}{x}.$$

Solution The function involved is rational. One way to find such a limit is to use an input–output table, as follows, using progressively larger x-values.

Inputs, x	1	10	50	100	2000
Outputs, $\dfrac{3x - 1}{x}$	2.0	2.9	2.98	2.99	2.9995

As the inputs get larger and larger without bound, the outputs get closer and closer to 3. Thus,

$$\lim_{x \to \infty} \frac{3x - 1}{x} = 3.$$

Another way to find this limit is to use some algebra and the fact that

$$\text{as } x \to \infty, \qquad \frac{b}{ax^n} \to 0,$$

for any positive integer n and any constants a and b, $a \neq 0$. We multiply by 1, using $(1/x) \div (1/x)$. This amounts to dividing both the numerator and the denominator by x:

$$\lim_{x \to \infty} \frac{3x - 1}{x} = \lim_{x \to \infty} \frac{3x - 1}{x} \cdot \frac{(1/x)}{(1/x)}$$

$$= \lim_{x \to \infty} \frac{(3x - 1)\dfrac{1}{x}}{x \cdot \dfrac{1}{x}}$$

$$= \lim_{x \to \infty} \frac{3x \cdot \dfrac{1}{x} - 1 \cdot \dfrac{1}{x}}{1}$$

$$= \lim_{x \to \infty} \left(3 - \frac{1}{x} \right)$$

$$= 3 - 0$$

$$= 3.$$

TECHNOLOGY CONNECTION

1. Verify the limit

$$\lim_{x \to \infty} \frac{3x - 1}{x} = 3$$

by using the TABLE feature with larger and larger x-values.

X	Y1	
50	2.98	
150	2.9933	
250	2.996	
350	2.9971	
450	2.9978	
550	2.9982	
650	2.9985	
X = 50		

X	Y1	
500	2.998	
1500	2.9993	
2500	2.9996	
3500	2.9997	
4500	2.9998	
5500	2.9998	
6500	2.9998	
X = 500		

2. Graph the function

$$f(x) = \frac{3x - 1}{x}$$

in DOT mode. Then use the TRACE feature, moving the cursor along the graph from left to right, and observe the behavior of the y-coordinates.

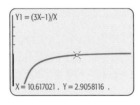

Y1 = (3X–1)/X

X = 10.617021 . Y = 2.9058116 .

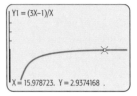

Y1 = (3X–1)/X

X = 15.978723 . Y = 2.9374168 .

Exercises

Consider $\lim\limits_{x \to \infty} \dfrac{2x + 5}{x}$.

1. Use the TABLE feature to find the limit.
2. Graph the function in DOT mode and use the TRACE feature to find the limit.

Example 2 Find

$$\lim_{x \to \infty} \frac{3x^2 - 7x + 2}{7x^2 + 5x + 1}.$$

Solution This function is rational. It involves a quotient of two polynomials. It involves a quotient of two polynomials. The *degree* of a polynomial is its highest power. Note that the degree of each polynomial above is 2.

The highest power of x in the denominator is x^2. We divide both the numerator and the denominator by x^2:

$$\lim_{x \to \infty} \frac{3x^2 - 7x + 2}{7x^2 + 5x + 1} = \lim_{x \to \infty} \frac{3 - \dfrac{7}{x} + \dfrac{2}{x^2}}{7 + \dfrac{5}{x} + \dfrac{1}{x^2}} = \frac{3 - 0 + 0}{7 + 0 + 0} = \frac{3}{7}. \qquad \blacktriangleleft$$

Example 3 Find

$$\lim_{x \to \infty} \frac{5x^2 + 7x + 9}{3x^3 + 2x - 4}.$$

Solution The highest power of x in the denominator is x^3. We divide the numerator and the denominator by x^3:

$$\lim_{x \to \infty} \frac{5x^2 + 7x + 9}{3x^3 + 2x - 4} = \lim_{x \to \infty} \frac{\dfrac{5}{x} + \dfrac{7}{x^2} + \dfrac{9}{x^3}}{3 + \dfrac{2}{x^2} - \dfrac{4}{x^3}} = \frac{0 + 0 + 0}{3 + 0 - 0} = \frac{0}{3} = 0. \qquad \blacktriangleleft$$

Exercises

Use the TABLE and TRACE features to find each of the following limits. Then check your work using the analytic procedure of Examples 1–4.

1. $\displaystyle\lim_{x\to\infty} \frac{2x^2 + x - 7}{3x^2 - 4x + 1}$

2. $\displaystyle\lim_{x\to\infty} \frac{5x + 4}{2x^3 - 3}$

3. $\displaystyle\lim_{x\to\infty} \frac{5x^2 - 2}{4x + 5}$

4. $\displaystyle\lim_{x\to\infty} \frac{x^2 - 1}{x^2 + x - 6}$

Example 4 Find

$$\lim_{x\to\infty} \frac{7x^5 - 8x - 6}{3x^2 + 5x + 2}.$$

Solution The highest power of x in the denominator is x^2. We divide the numerator and the denominator by x^2:

$$\lim_{x\to\infty} \frac{7x^5 - 8x - 6}{3x^2 + 5x + 2} = \lim_{x\to\infty} \frac{7x^3 - \dfrac{8}{x} - \dfrac{6}{x^2}}{3 + \dfrac{5}{x} + \dfrac{2}{x^2}} = \frac{\displaystyle\lim_{x\to\infty} 7x^3 - 0 - 0}{3 + 0 + 0} = \infty.$$

In this case, the numerator increases without bound positively while the denominator approaches 3. This can be checked with an input–output table. Thus the limit is ∞.

Graphs of Rational Functions

Figure 1 shows the graph of the rational function

$$f(x) = \frac{x^2 - 1}{x^2 + x - 6} = \frac{(x - 1)(x + 1)}{(x - 2)(x + 3)}.$$

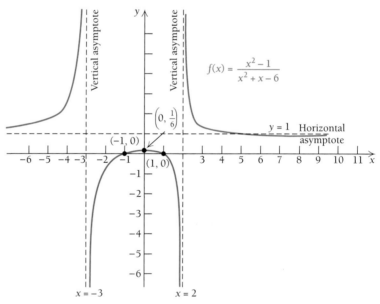

FIGURE 1

Let's make some observations about this graph.

First, note that as x gets closer to 2 from the left, the function values get smaller and smaller negatively, approaching $-\infty$. As x gets closer to 2 from the right, the function values get larger and larger positively. Thus,

$$\lim_{x\to 2^-} f(x) = -\infty \quad \text{and} \quad \lim_{x\to 2^+} f(x) = \infty.$$

For this graph, we can think of the line $x = 2$ as a "limiting line" called a *vertical asymptote*. Similarly, the line $x = -3$ is a vertical asymptote.

Definition

The line $x = a$ is a **vertical asymptote** if any of the following limit statements is true:

$$\lim_{x \to a^-} f(x) = \infty \quad \text{or} \quad \lim_{x \to a^-} f(x) = -\infty \quad \text{or}$$

$$\lim_{x \to a^+} f(x) = \infty \quad \text{or} \quad \lim_{x \to a^+} f(x) = -\infty.$$

The graph of a rational function *never* crosses a vertical asymptote. If the expression that defines the rational expression is simplified, meaning it has no common factor other than -1 or 1, then if a is an input that makes the denominator 0, the line $x = a$ is a vertical asymptote.

For example,

$$f(x) = \frac{x^2 - 1}{x - 1} = \frac{(x - 1)(x + 1)}{x - 1}$$

does not have a vertical asymptote at $x = 1$, even though 1 is an input that makes the denominator 0. This is the case because $(x^2 - 1)/(x - 1)$ is not simplified, that is, it has $x - 1$ as a common factor of the numerator and the denominator. On the other hand,

$$g(x) = \frac{x^2 - 1}{x^2 + x - 6} = \frac{(x + 1)(x - 1)}{(x - 2)(x + 3)}$$

is simplified and has $x = 2$ and $x = -3$ as vertical asymptotes.

Figure 2 shows four ways in which a vertical asymptote can occur.

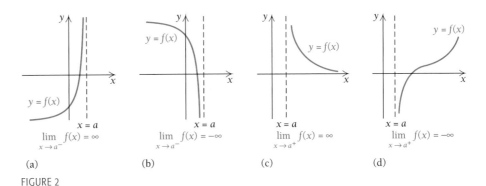

FIGURE 2

Look again at the graph in Fig. 1. Note that function values get closer and closer to 1 as x approaches $-\infty$, meaning that $f(x) \to 1$ as $x \to -\infty$. Also, function values get closer and closer to 1 as x approaches ∞, meaning that $f(x) \to 1$ as $x \to \infty$. (The graph actually crosses the horizontal asymptote $y = 1$ and then moves back to it.) Thus,

$$\lim_{x \to -\infty} f(x) = 1 \quad \text{and} \quad \lim_{x \to \infty} f(x) = 1.$$

The line $y = 1$ is called a *horizontal asymptote.*

TECHNOLOGY CONNECTION

Use the **GRAPH**, **TABLE**, and **TRACE** features of a grapher to verify that

$$\lim_{x \to \infty} \frac{x^2 - 1}{(x - 2)(x + 3)} = 1.$$

Definition

The line $y = b$ is a **horizontal asymptote** if either or both of the following limit statements is true:

$$\lim_{x \to -\infty} f(x) = b \quad \text{or} \quad \lim_{x \to \infty} f(x) = b.$$

The graph of a rational function may or may not cross a horizontal asymptote. Horizontal asymptotes occur when the degree of the numerator is less than or equal to the degree of the denominator.

In Figs. 3–5, we see three ways in which horizontal asymptotes can occur.

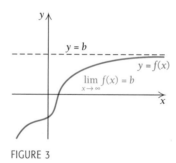

FIGURE 3

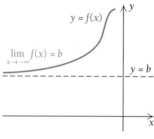

FIGURE 4

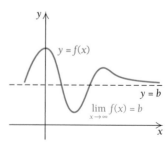

FIGURE 5

TECHNOLOGY CONNECTION

Asymptotes

Our discussion now allows us to attach the term "vertical asymptote" to those mysterious vertical lines that appear with the graphs of rational functions when the grapher is in **CONNECTED** mode. For example, consider the graph of $f(x) = 8/(x^2 - 4)$, using the viewing window $[-6, 6, -8, 8]$. Vertical asymptotes occur at $x = -2$ and $x = 2$. These lines are not part of the graph.

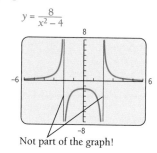

Not part of the graph!

Exercises

Graph each of the following in both **DOT** and **CONNECTED** modes. Try to locate the vertical asymptotes visually. Then verify your results using the methods of Examples 1 and 2. You may need to try different viewing windows.

1. $f(x) = \dfrac{x^2 + 7x + 10}{x^2 + 3x - 28}$

2. $f(x) = \dfrac{x^2 + 5}{x^3 - x^2 - 6x}$

Occurrences of Asymptotes

It is important in graphing rational functions to determine where the asymptotes, if any, occur. Vertical asymptotes are easy to locate when the expression is in simplified form and a denominator can be factored. The x-inputs that make a denominator 0 give us the vertical asymptotes.

Example 5   Determine the vertical asymptotes:

$$f(x) = \frac{3x - 2}{x(x - 5)(x + 3)}.$$

Solution   The expression is in simplified form. The vertical asymptotes are the lines $x = 0$, $x = 5$, and $x = -3$. ◄

Example 6   Determine the vertical asymptotes:

$$f(x) = \frac{x - 2}{x^3 - x} = \frac{x - 2}{x(x - 1)(x + 1)}.$$

Solution   The expression is in simplified form. The vertical asymptotes are the lines $x = 0$, $x = -1$, and $x = 1$. ◄

Example 7   Find the horizontal asymptotes:

$$f(x) = \frac{2x + 3}{x^3 - 2x^2 + 4}.$$

Solution   Since the degree of the numerator is less than the degree of the denominator, there is a horizontal asymptote. We then divide the numerator and the denominator by x^3 and find the limits as x approaches $-\infty$ and ∞; that is, as $|x|$ gets larger and larger:

$$f(x) = \frac{2x + 3}{x^3 - 2x^2 + 4} = \frac{\dfrac{2}{x^2} + \dfrac{3}{x^3}}{1 - \dfrac{2}{x} + \dfrac{4}{x^3}}.$$

As x gets smaller and smaller negatively, $|x|$ gets larger and larger. Similarly, as x gets larger and larger positively, $|x|$ gets larger and larger. Thus, as $|x|$ becomes very large, each expression with x or some power of x in the denominator takes on values ever closer to 0. Thus the numerator approaches 0 and the denominator approaches 1; hence the entire expression takes on values ever closer to 0. We have

$$f(x) \approx \frac{0 + 0}{1 - 0 + 0},$$

so $\displaystyle\lim_{x \to -\infty} f(x) = 0$ and $\displaystyle\lim_{x \to \infty} f(x) = 0,$

and the x-axis, the line $y = 0$, is a horizontal asymptote.   ◄

Example 8 Find the horizontal asymptotes:

$$f(x) = \frac{3x^2 + 2x - 4}{2x^2 - x + 1}.$$

Solution The numerator and the denominator have the same degree, so there is a horizontal asymptote. We divide the numerator and the denominator by x^2 and find the limit as $|x|$ gets larger and larger:

$$f(x) = \frac{3x^2 + 2x - 4}{2x^2 - x + 1} = \frac{3 + \dfrac{2}{x} - \dfrac{4}{x^2}}{2 - \dfrac{1}{x} + \dfrac{1}{x^2}}.$$

As $|x|$ gets very large, the numerator approaches 3 and the denominator approaches 2. Therefore, the function gets very close to $\frac{3}{2}$. Thus,

$$\lim_{x \to -\infty} f(x) = \frac{3}{2} \quad \text{and} \quad \lim_{x \to \infty} f(x) = \frac{3}{2}.$$

The line $y = \frac{3}{2}$ is a horizontal asymptote.

TECHNOLOGY CONNECTION

Exercises

Graph each of the following. Try to locate the horizontal asymptotes using the **TABLE** and **TRACE** features. Verify your results using the methods of Examples 7 and 8.

1. $f(x) = \dfrac{x^2 + 5}{x^3 - x^2 - 6x}$

2. $f(x) = \dfrac{9x^4 - 7x^2 - 9}{3x^4 + 7x^2 + 9}$

3. $f(x) = \dfrac{135x^5 - x^2}{x^7}$

4. $f(x) = \dfrac{3x^2 - 4x + 3}{6x^2 + 2x - 5}$

When the degree of the numerator is less than the degree of the denominator, the x-axis, or the line $y = 0$, is a horizontal asymptote.

When the degree of the numerator is the same as the degree of the denominator, the line $y = a/b$ is a horizontal asymptote, where a is the leading coefficient of the numerator and b is the leading coefficient of the denominator.

Oblique Asymptotes

There are asymptotes that are neither vertical nor horizontal. For example, in the graph of

$$f(x) = \frac{x^2 - 4}{x - 1},$$

shown at right, the line $x = 1$ is a vertical asymptote. Note that as $|x|$ gets larger and larger, the curve gets closer and closer to $y = x + 1$. The line $y = x + 1$ is called an *oblique asymptote*.

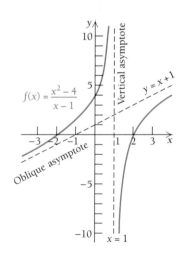

> **Definition**
>
> The line $y = mx + b$ is an **oblique asymptote** of the rational function $f(x) = P(x)/Q(x)$ if $f(x)$ can be expressed as
>
> $$f(x) = (mx + b) + g(x),$$
>
> where $g(x)$ approaches 0 as $|x|$ approaches ∞. Oblique asymptotes occur when the degree of the numerator is exactly 1 more than the degree of the denominator. A graph can cross an oblique asymptote.

How can we find an oblique asymptote? One way is by division.

TECHNOLOGY CONNECTION

Exercises

Graph each of the following. Try to visually locate the oblique asymptotes. Then use the method in Example 9 to find the oblique asymptote and graph it along with the original function.

1. $f(x) = \dfrac{3x^2 - 7x + 8}{x - 2}$

2. $f(x) = \dfrac{5x^3 + 2x + 1}{x^2 - 4}$

Example 9   Find the oblique asymptotes:

$$f(x) = \frac{x^2 - 4}{x - 1}.$$

Solution   When we divide the numerator by the denominator, we obtain a quotient of $x + 1$ and a remainder of -3. Thus,

$$f(x) = \frac{x^2 - 4}{x - 1} = (x + 1) + \frac{-3}{x - 1}.$$

$$
\begin{array}{r}
x + 1 \\
x - 1 \overline{)x^2 - 4} \\
\underline{x^2 - x} \\
x - 4 \\
\underline{x - 1} \\
-3
\end{array}
$$

Now we can see that when $|x|$ gets very large, $-3/(x - 1)$ approaches 0. Thus, for very large $|x|$, the expression $x + 1$ is the dominant part of

$$(x + 1) + \frac{-3}{x - 1}.$$

Thus, $y = x + 1$ is an oblique asymptote.

Intercepts

If they exist, the **x-intercepts** of a function occur at those values of x for which $y = f(x) = 0$ and give us points at which the graph crosses the x-axis. If it exists, the **y-intercept** of a function occurs at the value of y for which $x = 0$ and gives us the point at which the graph crosses the y-axis.

Example 10   Find the intercepts of

$$f(x) = \frac{x^3 - x^2 - 6x}{x^2 - 3x + 2}.$$

Solution   We factor the numerator and the denominator:

$$f(x) = \frac{x(x + 2)(x - 3)}{(x - 1)(x - 2)}.$$

Exercises
Graph each of the following. Use the ZERO feature and a table in ASK mode to find the *x*- and *y*-intercepts.

1. $f(x) = \dfrac{x(x - 3)(x + 5)}{(x + 2)(x - 4)}$

2. $f(x) = \dfrac{x^3 + 2x^2 - 3x}{x^2 + 5}$

To find the *x*-intercepts, we solve the equation $f(x) = 0$. Such values occur when the numerator is 0 but the denominator is not. Thus we solve the equation

$$x(x + 2)(x - 3) = 0.$$

The *x*-values making the numerator 0 are 0, -2, and 3. Since none of these makes the denominator 0, they yield the *x*-intercepts of the function: $(0, 0)$, $(-2, 0)$, and $(3, 0)$.

To find the *y*-intercept, we let $x = 0$:

$$f(0) = \frac{0^3 - 0^2 - 6(0)}{0^2 - 3(0) + 2} = 0.$$

In this case, the *y*-intercept is also an *x*-intercept, $(0, 0)$.

Sketching Graphs
We can now refine our analytic strategy for graphing.

Strategy for Sketching Graphs
a) *Intercepts.* Find the *x*-intercept(s) and the *y*-intercept of the graph.
b) *Asymptotes.* Find the vertical, horizontal, and oblique asymptotes.
c) *Derivatives.* Find $f'(x)$ and $f''(x)$.
d) *Undefined values and critical points of f.* Find the inputs for which the function is not defined, giving denominators of 0. Find also the critical points of *f*.
e) *Increasing and/or decreasing relative extrema.* Use the points found in step (d) to determine intervals over which the function *f* is increasing or decreasing. Use this information and/or the second derivative to determine the relative maxima and minima. A relative extremum can occur only at a point *c* for which $f(c)$ exists.
f) *Inflection points.* Determine candidates for inflection points by finding points *x* where $f''(x)$ does not exist or where $f''(x) = 0$. Find the function values at these points. If a function value $f(x)$ does not exist, then the function does not have an inflection point at *x*.
g) *Concavity.* Use the values *c* from step (f) as endpoints of intervals. Determine the concavity by checking to see where f' is increasing— that is, $f''(x) > 0$—and where f' is decreasing—that is, $f''(x) < 0$. Do this by selecting test points and substituting into $f''(x)$.
h) *Sketch the graph.* Use the information from steps (a) through (g) to sketch the graph, plotting extra points (computing them with your calculator) as needed.

Example 11 Sketch the graph of $f(x) = \dfrac{8}{x^2 - 4}$.

Solution

a) *Intercepts.* The x-intercepts occur at the points where the numerator is 0 but the denominator is not. Since in this case the numerator is the constant 8, there are no x-intercepts. To find the y-intercept, we compute $f(0)$:

$$f(0) = \frac{8}{0^2 - 4} = \frac{8}{-4} = -2.$$

This gives us one point on the graph, $(0, -2)$.

b) *Asymptotes.*

   Vertical: The denominator $x^2 - 4 = (x + 2)(x - 2)$. It is 0 for x-values of -2 and 2. Thus the graph has the lines $x = -2$ and $x = 2$ as vertical asymptotes. We draw them using dashed lines (they are *not* part of the actual graph).

   Horizontal: The degree of the numerator is less than the degree of the denominator, so the x-axis, or the line $y = 0$, is a horizontal asymptote. It is already drawn as an axis.

   Oblique: There is no oblique asymptote since the degree of the numerator is not 1 more than the degree of the denominator.

c) *Derivatives.* Find $f'(x)$ and $f''(x)$. Using the Quotient Rule, we get

$$f'(x) = \frac{-16x}{(x^2 - 4)^2} \quad \text{and} \quad f''(x) = \frac{16(3x^2 + 4)}{(x^2 - 4)^3}.$$

d) *Undefined values and critical points of f.* The domain of the original function is all real numbers except -2 and 2, where the vertical asymptotes occur. We find the critical points of f by looking for values of x where $f'(x) = 0$ or where $f'(x)$ does not exist. Now $f'(x) = 0$ for values of x for which $-16x = 0$, but the denominator is not 0. The only such number is 0 itself. The derivative $f'(x)$ does not exist at -2 and 2. Thus the undefined values and the critical points are -2, 0, and 2.

e) *Increasing and/or decreasing relative extrema.* Use the undefined values and the critical points to determine the intervals over which f is increasing and the intervals over which f is decreasing. The points to consider are -2, 0, and 2. These divide the real-number line into four intervals. We choose a test point in each interval and make a substitution into the derivative f':

A: Test -3, $f'(-3) = \dfrac{-16(-3)}{[(-3)^2 - 4]^2} = \dfrac{48}{25} > 0$;

B: Test -1, $f'(-1) = \dfrac{-16(-1)}{[(-1)^2 - 4]^2} = \dfrac{16}{9} > 0$;

C: Test 1, $f'(1) = \dfrac{-16(1)}{[(1)^2 - 4]^2} = \dfrac{-16}{9} < 0$;

D: Test 3, $f'(3) = \dfrac{-16(3)}{[(3)^2 - 4]^2} = \dfrac{-48}{25} < 0$.

Interval	$(-\infty, -2)$	$(-2, 0)$	$(0, 2)$	$(2, \infty)$
Test Value	$x = -3$	$x = -1$	$x = 1$	$x = 3$
Sign of $f'(x)$	$f'(-3) > 0$	$f'(-1) > 0$	$f'(1) < 0$	$f'(3) < 0$
Result	f is increasing	f is increasing	f is decreasing	f is decreasing

└── No change ──┘└── Change ──┘└── No change ──┘
indicates a
relative maximum
at 0.

Now $f(0) = -2$, so there is a relative maximum at $(0, -2)$.

f) *Inflection points.* Determine candidates for inflection points by finding where $f''(x)$ does not exist and where $f''(x) = 0$. Now $f(x)$ does not exist at -2 and 2, so $f''(x)$ does not exist at -2 and 2. We then determine where $f''(x) = 0$, or

$$16(3x^2 + 4) = 0.$$

But $16(3x^2 + 4) > 0$ for all real numbers x, so there are no points of inflection since $f(-2)$ and $f(2)$ do not exist.

g) *Concavity.* Use the values found in step (f) as endpoints of intervals. Determine the concavity by checking to see where f' is increasing and decreasing. The points -2 and 2 divide the real-number line into three intervals. We choose test points in each interval and make a substitution into f'':

A: Test -3, $f''(-3) = \dfrac{16[3(-3)^2 + 4]}{[(-3)^2 - 4]^3} > 0;$

B: Test 0, $f''(0) = \dfrac{16[3(0)^2 + 4]}{[(0)^2 - 4]^3} < 0;$

C: Test 3, $f''(3) = \dfrac{16[3(3)^2 + 4]}{[(3)^2 - 4]^3} > 0.$

TECHNOLOGY CONNECTION

Check the results of Example 11 using a grapher.

Interval	$(-\infty, -2)$	$(-2, 2)$	$(2, \infty)$
Test Value	$x = -3$	$x = 0$	$x = 3$
Sign of $f''(x)$	$f''(-3) > 0$	$f''(0) < 0$	$f''(3) > 0$
Result	f' is increasing; f is concave up.	f' is decreasing; f is concave down.	f' is increasing; f is concave up.

└── Change does ──┘└── Change does ──┘
not indicate a   not indicate a
point of inflection   point of inflection
since $f(-2)$ since $f(2)$
does not exist.   does not exist.

The function is concave up over the intervals $(-\infty, -2)$ and $(2, \infty)$. The function is concave down over the interval $(-2, 2)$.

h) *Sketch the graph.* Sketch the graph using the information in the following table, plotting extra points by computing values from your calculator as needed. The graph follows.

x	$f(x)$, approximately
-5	0.38
-4	0.67
-3	1.6
-1	-2.67
0	-2
1	-2.67
3	1.6
4	0.67
5	0.38

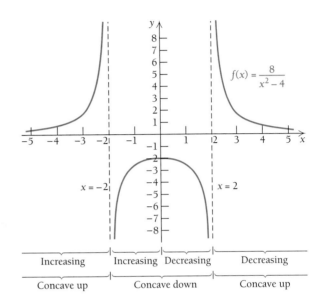

Example 12   Sketch the graph of $f(x) = \dfrac{x^2 + 4}{x}$.

Solution

a) *Intercepts.* The equation $f(x) = 0$ has no real-number solution. Thus there are no x-intercepts. The number 0 is not in the domain of the function. Thus there is no y-intercept.

b) *Asymptotes.*

Vertical: The denominator is x. Since its replacement with 0 makes the denominator 0, the line $x = 0$ is a vertical asymptote.

Horizontal: The degree of the numerator is greater than the degree of the denominator, so there are no horizontal asymptotes.

Oblique: The degree of the numerator is 1 greater than the degree of the denominator, so there is an oblique asymptote. We do the division and express the function in the form

$$f(x) = x + \frac{4}{x}. \qquad x\overline{)\begin{array}{l} x \\ x^2 + 4 \\ \underline{x^2} \\ 4 \end{array}}$$

As $|x|$ gets larger, the term $4/x$ approaches 0, so the line $y = x$ is an oblique asymptote.

c) *Derivatives.* Find $f'(x)$ and $f''(x)$:

$$f'(x) = 1 - 4x^{-2} = 1 - \frac{4}{x^2};$$

$$f''(x) = 8x^{-3} = \frac{8}{x^3}.$$

d) *Undefined values and critical points of f.* The number 0 is not in the domain of f. The derivative exists for all values of x except 0. Thus, to find critical points, we solve $f'(x) = 0$, looking for solutions other than 0:

$$1 - \frac{4}{x^2} = 0$$

$$1 = \frac{4}{x^2}$$

$$x^2 = 4$$

$$x = \pm 2.$$

Thus, -2 and 2 are critical points. Now $f(0)$ does not exist, but $f(-2) = -4$ and $f(2) = 4$. These give the points $(-2, -4)$ and $(2, 4)$ on the graph.

e) *Increasing and/or decreasing relative extrema.* Use the points found in step (d) to find intervals over which f is increasing and intervals over which f is decreasing. The points to consider are -2, 0, and 2. These divide the real-number line into four intervals. We choose test points in each interval and make a substitution into f':

A: Test -3, $f'(-3) = 1 - \dfrac{4}{(-3)^2} = \dfrac{5}{9} > 0;$

B: Test -1, $f'(-1) = 1 - \dfrac{4}{(-1)^2} = -3 < 0;$

C: Test 1, $f'(1) = 1 - \dfrac{4}{1^2} = -3 < 0;$

D: Test 3, $f'(3) = 1 - \dfrac{4}{3^2} = \dfrac{5}{9} > 0.$

Interval	$(-\infty, -2)$	$(-2, 0)$	$(0, 2)$	$(2, \infty)$
Test Value	$x = -3$	$x = -1$	$x = 1$	$x = 3$
Sign of $f'(x)$	$f'(-3) > 0$	$f'(-1) < 0$	$f'(1) < 0$	$f'(3) > 0$
Result	f is increasing	f is decreasing	f is decreasing	f is increasing

└── Change ──↑ └── No change ──↑ └── Change ──↑
indicates a relative   indicates a relative
maximum at -2.   minimum at 2.

TECHNOLOGY
CONNECTION

Check the result of
Example 12 using a
grapher.

Now $f(-2) = -4$ and $f(2) = 4$. There is a relative maximum at $(-2, -4)$ and a relative minimum at $(2, 4)$.

f) *Inflection points.* Determine candidates for inflection points by finding where $f''(x)$ does not exist or where $f''(x) = 0$. Now $f''(0)$ does not exist, but because $f(0)$ does not exist, there cannot be an inflection point at 0. Then look for values of x for which $f''(x) = 0$:

$$\frac{8}{x^3} = 0.$$

But this equation has no solution. Thus there are no points of inflection.

g) *Concavity.* Use the values found in step (f) as endpoints of intervals. Determine the concavity by checking to see where f' is increasing and decreasing. The number 0 divides the real-number line into two intervals, $(-\infty, 0)$ and $(0, \infty)$. We could choose a test value in each interval and make a substitution into f''. But note that for any $x < 0$, $x^3 < 0$, so

$$f''(x) = \frac{8}{x^3} < 0,$$

and for any $x > 0$, $x^3 > 0$, so

$$f''(x) = \frac{8}{x^3} > 0.$$

Thus, f is concave down over the interval $(-\infty, 0)$ and concave up over the interval $(0, \infty)$.

h) *Sketch the graph.* Sketch the graph using the preceding information and any additional computed values of f as needed. The graph follows.

x	$f(x)$, approximately
-6	-6.67
-5	-5.8
-4	-5
-3	-4.3
-2	-4
-1	-5
-0.5	-8.5
0.5	8.5
1	5
2	4
3	4.3
4	5
5	5.8
6	6.67

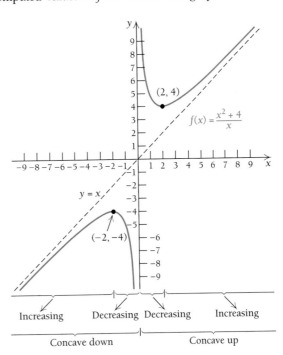

EXERCISE SET 3.3

Find the limit, if it exists.

1. $\lim\limits_{x \to \infty} \dfrac{2x - 4}{5x}$

2. $\lim\limits_{h \to 0} \dfrac{-5}{x(x + h)}$

3. $\lim\limits_{x \to \infty} \left(5 - \dfrac{2}{x}\right)$

4. $\lim\limits_{x \to \infty} \dfrac{3x + 1}{4x}$

5. $\lim\limits_{x \to \infty} \dfrac{2x - 5}{4x + 3}$

6. $\lim\limits_{x \to \infty} \left(7 + \dfrac{3}{x}\right)$

7. $\lim\limits_{x \to \infty} \dfrac{2x^2 - 5}{3x^2 - x + 7}$

8. $\lim\limits_{x \to \infty} \dfrac{6x + 1}{5x - 2}$

9. $\lim\limits_{x \to \infty} \dfrac{4 - 3x}{5 - 2x^2}$

10. $\lim\limits_{x \to \infty} \dfrac{4 - 3x - 12x^2}{1 + 5x + 3x^2}$

11. $\lim\limits_{x \to \infty} \dfrac{8x^4 - 3x^2}{5x^2 + 6x}$

12. $\lim\limits_{x \to \infty} \dfrac{6x^2 - x}{4x^4 - 3x^3}$

13. $\lim\limits_{x \to \infty} \dfrac{6x^4 - 5x^2 + 7}{8x^6 + 4x^3 - 8x}$

14. $\lim\limits_{x \to \infty} \dfrac{6x^4 - x^3}{4x^2 - 3x^3}$

15. $\lim\limits_{x \to \infty} \dfrac{11x^5 + 4x^3 - 6x + 2}{6x^3 + 5x^2 + 3x - 1}$

16. $\lim\limits_{x \to \infty} \dfrac{7x^9 - 6x^3 + 2x^2 - 10}{2x^6 + 4x^2 - x + 23}$

Sketch a graph of the function.

17. $f(x) = \dfrac{4}{x}$

18. $f(x) = -\dfrac{5}{x}$

19. $f(x) = \dfrac{-2}{x - 5}$

20. $f(x) = \dfrac{1}{x - 5}$

21. $f(x) = \dfrac{1}{x - 3}$

22. $f(x) = \dfrac{1}{x + 2}$

23. $f(x) = \dfrac{-2}{x + 5}$

24. $f(x) = \dfrac{-3}{x - 3}$

25. $f(x) = \dfrac{2x + 1}{x}$

26. $f(x) = \dfrac{3x - 1}{x}$

27. $f(x) = x + \dfrac{9}{x}$

28. $f(x) = x + \dfrac{2}{x}$

29. $f(x) = \dfrac{2}{x^2}$

30. $f(x) = \dfrac{-1}{x^2}$

31. $f(x) = \dfrac{x}{x - 3}$

32. $f(x) = \dfrac{x}{x + 2}$

33. $f(x) = \dfrac{1}{x^2 + 3}$

34. $f(x) = \dfrac{-1}{x^2 + 2}$

35. $f(x) = \dfrac{x - 1}{x + 2}$

36. $f(x) = \dfrac{x - 2}{x + 1}$

37. $f(x) = \dfrac{x^2 - 4}{x + 3}$

38. $f(x) = \dfrac{x^2 - 9}{x + 1}$

39. $f(x) = \dfrac{x - 1}{x^2 - 2x - 3}$

40. $f(x) = \dfrac{x + 2}{x^2 + 2x - 15}$

41. $f(x) = \dfrac{2x^2}{x^2 - 16}$

42. $f(x) = \dfrac{x^2 + x - 2}{2x^2 + 1}$

43. $f(x) = \dfrac{1}{x^2 - 1}$

44. $f(x) = \dfrac{10}{x^2 + 4}$

45. $f(x) = \dfrac{x^2 + 1}{x}$

46. $f(x) = \dfrac{x^3}{x^2 - 1}$

Applications
BUSINESS AND ECONOMICS

47. *Depreciation.* Suppose that the value V of a certain product decreases, or depreciates, with time t, in months, where

$$V(t) = 50 - \dfrac{25t^2}{(t + 2)^2}.$$

a) Find $V(0)$, $V(5)$, $V(10)$, and $V(70)$.
b) Find the maximum value of the product over the interval $[0, \infty)$.
c) Sketch a graph of V.
d) Find $\lim\limits_{t \to \infty} V(t)$.
 e) Does there seem to be a value below which V will never fall? Explain.

48. *Average Cost.* The total-cost function for Acme, Inc., to produce x units of a product is given by

$$C(x) = 3x^2 + 80.$$

a) The *average cost* is given by $A(x) = C(x)/x$. Find $A(x)$.
b) Graph the average cost.
c) Find the oblique asymptote for the graph of $y = A(x)$ and interpret it.

49. *Cost of Pollution Control.* Cities and companies find the cost of pollution control to increase tremendously with respect to the percentage of pollutants to be removed from a situation. Suppose

that the cost C of removing $p\%$ of the pollutants from a chemical dumping site is given by

$$C(p) = \frac{\$48,000}{100 - p}.$$

a) Find $C(0)$, $C(20)$, $C(80)$, and $C(90)$.
b) Find $\lim_{p \to 100^-} C(p)$.
tw c) Explain the meaning of the limit found in part (b).
d) Sketch a graph of C.
tw e) Can the company afford to remove 100% of the pollutants? Explain.

LIFE AND PHYSICAL SCIENCES

50. *Medication in the Bloodstream.* After an injection, the amount of a medication A in the bloodstream decreases after time t, in hours. Suppose that under certain conditions A is given by

$$A(t) = \frac{A_0}{t^2 + 1},$$

where A_0 is the initial amount of the medication given. Assume that an initial amount of 100 cc is injected.

a) Find $A(0)$, $A(1)$, $A(2)$, $A(7)$, and $A(10)$.
b) Find $\lim_{t \to \infty} A(t)$.
c) Find the maximum value of the injection over the interval $[0, \infty)$.
d) Sketch a graph of the function.
tw e) According to this function, does the medication ever completely leave the bloodstream? Explain your answer.

GENERAL INTEREST

51. *Baseball: Earned-Run Average.* A pitcher's *earned-run average* (the average number of runs given up every 9 innings, or 1 game) is given by

$$E = 9 \cdot \frac{n}{i},$$

where n is the number of earned runs allowed and i is the number of innings pitched. Suppose that we fix the number of earned runs allowed at 4 and let i vary. We get a function given by

$$E(i) = 9 \cdot \frac{4}{i}.$$

a) Complete the following table, rounding to two decimal places.

Innings Pitched (i)	Earned-Run Average (E)
9	
8	
7	
6	
5	
4	
3	
2	
1	
$\frac{2}{3}$ (2 outs)	
$\frac{1}{3}$ (1 out)	

b) Find $\lim_{i \to 0^+} E(i)$.
tw c) On the basis of parts (a) and (b), determine a pitcher's earned run average if 4 runs were allowed and there were 0 outs.

Synthesis

tw **52.** Explain why a vertical asymptote cannot be part of the graph of a function.

tw **53.** Using graphs and limits, explain the idea of an asymptote to the graph of a function. Describe three types of asymptotes.

Find the limit, if it exists.

54. $\lim\limits_{x \to -\infty} \dfrac{-3x^2 + 5}{2 - x}$ **55.** $\lim\limits_{x \to 0} \dfrac{|x|}{x}$

56. $\lim\limits_{x \to -2} \dfrac{x^3 + 8}{x^2 - 4}$ **57.** $\lim\limits_{x \to \infty} \dfrac{-6x^3 + 7x}{2x^2 - 3x - 10}$

58. $\lim\limits_{x \to -\infty} \dfrac{-6x^3 + 7x}{2x^2 - 3x - 10}$ **59.** $\lim\limits_{x \to 1} \dfrac{x^3 - 1}{x^2 - 1}$

60. $\lim\limits_{x \to -\infty} \dfrac{7x^5 + x - 9}{6x + x^3}$ **61.** $\lim\limits_{x \to -\infty} \dfrac{2x^4 + x}{x + 1}$

TECHNOLOGY CONNECTION

Graph the function.

62. $f(x) = x^2 + \dfrac{1}{x^2}$

63. $f(x) = \dfrac{x}{\sqrt{x^2 + 1}}$

64. $f(x) = \dfrac{x^3 + 4x^2 + x - 6}{x^2 - x - 2}$

65. $f(x) = \dfrac{x^3 + 2x^2 - 15x}{x^2 - 5x - 14}$

66. $f(x) = \dfrac{x^3 + 2x^2 - 3x}{x^2 - 25}$

67. $f(x) = \left| \dfrac{1}{x} - 2 \right|$

68. Graph the function

$$f(x) = \dfrac{x^2 - 3}{2x - 4}.$$

Using only the TRACE and ZOOM features:

a) Find all the x-intercepts.
b) Find the y-intercept.
c) Find all the asymptotes.

tw **69.** Graph the function

$$f(x) = \dfrac{\sqrt{x^2 + 3x + 2}}{x - 3}.$$

a) Estimate $\lim_{x \to \infty} f(x)$ and $\lim_{x \to -\infty} f(x)$ using the graph and input–output tables as needed to refine your estimates.
b) Describe the outputs of the function over the interval $(-2, -1)$.
c) What appears to be the domain of the function? Explain.
d) Find $\lim_{x \to -2^-} f(x)$ and $\lim_{x \to -1^+} f(x)$.

3.4

OBJECTIVES

➤ Find absolute extrema using Maximum–Minimum Principle 1.
➤ Find absolute extrema using Maximum–Minimum Principle 2.

Using Derivatives to Find Absolute Maximum and Minimum Values

Absolute Maximum and Minimum Values

A relative minimum may or may not be an absolute minimum, meaning the smallest value of the function over its entire domain. Similarly, a relative maximum may or may not be an absolute maximum, meaning the greatest value of a function over its entire domain.

The function in the following graph has relative minima at interior points c_1 and c_3 of the closed interval $[a, b]$.

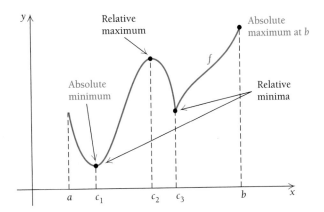

The relative minimum at c_1 also happens to be an absolute minimum. The function has a relative maximum at c_2 but it is not an absolute maximum. The absolute maximum occurs at the endpoint b.

Definition

Suppose that f is a function whose value $f(c)$ exists at input c in an interval I in the domain of f. Then:

$f(c)$ is an **absolute minimum** if $f(c) \leq f(x)$ for all x in I.

$f(c)$ is an **absolute maximum** if $f(x) \leq f(c)$ for all x in I.

Finding Absolute Maximum and Minimum Values Over Closed Intervals

We first consider a continuous function over a closed interval. To do so, look at these graphs and try to determine the points over the closed interval at which the absolute maxima and minima (extrema) occur.

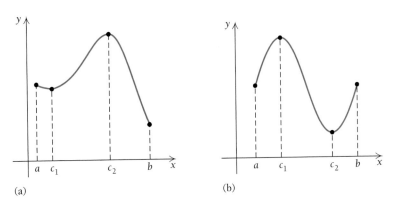

(a)   (b)

You may have discovered two theorems. Each of the functions did indeed have an absolute maximum value and an absolute minimum value. This leads us to one of the theorems.

Theorem 7

The Extreme-Value Theorem

A continuous function f defined over a closed interval $[a, b]$ must have an absolute maximum value and an absolute minimum value at points in $[a, b]$.

Look carefully at the preceding graphs and consider the critical points and the endpoints. In part (a), the graph starts at $f(a)$ and falls to $f(c_1)$. Then it rises from $f(c_1)$ to $f(c_2)$. From there it falls to $f(b)$. In part (b), the graph starts at $f(a)$ and rises to $f(c_1)$. Then it falls from $f(c_1)$ to $f(c_2)$. From there it rises to $f(b)$. It seems reasonable that whatever the maximum and minimum values are, they occur among the function values $f(a), f(c_1), f(c_2)$, and $f(b)$. This leads us to a procedure for determining *absolute extrema*.

Theorem 8

Maximum–Minimum Principle 1

Suppose that f is a continuous function over a closed interval $[a, b]$. To find the absolute maximum and minimum values of the function over $[a, b]$:

a) First find $f'(x)$.

b) Then determine the critical points of f in $[a, b]$. That is, find all points c for which

$$f'(c) = 0 \quad \text{or} \quad f'(c) \text{ does not exist.}$$

c) List the critical points of f and the endpoints of the interval:

$$a, c_1, c_2, \ldots, c_n, b.$$

d) Find the function values at the points in part (c):

$$f(a), f(c_1), f(c_2), \ldots, f(c_n), f(b).$$

The largest of these is the **absolute maximum** of f over the interval $[a, b]$. The smallest of these is the **absolute minimum** of f over the interval $[a, b]$.

Example 1   Find the absolute maximum and minimum values of

$$f(x) = x^3 - 3x + 2$$

over the interval $\left[-2, \frac{3}{2}\right]$.

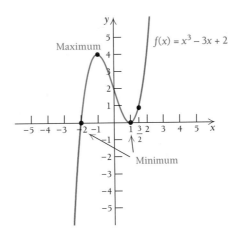

Solution   In each of Examples 1–6 of this section, we show the related graph (as here) so that you can see the absolute extrema. The procedures we use do not require the drawing of a graph. Keep in mind that we are considering only the interval $\left[-2, \frac{3}{2}\right]$.

a) Find $f'(x)$:

$$f'(x) = 3x^2 - 3.$$

b) Find the critical points. The derivative exists for all real numbers. Thus we merely solve $f'(x) = 0$:

$$3x^2 - 3 = 0$$
$$3x^2 = 3$$
$$x^2 = 1$$
$$x = \pm 1.$$

c) List the critical points and the endpoints. These points are -2, -1, 1, and $\frac{3}{2}$.

d) Find the function values at the points in step (c):

$$f(-2) = (-2)^3 - 3(-2) + 2 = -8 + 6 + 2 = 0; \longrightarrow \text{Minimum}$$
$$f(-1) = (-1)^3 - 3(-1) + 2 = -1 + 3 + 2 = 4; \longrightarrow \text{Maximum}$$
$$f(1) = (1)^3 - 3(1) + 2 = 1 - 3 + 2 = 0; \longrightarrow \text{Minimum}$$
$$f\left(\tfrac{3}{2}\right) = \left(\tfrac{3}{2}\right)^3 - 3\left(\tfrac{3}{2}\right) + 2 = \tfrac{27}{8} - \tfrac{9}{2} + 2 = \tfrac{7}{8}.$$

The largest of these values, 4, is the maximum. It occurs at $x = -1$. The smallest of these values is 0. It occurs twice: at $x = -2$ and $x = 1$. Thus over the interval $\left[-2, \frac{3}{2}\right]$ the

$$\text{absolute maximum} = 4 \text{ at } x = -1$$

and the

$$\text{absolute minimum} = 0 \text{ at } x = -2 \text{ and } x = 1.$$

Note that an absolute maximum or minimum value can occur at more than one point.

TECHNOLOGY CONNECTION

Finding Absolute Extrema

How can we use a grapher to find absolute extrema? Let's first consider Example 1. We can use any of the methods described in the Technology Connection on pp. 195–196. In this case, we adapt Methods 3 and 4.

Method 3

Method 3 is selected because there are relative extrema in the interval $\left[-2, \frac{3}{2}\right]$. This method gives us approximations for the relative extrema.

$f(x) = x^2 - 3x + 2$

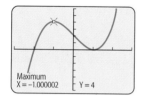

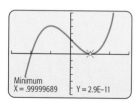

Then we check function values at these *x*-values and at the endpoints, using Maximum–Minimum Principle 1 to determine the absolute maximum and minimum values:

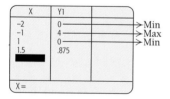

Method 4

Example 2 considers the same function as in Example 1, but over a different interval. This time we use fMax and fMin features from the MATH menu. Because there are no relative extrema, we can use fMax and fMin. The minimum and maximum values occur at the endpoints.

fMin(Y1,X,–3,–1.5)
     –2.999994692
Y1(–2.999994692)
     –15.99987261

fMax(Y1,X,–3,–1.5)
     –1.500005458
Y1(–1.500005458)
     3.124979532

Exercise

1. Use a grapher to estimate the absolute maximum and minimum values of $f(x) = x^3 - x^2 - x + 2$ first over the interval $[-2, 1]$ and then over the interval $[-1, 2]$. Then check your work using the methods of Examples 1 and 2.

Example 2   Find the absolute maximum and minimum values of

$$f(x) = x^3 - 3x + 2$$

over the interval $\left[-3, -\frac{3}{2}\right]$.

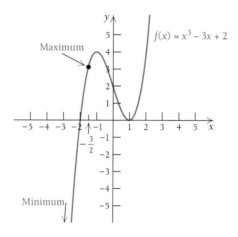

Solution   As in Example 1, the derivative is 0 at -1 and 1. But neither -1 nor 1 is in the interval $\left[-3, -\frac{3}{2}\right]$, so there are no critical points in this interval. Thus the maximum and minimum values occur at the endpoints:

$$f(-3) = (-3)^3 - 3(-3) + 2$$
$$= -27 + 9 + 2 = -16; \longrightarrow \text{Minimum}$$
$$f\left(-\frac{3}{2}\right) = \left(-\frac{3}{2}\right)^3 - 3\left(-\frac{3}{2}\right) + 2$$
$$= -\frac{27}{8} + \frac{9}{2} + 2 = \frac{25}{8} = 3\frac{1}{8}. \longrightarrow \text{Maximum}$$

Thus, over the interval $\left[-3, -\frac{3}{2}\right]$, the

$$\text{absolute maximum} = 3\frac{1}{8} \text{ at } x = -\frac{3}{2}$$

and the

$$\text{absolute minimum} = -16 \text{ at } x = -3.$$

Finding Absolute Maximum and Minimum Values Over Other Intervals

When there is only one critical point c in I, we may not need to check endpoint values to determine whether the function has an absolute maximum or minimum value at that point.

> **Theorem 9**
>
> Maximum–Minimum Principle 2
>
> Suppose that f is a function such that $f'(x)$ exists for every x in an interval I, and that there is *exactly one* (critical) point c, interior to I, for which $f'(c) = 0$. Then
>
> $$f(c) \text{ is the absolute maximum value over } I \text{ if } f''(c) < 0$$
>
> or
>
> $$f(c) \text{ is the absolute minimum value over } I \text{ if } f''(c) > 0.$$

This theorem holds no matter what the interval I is—whether open, closed, or extending to infinity. If $f''(c) = 0$, either we must use Maximum–Minimum Principle 1 or we must know more about the behavior of the function over the given interval.

TECHNOLOGY CONNECTION

Finding Absolute Extrema

Let's do Example 3 using a grapher. Again we adapt the methods of the Technology Connection on pp. 195–196. Strictly speaking, we cannot use the fMin or fMax features or the **MAXIMUM** or **MINIMUM** features from the **CALC** menu since we do not have a closed interval.

Methods 1 and 2

We create a graph, examine its shape, and use the **TRACE** and/or **TABLE** features. This procedure leads us to see that there is indeed no absolute minimum. We do get an absolute maximum $f(x) = 4$ at $x = 2$.

Exercise

1. Use a grapher to estimate the absolute maximum and minimum values of $f(x) = x^2 - 4x$. Then check your work using the method of Example 3.

Example 3   Find the absolute maximum and minimum values of

$$f(x) = 4x - x^2.$$

Solution   When no interval is specified, we consider the entire domain of the function. In this case, the domain is the set of all real numbers.

a) Find $f'(x)$:

$$f'(x) = 4 - 2x.$$

b) Find the critical points. The derivative exists for all real numbers. Thus we merely solve $f'(x) = 0$:

$$4 - 2x = 0$$
$$-2x = -4$$
$$x = 2.$$

c) Since there is only one critical point, we can apply Maximum–Minimum Principle 2 using the second derivative:

$$f''(x) = -2.$$

The second derivative is constant. Thus, $f''(2) = -2$, and since this is negative, we have the

$$\text{absolute maximum} = f(2) = 4 \cdot 2 - 2^2$$
$$= 8 - 4 = 4 \text{ at } x = 2.$$

The function has no minimum, as the graph at the top of the following page indicates.

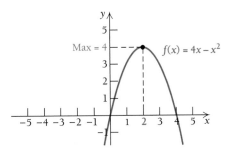

Example 4 Find the absolute maximum and minimum values of $f(x) = 4x - x^2$ over the interval $[0, 4]$.

Solution By the reasoning in Example 3, we know that the absolute maximum of f on $(-\infty, \infty)$ is $f(2)$, or 4. Since 2 is in the interval $[0, 4]$, we know that the absolute maximum of f over $[0, 4]$ will occur at 2. To find the absolute minimum, we need to check the endpoints:

$$f(0) = 4 \cdot 0 - 0^2 = 0$$

and

$$f(4) = 4 \cdot 4 - 4^2 = 0.$$

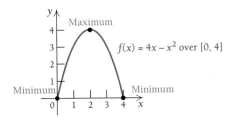

We see in the figure that the minimum is 0. It occurs twice, at $x = 0$ and $x = 4$. Thus the

$$\text{absolute maximum} = 4 \text{ at } x = 2$$

and the

$$\text{absolute minimum} = 0 \text{ at } x = 0 \text{ and } x = 4.$$

We have thus far restricted the use of Maximum–Minimum Principle 2 to intervals with one critical point. Suppose that a closed interval contains two critical points. Then we could break the interval up into two subintervals, consider maximum and minimum values over those subintervals, and compare. But we would need to consider values at the endpoints, and since we would, in effect, be using Maximum–Minimum Principle 1, we may as well use it at the outset.

A Strategy for Finding Maximum and Minimum Values

The following general strategy can be used when finding maximum and minimum values of continuous functions.

A Strategy for Finding Absolute Maximum and Minimum Values

To find absolute maximum and minimum values of a continuous function over an interval:

a) Find $f'(x)$.

b) Find the critical points.

c) If the interval is closed and there is more than one critical point, use Maximum–Minimum Principle 1.

d) If the interval is closed and there is exactly one critical point, use either Maximum–Minimum Principle 1 or Maximum–Minimum Principle 2. If the function is easy to differentiate, use Maximum–Minimum Principle 2.

e) If the interval is not closed, does not have endpoints, or does not contain its endpoints, such as $(-\infty, \infty)$, $(0, \infty)$, or (a, b), and the function has only one critical point, use Maximum–Minimum Principle 2. In such a case, if the function has a maximum, it will have no minimum; and if it has a minimum, it will have no maximum.

The case of finding absolute maximum and minimum values when more than one critical point occurs in an interval described in step (e) above must be dealt with by a detailed graph or by techniques beyond the scope of this book.

Example 5   Find the absolute maximum and minimum values of

$$f(x) = (x - 2)^3 + 1.$$

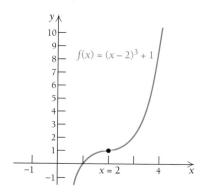

Solution

a) Find $f'(x)$:

$$f'(x) = 3(x - 2)^2.$$

b) Find the critical points. The derivative exists for all real numbers. Thus we solve $f'(x) = 0$:

$$3(x - 2)^2 = 0$$
$$(x - 2)^2 = 0$$
$$x - 2 = 0$$
$$x = 2.$$

c) Since there is only one critical point and there are no endpoints, we can try to apply Maximum–Minimum Principle 2 using the second derivative:

$$f''(x) = 6(x - 2).$$

Now

$$f''(2) = 6(2 - 2) = 0,$$

so Maximum–Minimum Principle 2 fails. We cannot use Maximum–Minimum Principle 1 because there are no endpoints. But note that $f'(x) = 3(x - 2)^2$ is never negative. Thus, $f(x)$ is increasing everywhere except at $x = 2$, so there is no maximum and no minimum. For $x < 2$, $x - 2 < 0$, so $f''(x) = 6(x - 2) < 0$. Similarly, for $x > 2$, $x - 2 > 0$, so $f''(x) = 6(x - 2) > 0$. Thus, at $x = 2$, the function has a *point of inflection*.   ◄

Example 6   Find the absolute maximum and minimum values of

$$f(x) = 5x + \frac{35}{x}$$

over the interval $(0, \infty)$.

Solution

a) Find $f'(x)$. We first express $f(x)$ as

$$f(x) = 5x + 35x^{-1}.$$

Then

$$f'(x) = 5 - 35x^{-2}$$

$$= 5 - \frac{35}{x^2}.$$

b) Find the critical points. Now $f'(x)$ exists for all values of x in $(0, \infty)$. Thus the only critical points are those for which $f'(x) = 0$:

$$5 - \frac{35}{x^2} = 0$$

$$5 = \frac{35}{x^2}$$

$$5x^2 = 35 \qquad \text{Multiplying by } x^2, \text{ since } x \neq 0$$

$$x^2 = 7$$

$$x = \pm\sqrt{7} \approx \pm 2.646.$$

e) The interval is not closed and is $(0, \infty)$. The only critical point is $\sqrt{7}$. Therefore, we can apply Maximum–Minimum Principle 2 using the second derivative,

$$f''(x) = 70x^{-3}$$

$$= \frac{70}{x^3},$$

to determine whether we have a maximum or a minimum. Now $f''(x)$ is positive for all values of x in $(0, \infty)$, so $f''(\sqrt{7}) > 0$, and the

$$\text{absolute minimum} = f(\sqrt{7})$$

$$= 5 \cdot \sqrt{7} + \frac{35}{\sqrt{7}}$$

$$= 5\sqrt{7} + \frac{35}{\sqrt{7}} \cdot \frac{\sqrt{7}}{\sqrt{7}}$$

$$= 5\sqrt{7} + \frac{35\sqrt{7}}{7}$$

$$= 5\sqrt{7} + 5\sqrt{7}$$

$$= 10\sqrt{7} \approx 26.458$$

at $x = \sqrt{7}$.

The function has no maximum value.

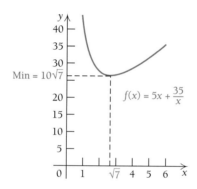

TECHNOLOGY CONNECTION

Finding Absolute Extrema

Let's do Example 6 using the **MAXIMUM** and **MINIMUM** features from the **CALC** menu. The shape of the graph leads us to see that there is no absolute maximum, but there is an absolute minimum.

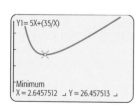

[0, 10, 0, 50]

Note that

$$\sqrt{7} \approx 2.65 \quad \text{and} \quad 10\sqrt{7} \approx 26.458,$$

which confirms the analytic solution.

Exercise

1. Use a grapher to estimate the absolute maximum and minimum values of $f(x) = 10x + 1/x$ over the interval $(0, \infty)$. Then check your work using the analytic method of Example 6.

EXERCISE   SET   3.4

Gasoline Mileage. The curves on the graph below show the gasoline mileage obtained when traveling at a constant speed for an average-size car and for a compact car.

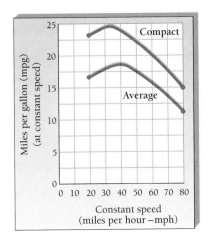

1. Consider the graph for the average-size car over the interval [20, 80].

   a) Estimate the speed at which the absolute maximum gasoline mileage is obtained.
   b) Estimate the speed at which the absolute minimum gasoline mileage is obtained.
   c) What is the mileage obtained at 70 mph?
   d) What is the mileage obtained at 55 mph?
   e) What percent increase in mileage is there by traveling at 55 mph rather than at 70 mph?

2. Answer the questions in Exercise 1 for a compact car.

Find the absolute maximum and minimum values of the function, if they exist, over the indicated interval.

3. $f(x) = 5 + x - x^2$; [0, 2]

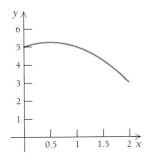

4. $f(x) = 4 + x - x^2$; [0, 2]

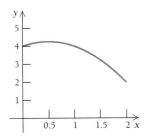

5. $f(x) = x^3 - x^2 - x + 2$; [0, 2]

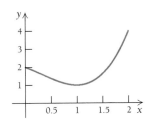

6. $f(x) = x^3 + \frac{1}{2}x^2 - 2x + 5$; [0, 1]

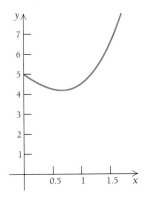

7. $f(x) = x^3 - x^2 - x + 2$; [-1, 0]
8. $f(x) = x^3 + \frac{1}{2}x^2 - 2x + 5$; [-2, 0]
9. $f(x) = 3x - 2$; [-1, 1]
10. $f(x) = 2x + 4$; [-1, 1]
11. $f(x) = 7 - 4x$; [-2, 5]
12. $f(x) = -2 + 8x$; [-10, 10]
13. $f(x) = -5$; [-1, 1]
14. $g(x) = 24$; [4, 13]
15. $f(x) = x^2 - 6x - 3$; [-1, 5]
16. $f(x) = x^2 - 4x + 5$; [-1, 3]

17. $f(x) = 3 - 2x - 5x^2$; $[-3, 3]$
18. $f(x) = 1 + 6x - 3x^2$; $[0, 4]$
19. $f(x) = x^3 - 3x^2$; $[0, 5]$
20. $f(x) = x^3 - 3x + 6$; $[-1, 3]$
21. $f(x) = x^3 - 3x$; $[-5, 1]$
22. $f(x) = 3x^2 - 2x^3$; $[-5, 1]$
23. $f(x) = 1 - x^3$; $[-8, 8]$
24. $f(x) = 2x^3$; $[-10, 10]$
25. $f(x) = 12 + 9x - 3x^2 - x^3$; $[-3, 1]$
26. $f(x) = x^3 - 6x^2 + 10$; $[0, 4]$
27. $f(x) = x^4 - 2x^3$; $[-2, 2]$
28. $f(x) = x^3 - x^4$; $[-1, 1]$
29. $f(x) = x^4 - 2x^2 + 5$; $[-2, 2]$
30. $f(x) = x^4 - 8x^2 + 3$; $[-3, 3]$
31. $f(x) = (x + 3)^{2/3} - 5$; $[-4, 5]$
32. $f(x) = 1 - x^{2/3}$; $[-8, 8]$

33. $f(x) = x + \dfrac{1}{x}$; $[1, 20]$

34. $f(x) = x + \dfrac{4}{x}$; $[-8, -1]$

35. $f(x) = \dfrac{x^2}{x^2 + 1}$; $[-2, 2]$

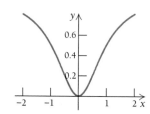

36. $f(x) = \dfrac{4x}{x^2 + 1}$; $[-3, 3]$

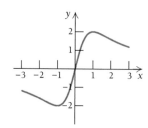

37. $f(x) = (x + 1)^{1/3}$; $[-2, 26]$
38. $f(x) = \sqrt[3]{x}$; $[8, 64]$

39.–48. Check Exercises 3, 5, 9, 13, 19, 23, 33, 35, 37, and 38 using a grapher.

Find the absolute maximum and minimum values of the function, if they exist, over the indicated interval. When no interval is specified, use the real line $(-\infty, \infty)$.

49. $f(x) = x(70 - x)$
50. $f(x) = x(50 - x)$
51. $f(x) = 2x^2 - 40x + 400$
52. $f(x) = 2x^2 - 20x + 100$
53. $f(x) = x - \frac{4}{3}x^3$; $(0, \infty)$
54. $f(x) = 16x - \frac{4}{3}x^3$; $(0, \infty)$
55. $f(x) = 17x - x^2$
56. $f(x) = 27x - x^2$
57. $f(x) = \frac{1}{3}x^3 - 3x$; $[-2, 2]$
58. $f(x) = \frac{1}{3}x^3 - 5x$; $[-3, 3]$
59. $f(x) = -0.001x^2 + 4.8x - 60$
60. $f(x) = -0.01x^2 + 1.4x - 30$
61. $f(x) = -\frac{1}{3}x^3 + 6x^2 - 11x - 50$; $(0, 3)$
62. $f(x) = -x^3 + x^2 + 5x - 1$; $(0, \infty)$
63. $f(x) = 15x^2 - \frac{1}{2}x^3$; $[0, 30]$
64. $f(x) = 4x^2 - \frac{1}{2}x^3$; $[0, 8]$

65. $f(x) = 2x + \dfrac{72}{x}$; $(0, \infty)$

66. $f(x) = x + \dfrac{3600}{x}$; $(0, \infty)$

67. $f(x) = x^2 + \dfrac{432}{x}$; $(0, \infty)$

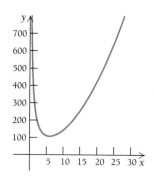

68. $f(x) = x^2 + \dfrac{250}{x}$; $(0, \infty)$

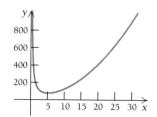

69. $f(x) = 2x^4 - x$; $[-1, 1]$
70. $f(x) = 2x^4 + x$; $[-1, 1]$
71. $f(x) = \sqrt[3]{x}$; $[0, 8]$
72. $f(x) = \sqrt{x}$; $[0, 4]$
73. $f(x) = (x + 1)^3$
74. $f(x) = (x - 1)^3$
75. $f(x) = 2x - 3$; $[-1, 1]$
76. $f(x) = 9 - 5x$; $[-10, 10]$
77. $f(x) = 2x - 3$
78. $f(x) = 9 - 5x$
79. $f(x) = x^{2/3}$; $[-1, 1]$
80. $g(x) = x^{2/3}$
81. $f(x) = \frac{1}{3}x^3 - x + \frac{2}{3}$
82. $f(x) = \frac{1}{3}x^3 - \frac{1}{2}x^2 - 2x + 1$
83. $f(x) = \frac{1}{3}x^3 - 2x^2 + x$; $[0, 4]$
84. $g(x) = \frac{1}{3}x^3 + 2x^2 + x$; $[-4, 0]$
85. $t(x) = x^4 - 2x^2$

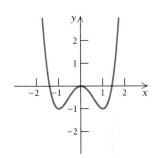

86. $f(x) = 2x^4 - 4x^2 + 2$

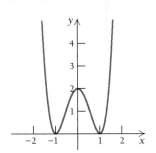

87.–96. Check Exercises 49, 51, 53, 57, 61, 65, 67, 69, 73, and 85 using a grapher.

Applications ....

BUSINESS AND ECONOMICS

97. *Monthly Productivity.* An employee's monthly productivity M, in number of units produced, is found to be a function of the number t of years of service. For a certain product, a productivity function is given by

$$M(t) = -2t^2 + 100t + 180, \quad 0 \le t \le 40.$$

Find the maximum productivity and the year in which it is achieved.

98. *Advertising.* A firm estimates that it will sell N units of a product after spending a dollars on advertising, where

$$N(a) = -a^2 + 300a + 6, \quad 0 \le a \le 300,$$

and a is in thousands of dollars. Find the maximum number of units that can be sold and the amount that must be spent on advertising in order to achieve that maximum.

99. *Maximizing Profit.* A firm determines that its total profit in dollars from the production and sale of x units of a product is given by

$$P(x) = \frac{1500}{x^2 - 6x + 10}.$$

Find the number of units x for which the total profit is a maximum.

LIFE AND PHYSICAL SCIENCES

100. *Blood Pressure.* For a dosage of x cubic centimeters (cc) of a certain drug, the resulting blood pressure B is approximated by

$$B(x) = 0.05x^2 - 0.3x^3, \quad 0 \le x \le 0.16.$$

Find the maximum blood pressure and the dosage at which it occurs.

101. *Minimizing Automobile Accidents.* At travel speed (constant velocity) x, in miles per hour, there are y accidents at nighttime for every 100 million miles of travel, where y is given by

$$y = -6.1x^2 + 752x + 22{,}620.$$

At what travel speed does the greatest number of accidents occur?

102. *Temperature During an Illness.* The temperature T of a person during an illness is given by

$$T(t) = -0.1t^2 + 1.2t + 98.6, \quad 0 \le t \le 12,$$

where T is the temperature (°F) at time t, in days. Find the maximum value of the temperature and when it occurs.

Synthesis ....

Find the absolute maximum and minimum values of the function, if they exist, over the indicated interval.

103. $g(x) = x\sqrt{x + 3};$ $[-3, 3]$

104. $h(x) = x\sqrt{1 - x};$ $[0, 1]$

105. *Business: Total Cost.* Several costs in a business environment can be separated into two components: those that increase with volume and those that decrease with volume. Although the quality of customer service becomes more expensive as it is increased, part of the increased cost is offset by customer goodwill. A firm has determined that its cost of service is given by the following function of "quality units,"

$$C(x) = (2x + 4) + \left(\frac{2}{x - 6}\right), \quad x > 6.$$

Find the number of "quality units" that the firm should use in order to minimize its total cost of service.

106. Let

$$y = (x - a)^2 + (x - b)^2.$$

For what value of x is y a minimum?

 107. Explain the usefulness of the first derivative in finding the absolute extrema of a function.

 108. Explain the usefulness of the second derivative in finding the absolute extrema of a function.

TECHNOLOGY   CONNECTION

Use a grapher to graph each function over the given interval. Visually estimate where absolute maximum and minimum values occur. Then use the **TABLE** feature to refine your estimate.

109. $f(x) = x^4 - 4x^3 + 10;$ $[0, 4]$

110. $f(x) = x^{2/3}(x - 5);$ $[1, 4]$

111. $f(x) = \frac{3}{4}(x^2 - 1)^{2/3}$

112. $f(x) = x\left(\frac{x}{2} - 5\right)^4$

113. *Life and Physical Sciences: Contractions During Pregnancy.* The following table and graph give the size of a pregnant woman's contractions as a function of time.

Time, t (in minutes)	Pressure (in millimeters of mercury)
0	10
1	8
2	9.5
3	15
4	12
5	14
6	14.5

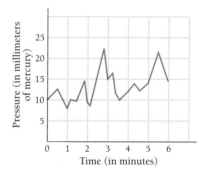

Use a grapher that has a **REGRESSION** feature.

a) Fit a linear equation to the data. Predict the size of the contraction after 7 min.

b) Fit a quartic polynomial to the data. Predict the size of the contraction after 7 min. Find the smallest contraction over the interval $[0, 10]$.

 114. *Business and Economics: Diet Cola Consumption.* The data in the following table show the rate of consumption of diet cola over the period from 1991 to 1995.

Years after 1990	Consumption of Diet Cola (in gallons per person)
1	11.7
2	11.6
3	11.9
4	12.3
5	12.3
6	?
7	?
8	?

Source: U.S. Department of Agriculture Economic Research Service

a) Use the REGRESSION feature on a grapher to fit linear, quadratic, cubic, and quartic polynomials to the data.
b) Decide which type of function best fits the data and explain your reasons.
c) Explain your reasons why a cubic might be a better fit than a quartic.
d) Find the absolute maximum and minimum values of the "best-fit" function over the interval [0, 6]; over the interval [0, 7].
e) Predict the consumption of diet cola in the years 1996, 1997, 1998, 1999, and 2000.
f) How might you verify your predictions? How might you use the new data to find another "best-fit" function?

3.5

Maximum–Minimum Problems; Business and Economics Applications

One very important application of differential calculus is the solving of maximum–minimum problems, that is, finding the absolute maximum or minimum value of some varying quantity Q and the point at which that maximum or minimum occurs.

Example 1 *Maximizing Area.* A hobby store has 20 ft of fencing to fence off a rectangular area for an electric train in one corner of its display room. The two sides up against the wall require no fence. What dimensions of the rectangle will maximize the area? What is the maximum area?

Solution At first glance, we might think that it does not matter what dimensions we use: They will all yield the same area. This is not the case. Let's first make a drawing and express the area in terms of one variable. If we let $x =$ the length of one side and $y =$ the length of the other, then since the sum of the lengths must be 20 ft, we have

$$x + y = 20 \quad \text{and} \quad y = 20 - x.$$

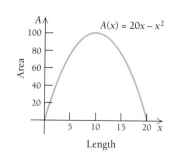

Thus the area is given by

$$A = xy$$
$$= x(20 - x)$$
$$= 20x - x^2.$$

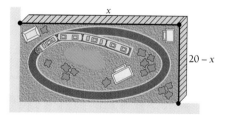

We are trying to find the maximum value of

$$A(x) = 20x - x^2 \quad \text{over the interval} \quad (0, 20).$$

We consider the interval $(0, 20)$ because x is the length of one side and cannot be negative or 0. Since there is only 20 ft of fencing, x cannot be greater than 20. Also, x cannot be 20 because then the length of y would be 0.

TECHNOLOGY CONNECTION

Exploratory Exercises

1. Complete this table using a grapher.

x	$y = 20 - x$	$A = x(20 - x)$
0		
4		
6.5		
8		
10		
12		
13.2		
20		

2. Graph $A(x) = x(20 - x)$ over the interval $[0, 20]$.
3. Estimate a maximum value and where it would occur.

a) We first find $A'(x)$:

$$A'(x) = 20 - 2x.$$

b) This derivative exists for all values of x in $(0, 20)$. Thus the only critical points are where

$$A'(x) = 20 - 2x = 0$$
$$-2x = -20$$
$$x = 10.$$

Since there is only one critical point in the interval, we can use the second derivative to determine whether we have a maximum. Note that

$$A''(x) = -2,$$

which is a constant. Thus, $A''(10)$ is negative, so $A(10)$ is a maximum. Now

$$A(10) = 10(20 - 10)$$
$$= 10 \cdot 10$$
$$= 100.$$

Thus the maximum area of 100 ft^2 is obtained using 10 ft for the length of one side and $20 - 10$, or 10 ft for the other. Note that $A(5) = 75$, $A(16) = 64$, and $A(12) = 96$; so length does affect area. ◄

Here is a general strategy for solving maximum–minimum problems. Although it may not guarantee success, it should certainly improve your chances.

A Strategy for Solving Maximum–Minimum Problems

1. Read the problem carefully. If relevant, make a drawing.
2. Label the picture with appropriate variables and constants, noting what varies and what stays fixed.
3. Translate the problem to an equation involving a quantity Q to be maximized or minimized. Try to represent Q in terms of the variables of step (2).
4. Try to express Q as a function of *one* variable. Use the procedures developed in Sections 3.1, 3.2, and 3.4 to determine the maximum or minimum values and the points at which they occur.

Example 2 *Maximizing Volume.* From a thin piece of cardboard 8 in. by 8 in., square corners are cut out so that the sides can be folded up to make a box. What dimensions will yield a box of maximum volume? What is the maximum volume?

Solution   We might again think at first that it does not matter what the dimensions are, but our experience with Example 1 should lead us to think otherwise. We make a drawing.

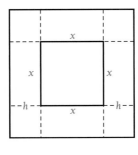 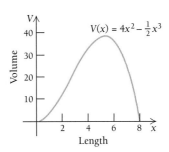

When squares of length h on a side are cut out of the corners, we are left with a square base with sides of length x. The volume of the resulting box is

$$V = lwh = x \cdot x \cdot h.$$

We want to express V in terms of one variable. Note that the overall length of a side of the cardboard is 8 in. We see from the figure that

$$h + x + h = 8,$$

or

$$x + 2h = 8.$$

Solving for h, we get

$$2h = 8 - x$$
$$h = \tfrac{1}{2}(8 - x)$$
$$= \tfrac{1}{2} \cdot 8 - \tfrac{1}{2}x$$
$$= 4 - \tfrac{1}{2}x.$$

Thus,

$$V = x \cdot x \cdot \left(4 - \tfrac{1}{2}x\right)$$
$$= x^2\left(4 - \tfrac{1}{2}x\right)$$
$$= 4x^2 - \tfrac{1}{2}x^3.$$

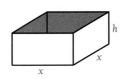

We are trying to find the maximum value of

$$V(x) = 4x^2 - \tfrac{1}{2}x^3 \quad \text{over the interval} \quad (0, 8).$$

TECHNOLOGY
CONNECTION

Exploratory Exercises

1. Complete this table using a grapher.

x	$h = 4 - \frac{1}{2}x$	$V = 4x^2 - \frac{1}{2}x^3$
0		
1		
2		
3		
4		
4.6		
5		
6		
6.8		
7		
8		

2. Graph $V(x) = 4x^2 - \frac{1}{2}x^3$ over the interval $[0, 8]$.

3. Estimate a maximum value and where it would occur.

We first find $V'(x)$:

$$V'(x) = 8x - \frac{3}{2}x^2.$$

Now $V'(x)$ exists for all x in the interval $(0, 8)$, so we set it equal to 0 to find the critical values:

$$V'(x) = 8x - \frac{3}{2}x^2 = 0$$
$$x\left(8 - \frac{3}{2}x\right) = 0$$
$$x = 0 \quad or \quad 8 - \frac{3}{2}x = 0$$
$$x = 0 \quad or \quad -\frac{3}{2}x = -8$$
$$x = 0 \quad or \quad x = -\frac{2}{3}(-8) = \frac{16}{3}.$$

The only critical point in $(0, 8)$ is $\frac{16}{3}$. Thus we can use the second derivative,

$$V''(x) = 8 - 3x,$$

to determine whether we have a maximum. Since

$$V''\left(\frac{16}{3}\right) = 8 - 3 \cdot \frac{16}{3}$$
$$= -8,$$

$V''\left(\frac{16}{3}\right)$ is negative, so $V\left(\frac{16}{3}\right)$ is a maximum, and

$$V\left(\frac{16}{3}\right) = 4 \cdot \left(\frac{16}{3}\right)^2 - \frac{1}{2}\left(\frac{16}{3}\right)^3$$
$$= \frac{1024}{27} = 37\frac{25}{27}.$$

The maximum volume is $37\frac{25}{27}$ in³. The dimensions that yield this maximum volume are

$$x = \frac{16}{3} = 5\frac{1}{3} \text{ in.}, \quad by \ x = 5\frac{1}{3} \text{ in.},$$
$$by \ h = 4 - \frac{1}{2}\left(\frac{16}{3}\right) = 1\frac{1}{3} \text{ in.} \qquad \blacktriangleleft$$

In the following problem, an open-top container of fixed volume is to be constructed. We want to determine the dimensions that will allow it to be built with the least amount of material. Such a problem could be important from an ecological standpoint.

Example 3 *Minimizing Surface Area.* A container firm is designing an open-top rectangular box, with a square base, that will hold 108 cubic centimeters (cc). What dimensions yield the minimum surface area? What is the minimum surface area?

Solution We first make a drawing. The surface area of the box is given by the area of the base plus the area of the four sides, or

$$S = x^2 + 4xy.$$

The volume must be 108 cc, and is given by

$$V = x^2y = 108.$$

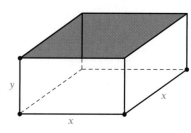

To express S in terms of one variable, we solve $x^2 y = 108$ for y:

$$y = \frac{108}{x^2}.$$

Then

$$S(x) = x^2 + 4x\left(\frac{108}{x^2}\right)$$

$$= x^2 + \frac{432}{x}.$$

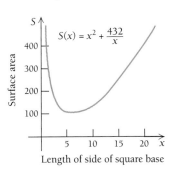

$S(x) = x^2 + \frac{432}{x}$

Length of side of square base

The nature of the application tells us that the domain of S is all positive numbers. Thus we are minimizing S over the interval $(0, \infty)$. We first find dS/dx:

$$\frac{dS}{dx} = 2x - \frac{432}{x^2}.$$

Since dS/dx exists for all x in $(0, \infty)$, the only critical points occur where $dS/dx = 0$. Thus we solve the following equation:

$$2x - \frac{432}{x^2} = 0$$

$$x^2\left(2x - \frac{432}{x^2}\right) = x^2 \cdot 0 \qquad \text{We multiply by } x^2 \text{ to clear the fractions.}$$

$$2x^3 - 432 = 0$$

$$2x^3 = 432$$

$$x^3 = 216$$

$$x = 6.$$

This is the only critical point, so we can use the second derivative to determine whether we have a minimum:

$$\frac{d^2S}{dx^2} = 2 + \frac{864}{x^3}.$$

Note that this is positive for all positive values of x. Thus we have a minimum at $x = 6$. When $x = 6$, it follows that $y = 3$:

$$y = \frac{108}{6^2} = \frac{108}{36} = 3.$$

Thus the surface area is minimized when $x = 6$ cm (centimeters) and $y = 3$ cm. The minimum surface area is

$$S = 6^2 + 4 \cdot 6 \cdot 3$$

$$= 108 \text{ cm}^2.$$

By coincidence, this is the same number as the fixed volume.

Business and Economics Applications: Marginal Analysis

Example 4   *Business: Maximizing Revenue.* A stereo manufacturer determines that in order to sell x units of a new stereo, the price per unit must be

$$p = 1000 - x.$$

The manufacturer also determines that the total cost of producing x units is given by

$$C(x) = 3000 + 20x.$$

a) Find the total revenue $R(x)$.

b) Find the total profit $P(x)$.

c) How many units must the company produce and sell in order to maximize profit?

d) What is the maximum profit?

e) What price per unit must be charged in order to make this maximum profit?

Solution

a) $R(x)$ = Total revenue = (Number of units) · (Price per unit)

$$= \quad x \quad · \quad p$$

$$= x(1000 - x) = 1000x - x^2$$

b) $P(x) = R(x) - C(x) = (1000x - x^2) - (3000 + 20x)$

$$= -x^2 + 980x - 3000$$

c)   To find the maximum value of $P(x)$, we first find $P'(x)$:

$$P'(x) = -2x + 980.$$

This is defined for all real numbers (actually we are interested in numbers x in $[0, \infty)$ only, since we cannot produce a negative number of stereos). Thus we solve:

$$P'(x) = -2x + 980 = 0$$

$$-2x = -980$$

$$x = 490.$$

Since there is only one critical point, we can try to use the second derivative to determine whether we have a maximum. Note that

$$P''(x) = -2, \quad \text{a constant.}$$

Thus, $P''(490)$ is negative, so $P(490)$ is a maximum.

d)   The maximum profit is given by

$$P(490) = -(490)^2 + 980 · 490 - 3000$$

$$= \$237,100.$$

Thus the stereo manufacturer makes a maximum profit of $237,100 by producing and selling 490 stereos.

e) The price per unit needed to make the maximum profit is

$$p = 1000 - 490 = \$510.$$

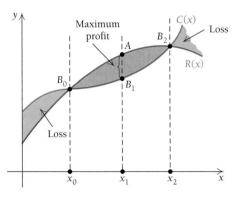

FIGURE 1

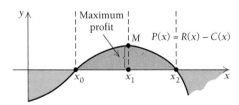

FIGURE 2

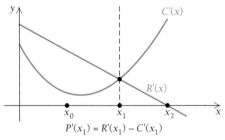

FIGURE 3

Let's take a general look at the total-profit function and its related functions.

Figure 1 shows an example of total-cost and total-revenue functions. We can estimate what the maximum profit might be by looking for the widest gap between $R(x)$ and $C(x)$. Points B_0 and B_2 are break-even points.

Figure 2 shows an example of total-profit function. Note that when production is too low ($< x_0$), there is a loss because of high fixed or initial costs and low revenue. When production is too high ($> x_2$), there is also a loss due to high marginal costs and low marginal revenues (as shown in Fig. 3).

The business operates at a profit everywhere between x_0 and x_2. Note that maximum profit occurs at a critical point x_1 of $P(x)$. If we assume that $P'(x)$ exists for all x in some interval, usually $[0, \infty)$, this critical point occurs at some number x such that

$$P'(x) = 0$$

and

$$P''(x) < 0.$$

Since $P(x) = R(x) - C(x)$, it follows that

$$P'(x) = R'(x) - C'(x)$$

and

$$P''(x) = R''(x) - C''(x).$$

Thus the maximum profit occurs at some number x such that

$$P'(x) = R'(x) - C'(x) = 0$$

and

$$P''(x) = R''(x) - C''(x) < 0,$$

or

$$R'(x) = C'(x)$$

and

$$R''(x) < C''(x).$$

In summary, we have the following theorem.

> **Theorem 10**
>
> Maximum profit is achieved when marginal revenue equals marginal cost and the rate of change of marginal revenue is less than the rate of change of marginal cost:
>
> $$R'(x) = C'(x) \quad \text{and} \quad R''(x) < C''(x).$$

Example 5   *Business: Determining a Ticket Price.* Fight promoters ride a thin line between profit and loss, especially in determining the price to charge for admission to closed-circuit television showings in local theaters. By keeping records, a theater determines that if the admission price is $20, it averages 1000 people in attendance. But for every increase of $1, it loses 100 customers from the average number. Every customer spends an average of $1.80 on concessions. What admission price should the theater charge in order to maximize total revenue?

Solution   Let x = the amount by which the price of $20 should be increased. (If x is negative, the price is decreased.) We first express the total revenue R as a function of x. Note that

$$
\begin{aligned}
R(x) &= (\text{Revenue from tickets}) + (\text{Revenue from concessions}) \\
&= (\text{Number of people}) \cdot (\text{Ticket price}) + \$1.80(\text{Number of people}) \\
&= (1000 - 100x)(20 + x) + 1.80(1000 - 100x) \\
&= 20{,}000 - 2000x + 1000x - 100x^2 + 1800 - 180x \\
R(x) &= -100x^2 - 1180x + 21{,}800.
\end{aligned}
$$

We are trying to find the maximum value of R over the set of all real numbers. To find x such that $R(x)$ is a maximum, we first find $R'(x)$:

$$R'(x) = -200x - 1180.$$

This derivative exists for all real numbers x. Thus the only critical points are where $R'(x) = 0$, so we solve that equation:

$$
\begin{aligned}
-200x - 1180 &= 0 \\
-200x &= 1180 \\
x &= -5.9 = -\$5.90.
\end{aligned}
$$

Since this is the only critical point, we can use the second derivative,

$$R''(x) = -200,$$

to determine whether we have a maximum. Since $R''(-5.9)$ is negative, $R(-5.9)$ is a maximum. Therefore, in order to maximize revenue, the theater should charge

$$\$20 + (-\$5.90), \quad \text{or} \quad \$14.10 \text{ per ticket.}$$

That is, this reduced ticket price will attract more people to the movie theater,

$$1000 - 100(-5.9), \quad \text{or} \quad 1590,$$

and will result in maximum revenue.

How can inventory costs be minimized?

Minimizing Inventory Costs

A retail outlet of a business is concerned about inventory costs. Suppose, for example, that an appliance store sells 2500 television sets per year. It *could* operate by ordering all the sets at once. But then the owners would face the carrying costs (insurance, building space, and so on) of storing them all. Thus they might make several smaller orders, say 5, so the largest number they would ever have to store is 500. On the other hand, each time they reorder, there are costs for paperwork, delivery charges, manpower, and so on. It seems, therefore, that there must be some balance between carrying costs and reorder costs. Let's see how calculus can help to determine what that balance might be. We are trying to minimize the following function:

$$\text{Total inventory costs} = \begin{pmatrix} \text{Yearly carrying} \\ \text{costs} \end{pmatrix} + \begin{pmatrix} \text{Yearly reorder} \\ \text{costs} \end{pmatrix}.$$

The *lot size x* refers to the largest amount ordered each reordering period. If x is ordered each period, then during that time there is somewhere between 0 and x units in stock. To have a representative expression for the amount in stock at any one time in the period, we can use the average, $x/2$. This represents the average amount held in stock over the course of the year.

Refer to the following graphs. If the lot size is 2500, then during the period between orders, there is somewhere between 0 and 2500 units in stock. On the average, there is 2500/2, or 1250 units in stock. If the lot size is 1250, then during the period between orders, there is somewhere between 0 and 1250 units in stock. On the average, there is 1250/2, or 625 units in stock.

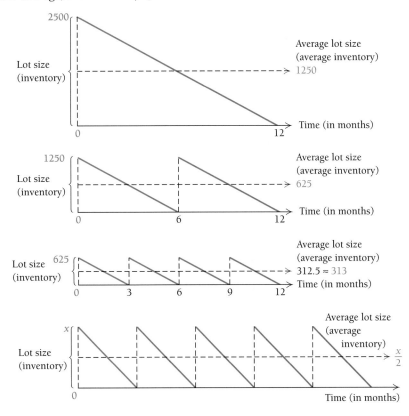

Example 6 *Business: Minimizing Inventory Costs.* A retail appliance store sells 2500 television sets per year. It costs $10 to store one set for a year. To reorder, there is a fixed cost of $20, plus $9 for each set. How many times per year should the store reorder, and in what lot size, in order to minimize inventory costs?

Solution   Let x = the lot size. Inventory costs are given by

$$C(x) = \text{(Yearly carrying costs)} + \text{(Yearly reorder costs)}.$$

We consider each separately.

a) *Yearly carrying costs.* The average amount held in stock is $x/2$, and it costs $10 per set for storage. Thus

$$\text{Yearly carrying costs} = \left(\begin{array}{c}\text{Yearly cost}\\\text{per item}\end{array}\right) \cdot \left(\begin{array}{c}\text{Average number}\\\text{of items}\end{array}\right)$$

$$= 10 \cdot \frac{x}{2}.$$

b) *Yearly reorder costs.* We know that x = the lot size, and suppose that there are N reorders each year. Then $Nx = 2500$, and $N = 2500/x$. Thus,

$$\text{Yearly reorder costs} = \left(\begin{array}{c}\text{Cost of each}\\\text{order}\end{array}\right) \cdot \left(\begin{array}{c}\text{Number of}\\\text{reorders}\end{array}\right)$$

$$= (20 + 9x)\frac{2500}{x}.$$

c) Hence

$$C(x) = 10 \cdot \frac{x}{2} + (20 + 9x)\frac{2500}{x}$$

$$= 5x + \frac{50{,}000}{x} + 22{,}500.$$

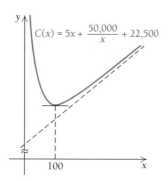

$$C(x) = 5x + \frac{50{,}000}{x} + 22{,}500$$

d) We want to find a minimum value of C over the interval $[1, 2500]$. (See the graph above.) We first find $C'(x)$:

$$C'(x) = 5 - \frac{50{,}000}{x^2}.$$

TECHNOLOGY   CONNECTION   . . . .

Exploratory Exercises

Many graphers have the ability to make tables and/or spreadsheets of function values. In reference to Example 6, without a knowledge of calculus, one might make a rough estimate of the lot size that will minimize total inventory costs by using a table like the following. Complete the table and make an estimate.

Lot size, x	Number of Reorders, $\dfrac{2500}{x}$	Average Inventory, $\dfrac{x}{2}$	Carrying Costs, $10 \cdot \dfrac{x}{2}$	Cost of Each Order, $20 + 9x$	Reorder Costs, $(20 + 9x)\dfrac{2500}{x}$	Total Inventory Costs, $C(x) = 10 \cdot \dfrac{x}{2} + (20 + 9x)\dfrac{2500}{x}$
2500	1	1250	$12,500	$22,200	$22,520	$35,020
1250	2	625	6,250	11,270	22,540	
500	5	250	2,500	4,520		
250	10	125				
167	15	84				
125	20					
100	25					
90	28					
50	50					

1. Use a grapher to graph $C(x)$ over the interval $[0, 2500]$.
2. With a grapher, estimate a minimum value and where it occurs.

e) $C'(x)$ exists for all x in $[1, 2500]$, so the only critical points are those x such that $C'(x) = 0$. We solve $C'(x) = 0$:

$$5 - \frac{50{,}000}{x^2} = 0$$

$$5 = \frac{50{,}000}{x^2}$$

$$5x^2 = 50{,}000$$

$$x^2 = 10{,}000$$

$$x = \pm 100.$$

Since there is only one critical point in the interval $[1, 2500]$, that is, $x = 100$, we can use the second derivative to see whether we have a maximum or a minimum:

$$C''(x) = \frac{100{,}000}{x^3}.$$

$C''(x)$ is positive for all x in $[1, 2500]$, so we do have a minimum at $x = 100$. Thus in order to minimize inventory costs, the store should order sets (2500/100), or 25 times per year. The lot size is 100.   ◄

What happens in such problems when the answer is not a whole number? For those functions, we consider the two whole numbers closest to the answer and substitute them into $C(x)$. The value that yields the smaller $C(x)$ is the lot size.

Example 7   *Business: Minimizing Inventory Costs.* Let's repeat Example 6 using all the data given, but change the $10 storage cost to $20. How many times per year should the store reorder television sets, and in what lot size, in order to minimize inventory costs?

Solution   Comparing this with Example 6, we find that the inventory cost function becomes

$$C(x) = 20 \cdot \frac{x}{2} + (20 + 9x)\frac{2500}{x}$$

$$= 10x + \frac{50{,}000}{x} + 22{,}500.$$

Then we find $C'(x)$, set it equal to 0, and solve for x:

$$C'(x) = 10 - \frac{50{,}000}{x^2} = 0$$

$$10 = \frac{50{,}000}{x^2}$$

$$10x^2 = 50{,}000$$

$$x^2 = 5000$$

$$x = \sqrt{5000}$$

$$\approx 70.7.$$

Since it does not make sense to reorder 70.7 sets each time, we consider the two numbers closest to 70.7, which are 70 and 71. Now

$$C(70) \approx \$23{,}914.29 \quad \text{and} \quad C(71) \approx \$23{,}914.23.$$

It follows that the lot size that will minimize cost is 71, although the difference, $0.06, is not significant. (*Note*: Such a procedure will not work for all types of functions, but will work for the type we are considering here.) The number of times that an order should be placed is 2500/71 ≈ 35, so there is still some estimating involved.   ◄

The lot size that minimizes total inventory costs is often referred to as the *economic ordering quantity.* There are three assumptions made in using the preceding method to determine the economic ordering quantity. The first is that the demand for the product is the same throughout the year. For television sets this may be reasonable, but for seasonal items such as clothing or skis, this assumption may not be reasonable. The second assumption is that the time between the placing of an order and its receipt will be consistent throughout the year. The third assumption is that the various costs involved, such as storage, shipping charges, and so on, do not vary. This may not be reasonable in a time of inflation, although one may account for them by anticipating what they might be and using average costs. Nevertheless, the model described above can be useful, and it allows us to analyze a seemingly difficult problem using calculus.

EXERCISE SET 3.5

1. Of all numbers whose sum is 50, find the two that have the maximum product. That is, maximize $Q = xy$, where $x + y = 50$.

2. Of all numbers whose sum is 70, find the two that have the maximum product. That is, maximize $Q = xy$, where $x + y = 70$.

3. In Exercise 1, can there be a minimum product? Explain.

4. In Exercise 2, can there be a minimum product? Explain.

5. Of all numbers whose difference is 4, find the two that have the minimum product.

6. Of all numbers whose difference is 6, find the two that have the minimum product.

7. Maximize $Q = xy^2$, where x and y are positive numbers, such that $x + y^2 = 1$.

8. Maximize $Q = xy^2$, where x and y are positive numbers, such that $x + y^2 = 4$.

9. Minimize $Q = x^2 + y^2$, where $x + y = 20$.

10. Minimize $Q = x^2 + y^2$, where $x + y = 10$.

11. Maximize $Q = xy$, where x and y are positive numbers, such that $\frac{4}{3}x^2 + y = 16$.

12. Maximize $Q = xy$, where x and y are positive numbers, such that $x + \frac{4}{3}y^2 = 1$.

13. *Maximizing Area.* A rancher wants to build a rectangular fence next to a river, using 120 yd of

fencing. What dimensions of the rectangle will maximize the area? What is the maximum area? (Note that the rancher need not fence in the side next to the river.)

14. *Maximizing Area.* A rancher wants to enclose two rectangular areas near a river, one for sheep and one for cattle. There is 240 yd of fencing available. What is the largest total area that can be enclosed?

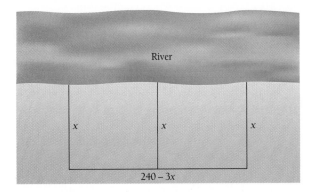

15. *Maximizing Area.* A carpenter is building a rectangular room with a fixed perimeter of 54 ft. What are the dimensions of the largest room that can be built? What is its area?

16. *Maximizing Area.* Of all rectangles that have a perimeter of 34 ft, find the dimensions of the one with the largest area. What is its area?

17. *Maximizing Volume.* From a thin piece of cardboard 30 in. by 30 in., square corners are cut out so that the sides can be folded up to make a box. What dimensions will yield a box of maximum volume? What is the maximum volume?

18. *Maximizing Volume.* From a thin piece of cardboard 20 in. by 20 in., square corners are cut out so that the sides can be folded up to make a box. What dimensions will yield a box of maximum volume? What is the maximum volume?

19. *Minimizing Surface Area.* A container company is designing an open-top, square-based, rectangular box that will have a volume of 62.5 in³. What dimensions yield the minimum surface area? What is the minimum surface area?

20. *Minimizing Surface Area.* A soup company is constructing an open-top, square-based, rectangular metal tank that will have a volume of 32 ft³. What dimensions yield the minimum surface area? What is the minimum surface area?

Applications ....
BUSINESS AND ECONOMICS

Maximizing Profit. Find the maximum profit and the number of units that must be produced and sold in order to yield the maximum profit.

21. $R(x) = 50x - 0.5x^2$, $C(x) = 4x + 10$

22. $R(x) = 50x - 0.5x^2$, $C(x) = 10x + 3$

23. $R(x) = 2x$, $C(x) = 0.01x^2 + 0.6x + 30$

24. $R(x) = 5x$, $C(x) = 0.001x^2 + 1.2x + 60$

25. $R(x) = 9x - 2x^2$, $C(x) = x^3 - 3x^2 + 4x + 1$; $R(x)$ and $C(x)$ are in thousands of dollars, and x is in thousands of units.

26. $R(x) = 100x - x^2$, $C(x) = \frac{1}{3}x^3 - 6x^2 + 89x + 100$; $R(x)$ and $C(x)$ are in thousands of dollars, and x is in thousands of units.

27. *Maximizing Profit.* Raggs, Ltd., a clothing firm, determines that in order to sell x suits, the price per suit must be

$$p = 150 - 0.5x.$$

It also determines that the total cost of producing x suits is given by

$$C(x) = 4000 + 0.25x^2.$$

a) Find the total revenue $R(x)$.
b) Find the total profit $P(x)$.

c) How many suits must the company produce and sell in order to maximize profit?
d) What is the maximum profit?
e) What price per suit must be charged in order to make this maximum profit?

28. *Maximizing Profit.* An appliance firm is marketing a new refrigerator. It determines that in order to sell x refrigerators, the price per refrigerator must be

$$p = 280 - 0.4x.$$

It also determines that the total cost of producing x refrigerators is given by

$$C(x) = 5000 + 0.6x^2.$$

a) Find the total revenue $R(x)$.
b) Find the total profit $P(x)$.
c) How many refrigerators must the company produce and sell in order to maximize profit?
d) What is the maximum profit?
e) What price per refrigerator must be charged in order to make this maximum profit?

29. *Maximizing Revenue.* A university is trying to determine what price to charge for football tickets. At a price of $6 per ticket, it averages 70,000 people per game. For every increase of $1, it loses 10,000 people from the average number. Every person at the game spends an average of $1.50 on concessions. What price per ticket should be charged in order to maximize revenue? How many people will attend at that price?

30. *Maximizing Revenue.* Suppose that you are the owner of a 30-unit motel. All units are occupied when you charge $20 a day per unit. For every increase of x dollars in the daily rate, there are x units vacant. Each occupied room costs $2 per day to service and maintain. What should you charge per unit in order to maximize profit?

31. *Maximizing Yield.* An apple farm yields an average of 30 bushels of apples per tree when 20 trees are planted on an acre of ground. Each time 1 more tree is planted per acre, the yield decreases 1 bu per tree due to the extra congestion. How many trees should be planted in order to get the highest yield?

32. *Maximizing Revenue.* When a theater owner charges $3 for admission, there is an average attendance of 100 people. For every $0.10 increase in admission, there is a loss of 1 customer from the average number. What admission should be charged in order to maximize revenue?

33. *Minimizing Costs.* A rectangular box with a volume of 320 ft³ is to be constructed with a square base and top. The cost per square foot for the bottom is 15¢, for the top is 10¢, and for the sides is 2.5¢. What dimensions will minimize the cost?

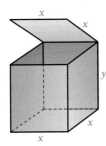

tw 34. A merchant who was purchasing a display sign from a salesclerk said, "I want a sign 10 ft by 10 ft." The salesclerk responded, "That's just what we'll give you; only to make it more aesthetically pleasing, why don't we change it to 7 ft by 13 ft?" Comment.

35. *Maximizing Profit.* The amount of money deposited in a financial institution in savings accounts is directly proportional to the interest rate that the financial institution pays on the money. Suppose that a financial institution can loan *all* the money it takes in on its savings accounts at an interest rate of 18%. What interest rate should it pay on its savings accounts in order to maximize profit?

36. *Maximizing Area.* A page in this book measures 73.125 in². On the average, there is a 0.75-in. margin at the top and at the bottom of each page and a 0.5-in. margin on each of the sides. What should the outside dimensions of each page be so that the printed area is a maximum? Measure the outside dimensions to see whether the actual dimensions maximize the printed area.

37. *Minimizing Inventory Costs.* A sporting goods store sells 100 pool tables per year. It costs $20 to store one pool table for one year. To reorder, there is a fixed cost of $40, plus $16 for each pool table. How many times per year should the store order pool tables, and in what lot size, in order to minimize inventory costs?

38. *Minimizing Inventory Costs.* A pro shop in a bowling center sells 200 bowling balls per year. It costs $4 to store one bowling ball for one year. To reorder, there is a fixed cost of $1, plus $0.50 for each bowling ball. How many times per year should the shop order bowling balls, and in what lot size, in order to minimize inventory costs?

39. *Minimizing Inventory Costs.* A retail outlet for Boxowitz Calculators sells 360 calculators per year. It costs $8 to store one calculator for one year. To reorder, there is a fixed cost of $10, plus $8 for each calculator. How many times per year should the store order calculators, and in what lot size, in order to minimize inventory costs?

40. *Minimizing Inventory Costs.* A sporting goods store in southern California sells 720 surfboards per year. It costs $2 to store one surfboard for one year. To reorder, there is a fixed cost of $5, plus $2.50 for each surfboard. How many times per year should the store order surfboards, and in what lot size, in order to minimize inventory costs?

41. *Minimizing Inventory Costs.* Repeat Exercise 39 using all the data given, but change the $8 storage charge to $9.

42. *Minimizing Inventory Costs.* Repeat Exercise 40 using all the data given, but change the $5 fixed cost to $4.

GENERAL INTEREST

43. *Maximizing Volume.* The postal service places a limit of 84 in. on the combined length and girth (distance around) of a package to be sent parcel post. What dimensions of a rectangular box with square cross-section will contain the largest volume that can be mailed? (*Hint:* There are two different girths.)

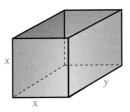

44. *Minimizing Cost.* A rectangular play area is to be fenced off in a person's yard and is to contain 48 yd². The neighbor agrees to pay half the cost of the fence on the side of the play area that lines the lot. What dimensions will minimize the cost of the fence?

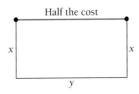

45. *Maximizing Light.* A Norman window is a rectangle with a semicircle on top. Suppose that the perimeter of a particular Norman window is to be 24 ft. What should its dimensions be in order to allow the maximum amount of light to enter through the window?

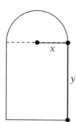

46. *Maximizing Light.* Repeat Exercise 45, but assume that the semicircle is to be stained glass, which transmits only half as much light as the semicircle in Exercise 45.

Synthesis

47. For what positive number is the sum of its reciprocal and five times its square a minimum?

48. For what positive number is the sum of its reciprocal and four times its square a minimum?

49. *Business: Minimizing Inventory Costs—a general solution.* A store sells Q units of a product per year. It costs a dollars to store one unit for one year. To reorder, there is a fixed cost of b dollars, plus c dollars for each unit. How many times per year should the store reorder, and in what lot size, in order to minimize inventory costs?

50. *Business: Minimizing Inventory Costs.* Use the general solution found in Exercise 49 to find how many times per year a store should reorder, and in what lot size, when $Q = 2500$, $a = \$10$, $b = \$20$, and $c = \$9$.

51. A 24-in. piece of string is cut in two pieces. One piece is used to form a circle and the other to form

a square. How should the string be cut so that the sum of the areas is a minimum? a maximum?

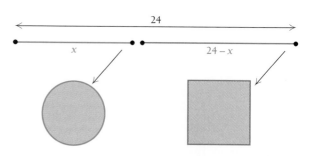

52. *Business: Minimizing Costs.* A power line is to be constructed from a power station at point A to an island at point C, which is 1 mi directly out in the water from a point B on the shore. Point B is 4 mi downshore from the power station at A. It costs \$5000 per mile to lay the power line under water and \$3000 per mile to lay the line under ground. At what point S downshore from A should the line come to the shore in order to minimize cost? Note that S could very well be B or A. (*Hint*: The length of CS is $\sqrt{1 + x^2}$.)

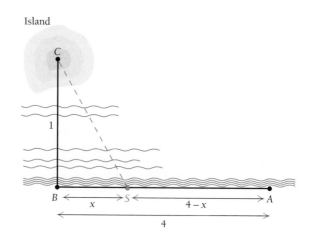

53. *Life Science: Flights of Homing Pigeons.* It is known that homing pigeons tend to avoid flying over water in the daytime, perhaps because the downdrafts of air over water make flying difficult. Suppose a homing pigeon is released on an island at point C, which is 3 mi directly out in the water from a point B on shore. Point B is 8 mi downshore from the pigeon's home loft at point A. Assume that a pigeon requires 1.28 times the rate of energy over land to fly over water. Toward what point S downshore from A should the pigeon fly in order to minimize the total energy required to get to home loft A? Assume that

(Total energy)

= (Energy rate over water) · (Distance over water)

+ (Energy rate over land) · (Distance over land)

Island

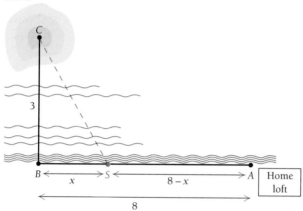

54. *Business: Minimizing Distance.* A road is to be built between two cities C_1 and C_2, which are on opposite sides of a river of uniform width r. Because of the river, a bridge must be built. C_1 is a units from the river, and C_2 is b units from the river; $a \leq b$. Where should the bridge be located in order to minimize the total distance between the cities? Give a general solution using the constants a, b, p, and r in the figure shown here.

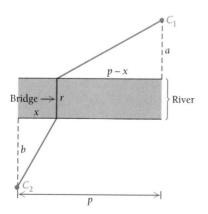

55. *Business: Minimizing Cost.* The total-cost function for producing x units of a certain product is given by

$$C(x) = 8x + 20 + \frac{x^3}{100}.$$

a) Find the marginal cost $C'(x)$.
b) Find the average cost $A(x) = C(x)/x$.
c) Find the marginal average cost $A'(x)$.
d) Find the minimum of $A(x)$ and the value x_0 at which it occurs. Find the marginal cost at x_0.
e) Compare $A(x_0)$ and $C'(x_0)$.

56. *Business: Minimizing Cost.* Consider $A(x) = C(x)/x$.

a) Find $A'(x)$ in terms of $C'(x)$ and $C(x)$.
b) Show that $A(x)$ has a minimum at that value of x_0 such that

$$C'(x_0) = A(x_0)$$
$$= \frac{C(x_0)}{x_0}.$$

This shows that when marginal cost and average cost are the same, a product is being produced at the least average cost.

57. Minimize $Q = x^3 + 2y^3$, where x and y are positive numbers, such that $x + y = 1$.

58. Minimize $Q = 3x + y^3$, where $x^2 + y^2 = 2$.

tw 59. Explain how marginal revenue, cost, and profit can be used to find maximum profit.

3.6

Differentials

In this section, we consider ways of using calculus to make linear approximations. Suppose, for example, that at a certain time in an illness, a patient's temperature is 102° and the instantaneous rate of change of temperature is $-2°$ per minute. It would seem reasonable to *estimate* the patient's temperature 1 min later to be about 100°, especially if the temperature were following some particular type of function formula.

Delta Notation

Recall the difference quotient

$$\frac{f(x + h) - f(x)}{h},$$

illustrated in this graph.

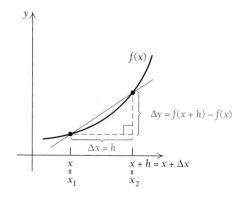

The difference quotient is used to define the derivative of a function at x. The number h is considered to be a *change* in x. Another notation for such a change is Δx, read "delta x" and called **delta notation.** The expression Δx is *not* the product of Δ and x, but is its own entity; that is, it is a new type of variable that represents the *change* in the value of x from a *first* value to a *second*. Thus,

$$\Delta x = (x + h) - x = h.$$

If subscripts are used for the first and second values of x, we have

$$\Delta x = x_2 - x_1, \quad \text{or} \quad x_2 = x_1 + \Delta x.$$

Δx can be positive or negative.

Example 1

a) If $x_1 = 4$ and $\Delta x = 0.7$, then $x_2 = 4.7$.

b) If $x_1 = 4$ and $\Delta x = -0.7$, then $x_2 = 3.3$.   ◄

We generally omit the subscripts and use x and $x + \Delta x$.

Now suppose we have a function given by $y = f(x)$. A change in x from x to $x + \Delta x$ yields a change in y from $f(x)$ to $f(x + \Delta x)$. The change in y is given by

$$\Delta y = f(x + \Delta x) - f(x).$$

Example 2   For $y = x^2$, $x = 4$, and $\Delta x = 0.1$, find Δy.

Solution   We have

$$\Delta y = (4 + 0.1)^2 - 4^2$$
$$= (4.1)^2 - 4^2 = 16.81 - 16 = 0.81.$$

Example 3   For $y = x^3$, $x = 2$, and $\Delta x = -0.1$, find Δy.

Solution   We have

$$\Delta y = [2 + (-0.1)]^3 - 2^3$$
$$= (1.9)^3 - 2^3 = 6.859 - 8 = -1.141.$$

Let's see how we can use calculus to make estimates of function values (to make predictions). If delta notation is used, the difference quotient

$$\frac{f(x + h) - f(x)}{h}$$

becomes

$$\frac{f(x + \Delta x) - f(x)}{\Delta x} = \frac{\Delta y}{\Delta x}.$$

We can then express the derivative as

$$\frac{dy}{dx} = \lim_{\Delta x \to 0} \frac{\Delta y}{\Delta x}.$$

Note that the delta notation resembles the Leibniz notation (see p. 138).

For values of Δx close to 0, we have the approximation

$$\frac{dy}{dx} \approx \frac{\Delta y}{\Delta x}, \quad \text{or} \quad f'(x) \approx \frac{\Delta y}{\Delta x}.$$

Multiplying both sides of the second expression by Δx gives us

$$\Delta y \approx f'(x)\, \Delta x.$$

We can see this in the graph at right.

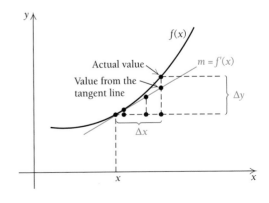

From this graph, it seems reasonable to assume that average rates of change $\Delta y/\Delta x$ of the function are approximately the same as the slope of the tangent line and the tangent line can be used to estimate a function value.

Let's illustrate this idea by considering the square-root function, $f(x) = \sqrt{x}$. We know how to approximate $\sqrt{27}$ using a calculator. But suppose we didn't. We could begin with $\sqrt{25}$ and use as a change in input $\Delta x = 2$. We would use the change in y, $\Delta y \approx f'(x) \, \Delta x$, to estimate $\sqrt{27}$.

Example 4   Approximate $\sqrt{27}$ using $\Delta y \approx f'(x) \, \Delta x$.

Solution   We first think of the number closest to 27 that is a perfect square. This is 25. What we will do is approximate how y, or $\sqrt{x}$, changes when 25 changes by $\Delta x = 2$. Let

$$y = f(x) = \sqrt{x}.$$

Then

$$\Delta y = \sqrt{x + \Delta x} - \sqrt{x} = \sqrt{x + \Delta x} - y,$$

so

$$y + \Delta y = \sqrt{x + \Delta x}.$$

Now

$$\Delta y \approx f'(x) \, \Delta x = \frac{1}{2}x^{-1/2} \, \Delta x = \frac{1}{2\sqrt{x}} \, \Delta x.$$

If we let $x = 25$ and $\Delta x = 2$, then

$$\Delta y \approx f'(x) \, \Delta x = \frac{1}{2\sqrt{25}} \cdot 2$$

$$= \frac{1}{\sqrt{25}} = \frac{1}{5} = 0.2.$$

We now have

$$\sqrt{27} = \sqrt{25 + 2} = \sqrt{x + \Delta x} = y + \Delta y$$
$$= \sqrt{25} + \Delta y \qquad y = \sqrt{x}; x = 25$$
$$= 5 + \Delta y$$
$$\approx 5 + 0.2$$
$$\approx 5.2.$$

To five decimal places, $\sqrt{27} = 5.19615$. Thus our approximation is fairly accurate.

◄

Business and Economics Applications: Marginal Analysis

Suppose we have a total-cost function $C(x)$. When $\Delta x = 1$, we have

$$\Delta C \approx C'(x). \qquad C'(x)\,\Delta x = C'(x) \cdot 1 = C'(x)$$

Whether this is a good approximation depends on the function and on the values of x. Let's consider an example.

Example 5   Consider the total-cost function

$$C(x) = 2x^3 - 12x^2 + 30x + 200.$$

a) Find ΔC and $C'(x)$ when $x = 2$ and $\Delta x = 1$.

b) Find ΔC and $C'(x)$ when $x = 100$ and $\Delta x = 1$.

Solution

a) We have

$$\Delta C = C(2 + 1) - C(2) = C(3) - C(2)$$
$$= \$236 - \$228 = \$8.$$

Recall that $C(2)$ is the total cost of producing 2 units, and $C(3)$ is the total cost of producing 3 units, so $C(3) - C(2)$, or \$8, is the cost of the third unit. Now

$$C'(x) = 6x^2 - 24x + 30, \quad \text{so} \quad C'(2) = \$6.$$

b) We have

$$\Delta C = C(100 + 1) - C(100)$$
$$= C(101) - C(100) = \$58,220.$$

Note that this is the cost of the 101st unit. Now

$$C'(100) = \$57,630.$$

Note that in part (a), the approximation of \$6 is \$2 less than the "correct" value of \$8; this is a percentage difference of \$2/\$8, or 25%, from the correct value. In part (b), the approximation of \$57,630 is \$590 less than the "correct" value of \$58,220; this is a percentage difference of \$590/\$58,220, or about 1% from the correct value, so it is a very close approximation.   ◄

We purposely used $\Delta x = 1$ in Example 5 to illustrate the following.

$$C'(x) \approx C(x + 1) - C(x)$$

Marginal cost is (approximately) the cost of the $(x + 1)$st, or next, unit.

This is the historical definition that economists have given to marginal cost. Similarly, the following is true.

$$R'(x) \approx R(x + 1) - R(x)$$

Marginal revenue is (approximately) the revenue from the sale of the $(x + 1)$st, or next, unit.

And

$$P'(x) \approx P(x + 1) - P(x)$$

Marginal profit is (approximately) the profit from the production and sale of the $(x + 1)$st, or next, unit.

Differentials

Up to now we have not defined the symbols dy and dx as separate entities, and we have treated dy/dx as one symbol. We now define dy and dx. These symbols are called **differentials.**

Definition

For $y = f(x)$, we define

dx, called the **differential of x,** by $dx = \Delta x$

and

dy, called the **differential of y,** by $dy = f'(x)\, dx$.

We can illustrate dx and dy as shown below. Note that $dx = \Delta x$, but $dy \neq \Delta y$, though $dy \approx \Delta y$, for small values of dx.

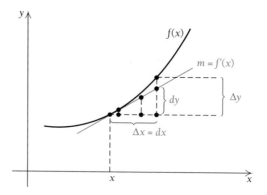

Example 6 For $y = x(4 - x)^3$:

a) Find dy.

b) Find dy when $x = 5$ and $dx = 0.2$.

Solution

a) First, we find dy/dx:

$$\frac{dy}{dx} = -x[3(4 - x)^2] + (4 - x)^3 \qquad \text{Using the Product Rule and the Extended Power Rule}$$

$$= (4 - x)^2[-3x + (4 - x)]$$

$$= (4 - x)^2[-4x + 4] \qquad \text{Simplifying}$$

$$= -4(4 - x)^2(x - 1).$$

Then

$$dy = -4(4 - x)^2(x - 1)\ dx.$$

Note that the expression for dy contains *two* variables, x and dx.

b) When $x = 5$ and $dx = 0.2$,

$$dy = -4(4 - 5)^2(5 - 1)0.2 = -4(-1)^2(4)(0.2) = -3.2.$$

EXERCISE SET 3.6

Find Δy and $f'(x)\ \Delta x$.

1. For $y = f(x) = x^2$, $x = 2$, and $\Delta x = 0.01$
2. For $y = f(x) = x^3$, $x = 2$, and $\Delta x = 0.01$
3. For $y = f(x) = x + x^2$, $x = 3$, and $\Delta x = 0.04$
4. For $y = f(x) = x - x^2$, $x = 3$, and $\Delta x = 0.02$
5. For $y = f(x) = 1/x^2$, $x = 1$, and $\Delta x = 0.5$
6. For $y = f(x) = 1/x$, $x = 1$, and $\Delta x = 0.2$
7. For $y = f(x) = 3x - 1$, $x = 4$, and $\Delta x = 2$
8. For $y = f(x) = 2x - 3$, $x = 8$, and $\Delta x = 0.5$

9. For the total-cost function

$$C(x) = 0.01x^2 + 0.6x + 30,$$

find ΔC and $C'(x)$ when $x = 70$ and $\Delta x = 1$.

10. For the total-cost function

$$C(x) = 0.01x^2 + 1.6x + 100,$$

find ΔC and $C'(x)$ when $x = 80$ and $\Delta x = 1$.

11. For the total-revenue function

$$R(x) = 2x,$$

find ΔR and $R'(x)$ when $x = 70$ and $\Delta x = 1$.

12. For the total-revenue function

$$R(x) = 3x,$$

find ΔR and $R'(x)$ when $x = 80$ and $\Delta x = 1$.

13. a) Using $C(x)$ of Exercise 9 and $R(x)$ of Exercise 11, find the total profit $P(x)$.
    b) Find ΔP and $P'(x)$ when $x = 70$ and $\Delta x = 1$.

14. a) Using $C(x)$ of Exercise 10 and $R(x)$ of Exercise 12, find the total profit $P(x)$.
    b) Find ΔP and $P'(x)$ when $x = 80$ and $\Delta x = 1$.

Approximate using $\Delta y \approx f'(x)\ \Delta x$.

15. $\sqrt{19}$
16. $\sqrt{10}$
17. $\sqrt{102}$
18. $\sqrt{103}$
19. $\sqrt[3]{10}$
20. $\sqrt[3]{28}$

Find dy.

21. $y = (2x^3 + 1)^{3/2}$
22. $y = x^3(2x + 5)^2$
23. $y = \sqrt[5]{x + 27}$
24. $y = \dfrac{x^3 + x + 2}{x^2 + 3}$

25. $y = x^4 - 2x^3 + 5x^2 + 3x - 4$

26. $y = (7 - x)^8$

27. In Exercise 25, find dy when $x = 2$ and $dx = 0.1$.

28. In Exercise 26, find dy when $x = 1$ and $dx = 0.01$.

Applications

BUSINESS AND ECONOMICS

29. *Average Cost.* The average cost of a company to produce x units of a product is given by the function

$$A(x) = \frac{13x + 100}{x}.$$

By approximately how much does the average cost change as production goes from 100 units to 101 units?

30. *Supply.* A supply function for a certain product is given by

$$S(p) = 0.08p^3 + 2p^2 + 10p + 11.$$

Approximately how many more units will a seller supply when the price changes from $18.00 per unit to $18.20 per unit?

31. *Advertising.* A firm estimates that it will sell N units of a product after spending a dollars on advertising, where

$$N(a) = -a^2 + 300a + 6,$$

and a is in thousands of dollars. Approximately how many more products will a company sell by increasing its advertising expenditure from $100 thousand to $101 thousand?

LIFE AND PHYSICAL SCIENCES

32. *Healing Wound.* The circular area of a healing wound is given by

$$A = \pi r^2,$$

where r is the radius, in centimeters. By approximately how much does the area decrease when the radius is decreased from 2 cm to 1.9 cm? Use 3.14 for π.

33. *Tumor Growth.* The spherical volume of a tumor is given by

$$V = \tfrac{4}{3}\pi r^3,$$

where r is the radius, in centimeters. By approximately how much does the volume increase when the radius is increased from 1 cm to 1.2 cm? Use 3.14 for π.

34. *Medical Dosage.* The function

$$N(t) = \frac{0.8t + 1000}{5t + 4}$$

gives the bodily concentration $N(t)$, in parts per million, of a dosage of medication after time t, in hours. By approximately how much does the concentration change as time changes from 2.8 hr to 2.9 hr?

GENERAL INTEREST

35. Suppose that a rope surrounds the earth at the equator. The rope is lengthened by 10 ft. By about how much is the rope raised above the earth?

Synthesis

tw 36. Look up the idea of a differential in a book on the history of mathematics. Write a short paragraph.

tw 37. Explain the uses of the differential.

3.7

Implicit Differentiation and Related Rates*

Implicit Differentiation

Consider the equation

$$y^3 = x.$$

This equation *implies* that y is a function of x, for if we solve for y, we get

$$y = \sqrt[3]{x}$$
$$= x^{1/3}.$$

We know from our work in this chapter that

$$\frac{dy}{dx} = \frac{1}{3}x^{-2/3}. \tag{1}$$

A method known as **implicit differentiation** allows us to find dy/dx *without* solving for y. We use the Chain Rule, treating y as a function of x. We use the Extended Power Rule and differentiate both sides of

$$y^3 = x$$

with respect to x:

$$\frac{d}{dx}y^3 = \frac{d}{dx}x.$$

The derivative on the left side is found using the Extended Power Rule:

$$3y^2\frac{dy}{dx} = 1.$$

Then

$$\frac{dy}{dx} = \frac{1}{3y^2}, \quad \text{or} \quad \frac{1}{3}y^{-2}.$$

We can show that this indeed gives us the same answer as equation (1) by replacing y with $x^{1/3}$:

$$\frac{dy}{dx} = \frac{1}{3}y^{-2} = \frac{1}{3}(x^{1/3})^{-2} = \frac{1}{3}x^{-2/3}.$$

Often, it is difficult or impossible to solve for y, obtaining an explicit expression in terms of x. For example, the equation

$$y^3 + x^2y^5 - x^4 = 27$$

---

*This section can be omitted without loss of continuity.

determines y as a function of x, but it would be difficult to solve for y. We can nevertheless find a formula for the derivative of y *without* solving for y. This involves computing $\dfrac{d}{dx}y^n$ for various integers n, and hence involves the Extended Power Rule in the form

$$\frac{d}{dx}y^n = ny^{n-1} \cdot \frac{dy}{dx}.$$

Example 1   For $y^3 + x^2y^5 - x^4 = 27$:

a) Find dy/dx using implicit differentiation.

b) Find the slope of the tangent line to the curve at the point $(0, 3)$.

Solution

a) We differentiate the term x^2y^5 using the Product Rule. Note that whenever an expression involving y is differentiated, dy/dx must be a factor of the answer. When an expression involving just x is differentiated, there is no factor dy/dx.

$$\frac{d}{dx}(y^3 + x^2y^5 - x^4) = \frac{d}{dx}(27)$$

$$\frac{d}{dx}y^3 + \frac{d}{dx}x^2y^5 - \frac{d}{dx}x^4 = 0$$

$$3y^2 \cdot \frac{dy}{dx} + x^2 \cdot 5y^4 \cdot \frac{dy}{dx} + 2x \cdot y^5 - 4x^3 = 0.$$

Then

$$3y^2 \cdot \frac{dy}{dx} + 5x^2y^4 \cdot \frac{dy}{dx} = 4x^3 - 2xy^5 \qquad \text{Only those terms involving } dy/dx \text{ should appear on one side.}$$

$$(3y^2 + 5x^2y^4)\frac{dy}{dx} = 4x^3 - 2xy^5$$

$$\frac{dy}{dx} = \frac{4x^3 - 2xy^5}{3y^2 + 5x^2y^4} \qquad \text{Solving for } dy/dx. \text{ Leave the answer in terms of } x \text{ and } y.$$

b) To find the slope of the tangent line to the curve at $(0, 3)$, we replace x with 0 and y with 3:

$$\frac{dy}{dx} = \frac{4 \cdot 0^3 - 2 \cdot 0 \cdot 3^5}{3 \cdot 3^2 + 5 \cdot 0^2 \cdot 3^4} = 0.$$

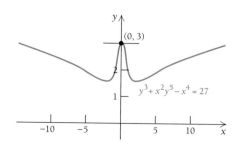

The demand function for a product (see Sections 1.5 and 2.6) is often given implicitly.

Example 2 For the demand equation $x = \sqrt{200 - p^3}$, differentiate implicitly to find dp/dx.

Solution

$$\frac{d}{dx}x = \frac{d}{dx}\sqrt{200 - p^3}$$

$$1 = \frac{1}{2}(200 - p^3)^{-1/2} \cdot (-3p^2) \cdot \frac{dp}{dx}$$

$$1 = \frac{-3p^2}{2\sqrt{200 - p^3}} \cdot \frac{dp}{dx}$$

$$\frac{2\sqrt{200 - p^3}}{-3p^2} = \frac{dp}{dx}$$

Related Rates

Suppose that y is a function of x, say

$$y = f(x),$$

and x varies with time t (as a function of time t). Since y depends on x and x depends on t, y also depends on t. That is, y is also a function of time t. The Chain Rule gives the following:

$$\frac{dy}{dt} = \frac{dy}{dx} \cdot \frac{dx}{dt}.$$

Thus the rate of change of y is *related* to the rate of change of x. Let's see how this comes up in problems. It helps to keep in mind that any variable can be thought of as a function of time t, even though a specific expression in terms of t may not be given.

Example 3 *Business: Service Area.* A restaurant supplier services the restaurants in a circular area in such a way that the radius r is increasing at the rate of 2 mi per year at the moment when r goes through the value $r = 5$ mi. At that moment, how fast is the area increasing?

Solution The area A and the radius r are always related by the equation for the area of a circle:

$$A = \pi r^2.$$

We take the derivative of both sides with respect to t:

$$\frac{dA}{dt} = 2\pi r \cdot \frac{dr}{dt}.$$

At the moment in question, $dr/dt = 2$ mi/yr (miles per year) and $r = 5$ mi, so

$$\frac{dA}{dt} = 2\pi(5 \text{ mi})\left(2\frac{\text{mi}}{\text{yr}}\right)$$

$$= 20\pi \frac{\text{mi}^2}{\text{yr}}$$

$$\approx 63 \text{ square miles per year.}$$

Example 4 *Business: Rates of Change of Revenue, Cost, and Profit.* For a company making stereos, the total revenue from the sale of x stereos is given by

$$R(x) = 1000x - x^2,$$

and the total cost is given by

$$C(x) = 3000 + 20x.$$

Suppose that the company is producing and selling stereos at the rate of 10 stereos per day at the moment when the 400th stereo is produced. At that same moment, what is the rate of change of **(a)** total revenue? **(b)** total cost? **(c)** total profit?

Solution

a) $\dfrac{dR}{dt} = 1000 \cdot \dfrac{dx}{dt} - 2x \cdot \dfrac{dx}{dt}$ Differentiating with respect to time

$= 1000 \cdot 10 - 2(400)10$ Substituting 10 for dx/dt and 400 for x

$= \$2000$ per day

b) $\dfrac{dC}{dt} = 20 \cdot \dfrac{dx}{dt}$ Differentiating with respect to time

$= 20(10)$

$= \$200$ per day

c) Since $P = R - C$,

$$\frac{dP}{dt} = \frac{dR}{dt} - \frac{dC}{dt}$$

$= \$2000$ per day $- \$200$ per day

$= \$1800$ per day.

EXERCISE SET 3.7

Differentiate implicitly to find dy/dx. Then find the slope of the curve at the given point.

1. $xy - x + 2y = 3$; $\left(-5, \dfrac{2}{3}\right)$

2. $xy + y^2 - 2x = 0$; $(1, -2)$

3. $x^2 + y^2 = 1$; $\left(\dfrac{1}{2}, \dfrac{\sqrt{3}}{2}\right)$

4. $x^2 - y^2 = 1$; $(\sqrt{3}, \sqrt{2})$

5. $x^2 y - 2x^3 - y^3 + 1 = 0$; $(2, -3)$

6. $4x^3 - y^4 - 3y + 5x + 1 = 0$; $(1, -2)$

Differentiate implicitly to find dy/dx.

7. $2xy + 3 = 0$

8. $x^2 + 2xy = 3y^2$

9. $x^2 - y^2 = 16$

10. $x^2 + y^2 = 25$

11. $y^5 = x^3$

12. $y^3 = x^5$

13. $x^2 y^3 + x^3 y^4 = 11$

14. $x^3 y^2 - x^5 y^3 = -19$

For the given demand equation, differentiate implicitly to find dp/dx.

15. $p^2 + p + 2x = 40$

16. $xp^3 = 24$

17. $(p + 4)(x + 3) = 48$

18. $1000 - 300p + 25p^2 = x$

19. Two variable quantities A and B are found to be related by the equation

$$A^3 + B^3 = 9.$$

What is the rate of change dA/dt at the moment when $A = 2$ and $dB/dt = 3$?

20. Two variable quantities G and H, nonnegative, are found to be related by the equation

$$G^2 + H^2 = 25.$$

What is the rate of change dH/dt when $dG/dt = 3$ and $G = 0$? $G = 1$? $G = 3$?

Applications

BUSINESS AND ECONOMICS

Rates of Change of Total Revenue, Cost, and Profit. Find the rates of change of total revenue, cost, and profit for each of the following.

21. $R(x) = 50x - 0.5x^2$,
$C(x) = 4x + 10$,
when $x = 30$ and $dx/dt = 20$ units per day

22. $R(x) = 50x - 0.5x^2$,
$C(x) = 10x + 3$,
when $x = 10$ and $dx/dt = 5$ units per day

23. $R(x) = 2x$,
$C(x) = 0.01x^2 + 0.6x + 30$,
when $x = 20$ and $dx/dt = 8$ units per day

24. $R(x) = 280x - 0.4x^2$,
$C(x) = 5000 + 0.6x^2$,
when $x = 200$ and $dx/dt = 300$ units per day

LIFE AND PHYSICAL SCIENCES

25. *Rate of Change of a Tumor.* The volume of a tumor is given by

$$V = \tfrac{4}{3}\pi r^3.$$

The radius is increasing at the rate of 0.03 centimeter per day (cm/day) at the moment when $r = 1.2$ cm. How fast is the volume changing at that moment?

26. *Rate of Change of a Healing Wound.* The area of a healing wound is given by

$$A = \pi r^2.$$

The radius is decreasing at the rate of 1 millimeter per day (-1 mm/day) at the moment when $r = 25$ mm. How fast is the area decreasing at that moment?

Poiseuille's Law. The flow of blood in a blood vessel is faster toward the center of the vessel and slower toward the outside. The speed of the blood V is given by

$$V = \frac{p}{4Lv}(R^2 - r^2),$$

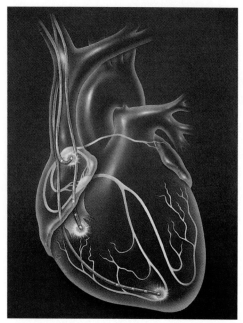

The flow of blood in a blood vessel can be modeled by Poiseuille's Law.

where R is the radius of the blood vessel, r is the distance of the blood from the center of the vessel, and p, L, and v are physical constants related to pressure, length, and viscosity of the blood vessels, respectively. Use this formula for Exercises 27 and 28.

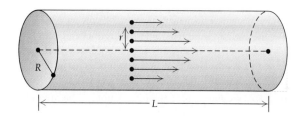

27. Assume that r is a constant as well as p, L, and v.

    a) Find the rate of change dV/dt in terms of R and dR/dt when $L = 1$ mm, $p = 100$, and $v = 0.05$.

    b) A person goes out into the cold to shovel snow. Cold air has the effect of contracting blood vessels far from the heart. Suppose that a blood vessel contracts at a rate of

    $$\frac{dR}{dt} = -0.0015 \text{ mm/min}$$

    at a place in the blood vessel where the radius $R = 0.0075$ mm. Find the rate of change dV/dt at that location.

28. Assume that r is a constant as well as p, L, and v.

    a) Find the rate of change dV/dt in terms of R and dR/dt when $L = 1$ mm, $p = 100$, and $v = 0.05$.

    b) When shoveling snow in cold air, a person with a history of heart trouble can develop angina (chest pains) due to contracting blood vessels. To counteract this, he or she may take a nitroglycerin tablet, which dilates the blood vessels. Suppose that after a nitroglycerin tablet is taken, a blood vessel dilates at a rate of

    $$\frac{dR}{dt} = 0.0025 \text{ mm/min}$$

    at a place in the blood vessel where the radius $R = 0.02$ mm. Find the rate of change dV/dt.

GENERAL INTEREST

29. Two cars start from the same point at the same time. One travels north at 25 mph, and the other travels east at 60 mph. How fast is the distance between them increasing at the end of 1 hr? (*Hint:* $D^2 = x^2 + y^2$. To find D after 1 hr, solve $D^2 = 25^2 + 60^2$.)

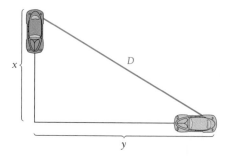

30. A ladder 26 ft long leans against a vertical wall. If the lower end is being moved away from the wall at the rate of 5 ft/sec, how fast is the height of the top decreasing (this will be a negative rate) when the lower end is 10 ft from the wall?

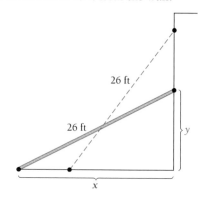

Synthesis

Differentiate implicitly to find dy/dx.

31. $\sqrt{x} + \sqrt{y} = 1$ **32.** $\dfrac{1}{x^2} + \dfrac{1}{y^2} = 5$

33. $y^3 = \dfrac{x - 1}{x + 1}$ **34.** $y^2 = \dfrac{x^2 - 1}{x^2 + 1}$

35. $x^{3/2} + y^{2/3} = 1$

36. $(x - y)^3 + (x + y)^3 = x^5 + y^5$

Differentiate implicitly to find dy/dx and d^2y/dx^2.

37. $xy + x - 2y = 4$ **38.** $y^2 - xy + x^2 = 5$

39. $x^2 - y^2 = 5$ **40.** $x^3 - y^3 = 8$

tw **41.** Explain the usefulness of implicit differentiation.

tw **42.** Look up the word "implicit" in a dictionary. Explain how that definition can be related to the concept of a function that is defined "implicitly."

TECHNOLOGY CONNECTION

Use a grapher to graph each of the following equations. On most graphers, equations must be solved for y before they can be entered.

43. $x^2 + y^2 = 4$ (*Note*: You will probably need to sketch the graph in two parts:

$$y = \sqrt{4 - x^2} \quad \text{and} \quad y = -\sqrt{4 - x^2}.)$$

Then graph the tangent line to the graph at the point $(-1, \sqrt{3})$.

44. $x^4 = y^2 + x^6$

Then graph the tangent line to the graph at the point $(-0.8, 0.384)$.

45. $y^4 = y^2 - x^2$

46. $x^3 = y^2(2 - x)$

47. $y^2 = x^3$

CHAPTER 3 SUMMARY AND REVIEW

Terms to Know

Increasing function, p. 187
Decreasing function, p. 187
Critical point, p. 188
Relative maximum, p. 189
Relative minimum, p. 189
Relative extrema, p. 189
First-Derivative Test, p. 192
Concavity, pp. 202, 203
Second-Derivative Test, p. 204

Point of inflection, p. 208
Rational function, p. 217
Vertical asymptote, p. 221
Horizontal asymptote, p. 222
Oblique asymptote, p. 225
x-intercepts, p. 225
y-intercept, p. 225
Absolute minimum, p. 235
Absolute maximum, p. 235

Maximum–Minimum Principle 1, p. 236
Maximum–Minimum Principle 2, p. 240
Delta notation, p. 266
Differential, p. 270
Implicit differentiation, p. 273
Related rates, p. 275

Review Exercises

These exercises are for test preparation. They can also be used as a lengthened practice test. Answers are at the back of the book. The answers also contain bracketed section references, which tell you where to restudy if your answer is incorrect.

Find the relative extrema of the function. List your answers in terms of ordered pairs. Then sketch a graph

of the function.

1. $f(x) = 3 - 2x - x^2$

2. $f(x) = x^4 - 2x^2 + 3$

3. $f(x) = \dfrac{-8x}{x^2 + 1}$

4. $f(x) = 4 + (x - 1)^3$

5. $f(x) = x^3 + x^2 - x + 3$

6. $f(x) = 3x^{2/3}$

7. $f(x) = 2x^3 - 3x^2 - 12x + 10$

8. $f(x) = x^3 - 3x + 2$

Sketch a graph of the function.

9. $f(x) = \dfrac{-4}{x - 3}$

10. $f(x) = \dfrac{1}{2}x + \dfrac{1}{x}$

11. $f(x) = \dfrac{x^2 - 2x + 2}{x - 1}$

12. $f(x) = \dfrac{x}{x - 2}$

Find the absolute maximum and minimum values of the function, if they exist, over the indicated interval. Where no interval is specified, use the real line.

13. $f(x) = \dfrac{1}{3}x^3 + 3x^2 + 9x + 2$

14. $f(x) = x^2 - 10x + 8;\quad [-2, 6]$

15. $f(x) = 4x^3 - 6x^2 - 24x + 5;\quad [-2, 3]$

16. $f(x) = 5x - 7;\quad [-1, 1]$

17. $f(x) = x^2 - \dfrac{2}{x};\quad (-\infty, 0)$

18. $f(x) = 5x - 7$

19. $f(x) = 3x^4 + 2x^3 - 3x^2 + 1;\quad [-3, 0]$

20. $f(x) = 5x^2 + \dfrac{5}{x^2};\quad (0, \infty)$

21. $f(x) = 5x^4 - x^5;\quad [-1, 5]$

22. $f(x) = -x^2 + 5x + 7$

23. Of all numbers whose sum is 60, find the two that have the maximum product.

24. Find the minimum value of $Q = x^2 - 2y^2$, where $x - 2y = 1$.

25. *Business: Maximizing Profit.* If

$$R(x) = 52x - 0.5x^2 \quad \text{and} \quad C(x) = 22x - 1,$$

find the maximum profit and the number of units that must be produced and sold in order to yield this maximum profit.

26. A rectangular box with a square base and a cover is to contain 2500 ft³. If the cost per square foot for the bottom is $2, for the top is $3, and for the sides is $1, what should the dimensions be in order to minimize the cost?

27. *Business: Minimizing Inventory Cost.* A store in California sells 360 multispeed bicycles per year. It costs $8 to store one bicycle for one year. To reorder, there is a fixed cost of $10, plus $2 for each bicycle. How many times per year should the store order bicycles, and in what lot size, in order to minimize inventory costs?

Given $y = f(x) = x^3 - x$.

28. Find Δy and $f'(x)\,\Delta x$, given that $x = 3$ and $\Delta x = -0.5$.

29. a) Find dy.
    b) Find dy when $x = 2$ and $dx = 0.01$.

30. Approximate $\sqrt{69}$ using $\Delta y \approx f'(x)\,\Delta x$.

31. Differentiate the following implicitly to find dy/dx. Then find the slope of the curve at the given point.

$$2x^3 + 2y^3 = -9xy;\quad (-1, -2)$$

32. A ladder 25 ft long leans against a vertical wall. If the lower end is being moved away from the wall at the rate of 6 ft/sec, how fast is the height of the top decreasing when the lower end is 7 ft from the wall?

33. *Business: Total Revenue, Cost, and Profit.* Find the rates of change of total revenue, cost, and profit for

$$R(x) = 120x - 0.5x^2 \quad \text{and} \quad C(x) = 15x + 6,$$

when $x = 100$ and $dx/dt = 30$ units per day.

Synthesis

34. Find the absolute maximum and minimum values, if they exist, over the indicated interval.

$$f(x) = (x - 3)^{2/5};\quad (-\infty, \infty)$$

35. Differentiate implicitly to find dy/dx:

$$(x - y)^4 + (x + y)^4 = x^6 + y^6.$$

36. Find the relative maxima and minima of

$$y = x^4 - 8x^3 - 270x^2.$$

 TECHNOLOGY   CONNECTION

Use a grapher to estimate the relative extrema of the function.

37. $f(x) = 3.8x^5 - 18.6x^3$

38. $f(x) = \sqrt[3]{|9 - x^2|} - 1$

39. *Life and Physical Sciences: Incidence of Breast Cancer.* The following table provides data relating the incidence of breast cancer per 100,000 women of various ages.

Age	Incidence per 100,000
0	0
27	10
32	25
37	60
42	125
47	187
52	224
57	270
62	340
67	408
72	437
77	475
82	460
87	420

Source: National Cancer Institute

a) Use a grapher with a **REGRESSION** feature to fit linear, quadratic, cubic, and quartic functions to the data.

tw b) Decide which function best fits the data and explain your reasons.

tw c) Determine the domain of the function on the basis of the function and the problem situation and explain.

d) Determine the maximum value of the function on the domain. At what age is the incidence of breast cancer the greatest?

Note: The function used in Example 7 of Section 1.1 was found in this manner.

CHAPTER 3 TEST

Find the relative extrema of the function. List your answers in terms of ordered pairs. Then sketch a graph of the function.

1. $f(x) = x^2 - 4x - 5$
2. $f(x) = 2x^4 - 4x^2 + 1$
3. $f(x) = (x - 2)^{2/3} - 4$
4. $f(x) = \dfrac{16}{x^2 + 4}$
5. $f(x) = x^3 + x^2 - x + 1$
6. $f(x) = 4 + 3x - x^3$
7. $f(x) = (x + 2)^3$
8. $f(x) = x\sqrt{9 - x^2}$

Sketch a graph of the function.

9. $f(x) = \dfrac{2}{x - 1}$
10. $f(x) = \dfrac{-8}{x^2 - 4}$

11. $f(x) = \dfrac{x^2 - 1}{x}$
12. $f(x) = \dfrac{x + 2}{x - 3}$

Find the absolute maximum and minimum values of the function, if they exist, over the indicated interval. Where no interval is specified, use the real line.

13. $f(x) = x(6 - x)$
14. $f(x) = x^3 + x^2 - x + 1;\ \left[-2, \frac{1}{2}\right]$
15. $f(x) = -x^2 + 8.6x + 10$
16. $f(x) = -2x + 5;\ [-1, 1]$
17. $f(x) = -2x + 5$
18. $f(x) = 3x^2 - x - 1$
19. $f(x) = x^2 + \dfrac{128}{x};\ (0, \infty)$
20. Of all numbers whose difference is 8, find the two that have the minimum product.
21. Minimize $Q = x^2 + y^2$, where $x - y = 10$.

22. Business: Maximum Profit. Find the maximum profit and the number of units that must be produced and sold in order to yield the maximum profit.

$$R(x) = x^2 + 110x + 60,$$
$$C(x) = 1.1x^2 + 10x + 80$$

23. From a thin piece of cardboard 60 in. by 60 in., square corners are cut out so the sides can be folded up to make a box. What dimensions will yield a box of maximum volume? What is the maximum volume?

24. Business: Minimizing Inventory Costs. A sporting goods store sells 1225 tennis rackets per year. It costs $2 to store one tennis racket for one year. To reorder, there is a fixed cost of $1, plus $0.50 for each tennis racket. How many times per year should the sporting goods store order tennis rackets, and in what lot size, in order to minimize inventory costs?

25. For $y = f(x) = x^2 - 3$, $x = 5$, and $\Delta x = 0.1$, find Δy and $f'(x) \, \Delta x$.

26. Approximate $\sqrt{104}$ using $\Delta y \approx f'(x) \, \Delta x$.

27. For $y = \sqrt{x^2 + 3}$:

   a) Find dy.
   b) Find dy when $x = 2$ and $dx = 0.01$.

28. Differentiate the following implicitly to find dy/dx. Then find the slope of the curve at the given point.

$$x^3 + y^3 = 9; \quad (1, 2)$$

29. A board 13 ft long leans against a vertical wall. If the lower end is being moved away from the wall at the rate of 0.4 ft/sec, how fast is the upper end coming down when the lower end is 12 ft from the wall?

Synthesis

30. Find the absolute maximum and minimum values of the function, if they exist, over the indicated interval.

$$f(x) = \frac{x^2}{1 + x^3}; \quad [0, \infty)$$

31. Business: Minimizing Average Cost. The total cost of producing x units of a product is given by

$$C(x) = 100x + 100\sqrt{x} + \frac{\sqrt{x^3}}{100}.$$

   a) Find the average cost $A(x)$.
   b) Find the minimum value of $A(x)$.

 TECHNOLOGY CONNECTION

32. Use a grapher to estimate the relative extrema of the function.

$$f(x) = 5x^3 - 30x^2 + 45x + 5\sqrt{x}$$

33. Business: Advertising. The business of manufacturing and selling bowling balls is one of constant change. Companies must introduce new models to the market about every 3 or 4 months. Typically, a new model is created because of advances in technology such as new surface stock or a new way to place weight blocks in a ball. To decide how to use its advertising dollars best, companies track the sales of their bowling balls in relation to the amount spent on advertising. Suppose that a company has collected the following data from past sales.

Amount Spent on Advertising (in thousands)	Number of Bowling Balls Sold, N
$ 0	8
50	13,115
100	19,780
150	22,612
200	20,083
250	12,430
300	4

   a) Use a grapher with a **REGRESSION** feature to fit linear, quadratic, cubic, and quartic functions to the data.
   tw b) Decide which function best fits the data and explain your reasons.
   tw c) Determine the domain of the function on the basis of the function and the problem situation and explain.
   d) Determine the maximum value of the function on the domain. How much should the company spend on its next new model in order to maximize the number of bowling balls sold?

Note: The function used in Exercise 98 of Exercise Set 3.4 might have been found in this manner.

EXTENDED TECHNOLOGY APPLICATION

Maximum Sustainable Harvest

In certain situations, biologists are able to determine what is called a **reproduction curve.** This is a function

$$y = f(P)$$

such that if P is the population at a certain time t, then the population one year later, at time $t + 1$, is $f(P)$. Such a curve is shown below.

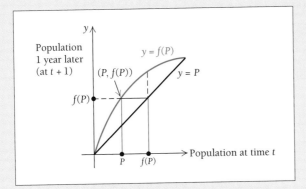

The line $y = P$ is significant for two reasons. First, if $y = P$ is a description of f, then we know that the population stays the same from year to year. But the graph of f above lies mostly above the line, indicating that, in this case, the population is increasing.

Too many deer in a forest can deplete the food supply and eventually cause the population to decrease for lack of food. Often in such cases and with some controversy, hunters are allowed to "harvest" some of the deer. Then with a greater food supply, the remaining deer population might actually prosper and increase.

We know that a population P will grow to a population $f(P)$ in one year. If this were a population of fur-bearing animals and the population were increasing, then one could "harvest" the amount

$$f(P) - P$$

each year without depleting the initial population P. If the population were remaining the same or decreasing, then such a harvest would deplete the population.

Suppose that we want to know the value of P_0 that would allow the harvest to be the largest. If we could determine that P_0, then we could let the population grow until it reached that level, and then begin harvesting year after year the amount $f(P_0) - P_0$.

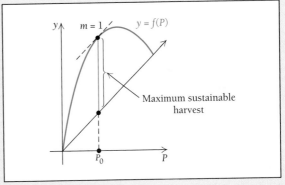

Let the harvest function H be given by

$$H(P) = f(P) - P.$$

Then $H'(P) = f'(P) - 1.$

Now, if we assume that $H'(P)$ exists for all values of P and that there is only one critical point, it follows that the **maximum sustainable harvest** occurs at that value P_0 such that

$$H'(P_0) = f'(P_0) - 1 = 0$$

and $H''(P_0) = f''(P_0) < 0.$

Or, equivalently, we have the following.

> ### Theorem
>
> The **maximum sustainable harvest** occurs at P_0 such that
>
> $$f'(P_0) = 1 \quad \text{and} \quad f''(P_0) < 0,$$
>
> and is given by
>
> $$H(P_0) = f(P_0) - P_0.$$

Exercises

For each reproduction curve in Exercises 1–3, do the following.

a) Graph the reproduction function, the function $y = P$, and the harvest function using the same set of axes or viewing window.
b) Find the population at which the maximum sustainable harvest occurs. Use both a graphical solution and a calculus solution.
c) Find the maximum sustainable harvest.

1. $f(P) = P(10 - P)$, where P is measured in thousands.

2. $f(P) = -0.025P^2 + 4P$, where P is measured in thousands. This is the reproduction curve in the Hudson bay area for the snowshoe hare, a fur-bearing animal.

3. $f(P) = -0.01P^2 + 2P$, where P is measured in thousands. This is the reproduction curve in the Hudson Bay area for the lynx, a fur-bearing animal.

For each reproduction curve in Exercises 4 and 5, do the following.

a) Graph the reproduction function, the function $y = P$, and the harvest function using the same set of axes or viewing window.
b) Find the population at which the maximum sustainable harvest occurs. Use just a graphical solution.
c) Find the maximum sustainable harvest.

4. $f(P) = 40\sqrt{P}$, where P is measured in thousands. Assume that this is the reproduction curve for the brown trout population in a large lake.

5. $f(P) = 0.237P\sqrt{2000 - P^2}$, where P is measured in thousands.

6. The following table lists data regarding the reproduction of a certain animal.

Population, P (in thousands)	Population, $f(P)$, One Year Later
10	9.7
20	23.1
30	37.4
40	46.2
50	42.6

a) Use the REGRESSION feature on a grapher to fit a cubic polynomial to these data.
b) Graph the reproduction function, the function $y = P$, and the harvest function using the same set of axes or viewing window.
c) Find the population at which the maximum sustainable harvest occurs. Use just a graphical solution.

4

Exponential and Logarithmic Functions

INTRODUCTION

In this chapter, we consider two types of functions that are closely related: *exponential functions* and *logarithmic functions*. We will learn to find derivatives of such functions. Both are rich in applications such as population growth, decay, interest compounded continuously, spread of rumors, and carbon dating.

AN APPLICATION

Titanic is one of the greatest box-office attractions of all time. Having opened in December 1997, it was still running in theaters in August 1998 when it went into videotape sales. The total U.S. box-office revenue R, in millions of dollars, after time t, in weeks, can be approximated by the logistic function

$$R(t) = \frac{596.423}{1 + 4.974e^{-0.1814t}}.$$

(*Source*: Exhibitor Relations Co, Inc.) Sketch a graph of the function.

This is a *logistic function*. The graph is as follows.

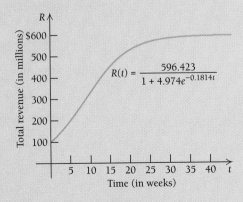

This problem appears as Example 8 in Section 4.3.

4.1

Exponential Functions

Graphs of Exponential Functions

Consider the following graphs. The rapid rise of the graphs indicates that they approximate *exponential functions*. We now consider such functions and many of their applications.

World Population Growth

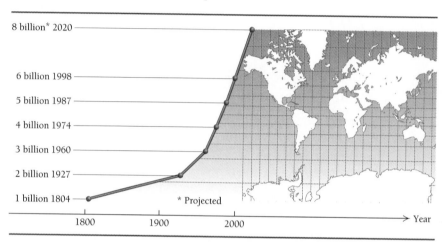

Source: U.S. Bureau of the Census

Let's review definitions of such expressions as a^x, where x is a rational number. For example,

$$a^{2.34} \quad \text{or} \quad a^{234/100}$$

means "raise a to the 234th power and then take the 100th root $\left(\sqrt[100]{a^{234}}\right)$."

What about expressions with irrational exponents, such as $2^{\sqrt{2}}$, 2^{π}, or $2^{-\sqrt{3}}$? An irrational number is a number named by an infinite, nonrepeating decimal. Let's consider 2^{π}. We know that π is irrational with infinite, nonrepeating decimal expansion:

$$3.141592653\ldots.$$

This means that π is approached as a limit by the rational numbers

$$3, \ 3.1, \ 3.14, \ 3.141, \ 3.1415, \ \ldots,$$

so it seems reasonable that 2^{π} should be approached as a limit by the rational powers

$$2^3, \ 2^{3.1}, \ 2^{3.14}, \ 2^{3.141}, \ 2^{3.1415}, \ \ldots.$$

Estimating each power with a calculator, we get the following:

$$8, \ 8.574188, \ 8.815241, \ 8.821353, \ 8.824411, \ \ldots.$$

In general, a^x is approximated by the values of a^r for rational numbers r near x;

We can approximate 2^π using the exponential keys on a grapher:

$2^\pi \approx 8.824977827$.

Exercises

Approximate.

1. 5^π 2. $5^{\sqrt{3}}$

3. $7^{-\sqrt{2}}$ 4. $18^{-\pi}$

a^x is the limit of a^r as r approaches x through rational values. In summary, for $a > 0$, the definition of a^x for rational numbers x can be extended to arbitrary real numbers x in such a way that the usual laws of exponents, such as

$$a^x \cdot a^y = a^{x+y}, \qquad a^x \div a^y = a^{x-y}, \qquad (a^x)^y = a^{xy}, \quad \text{and} \quad a^{-x} = \frac{1}{a^x},$$

still hold. Moreover, the function so obtained, $f(x) = a^x$, is continuous.

Definition

An **exponential function** f is given by

$$f(x) = a^x,$$

where x is any real number, $a > 0$, and $a \neq 1$. The number a is called the **base**.

The following are examples of exponential functions:

$$f(x) = 2^x, \qquad f(x) = \left(\tfrac{1}{2}\right)^x, \qquad f(x) = (0.4)^x.$$

Note that in contrast to power functions like $y = x^2$ or $y = x^3$, the variable in an exponential function is in the exponent, not the base. Exponential functions have extensive application. Let's consider their graphs.

Example 1   Graph: $y = f(x) = 2^x$.

Solution

a) First, we find some function values.

x	$y = f(x)$ (or 2^x)
0	1
$\frac{1}{2}$	1.4
1	2
2	4
3	8
-1	$\frac{1}{2}$
-2	$\frac{1}{4}$

Note: For:

$x = 0, \qquad y = 2^0 = 1;$

$x = \frac{1}{2}, \qquad y = 2^{1/2} = \sqrt{2} \approx 1.4;$

$x = 1, \qquad y = 2^1 = 2;$

$x = 2, \qquad y = 2^2 = 4;$

$x = 3, \qquad y = 2^3 = 8;$

$x = -1, \quad y = 2^{-1} = \frac{1}{2};$

$x = -2, \quad y = 2^{-2} = \frac{1}{2^2} = \frac{1}{4}.$

The curve comes very close to the x-axis, but does not touch or cross it. The x-axis is a horizontal asymptote.

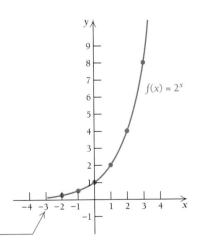

b) Next, we plot the points and connect them with a smooth curve. The graph is continuous, increasing, and concave up. We see too that the x-axis is a horizontal asymptote (see Section 3.3), that is,

$$\lim_{x \to -\infty} f(x) = 0 \quad \text{and} \quad \lim_{x \to \infty} f(x) = \infty.$$

Example 2   Graph: $y = f(x) = \left(\tfrac{1}{2}\right)^x$.

Solution

a) First, we find some function values. Before we do so, note that

$$y = f(x) = \left(\tfrac{1}{2}\right)^x$$
$$= (2^{-1})^x$$
$$= 2^{-x}.$$

This will ease our work.

x	0	$\frac{1}{2}$	1	2	-1	-2	-3
y	1	0.7	$\frac{1}{2}$	$\frac{1}{4}$	2	4	8

Note: For:

$$x = 0, \quad y = 2^{-0} = 1;$$
$$x = \frac{1}{2}, \quad y = 2^{-1/2} = \frac{1}{2^{1/2}}$$
$$= \frac{1}{\sqrt{2}} \approx \frac{1}{1.4} \approx 0.7;$$
$$x = 1, \quad y = 2^{-1} = \tfrac{1}{2};$$
$$x = 2, \quad y = 2^{-2} = \tfrac{1}{4};$$
$$x = -1, \quad y = 2^{-(-1)} = 2;$$
$$x = -2, \quad y = 2^{-(-2)} = 4;$$
$$x = -3, \quad y = 2^{-(-3)} = 8.$$

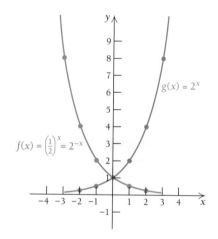

TECHNOLOGY
CONNECTION

Check the graphs of the functions in Examples 1 and 2. Then graph

$f(x) = 3^x$ and

$g(x) = \left(\tfrac{1}{3}\right)^x$

and look for patterns.

b) Next, we plot these points and connect them with a smooth curve, as shown by the red curve in the figure. The graph is continuous, decreasing, and concave up. We see too that the x-axis is a horizontal asymptote, that is,

$$\lim_{x \to \infty} f(x) = 0 \quad \text{and} \quad \lim_{x \to -\infty} f(x) = \infty.$$

The graph of $g(x) = 2^x$, the blue curve, is shown for comparison.

The following are some properties of the exponential function.

1. The function $f(x) = a^x$, where $a > 1$, is a positive, increasing, continuous function. As x gets smaller, a^x approaches 0. The graph is concave up, as shown at right. (If you studied Section 3.3, you also know that the x-axis is an asymptote.)

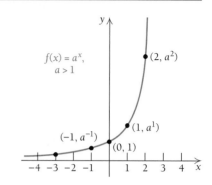

2. The function $f(x) = a^x$, where $0 < a < 1$, is a positive, decreasing, continuous function. As x gets larger, a^x approaches 0. The graph is concave up, as shown at right. (The x-axis is an asymptote.)

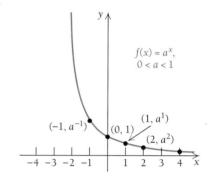

When $a = 1$, $f(x) = a^x = 1^x = 1$, and is a constant function. This is why we do not allow 1 to be the base of an exponential function.

The Derivative of a^x; the Number e

Let's consider finding the derivative of the exponential function

$$f(x) = a^x.$$

The derivative is given by

$$f'(x) = \lim_{h \to 0} \frac{f(x + h) - f(x)}{h} \qquad \text{Definition of the derivative}$$

$$= \lim_{h \to 0} \frac{a^{x+h} - a^x}{h} \qquad \text{Substituting } a^{x+h} \text{ for } f(x + h) \text{ and } a^x \text{ for } f(x)$$

$$= \lim_{h \to 0} \frac{a^x \cdot a^h - a^x \cdot 1}{h}$$

$$= \lim_{h \to 0} a^x \cdot \left(\frac{a^h - 1}{h} \right)$$

$$= a^x \cdot \lim_{h \to 0} \frac{a^h - 1}{h}. \qquad \text{Since the variable is } h \text{ and } h \text{ approaches 0, we treat } a^x \text{ as a constant, and the limit of a constant times a function is the constant times the limit.}$$

We get

$$f'(x) = a^x \cdot \lim_{h \to 0} \frac{a^h - 1}{h}. \tag{1}$$

In particular, for $g(x) = 2^x$,

$$g'(x) = 2^x \cdot \lim_{h \to 0} \frac{2^h - 1}{h}.$$

Note that the limit does not depend on the value of x at which we are evaluating the derivative. In order for $g'(x)$ to exist, we must determine whether

$$\lim_{h \to 0} \frac{2^h - 1}{h} \quad \text{exists.}$$

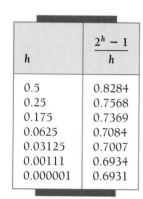

h	$\dfrac{2^h - 1}{h}$
0.5	0.8284
0.25	0.7568
0.175	0.7369
0.0625	0.7084
0.03125	0.7007
0.00111	0.6934
0.000001	0.6931

Let's investigate this question.

We choose a sequence of numbers h approaching 0 and compute $(2^h - 1)/h$, listing the results in a table, as shown at left. It seems reasonable to assume that $(2^h - 1)/h$ has a limit as h approaches 0 and that its approximate value is 0.7; thus,

$$g'(x) \approx (0.7)2^x.$$

In other words, the derivative is a constant times the function value 2^x. Similarly, for $t(x) = 3^x$,

$$t'(x) = 3^x \cdot \lim_{h \to 0} \frac{3^h - 1}{h}.$$

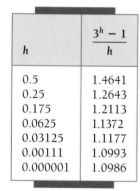

h	$\dfrac{3^h - 1}{h}$
0.5	1.4641
0.25	1.2643
0.175	1.2113
0.0625	1.1372
0.03125	1.1177
0.00111	1.0993
0.000001	1.0986

Again, we can find an approximation for the limit that does not depend on the value of x at which we are evaluating the derivative. Consider the table shown at left. Again, it seems reasonable to assume that $(3^h - 1)/h$ has a limit as h approaches 0 and that its approximate value is 1.1; thus

$$t'(x) \approx (1.1)3^x.$$

In other words, the derivative is a constant times the function value 3^x.

Let's now observe and analyze what we have done. We proved that

$$\text{if } f(x) = a^x, \quad \text{then } f'(x) = \lim_{h \to 0} \frac{a^h - 1}{h} \cdot a^x.$$

Consider $\displaystyle\lim_{h \to 0} \frac{a^h - 1}{h}$.

TECHNOLOGY CONNECTION

Exercises

Use a grapher, set in ASK mode, to produce the tables above.

1. For $a = 2$,

$$\lim_{h \to 0} \frac{a^h - 1}{h} = \lim_{h \to 0} \frac{2^h - 1}{h} \approx 0.7.$$

2. For $a = 3$,

$$\lim_{h \to 0} \frac{a^h - 1}{h} = \lim_{h \to 0} \frac{3^h - 1}{h} \approx 1.1.$$

It seems reasonable to assume that there exists some number between 2 and 3 such that

$$\lim_{h \to 0} \frac{a^h - 1}{h} = 1$$

and

$$f'(x) = a^x \cdot \lim_{h \to 0} \frac{a^h - 1}{h} = a^x \cdot 1 = f(x).$$

Stated in another way, there should exist some number between 2 and 3 such that the derivative of the function is the same as the function itself. We define that number to be e. We will soon see that the number is approximated as

$$e \approx 2.718281828459.$$

Definition
The number e is the unique positive real number for which

$$\lim_{h \to 0} \frac{e^h - 1}{h} = 1.$$

It follows that for the exponential function $f(x) = e^x$,

$$f'(x) = e^x \cdot \lim_{h \to 0} \frac{e^h - 1}{h}$$
$$= e^x \cdot 1$$
$$= e^x.$$

That is, the derivative of e^x is e^x.

Compound Interest and an Approximation for e

An application for e using the compound-interest formula leads to

$$A = P\left(1 + \frac{i}{n}\right)^{nt},$$

which we developed in Chapter 1, where A is the amount that an initial investment P will be worth after t years at interest rate i, compounded n times per year.

Suppose that \$1 is an initial investment at 100% interest for 1 yr (though obviously no financial institution would pay this). The formula becomes

$$A = \left(1 + \frac{1}{n}\right)^n.$$

Suppose we were to have the compounding periods n increase indefinitely. Let's

Exploratory.
Graph

$$y = \left(1 + \frac{1}{x}\right)^x$$

using the viewing
window $[0, 5000, 0, 5]$,
with $Xscl = 1000$ and
$Yscl = 1$. Trace along
the graph. Why does
the graph appear to be
horizontal? As you
trace to the right, note
the values of the
y-coordinate. Is it
approaching a
constant? What seems
to be its value?

investigate the behavior of the function A. We obtain the following table of values
and graph.

n	$A = \left(1 + \dfrac{1}{n}\right)^n$
1	$2.00000
2	$2.25000
3	$2.37037
4	$2.44141
12	$2.61304
52	$2.69260
365	$2.71457
8,760	$2.71813
525,600	$2.71828

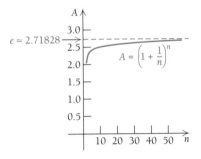

The amount in the investment grows at interest compounded continuously and approaches a limit. That limit and an approximation of e are given in the following
theorem.

Theorem 1

$$e = \lim_{n \to \infty} \left(1 + \frac{1}{n}\right)^n$$

$$\approx 2.718281828459$$

That is, e is that number which

$$\left(1 + \frac{1}{n}\right)^n$$

approaches as n gets larger without bound.

Finding Derivatives of Functions Involving e

We have established that for the function $f(x) = e^x$, we also have $f'(x) = e^x$, or,
simply, the following.

Theorem 2

$$\frac{d}{dx}e^x = e^x$$

E x p l o r a t o r y .   Graph $f(x) = e^x$ using the viewing window $[-5, 5, -1, 10]$. Trace to a point and observe the y-value. Then find the value of y' at that point and compare y and y'.

Repeat this process for three other values of x. What do you observe?

Let's compare the graphs of $g(x) = 2^x$ and $g'(x) \approx (0.7)2^x$, $t(x) = 3^x$ and $t'(x) \approx (1.1)3^x$, and $f(x) = e^x$ and $f'(x) = e^x$. Note that the graph of g' lies *below* the graph of g and the graph of t' lies *above* the graph of t, but the graph of f' is exactly the same as the graph of f.

Theorem 2 says that for the function $f(x) = e^x$, the derivative $f'(x) = e^x$ (the slope of the tangent line) is the same as the function value at any x.

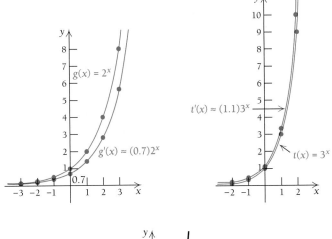

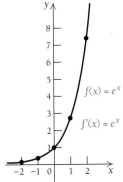

Let's find some other derivatives.

Example 3

$$\frac{d}{dx} 3e^x = 3\frac{d}{dx} e^x$$
$$= 3e^x$$

Example 4

$$\frac{d}{dx}(x^2 e^x) = x^2 \cdot e^x + 2x \cdot e^x \qquad \text{By the Product Rule}$$

$$= e^x(x^2 + 2x), \quad \text{or} \quad xe^x(x + 2) \qquad \text{Factoring}$$

Check the results of
Examples 3–5 using a
grapher. This assumes
that you have a
grapher that graphs f
and f'.

Then differentiate
$f(x) = e^x/x^2$ and
check your answer
using the grapher.

Example 5

$$\frac{d}{dx}\left(\frac{e^x}{x^3}\right) = \frac{x^3 \cdot e^x - 3x^2 \cdot e^x}{x^6} \qquad \text{By the Quotient Rule}$$

$$= \frac{x^2 e^x(x-3)}{x^6} \qquad \text{Factoring}$$

$$= \frac{e^x(x-3)}{x^4} \qquad \text{Simplifying}$$

Suppose that we have a more complicated function in the exponent, such as

$$h(x) = e^{x^2 - 5x}.$$

This is a composition of functions. In general, we have

$$h(x) = e^{f(x)} = g[f(x)], \quad \text{where} \quad g(x) = e^x.$$

Now $g'(x) = e^x$. Then by the Chain Rule (Section 2.8), we have

$$h'(x) = g'[f(x)] \cdot f'(x)$$
$$= e^{f(x)} \cdot f'(x).$$

For the case above, $f(x) = x^2 - 5x$, so $f'(x) = 2x - 5$. Then

$$h'(x) = g'[f(x)] \cdot f'(x)$$
$$= e^{f(x)} \cdot f'(x)$$
$$= e^{x^2 - 5x}(2x - 5).$$

The next rule, which we have proven using the Chain Rule, allows us to find derivatives of functions like the one above.

Theorem 3

$$\frac{d}{dx}e^{f(x)} = f'(x)\,e^{f(x)},$$

or

$$\frac{d}{dx}e^u = \frac{du}{dx} \cdot e^u.$$

The derivative of e to some power is the derivative of the power times e to the power.

The following gives us a way to remember this rule.

$$g(x) = e^{x^2 - 5x} \quad ①$$

$$g'(x) = (2x - 5)e^{x^2 - 5x}$$

$$②$$

① Take the derivative of the exponent.

② Multiply the derivative of the exponent by the original function.

Example 6

$$\frac{d}{dx}e^{5x} = 5e^{5x}$$

Example 7

$$\frac{d}{dx}e^{-x^2+4x-7} = (-2x + 4)e^{-x^2+4x-7}$$

Example 8

$$\frac{d}{dx}e^{\sqrt{x^2-3}} = \frac{d}{dx}e^{(x^2-3)^{1/2}}$$

$$= \frac{1}{2}(x^2 - 3)^{-1/2} \cdot 2x \cdot e^{(x^2-3)^{1/2}}$$

$$= x(x^2 - 3)^{-1/2} \cdot e^{\sqrt{x^2-3}}$$

$$= \frac{xe^{\sqrt{x^2-3}}}{\sqrt{x^2 - 3}}$$

Graphs of e^x, e^{-x}, and $1 - e^{-kx}$

Now that we know how to find the derivative of $f(x) = e^x$, let's look at the graph of $f(x) = e^x$ from the standpoint of calculus concepts and the curve-sketching techniques discussed in Section 3.2.

Example 9   Graph: $f(x) = e^x$. Analyze the graph using calculus.

Solution   If all we want is a quick graph, we might simply find some function values using a calculator, plot the points, and sketch the graph as shown below.

x	$f(x)$
-2	0.135
-1	0.368
0	1
1	2.718
2	7.389

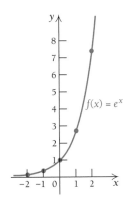

However, we can analyze the graph using calculus as follows.

a) *Derivatives.* Since $f(x) = e^x$, it follows that $f'(x) = e^x$, so $f''(x) = e^x$.

b) *Critical points of f.* Since $f'(x) = e^x > 0$ for all real numbers x, we know that the derivative exists for all real numbers and there is no solution of the equation $f'(x) = 0$. There are no critical points and therefore no maximum or minimum values.

c) *Increasing.* We have $f'(x) = e^x > 0$ for all real numbers x, so the function f is increasing over the entire real line, $(-\infty, \infty)$.

d) *Inflection points.* We have $f''(x) = e^x > 0$ for all real numbers x, so the equation $f''(x) = 0$ has no solution and there are no points of inflection.

e) *Concavity.* Since $f''(x) = e^x > 0$ for all real numbers x, the function f' is increasing and the graph is concave up over the entire real line. ◄

Example 10   Graph: $g(x) = e^{-x}$. Analyze the graph using calculus.

Solution   First, we find some function values using a calculator, plot the points, and sketch the graph as shown below.

x	$g(x)$
-2	7.389
-1	2.718
0	1
1	0.368
2	0.135

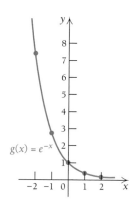

We can then analyze the graph using calculus as follows.

a) *Derivatives.* Since $g(x) = e^{-x}$, we have

$$g'(x) = (-1)e^{-x} = -e^{-x},$$

so

$$g''(x) = (-1)(-1)e^{-x} = e^{-x}.$$

b) *Critical points of g.* Since the expression $e^{-x} > 0$, the derivative $g'(x) = -e^{-x} < 0$ for all real numbers x. Thus the derivative exists for all real numbers, and the equation $g'(x) = 0$ has no solution. There are no critical points and therefore no maximum or minimum values.

c) *Decreasing.* Since the derivative $g'(x) = -e^{-x} < 0$ for all real numbers x, the function g is decreasing over the entire real line, $(-\infty, \infty)$.

d) *Inflection points.* We have $g''(x) = e^{-x} > 0$, so the equation $g''(x) = 0$ has no solution and there are no points of inflection.

e) *Concavity.* We also know that since $g''(x) = e^{-x} > 0$ for all real numbers x, the function g' is increasing and the graph is concave up over the entire real line. ◄

Functions of the type $f(x) = 1 - e^{kx}$, $x \geq 0$, also have important applications.

Example 11   Graph: $h(x) = 1 - e^{-2x}$, $x \geq 0$. Analyze the graph using calculus.

Solution   First, we find some function values using a calculator, plot the points, and sketch the graph as shown below.

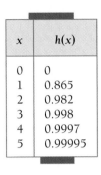

x	$h(x)$
0	0
1	0.865
2	0.982
3	0.998
4	0.9997
5	0.99995

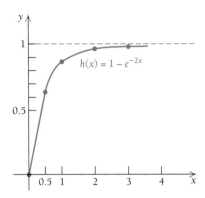

We can then analyze the graph using calculus as follows.

a) *Derivatives.* Since $h(x) = 1 - e^{-2x}$,

$$h'(x) = (-2)(-1)e^{-2x} = 2e^{-2x}$$

and

$$h''(x) = (-2)(2e^{-2x}) = -4e^{-2x}.$$

b) *Critical points.* Since the expression $e^{-2x} > 0$, the derivative $h'(x) = 2e^{-2x} > 0$ for all real numbers x. Thus the derivative exists for all real numbers, and the equation $h'(x) = 0$ has no solution. There are no critical points.

c) *Increasing.* Since the derivative $h'(x) = 2e^{-2x} > 0$ for all real numbers x, h is increasing over the entire real line.

d) *Inflection points.* Since $h''(x) = -4e^{-2x} < 0$, we know that the equation $h''(x) = 0$ has no solution, so there are no points of inflection.

e) *Concavity.* Since $h''(x) = -4e^{-2x} < 0$ for all real numbers x, the function h' is decreasing and the graph is concave down over the entire real line.   ◄

In general, the graph of $h(x) = 1 - e^{-kx}$, for $k > 0$ and $x \geq 0$, is increasing, which we expect since $h'(x) = ke^{-kx}$ is always positive. We also see that $h(x)$ approaches 1 as x approaches ∞; that is, $\lim_{x \to \infty} (1 - e^{-kx}) = 1$.

A word of caution! Functions of the type a^x (for example, 2^x, 3^x, and e^x) are different from functions of the type x^a (for example, x^2, x^3, $x^{1/2}$). For a^x, the variable is in the exponent. For x^a, the variable is in the base. The derivative of a^x is not xa^{x-1}. In particular, we have the following:

$$\frac{d}{dx}e^x \neq xe^{x-1}, \quad \text{but} \quad \frac{d}{dx}e^x = e^x.$$

EXERCISE SET 4.1

Graph.

1. $y = 4^x$
2. $y = 5^x$
3. $y = (0.4)^x$
4. $y = (0.2)^x$
5. $x = 4^y$
6. $x = 5^y$

Differentiate.

7. $f(x) = e^{3x}$
8. $f(x) = e^{2x}$
9. $f(x) = 5e^{-2x}$
10. $f(x) = 4e^{-3x}$
11. $f(x) = 3 - e^{-x}$
12. $f(x) = 2 - e^{-x}$
13. $f(x) = -7e^x$
14. $f(x) = -4e^x$
15. $f(x) = \frac{1}{2}e^{2x}$
16. $f(x) = \frac{1}{4}e^{4x}$
17. $f(x) = x^4 e^x$
18. $f(x) = x^5 e^x$
19. $f(x) = \dfrac{e^x}{x^4}$
20. $f(x) = \dfrac{e^x}{x^5}$
21. $f(x) = e^{-x^2+7x}$
22. $f(x) = e^{-x^2+8x}$
23. $f(x) = e^{-x^2/2}$
24. $f(x) = e^{x^2/2}$
25. $y = e^{\sqrt{x-7}}$
26. $y = e^{\sqrt{x-4}}$
27. $y = \sqrt{e^x - 1}$
28. $y = \sqrt{e^x + 1}$
29. $y = xe^{-2x} + e^{-x} + x^3$
30. $y = e^x + x^3 - xe^x$
31. $y = 1 - e^{-x}$
32. $y = 1 - e^{-3x}$
33. $y = 1 - e^{-kx}$
34. $y = 1 - e^{-mx}$

Graph the function. Then analyze the graph using calculus.

35. $f(x) = e^{2x}$
36. $f(x) = e^{(1/2)x}$
37. $f(x) = e^{-2x}$
38. $f(x) = e^{-(1/2)x}$
39. $f(x) = 3 - e^{-x}$, for nonnegative values of x
40. $f(x) = 2(1 - e^{-x})$, for nonnegative values of x

41.–46. For each of Exercises 35–40, graph the function and its first and second derivatives using a grapher.

47. Find the tangent line to the graph of $f(x) = e^x$ at the point $(0, 1)$.

48. Find the tangent line to the graph of $f(x) = 2e^{-3x}$ at the point $(0, 2)$.

49. and 50. For each of Exercises 47 and 48, graph the function and the tangent line using a grapher.

Applications ....

BUSINESS AND ECONOMICS

51. *Marginal Cost.* A company's total cost, in millions of dollars, is given by
$$C(t) = 100 - 50e^{-t},$$
where t is the time.

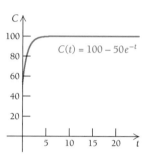

Find each of the following.

a) The marginal cost $C'(t)$
b) $C'(0)$
c) $C'(4)$
d) Find $\lim_{t \to \infty} C(t)$ and $\lim_{t \to \infty} C'(t)$. Why do you think the company's costs tend to level off as time increases?

52. *Marginal Cost.* A company's total cost, in millions of dollars, is given by
$$C(t) = 200 - 40e^{-t},$$
where t is the time.

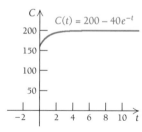

Find each of the following.

a) The marginal cost $C'(t)$
b) $C'(0)$
c) $C'(5)$

tw d) Find $\lim_{t\to\infty} C(t)$ and $\lim_{t\to\infty} C'(t)$. Why do you think the company's costs tend to level off as time increases?

53. *Marginal Demand.* The demand function for a certain kind of VCR is given by the function

$$x = D(p) = 480e^{-0.003p}.$$

a) How many VCRs will be sold when the price is $120? $180? $340?
b) Sketch a graph of $x = D(p)$ for $0 \le p \le 400$.
c) Find the marginal demand $D'(p)$.

tw d) Interpret the meaning of the derivative.

54. *Marginal Supply.* The supply function for the VCR in Exercise 53 is given by the function

$$x = S(p) = 150e^{0.004p}.$$

a) How many VCRs will the seller allow to be sold when the price is $120? $180? $340?
b) Sketch a graph of $x = S(p)$ for $0 \le p \le 400$.
c) Find the marginal supply $S'(p)$.

tw d) Interpret the meaning of the derivative.

LIFE AND PHYSICAL SCIENCES

55. *Medication Concentration.* The concentration C, in parts per million, of a medication in the body t hours after ingestion is given by the function

$$C(t) = 10t^2 e^{-t}.$$

a) Find the concentration after 0 hr; 1 hr; 2 hr; 3 hr; 10 hr.
b) Sketch a graph of the function for $0 \le t \le 10$.
c) Find the rate of change of the concentration $C'(t)$.
d) Find the maximum value of the concentration and where it occurs.

tw e) Interpret the meaning of the derivative.

SOCIAL SCIENCES

56. *Ebbinghaus Learning Model.* Suppose that you are given the task of learning 100% of a block of knowledge. Human nature tells us that we would retain only a percentage P of the knowledge t weeks after we have learned it. The *Ebbinghaus learning model* asserts that P is given by

$$P(t) = Q + (100\% - Q)e^{-kt},$$

where Q is the percentage that we would never forget and k is a constant that depends on the

knowledge learned. Suppose that $Q = 40\%$ and $k = 0.7$.

a) Find the percentage retained after 0 weeks; 1 week; 2 weeks; 6 weeks; 10 weeks.
b) Find $\lim_{t\to\infty} P(t)$.
c) Sketch a graph of P.
d) Find the rate of change of P with respect to time t, $P'(t)$.

tw e) Interpret the meaning of the derivative.

Synthesis

Differentiate.

57. $y = (e^{3x} + 1)^5$

58. $y = (e^{x^2} - 2)^4$

59. $y = \dfrac{e^{3t} - e^{7t}}{e^{4t}}$

60. $y = \sqrt[3]{e^{3t} + t}$

61. $y = \dfrac{e^x}{x^2 + 1}$

62. $y = \dfrac{e^x}{1 - e^x}$

63. $f(x) = e^{\sqrt{x}} + \sqrt{e^x}$

64. $f(x) = \dfrac{1}{e^x} + e^{1/x}$

65. $f(x) = e^{x/2} \cdot \sqrt{x - 1}$

66. $f(x) = \dfrac{xe^{-x}}{1 + x^2}$

67. $f(x) = \dfrac{e^x - e^{-x}}{e^x + e^{-x}}$

68. $f(x) = e^{e^x}$

Each of the following is an expression for e. Find the function values that are approximations for e. Round to five decimal places.

69. $e = \lim_{t\to 0} f(t)$; $f(t) = (1 + t)^{1/t}$
Find $f(1)$, $f(0.5)$, $f(0.2)$, $f(0.1)$, and $f(0.001)$.

70. $e = \lim_{t\to 1} g(t)$; $g(t) = t^{1/(t-1)}$
Find $g(0.5)$, $g(0.9)$, $g(0.99)$, $g(0.999)$, and $g(0.9998)$.

71. Find the maximum value of $f(x) = x^2 e^{-x}$ over $[0, 4]$.

72. Find the minimum value of $f(x) = xe^x$ over $[-2, 0]$.

tw 73. A student made the following error on a test:

$$\frac{d}{dx} e^x = xe^{x-1}.$$

Explain the error and how to correct it.

tw 74. Describe the differences in the graphs of $f(x) = 3^x$ and $g(x) = x^3$.

TECHNOLOGY CONNECTION

Graph each of the following and find the relative extrema.

75. $f(x) = x^2 e^{-x}$ **76.** $f(x) = e^{-x^2}$

For each of the following functions, graph f, f', and f''.

77. $f(x) = e^x$ **78.** $f(x) = e^{-x}$

79. $f(x) = 2e^{0.3x}$

80. $f(x) = 1000e^{-0.08x}$

81. Graph

$$f(x) = \left(1 + \frac{1}{x}\right)^x.$$

Use the **TABLE** feature and very large values of x to confirm that e is approached as a limit.

4.2

OBJECTIVES

➤ Convert between exponential and logarithmic equations.
➤ Solve exponential equations.
➤ Solve problems involving exponential and logarithmic functions.
➤ Differentiate functions involving natural logarithms.

Logarithmic Functions

Graphs of Logarithmic Functions

Suppose that we want to solve the equation

$$10^y = 1000.$$

We are trying to find that power of 10 that will give 1000. We can see that the answer is 3. The number 3 is called the "logarithm, base 10, of 1000."

Definition

A **logarithm** is defined as follows:

$$y = \log_a x \quad \text{means} \quad x = a^y, \quad a > 0, a \neq 1.$$

The number $\log_a x$ is the power y to which we raise a to get x. The number a is called the *logarithmic base*.

For logarithms base 10, $\log_{10} x$ is the power y such that $x = 10^y$. Therefore, a logarithm can be thought of as an exponent. We can convert from a logarithmic equation to an exponential equation, and conversely, as follows.

Logarithmic Equation	Exponential Equation
$\log_a M = N$	$a^N = M$
$\log_{10} 100 = 2$	$10^2 = 100$
$\log_{10} 0.01 = -2$	$10^{-2} = 0.01$
$\log_{49} 7 = \frac{1}{2}$	$49^{1/2} = 7$

In order to graph a logarithmic equation, we can graph its equivalent exponential equation.

Example 1   Graph: $y = \log_2 x$.

Solution   We first write the equivalent exponential equation:

$$x = 2^y.$$

We select values for y and find the corresponding values of 2^y. Then we plot points, remembering that x is still the first coordinate, and connect the points with a smooth curve.

TECHNOLOGY CONNECTION

Exploratory.
Graph $f(x) = 2^x$.
Then use the TABLE feature to find the coordinates of points on the graph. How can each ordered pair help you make a hand drawing of a graph of $g(x) = \log_2 x$?

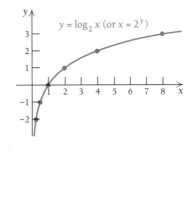

x, or 2^y	y
1	0
2	1
4	2
8	3
$\frac{1}{2}$	-1
$\frac{1}{4}$	-2

① Select y.

② Compute x.

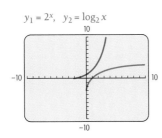

TECHNOLOGY CONNECTION

Graphing Logarithmic Functions

To graph $y = \log_2 x$, we first graph $y = 2^x$. We then use the DRAWINV feature. Both graphs are drawn together.

$y_1 = 2^x, \quad y_2 = \log_2 x$

Exercises
Graph.
1. $y = \log_3 x$ 2. $y = \log_5 x$
3. $f(x) = \log_e x$ 4. $f(x) = \log_{10} x$

The graphs of $f(x) = 2^x$ and $g(x) = \log_2 x$ are shown below using the same set of axes. Note that we can obtain the graph of g by reflecting the graph of f across the line $y = x$. Graphs obtained in this manner are known as *inverses* of each other.

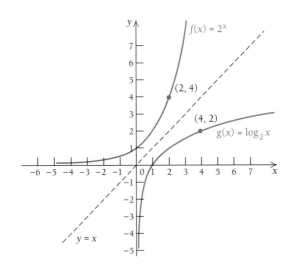

Although we cannot develop inverses in detail here, it is of interest to note that they "undo" each other. For example,

$$f(3) = 2^3 = 8 \qquad \text{The input 3 gives the output 8.}$$

and $\qquad g(8) = \log_2 8 = 3.$ The input 8 gets us back to 3.

Basic Properties of Logarithms

The following are some basic properties of logarithms. The proofs follow from properties of exponents.

Theorem 4

Properties of Logarithms

For any positive numbers M, N, and a, $a \neq 1$, and any real number k:

P1. $\log_a MN = \log_a M + \log_a N$

P2. $\log_a \dfrac{M}{N} = \log_a M - \log_a N$

P3. $\log_a M^k = k \cdot \log_a M$

P4. $\log_a a = 1$

P5. $\log_a a^k = k$

P6. $\log_a 1 = 0$

Proof of P1 and P2: Let $X = \log_a M$ and $Y = \log_a N$. Writing the equivalent exponential equations, we then have

$$M = a^X \quad \text{and} \quad N = a^Y.$$

Then by the properties of exponents (see Section 1.1), we have

$$MN = a^X \cdot a^Y = a^{X+Y},$$

so $\qquad \log_a MN = X + Y$

$$= \log_a M + \log_a N.$$

Also

$$\frac{M}{N} = a^X \div a^Y = a^{X-Y},$$

so $\qquad \log_a \dfrac{M}{N} = X - Y$

$$= \log_a M - \log_a N.$$

Proof of P3: Let $X = \log_a M$. Then

$$M = a^X,$$

so $\qquad M^k = (a^X)^k, \quad \text{or} \quad M^k = a^{Xk}.$

Thus,

$$\log_a M^k = Xk = kX = k \cdot \log_a M.$$

Proof of P4: $\log_a a = 1$ because $a^1 = a$.

Proof of P5: $\log_a (a^k) = k$ because $(a^k) = a^k$.

Proof of P6: $\log_a 1 = 0$ because $a^0 = 1$.

Let's illustrate these properties.

Example 2   Given

$$\log_a 2 = 0.301 \quad \text{and} \quad \log_a 3 = 0.477,$$

find each of the following.

a) $\log_a 6$ $\begin{aligned} \log_a 6 &= \log_a (2 \cdot 3) \\ &= \log_a 2 + \log_a 3 \qquad \text{By P1} \\ &= 0.301 + 0.477 \\ &= 0.778 \end{aligned}$

b) $\log_a \frac{2}{3}$ $\begin{aligned} \log_a \frac{2}{3} &= \log_a 2 - \log_a 3 \qquad \text{By P2} \\ &= 0.301 - 0.477 \\ &= -0.176 \end{aligned}$

c) $\log_a 81$ $\begin{aligned} \log_a 81 &= \log_a 3^4 \\ &= 4 \log_a 3 \qquad \text{By P3} \\ &= 4(0.477) \\ &= 1.908 \end{aligned}$

d) $\log_a \frac{1}{3}$ $\begin{aligned} \log_a \frac{1}{3} &= \log_a 1 - \log_a 3 \qquad \text{By P2} \\ &= 0 - 0.477 \qquad \text{By P6} \\ &= -0.477 \end{aligned}$

e) $\log_a \sqrt{a}$ $\log_a \sqrt{a} = \log_a a^{1/2} = \frac{1}{2} \qquad \text{By P5}$

f) $\log_a 2a$ $\begin{aligned} \log_a 2a &= \log_a 2 + \log_a a \qquad \text{By P1} \\ &= 0.301 + 1 \qquad \text{By P4} \\ &= 1.301 \end{aligned}$

g) $\log_a 5$ *No way to find using these properties.*
$(\log_a 5 \neq \log_a 2 + \log_a 3)$

h) $\dfrac{\log_a 3}{\log_a 2}$ $\dfrac{\log_a 3}{\log_a 2} = \dfrac{0.477}{0.301} \approx 1.58$
We simply divided and used none of the properties.

Common Logarithms

The number $\log_{10} x$ is called the **common logarithm** of x and is abbreviated $\log x$; that is:

> ### Definition
>
> For any positive number x,
>
> $$\log x = \log_{10} x.$$

Thus, when we write $\log x$ with no base indicated, base 10 is understood. Note the following comparison of common logarithms and powers of 10.

<table>
<tr><td>$1000 = 10^3$</td><td rowspan="7">The common logarithms at the right follow from the powers at the left.</td><td>$\log 1000 = 3$</td></tr>
<tr><td>$100 = 10^2$</td><td>$\log 100 = 2$</td></tr>
<tr><td>$10 = 10^1$</td><td>$\log 10 = 1$</td></tr>
<tr><td>$1 = 10^0$</td><td>$\log 1 = 0$</td></tr>
<tr><td>$0.1 = 10^{-1}$</td><td>$\log 0.1 = -1$</td></tr>
<tr><td>$0.01 = 10^{-2}$</td><td>$\log 0.01 = -2$</td></tr>
<tr><td>$0.001 = 10^{-3}$</td><td>$\log 0.001 = -3$</td></tr>
</table>

Exercise

1. Graph
   $f(x) = 10^x$,
   $y = x$, and
   $g(x) = \log_{10} x$
   using the same
   set of axes. Then
   find $f(3)$,
   $f(0.699)$, $g(5)$,
   and $g(1000)$.

Since $\log 100 = 2$ and $\log 1000 = 3$, it seems reasonable that $\log 500$ is somewhere between 2 and 3. Tables were originally used for such approximations, but with the advent of the calculator, that method of finding logarithms is used infrequently. Using a calculator with a 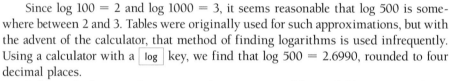 $\boxed{\log}$ key, we find that $\log 500 = 2.6990$, rounded to four decimal places.

Before calculators and computers became so readily available, common logarithms were used extensively to do certain kinds of computations. In fact, computation is the reason logarithms were developed. Since the standard notation we use for numbers is based on 10, it is logical that base-10, or common, logarithms were used for computations. Today, computations with common logarithms are mainly of historical interest; the logarithmic functions, base e, are of modern importance.

Natural Logarithms

The number e, which is approximately 2.718282, was developed in Section 4.1, and has extensive application in many fields. The number $\log_e x$ is called the **natural logarithm** of x and is abbreviated $\ln x$; that is:

> ### Definition
>
> For any positive number x,
>
> $$\ln x = \log_e x.$$

The following is a restatement of the basic properties of logarithms in terms of natural logarithms.

Theorem 5

P1. $\ln MN = \ln M + \ln N$

P2. $\ln \dfrac{M}{N} = \ln M - \ln N$

P3. $\ln a^k = k \cdot \ln a$

P4. $\ln e = 1$

P5. $\ln e^k = k$

P6. $\ln 1 = 0$

Let's illustrate these properties.

Example 3   Given

$$\ln 2 = 0.6931 \quad \text{and} \quad \ln 3 = 1.0986,$$

find each of the following.

a) $\ln 6$
$$\ln 6 = \ln (2 \cdot 3) = \ln 2 + \ln 3 \quad \text{By P1}$$
$$= 0.6931 + 1.0986$$
$$= 1.7917$$

b) $\ln 81$
$$\ln 81 = \ln (3^4)$$
$$= 4 \ln 3 \quad \text{By P3}$$
$$= 4(1.0986)$$
$$= 4.3944$$

c) $\ln \frac{2}{3}$
$$\ln \tfrac{2}{3} = \ln 2 - \ln 3 \quad \text{By P2}$$
$$= 0.6931 - 1.0986$$
$$= -0.4055$$

d) $\ln \frac{1}{3}$
$$\ln \tfrac{1}{3} = \ln 1 - \ln 3 \quad \text{By P2}$$
$$= 0 - 1.0986 \quad \text{By P6}$$
$$= -1.0986$$

e) $\ln 2e$
$$\ln 2e = \ln 2 + \ln e \quad \text{By P1}$$
$$= 0.6931 + 1 \quad \text{By P4}$$
$$= 1.6931$$

f) $\ln \sqrt{e^3}$
$$\ln \sqrt{e^3} = \ln e^{3/2}$$
$$= \tfrac{3}{2} \quad \text{By P5}$$

Finding Natural Logarithms Using a Calculator

You should have a calculator with a $\boxed{\ln}$ key. You can find natural logarithms directly using this key.

Example 4   Find each logarithm on your calculator. Round to six decimal places.

a) $\ln 5.24 = 1.656321$

b) $\ln 0.00001277 = -11.268412$

Exponential Equations

If an equation contains a variable in an exponent, we call the equation **exponential.** We can use logarithms to manipulate or solve exponential equations.

Example 5   Solve $e^t = 40$ for t.

Solution   We have

$$\ln e^t = \ln 40 \qquad \text{Taking the natural logarithm on both sides}$$
$$t = \ln 40 \qquad \text{By P5}$$
$$t = 3.688879$$
$$t \approx 3.7$$

It should be noted that this is an approximation for t even though an equals sign is often used.

Example 6   Solve $e^{-0.04t} = 0.05$ for t.

Solution   We have

$$\ln e^{-0.04t} = \ln 0.05 \qquad \text{Taking the natural logarithm on both sides}$$
$$-0.04t = \ln 0.05 \qquad \text{By P5}$$
$$t = \frac{\ln 0.05}{-0.04}$$
$$t = \frac{-2.995732}{-0.04}$$
$$t \approx 75.$$

For purposes of space and explanation, we have rounded the value of $\ln 0.05$ to -2.995732 in an intermediate step. When using your calculator, you should find

$$\frac{\ln 0.05}{-0.04}$$

directly, without rounding, as say,

$$\frac{-2.995732274}{-0.04}.$$

Then divide, and round at the end. Answers at the back of the book have been found in this manner. Remember, the number of places in a table or on a calculator may affect the accuracy of the answer. Usually, your answer should agree to at least three digits.

TECHNOLOGY   CONNECTION

Solving Exponential Equations

Let's solve the equation of Example 5, $e^t = 40$, using a grapher.

Method 1: The INTERSECT Feature

We change the variable to x and consider the system of equations $y_1 = e^x$ and $y_2 = 40$. We graph the equations in the viewing window $[-1, 8, -10, 70]$ in order to see the curvature and possible points of intersection.

$$y_1 = e^x, \quad y_2 = 40$$

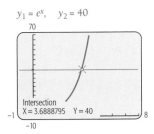

Then we use the **INTERSECT** feature to find the point of intersection, about $(3.7, 40)$. The x-coordinate, 3.7, is the solution of the original equation $e^t = 40$.

Method 2: The ZERO Feature

We change the variable to x and get a 0 on one side of the equation: $e^x - 40 = 0$. Then we graph $y = e^x - 40$ in the window $[-1, 8, -10, 10]$.

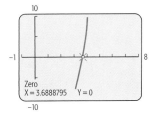

Using the **ZERO** feature, we see that the x-intercept is about $(3.7, 0)$, so 3.7 is the solution of the original equation $e^t = 40$.

Exercises

Solve using a grapher.

1. $e^t = 1000$
2. $e^{-x} = 60$
3. $e^{-0.04t} = 0.05$
4. $e^{0.23x} = 41{,}378$
5. $15e^{0.2x} = 34{,}785.13$

Graphs of Natural Logarithmic Functions

There are two ways in which we might obtain the graph of $y = f(x) = \ln x$. One is by writing its equivalent equation $x = e^y$. Then we select values for y and use a calculator to find the corresponding values of e^y. We then plot points, remembering that x is still the first coordinate. The graph follows.

x, or e^y	y
0.1	-2
0.4	-1
1.0	0
2.7	1
7.4	2
20.1	3

① Select y.

② Compute x.

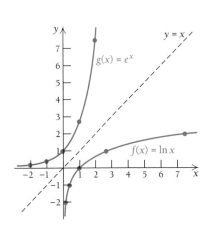

The figure on the preceding page shows the graph of $g(x) = e^x$ for comparison. Note again that the functions are inverses of each other. That is, the graph of $y = \ln x$, or $x = e^y$, is a reflection, or mirror image, across the line $y = x$ of the graph of $y = e^x$. Any ordered pair (a, b) on the graph of g yields an ordered pair (b, a) on f. Note too that $\lim_{x \to 0^-} \ln x = -\infty$ and the y-axis is a vertical asymptote.

The second method of graphing $y = \ln x$ is to use a calculator to find function values. For example, when $x = 2$, then $y = \ln 2 \approx 0.6931 \approx 0.7$. This gives the pair $(2, 0.7)$ on the graph, which follows.

TECHNOLOGY
CONNECTION

**Graphing
Logarithmic
Functions**

To graph $y = \ln x$, we can use the $\boxed{\text{LN}}$ key and enter the function as $Y_1 = \ln (X)$.

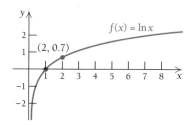

The following properties can be observed from the graph.

Theorem 6

$\ln x$ exists only for positive numbers x. The domain is $(0, \infty)$.

$\ln x < 0$ for $0 < x < 1$.

$\ln x = 0$ when $x = 1$.

$\ln x > 0$ for $x > 1$.

The range is the entire real line, $(-\infty, \infty)$.

TECHNOLOGY
CONNECTION

Exploratory.
Consider $f(x) = \ln x$. Graph f and f' using the viewing window $[0, 10, -3, 3]$. Make an input–output table for the functions. Compare a given x-value with its corresponding y-value for f and f'. What do you observe?

Derivatives of Natural Logarithmic Functions

Let's find the derivative of

$$f(x) = \ln x. \tag{1}$$

We first write its equivalent exponential equation:

$$e^{f(x)} = x. \qquad \text{$\ln x = \log_e x = f(x)$, so $e^{f(x)} = x$,} \tag{2}$$
$$\text{by the definition of logarithms.}$$

Now we differentiate on both sides of this equation:

$$\frac{d}{dx} e^{f(x)} = \frac{d}{dx} x$$

$$f'(x) \cdot e^{f(x)} = 1 \qquad \text{By the Chain Rule}$$

$$f'(x) \cdot x = 1 \qquad \text{Substituting x for $e^{f(x)}$ from equation (2)}$$

$$f'(x) = \frac{1}{x}.$$

Thus we have the following.

> **Theorem 7**
>
> For any positive number x,
> $$\frac{d}{dx}\ln x = \frac{1}{x}.$$

Theorem 7 asserts that for the function $f(x) = \ln x$, to find the slope of the tangent line at x, we need only take the reciprocal of x. This is true only for positive values of x, since $\ln x$ is defined only for positive numbers. (For negative numbers x, this derivative formula becomes

$$\frac{d}{dx}\ln |x| = \frac{1}{x},$$

but we will seldom consider such a case in this text.)

Let's find some derivatives.

Example 7

$$\frac{d}{dx}3\ln x = 3\frac{d}{dx}\ln x$$
$$= \frac{3}{x}$$

Example 8

$$\frac{d}{dx}(x^2\ln x + 5x) = x^2 \cdot \frac{1}{x} + 2x \cdot \ln x + 5 \qquad \text{Using the Product Rule on } x^2 \ln x$$
$$= x + 2x \cdot \ln x + 5, \qquad \text{Simplifying}$$
$$\text{or } x(1 + 2\ln x) + 5$$

Example 9

$$\frac{d}{dx}\frac{\ln x}{x^3} = \frac{x^3 \cdot (1/x) - (3x^2)(\ln x)}{x^6} \qquad \text{By the Quotient Rule}$$
$$= \frac{x^2 - 3x^2\ln x}{x^6}$$
$$= \frac{x^2(1 - 3\ln x)}{x^6} \qquad \text{Factoring}$$
$$= \frac{1 - 3\ln x}{x^4} \qquad \text{Simplifying}$$

Suppose that we want to differentiate a more complicated function, such as

$$h(x) = \ln (x^2 - 8x).$$

This is a composition of functions. In general, we have

$$h(x) = \ln f(x) = g[f(x)], \quad \text{where} \quad g(x) = \ln x.$$

Now $g'(x) = 1/x$. Then by the Chain Rule (Section 2.8), we have

$$h'(x) = g'[f(x)] \cdot f'(x) = \frac{1}{f(x)} \cdot f'(x).$$

For the above case, $f(x) = x^2 - 8x$, so $f'(x) = 2x - 8$. Then

$$h'(x) = \frac{1}{x^2 - 8x} \cdot (2x - 8) = \frac{2x - 8}{x^2 - 8x}.$$

The following rule, which we have proven using the Chain Rule, allows us to find derivatives of functions like the one above.

Theorem 8

$$\frac{d}{dx} \ln f(x) = f'(x) \cdot \frac{1}{f(x)} = \frac{f'(x)}{f(x)},$$

or

$$\frac{d}{dx} \ln u = \frac{du}{dx} \cdot \frac{1}{u}.$$

The derivative of the natural logarithm of a function is the derivative of the function divided by the function.

The following gives us a way of remembering this rule.

$$h(x) = \ln (x^2 - 8x)$$

①

$$h'(x) = \frac{2x - 8}{x^2 - 8x}$$
②

① Differentiate the "inside" function.

② Divide by the "inside" function.

Example 10

$$\frac{d}{dx} \ln 3x = \frac{3}{3x} = \frac{1}{x}.$$

Note that we could have done this another way, using Property 1:

$$\ln 3x = \ln 3 + \ln x;$$

then

$$\frac{d}{dx} \ln 3x = \frac{d}{dx} \ln 3 + \frac{d}{dx} \ln x = 0 + \frac{1}{x} = \frac{1}{x}.$$

◄

Example 11

$$\frac{d}{dx} \ln (x^2 - 5) = \frac{2x}{x^2 - 5}$$

◄

Example 12

$$\frac{d}{dx} \ln (\ln x) = \frac{1}{x} \cdot \frac{1}{\ln x} = \frac{1}{x \ln x}$$

◄

Example 13

$$\frac{d}{dx} \ln \left(\frac{x^3 + 4}{x}\right) = \frac{d}{dx} [\ln (x^3 + 4) - \ln x]$$ By P2. This avoids use of the Quotient Rule.

$$= \frac{3x^2}{x^3 + 4} - \frac{1}{x}$$

$$= \frac{3x^2}{x^3 + 4} \cdot \frac{x}{x} - \frac{1}{x} \cdot \frac{x^3 + 4}{x^3 + 4}$$

$$= \frac{(3x^2)x - (x^3 + 4)}{x(x^3 + 4)}$$

$$= \frac{3x^3 - x^3 - 4}{x(x^3 + 4)} = \frac{2x^3 - 4}{x(x^3 + 4)}$$

◄

TECHNOLOGY
CONNECTION

Check the results of Examples 10–13 using a grapher. Then differentiate

$$y = \ln\left(\frac{x^5 - 2}{x}\right)$$

and check your answer using the grapher.

Applications

Conduct your own memory experiment. Study this photograph carefully. Then put it aside and write down as many items as you can. Wait a half-hour and again write down as many as you can. Do this five more times. Make a graph of the number of items you remember versus the time. Does the graph appear to be logarithmic?

Example 14 *Social Science: Forgetting.* In a psychological experiment, students were shown a set of nonsense syllables, such as POK, RTZ, PDQ, and so on, and asked to recall them every second thereafter. The percentage $R(t)$ who retained the syllables after t seconds was found to be given by the logarithmic learning model

$$R(t) = 80 - 27 \ln t, \quad \text{for } t \geq 1.$$

Strictly speaking, the function is not continuous, but in order to use calculus, we "fill in" the graph with a smooth curve, considering $R(t)$ to be defined for any number $t \geq 1$. This is not unreasonable, since we could find the percentage who retained the syllables after $t = 3.417$ sec, instead of merely after integer values such as 1, 2, 3, 4, and so on.

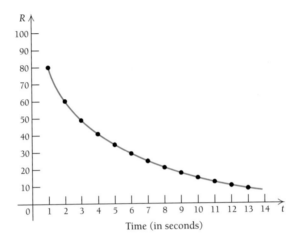

TECHNOLOGY CONNECTION

Graph
$y = 80 - 27 \ln x$
using the viewing window $[1, 14, -1, 100]$. Trace along the graph. Describe the meaning of each coordinate in an ordered pair.

a) What percentage retained the syllables after 1 sec?

b) Find $R'(t)$, the rate of change of R with respect to t.

c) Find the maximum and minimum values, if they exist.

Solution

a) $R(1) = 80 - 27 \cdot \ln 1 = 80 - 27 \cdot 0 = 80\%$

b) $R'(t) = \dfrac{-27}{t} = -\dfrac{27}{t}$

c) $R'(t)$ exists for all values of t in the interval $[1, \infty)$. Note that for $t \geq 1$, $-27/t < 0$. Thus there are no critical points and R is decreasing. Then R has a maximum value at the endpoint 1. This maximum value is $R(1)$, or 80%. There is no minimum value.

Example 15 *Business: An Advertising Model.* A company begins a radio advertising campaign in New York City to market a new product. The percentage of the "target market" that buys a product is normally a function of the length of the advertising campaign. The radio station estimates this percentage by using the model $f(t) = 1 - e^{-0.04t}$ for this type of product, where t is the number of days of the campaign. The target market is estimated to be 1,000,000 people and the price per unit is $0.50. The costs of advertising are $1000 per day. Find the length of the advertising campaign that will result in the maximum profit.

Solution That the percentage of the target market that buys the product can be modeled by $f(t) = 1 - e^{-0.04t}$ is justified if we look at its graph. The function increases from 0 (0%) toward 1 (100%). The longer the advertising campaign, the larger the percentage of the market that has bought the product.

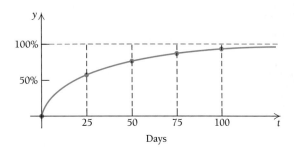

The total-profit function, here expressed in terms of time t, is given by

$$\text{Profit} = \text{Revenue} - \text{Cost}$$
$$P(t) = R(t) - C(t).$$

a) Find $R(t)$:

$$R(t) = (\text{Price per unit}) \cdot (\text{Target market}) \cdot (\text{Percentage buying})$$
$$R(t) = 0.5(1{,}000{,}000)(1 - e^{-0.04t}) = 500{,}000 - 500{,}000e^{-0.04t}.$$

b) Find $C(t)$:

$$C(t) = (\text{Advertising costs per day}) \cdot (\text{Number of days})$$
$$C(t) = 1000t.$$

c) Find $P(t)$ and take its derivative:

$$P(t) = R(t) - C(t)$$
$$P(t) = 500{,}000 - 500{,}000e^{-0.04t} - 1000t$$
$$P'(t) = (-0.04)(-500{,}000e^{-0.04t}) - 1000$$
$$P'(t) = 20{,}000e^{-0.04t} - 1000.$$

d) Set the first derivative equal to 0 and solve:

$$20{,}000e^{-0.04t} - 1000 = 0$$
$$20{,}000e^{-0.04t} = 1000$$
$$e^{-0.04t} = \frac{1000}{20{,}000} = 0.05$$
$$\ln e^{-0.04t} = \ln 0.05$$
$$-0.04t = \ln 0.05$$
$$t = \frac{\ln 0.05}{-0.04}$$
$$t = \frac{-2.995732}{-0.04}$$
$$t \approx 75.$$

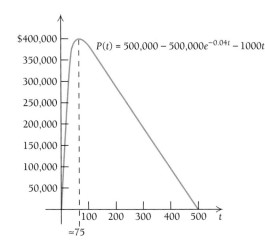

$P(t) = 500{,}000 - 500{,}000e^{-0.04t} - 1000t$

Graph the function P in Example 15 and verify that there is maximum profit if the length of the advertising campaign is about 75 days.

e) We have only one critical point, so we can use the second derivative to determine whether we have a maximum:

$$P''(t) = -0.04(20{,}000e^{-0.04t})$$
$$= -800e^{-0.04t}.$$

Since exponential functions are positive, $e^{-0.04t} > 0$ for all numbers t. Thus, since $-800e^{-0.04t} < 0$ for all numbers t, $P''(t)$ is less than 0 for $t = 75$ and we have a maximum.

The length of the advertising campaign must be 75 days in order to result in maximum profit.   ◄

EXERCISE SET 4.2

Write an equivalent exponential equation.

1. $\log_2 8 = 3$
2. $\log_3 81 = 4$
3. $\log_8 2 = \frac{1}{3}$
4. $\log_{27} 3 = \frac{1}{3}$
5. $\log_a K = J$
6. $\log_a J = K$
7. $\log_b T = v$
8. $\log_c Y = t$

Write an equivalent logarithmic equation.

9. $e^M = b$
10. $e^t = p$
11. $10^2 = 100$
12. $10^3 = 1000$
13. $10^{-1} = 0.1$
14. $10^{-2} = 0.01$
15. $M^p = V$
16. $Q^n = T$

Given $\log_b 3 = 1.099$ and $\log_b 5 = 1.609$, find each of the following.

17. $\log_b 15$
18. $\log_b \frac{3}{5}$
19. $\log_b \frac{1}{5}$
20. $\log_b \sqrt{b^3}$
21. $\log_b 5b$
22. $\log_b 75$

Given $\ln 4 = 1.3863$ and $\ln 5 = 1.6094$, find each of the following. Do not use a calculator.

23. $\ln 20$
24. $\ln \frac{5}{4}$
25. $\ln \frac{1}{4}$
26. $\ln 4e$
27. $\ln \sqrt{e^8}$
28. $\ln 100$

Find the logarithm. Round to six decimal places.

29. $\ln 5894$
30. $\ln 99{,}999$

31. $\ln 0.0182$
32. $\ln 0.00087$
33. $\ln 8100$
34. $\ln 0.011$

Solve for t.

35. $e^t = 100$
36. $e^t = 1000$
37. $e^t = 60$
38. $e^t = 90$
39. $e^{-t} = 0.1$
40. $e^{-t} = 0.01$
41. $e^{-0.02t} = 0.06$
42. $e^{0.07t} = 2$

Differentiate.

43. $y = -6 \ln x$
44. $y = -4 \ln x$
45. $y = x^4 \ln x - \frac{1}{2}x^2$
46. $y = x^5 \ln x - \frac{1}{4}x^4$
47. $y = \dfrac{\ln x}{x^4}$
48. $y = \dfrac{\ln x}{x^5}$
49. $y = \ln \dfrac{x}{4}$ $\left(Hint: \ln \dfrac{x}{4} = \ln x - \ln 4. \right)$
50. $y = \ln \dfrac{x}{2}$
51. $f(x) = \ln (5x^2 - 7)$
52. $f(x) = \ln (7x^3 + 4)$
53. $f(x) = \ln (\ln 4x)$
54. $f(x) = \ln (\ln 3x)$
55. $f(x) = \ln \left(\dfrac{x^2 - 7}{x} \right)$
56. $f(x) = \ln \left(\dfrac{x^2 + 5}{x} \right)$
57. $f(x) = e^x \ln x$

58. $f(x) = e^{2x} \ln x$ 59. $f(x) = \ln (e^x + 1)$

60. $f(x) = \ln (e^x - 2)$

61. $f(x) = (\ln x)^2$
    (*Hint*: Use the Extended Power Rule.)

62. $f(x) = (\ln x)^3$

Applications ....
BUSINESS AND ECONOMICS

63. *Advertising.* A model for advertising response is given by

$$N(a) = 1000 + 200 \ln a, \quad a \geq 1,$$

where $N(a)$ is the number of units sold and a is the amount spent on advertising, in thousands of dollars.

a) How many units were sold after spending $1000 ($a = 1$) on advertising?

b) Find $N'(a)$ and $N'(10)$.

c) Find the maximum and minimum values, if they exist.

tw d) Find $N'(a)$. Discuss $\lim_{a \to \infty} N'(a)$. Does it make sense to spend more and more dollars on advertising?

64. *Advertising.* A model for advertising response is given by

$$N(a) = 2000 + 500 \ln a, \quad a \geq 1,$$

where $N(a)$ is the number of units sold and a is the amount spent on advertising, in thousands of dollars.

a) How many units were sold after spending $1000 ($a = 1$) on advertising?

b) Find $N'(a)$ and $N'(10)$.

c) Find the maximum and minimum values, if they exist.

tw d) Find $\lim_{a \to \infty} N'(a)$. Discuss whether it makes sense to spend more and more dollars on advertising.

65. *An Advertising Model.* Solve Example 15 given that the costs of advertising are $2000 per day.

66. *An Advertising Model.* Solve Example 15 given that the costs of advertising are $4000 per day.

67. *Growth of a Stock.* The value V of a stock is modeled by

$$V(t) = \$58(1 - e^{-1.1t}) + \$20,$$

where V is the value of the stock after time t, in months.

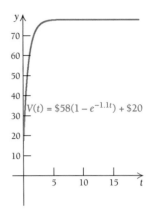

a) Find $V(1)$ and $V(12)$.

b) Find $V'(t)$.

c) After how many months will the value of the stock be $75?

tw d) Find $\lim_{t \to \infty} V(t)$. Discuss the value of the stock over a long period of time. Is this trend typical?

68. *Marginal Revenue.* The demand function for a certain product is given by

$$x = D(p) = 800e^{-0.125p}.$$

Recall that total revenue is given by $R(p) = pD(p)$.

a) Find $R(p)$.

b) Find the marginal revenue, $R'(p)$.

c) At what price per unit p will the revenue be maximum?

LIFE AND PHYSICAL SCIENCES

69. *Acceptance of a New Medicine.* The percentage P of doctors who accept a new medicine is given by

$$P(t) = 100(1 - e^{-0.2t}),$$

where t is the time, in months.

a) Find $P(1)$ and $P(6)$.

b) Find $P'(t)$.

c) How many months will it take for 90% of the doctors to accept the new medicine?

tw d) Find $\lim_{t \to \infty} P(t)$ and discuss its meaning.

70. *The Reynolds Number.* For many kinds of animals, the Reynolds number R is given by

$$R = A \ln r - Br,$$

where A and B are positive constants and r is the radius of the aorta. Find the maximum value of R.

SOCIAL SCIENCES

71. *Forgetting.* Students in college botany took a final exam. They took equivalent forms of the exam in monthly intervals thereafter. The average score $S(t)$, in percent, after t months was found to be given by

$$S(t) = 68 - 20 \ln (t + 1), \quad t \geq 0.$$

a) What was the average score when they initially took the test, $t = 0$?
b) What was the average score after 4 months?
c) What was the average score after 24 months?
d) What percentage of the initial score did they retain after 2 years (24 months)?
e) Find $S'(t)$.
f) Find the maximum and minimum values, if they exist.
tw g) Find $\lim\limits_{t \to \infty} S(t)$ and discuss its meaning.

72. *Forgetting.* Students in college zoology took a final exam. They took equivalent forms of the exam in monthly intervals thereafter. The average score $S(t)$, in percent, after t months was found to be given by

$$S(t) = 78 - 15 \ln (t + 1), \quad t \geq 0.$$

a) What was the average score when they initially took the test, $t = 0$?
b) What was the average score after 4 months?
c) What was the average score after 24 months?
d) What percentage of the initial score did they retain after 2 years (24 months)?
e) Find $S'(t)$.
f) Find the maximum and minimum values, if they exist.
tw g) Find $\lim\limits_{t \to \infty} S(t)$ and discuss its meaning.

73. *Walking Speed.* Bornstein and Bornstein found in a study that the average walking speed v of a person living in a city of population p, in thousands, is given by

$$v(p) = 0.37 \ln p + 0.05,$$

where v is in feet per second. (*Source: International Journal of Psychology*).

a) The population of Seattle is 531,000. What is the average walking speed of a person living in Seattle? [*Hint*: Find $v(531)$.]
b) The population of New York is 7,900,000. What is the average walking speed of a person living in New York?

c) Find $v'(p)$.
tw d) Interpret $v'(p)$ found in part (c).

74. *The Hullian Learning Model.* A typist learns to type W words per minute after t weeks of practice, where W is given by

$$W(t) = 100(1 - e^{-0.3t}).$$

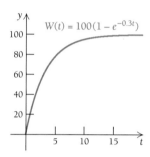

a) Find $W(1)$ and $W(8)$.
b) Find $W'(t)$.
c) After how many weeks will the typist's speed be 95 words per minute?
tw d) Find $\lim\limits_{t \to \infty} W(t)$ and discuss its meaning.

Synthesis ....

Differentiate.

75. $y = (\ln x)^{-4}$

76. $y = (\ln x)^n$

77. $f(t) = \ln (t^3 + 1)^5$

78. $f(t) = \ln (t^2 + t)^3$

79. $f(x) = [\ln (x + 5)]^4$

80. $f(x) = \ln [\ln (\ln 3x)]$

81. $f(t) = \ln [(t^3 + 3)(t^2 - 1)]$

82. $f(t) = \ln \dfrac{1 - t}{1 + t}$

83. $y = \ln \dfrac{x^5}{(8x + 5)^2}$

84. $y = \ln \sqrt{5 + x^2}$

85. $f(t) = \dfrac{\ln t^2}{t^2}$

86. $f(x) = \dfrac{1}{5}x^5 \left(\ln x - \dfrac{1}{5} \right)$

87. $y = \dfrac{x^{n+1}}{n + 1} \left(\ln x - \dfrac{1}{n + 1} \right)$

88. $y = \dfrac{x \ln x - x}{x^2 + 1}$

89. $y = \ln (t + \sqrt{1 + t^2})$

90. $f(x) = \ln \dfrac{1 + \sqrt{x}}{1 - \sqrt{x}}$

91. $f(x) = \ln [\ln x]^3$

92. $f(x) = \dfrac{\ln x}{1 + (\ln x)^2}$

93. Find $\displaystyle\lim_{h \to 0} \dfrac{\ln (1 + h)}{h}$.

Solve for t.

94. $P = P_0 e^{-kt}$ **95.** $P = P_0 e^{kt}$

Verify each of the following.

96. $\log x = \dfrac{\ln x}{\ln 10} \approx 0.4343 \ln x$

97. $\ln x = \dfrac{\log x}{\log e} \approx 2.3026 \log x$

tw 98. Explain how the graph of $y = \ln x$ could be used to find the graph of $y = e^x$.

tw 99. Consider the true statement

$$\frac{d}{dx} \ln 4x = \frac{1}{x}.$$

Explain what this means graphically.

TECHNOLOGY   CONNECTION

100. Which is larger, e^π or π^e?

101. Find $\sqrt[e]{e}$. Compare it to other expressions of the type $\sqrt[x]{x}$, $x > 0$. What can you conclude?

Use input–output tables to find the limit.

102. $\displaystyle\lim_{x \to 1} \ln x$ **103.** $\displaystyle\lim_{x \to \infty} \ln x$

Graph each function f and its derivative f'.

104. $f(x) = \ln x$

105. $f(x) = x \ln x$

106. $f(x) = x^2 \ln x$

107. $f(x) = \dfrac{\ln x}{x^2}$

Find the minimum value of the function.

108. $f(x) = x \ln x$

109. $f(x) = x^2 \ln x$

4.3

Applications: The Uninhibited Growth Model, $dP/dt = kP$

Exponential Growth

Consider the function

$$f(x) = 2e^{3x}.$$

Differentiating, we get

$$f'(x) = 3 \cdot 2e^{3x}$$
$$= 3 \cdot f(x).$$

Graphically, this says that the derivative, or slope of the tangent line, is simply the constant 3 times the function value.

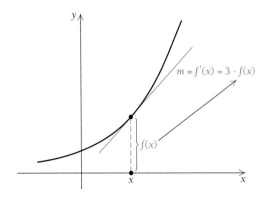

The exponential function $f(x) = ce^{kx}$ is the only function whose derivative is a constant times the function itself.

Theorem 9

A function $y = f(x)$ satisfies the equation

$$\frac{dy}{dx} = ky \qquad [f'(x) = k \cdot f(x)]$$

if and only if

$$y = ce^{kx} \qquad [f(x) = ce^{kx}]$$

for some constant c.

No matter what the variables, you should be able to write the function that satisfies what is called a *differential equation*.

Example 1   Find the function that satisfies the equation

$$\frac{dA}{dt} = kA.$$

Solution   The function is $A = ce^{kt}$, or $A(t) = ce^{kt}$.   ◅

Example 2   Find the function that satisfies the equation

$$\frac{dP}{dt} = kP.$$

Solution   The function is $P = ce^{kt}$, or $P(t) = ce^{kt}$.   ◅

Example 3   Find the function that satisfies the equation

$$f'(Q) = k \cdot f(Q).$$

Solution   The function is $f(Q) = ce^{kQ}$.   ◅

What will the world population be in 2010?

The equation

$$\frac{dP}{dt} = kP, \quad k > 0 \qquad [P'(t) = k \cdot P(t), \quad k > 0]$$

is the basic model of uninhibited population growth, whether it be a population of humans, a bacteria culture, or money invested at interest compounded continuously. Neglecting special inhibiting and stimulating factors, we know that a population normally reproduces itself at a rate proportional to its size, and this is exactly what the equation $dP/dt = kP$ says. The solution of the equation is

$$P(t) = ce^{kt}, \tag{1}$$

where t is the time. At $t = 0$, we have some "initial" population $P(0)$ that we will represent by P_0. We can rewrite equation (1) in terms of P_0 as

$$P_0 = P(0) = ce^{k \cdot 0} = ce^0 = c \cdot 1 = c.$$

Thus, $P_0 = c$, so we can express $P(t)$ as

$$P(t) = P_0 e^{kt}.$$

Its graph shows how uninhibited growth produces a "population explosion."

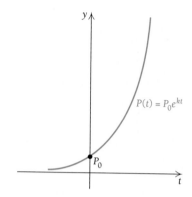

The constant k is called the **rate of exponential growth,** or simply the **growth rate.** This is not the rate of change of the population size, which is

$$\frac{dP}{dt} = kP,$$

but the constant by which P must be multiplied in order to get its rate of change. It is thus a different use of the word rate. It is like the *interest rate* paid by a bank. If the interest rate is 7%, or 0.07, we do not mean that your bank balance P is growing at the rate of 0.07 dollars per year, but at the rate of $0.07P$ dollars per year. We therefore express the rate as 7% per year, rather than 0.07 dollars per year. We could say that the rate is 0.07 dollars *per dollar* per year. When interest is compounded continuously, the interest rate is a true exponential growth rate.

 TECHNOLOGY CONNECTION

Exploratory Exercises: Growth

Use a sheet of $8\frac{1}{2}$-in. by 11-in. paper that has been cut into two equal pieces. Then cut these into four equal pieces. Then cut these into eight equal pieces, and so on, performing five cutting steps.

   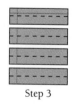

Start   Step 1   Step 2   Step 3

	t	$0.004 \cdot 2^t$
Start	0	$0.004 \cdot 2^0$, or 0.004
Step 1	1	$0.004 \cdot 2^1$, or 0.008
Step 2	2	$0.004 \cdot 2^2$, or 0.016
Step 3	3	
Step 4	4	
Step 5	5	

a) Place all the pieces in a stack and measure the thickness.

b) A piece of paper is typically 0.004 in. thick. Check the calculation in part (a) by completing the table.

c) Graph the function $f(t) = 0.004(2)^t$.

d) Compute the thickness of the paper (in miles) after 25 steps.

Example 4   *Business: Interest Compounded Continuously.* Suppose that an amount P_0 is invested in a savings account where interest is compounded continuously at 7% per year. That is, the balance P grows at the rate given by

$$\frac{dP}{dt} = 0.07P.$$

a) Find the function that satisfies the equation. List it in terms of P_0 and 0.07.

b) Suppose that $100 is invested. What is the balance after 1 yr?

c) After what period of time will an investment of $100 double itself?

Solution

a) $P(t) = P_0 e^{0.07t}$

b) $P(1) = 100 e^{0.07(1)} = 100 e^{0.07}$
$$= 100(1.072508)$$
$$\approx \$107.25$$

c) We are looking for that time T for which $P(T) = \$200$. The number T is called the **doubling time.** To find T, we solve the equation

$$200 = 100 e^{0.07 \cdot T}$$
$$2 = e^{0.07T}.$$

We use natural logarithms to solve this equation:

$$\ln 2 = \ln e^{0.07T}$$
$$\ln 2 = 0.07T \qquad \text{By P5: } \ln e^k = k$$
$$\frac{\ln 2}{0.07} = T$$
$$\frac{0.693147}{0.07} = T$$
$$9.9 \approx T.$$

Thus, $100 will double itself in 9.9 yr.

Let's consider another way of developing the formula

$$P(t) = P_0 e^{kt}$$

by starting with

$$A = P\left(1 + \frac{i}{n}\right)^{nt}$$

and compounding interest continuously. Let $P = P_0$ and $i = k$ to obtain

$$A = P_0\left(1 + \frac{k}{n}\right)^{nt}.$$

We are interested in what happens as n gets very large, that is, as n approaches ∞. To

Under ideal conditions, the growth rate of this rabbit population might be 11.7% per day. When will this population of rabbits double?

determine this limit, we first let

$$\frac{k}{n} = \frac{1}{q}.$$

Then

$$qk = n \quad \text{and} \quad q = \frac{n}{k},$$

which shows that since k is a positive constant, as n gets large, so does q. To find a formula for continuously compounded interest, we evaluate the following limit:

$$P(t) = \lim_{n \to \infty} \left[P_0 \left(1 + \frac{k}{n} \right)^{nt} \right] \qquad \text{Letting the number of compounding periods become infinite}$$

$$= P_0 \lim_{q \to \infty} \left[\left(1 + \frac{1}{q} \right)^{qkt} \right] \qquad \text{The limit of a constant times a function is the constant times the limit. We also substitute } 1/q \text{ for } k/n \text{ and } qk \text{ for } n. \text{ Also, } q \to \infty \text{ because } n \to \infty.$$

$$= P_0 \left[\lim_{q \to \infty} \left(1 + \frac{1}{q} \right)^{q} \right]^{kt} \qquad \text{The limit of a power is the power of the limit: a form of L2 in Section 2.2.}$$

$$= P_0 [e]^{kt}. \qquad \text{Theorem 1}$$

We can find a general expression relating the growth rate k and the doubling time T by solving the following equation:

$$2P_0 = P_0 e^{kT}$$
$$2 = e^{kT} \qquad \text{Dividing by } P_0$$
$$\ln 2 = \ln e^{kT}$$
$$\ln 2 = kT.$$

Theorem 10
The *growth rate* k and the *doubling time* T are related by

$$kT = \ln 2 = 0.693147,$$

or $\quad k = \dfrac{\ln 2}{T} = \dfrac{0.693147}{T},$

and $\quad T = \dfrac{\ln 2}{k} = \dfrac{0.693147}{k}.$

Note that this relationship between k and T does not depend on P_0.

Example 5   *Business: Internet Use.* Worldwide use of the Internet is increasing at

an exponential rate. Net traffic is doubling every 100 days. What is the exponential growth rate?

Solution   We have

$$k = \frac{\ln 2}{T} = \frac{0.693147}{100 \text{ days}}$$

To do the calculation on a grapher, simply enter it as $\frac{\ln 2}{100}$ without finding the logarithmic value first.

$$\approx 0.006931 \frac{1}{\text{day}} \approx 0.69\% \text{ per day}.$$

The exponential growth rate is approximately 0.69% per day.   ◄

The Rule of 70

The relationship between doubling time T and interest rate k is the basis of a rule often used in the investment world, called the **Rule of 70.** To estimate how long it will take to double your money at varying rates of return, divide 70 by the rate of return. To see how this works, let the interest rate $k = r\%$. Then

$$T = \frac{\ln 2}{k} = \frac{0.693147}{r\%}$$

$$= \frac{0.693147}{r \times 0.01} \cdot \frac{100}{100}$$

$$= \frac{69.3147}{r} \approx \frac{70}{r}.$$

Example 6   *Life Science: World Population Growth.* The population of the world was 6.0 billion in 1998 and it is projected to be 6.0400 billion in 2000. It has been estimated that the population is growing exponentially at the rate of 0.016, or 1.6% per year. (How was this estimate determined? The answer is in the model we develop in the following Technology Connection.) Thus,

$$\frac{dP}{dt} = 0.016P,$$

where t is the time, in years, after 2000. To simplify computations, we assume that the population was 6.0400 billion at the beginning of 2000.

a) Find the function that satisfies the equation. Assume that $P_0 = 6.0400$ and $k = 0.016$.

b) Estimate the world population in 2020 ($t = 20$).

c) After what period of time will the population be double that in 2000?

Solution

a) $P(t) = 6.0400e^{0.016t}$

b) $P(20) = 6.0400e^{0.016(20)} = 6.0400e^{0.32} \approx 8.3179$ billion

c) $T = \dfrac{\ln 2}{k} = \dfrac{\ln 2}{0.016} = 43.3$ yr

Thus, according to this model, the population in 2000 will double itself in 2043. (No wonder ecologists are alarmed!).   ◄

TECHNOLOGY   CONNECTION

Exponential Regression: Modeling World Population Growth

The table below shows data regarding world population growth. (A graph illustrating these data, along with the projected population in 2020, appeared in Section 4.1.)

Year	World Population (in billions)
1927	2
1960	3
1974	4
1987	5
1998	6

How were the data projected in 2020? Let's use the data from 1927 to 1998 to find an exponential model. The graph shows a rapidly growing population that can be modeled with an exponential function. We carry out the regression procedure very much as we did in Section 1.1, but here we choose ExpReg rather than LinReg.

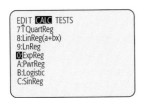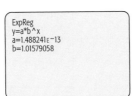

Note that this gives us an exponential model of the type $y = a \cdot b^x$, but the base is not the number *e*. We can make a conversion to an exponential function, base *e*, using the formula $b^x = e^{x(\ln b)}$, which we state without proof.

The exponential regression gives us the model

$$y = (1.488241 \cdot 10^{-13})(1.01579058)^x. \qquad (1)$$

Using $b^x = e^{x(\ln b)}$, we obtain

$$b^x = (1.01579058)^x = e^{x(\ln b)}$$
$$= e^{x(\ln 1.01579058)} \approx e^{x(0.0156672059)}$$
$$= e^{0.0156672059x}.$$

We have converted equation (1) to

$$y = (1.488241 \cdot 10^{-13})e^{0.0156672059x}. \qquad (2)$$

The advantage of this form is that we see the growth rate. Here the world population growth rate is 0.16, or 1.6%. To find world population in 2000, we can substitute 2000 for *x* in either equation (1) or (2). We choose equation (2):

$$y = (1.488241 \cdot 10^{-13})e^{0.0156672059(2000)}$$
$$\approx 6.0400 \text{ billion}.$$

Exercises

Use equation (1) or equation (2) to predict world population in each year.

1. 2005   **2.** 2010   **3.** 2020   **4.** 2050

Business: Projected College Costs. For Exercises 5 and 6, use the data regarding projected college costs listed in the table below.

School Year	Projected Cost at a State University
1989–1990	$ 6,172
1990–1991	6,543
1991–1992	6,935
1992–1993	7,351
1993–1994	7,792
1994–1995	8,260
1995–1996	8,756
1996–1997	9,281
1997–1998	9,838
1998–1999	10,424
1999–2000	11,054

Source: College Board, Senate Labor Committee

5. Use the **REGRESSION** feature to fit an exponential function $y = a \cdot b^x$ to the data. Then convert that formula to an exponential function, base *e*, and determine the exponential growth rate. Let 1989–1990 be represented by $x = 0$.

6. Use either of the exponential functions found in Exercise 5 to predict college costs in 2001, 2015, and 2040.

In the preceding Technology Connection, we found a way to use a procedure called *regression* to create an exponential model. There is another way to create such a model if regression is not an option. We do so by choosing two representative data points and using them to determine P_0 and k in $P(t) = P_0 e^{kt}$.

Modeling Other Phenomena

Example 7 Life Science: Alcohol Absorption and the Risk of Having an Accident. Extensive research has provided data relating the risk R (in percent) of having an automobile accident to the blood alcohol level b (in percent). Using two representative points (0, 1%) and (0.14, 20%), we can approximate the data with an exponential function. The modeling assumption is that the rate of change of the risk R with respect to the blood alcohol level b is given by

$$\frac{dR}{db} = kR.$$

a) Find the function that satisfies the equation. Assume that $R_0 = 1\%$.

b) Find k using the data point $R(0.14) = 20$.

c) Rewrite $R(b)$ in terms of k.

d) At what blood alcohol level will the risk of having an accident be 100%? Round to the nearest hundredth.

Some myths about alcohol. It's a fact—the blood alcohol concentration (BAC) in the human body is measurable. And there's no cure for its effect on the central nervous system except time. It takes time for the body's metabolism to recover. That means a cup of coffee, a cold shower, and fresh air can't erase the effect of several drinks.

There are variables, of course: a person's body weight, how many drinks have been consumed in a given time, how much has been eaten, and so on. These account for different BAC levels. But the myth that some people can "handle their liquor" better than others is a gross rationalization—especially when it comes to driving. Some people can act more sober than others. But an automobile doesn't act; it reacts.

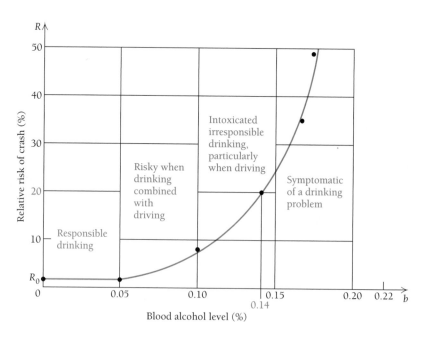

Blood alcohol level (%)

Solution

a) Since both R and b are percents, we omit the % symbol to simplify computation. The solution is

$$R(b) = e^{kb}, \quad \text{since } R_0 = 1.$$

We have made use of the data point (0, 1).

b) We now use the second data point, (0.14, 20), to determine k. We solve the equation

$$R(b) = e^{kb}, \quad \text{or} \quad 20 = e^{k(0.14)}$$

for k, using natural logarithms:

$$20 = e^{k(0.14)} = e^{0.14k}$$

$$\ln 20 = \ln e^{0.14k}$$

$$\ln 20 = 0.14k$$

$$\frac{\ln 20}{0.14} = k$$

$$\frac{2.995732}{0.14} = k$$

$$21.4 \approx k. \qquad \text{Rounded to the nearest tenth}$$

c) $R(b) = e^{21.4b}$

d) We substitute 100 for $R(b)$ and solve the equation for b:

$$100 = e^{21.4b}$$

$$\ln 100 = \ln e^{21.4b}$$

$$\ln 100 = 21.4b$$

$$\frac{\ln 100}{21.4} = b$$

$$\frac{4.605170}{21.4} = b$$

$$0.22 \approx b. \qquad \text{Rounded to the nearest hundredth}$$

The calculations done in this problem can be performed more conveniently on your calculator if you do not stop to round. For example, in part (b), we find $\ln 20$ and divide by 0.14, obtaining 21.39808767.... We then use that value for k in part (d). Answers will be found that way in the exercises. You may note some variance in the last one or two decimal places if you round as you go.

Thus when the blood alcohol level is 0.22%, according to this model, the risk of an accident is 100%. From the graph, we see that this would occur for a 160-lb man after 12 1-oz drinks of 86-proof whiskey. "Theoretically," the model tells us that after 12 drinks of whiskey, one is "sure" to have an accident. This might be questioned in reality, since a person who has had 12 drinks might not be able to drive at all. ◅

Models of Limited Growth

The growth model $P(t) = P_0 e^{kt}$ has many applications to unlimited population growth as we have seen thus far in this section. However, it seems reasonable that there can be factors that prevent a population from exceeding some limiting value L—perhaps a limitation on food, living space, or other natural resources. One

model of such growth is

$$P(t) = \frac{L}{1 + be^{-kt}},$$

which is called the *logistic equation*.

Example 8   *Business: Box-Office Revenue. Titanic* is one of the greatest box-office attractions of all time. Having opened in December 1997, it was still running in theaters in August 1998 when it went into videotape sales. The total U.S. box-office revenue R, in millions of dollars, after time t, in weeks, can be approximated by the logistic function

$$R(t) = \frac{596.423}{1 + 4.974e^{-0.1814t}}$$

(*Source*: Exhibitor Relations Co., Inc.).

a) Find the total revenue after 1 week; 2 weeks; 4 weeks; 5 weeks; 8 weeks; 26 weeks.

b) Find the rate of change $R'(t)$.

c) Sketch a graph of the function.

Solution

a) We use a calculator to find the function values, listing them in a table, as shown at left.

t	$R(t)$
1	115.8
2	133.7
4	175.0
5	198.3
8	275.4
26	571.0

b) We find the rate of change $R'(t)$ as follows, using the Quotient Rule:

$$R'(t) = \frac{(1 + 4.974e^{-0.1814t}) \cdot 0 - [(-0.1814)(4.974)e^{-0.1814t}]596.423}{(1 + 4.974e^{-0.1814t})^2}$$

$$\approx \frac{538.14e^{-0.1814t}}{(1 + 4.974e^{-0.1814t})^2}.$$

c) We consider the curve-sketching procedures discussed in Section 3.2. The derivative $R'(t)$ exists for all real numbers t; thus there are no critical points for which the derivative does not exist. The equation $R'(t) = 0$ holds only when the numerator is 0. But the numerator is positive for all real numbers t. Thus the function has no critical points and hence no relative extrema.

Since $e^{-0.1814t} > 0$ for all real numbers t, $R'(t)$ is positive

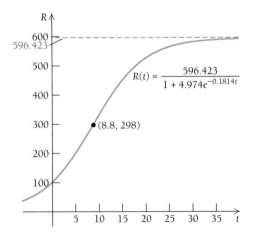

$$R(t) = \frac{596.423}{1 + 4.974e^{-0.1814t}}$$

for all real numbers t. Thus, P is increasing over the entire real line. Though we will not develop the procedure here, the second derivative can be used to show that the graph has a point of inflection at about (8.8, 298). The function is concave up (R' is increasing) over the interval (0, 8.8) and concave down (R' is decreasing) over the interval (8.8, ∞). The graph is the S-shaped curve shown above.

Another model of limited growth is provided by the function

$$P(t) = L(1 - e^{-kt}),$$

which is shown graphed below. This function also increases over the entire interval $[0, \infty)$.

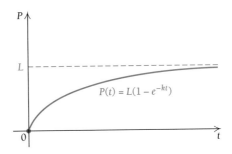

$$P(t) = L(1 - e^{-kt})$$

EXERCISE SET 4.3

1. Find the exponential function that satisfies the equation $dQ/dt = kQ$.

2. Find the exponential function that satisfies the equation $dR/dt = kR$.

Applications

BUSINESS AND ECONOMICS

3. *Franchise Expansion.* A national hamburger firm is selling franchises throughout the country. The marketing manager estimates that the number of franchises N will increase at the rate of 10% per year, that is,

$$\frac{dN}{dt} = 0.10N.$$

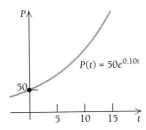

$$P(t) = 50e^{0.10t}$$

a) Find the function that satisfies the equation. Assume that the number of franchises at $t = 0$ is 50.

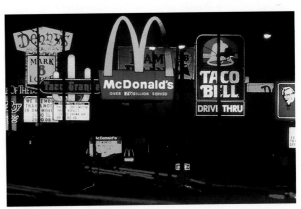

b) How many franchises will there be in 20 yr?

c) After what period of time will the initial number of 50 franchises double?

4. *Franchise Expansion.* Pizza, Unltd., a national

pizza firm, is selling franchises throughout the country. The CEO estimates that the number of franchises N will increase at the rate of 15% per year, that is,

$$\frac{dN}{dt} = 0.15N.$$

a) Find the function that satisfies the equation. Assume that the number of franchises at $t = 0$ is 40.
b) How many franchises will there be in 20 yr?
c) After what period of time will the initial number of 40 franchises double?

5. *Compound Interest.* Suppose that P_0 is invested in a savings account in which interest is compounded continuously at 6.5% per year. That is, the balance P grows at the rate given by

$$\frac{dP}{dt} = 0.065P.$$

a) Find the function that satisfies the equation. List it in terms of P_0 and 0.065.
b) Suppose that $1000 is invested. What is the balance after 1 yr? after 2 yr?
c) When will an investment of $1000 double itself?

6. *Compound Interest.* Suppose that P_0 is invested in a savings account in which interest is compounded continuously at 8% per year. That is, the balance P grows at the rate given by

$$\frac{dP}{dt} = 0.08P.$$

a) Find the function that satisfies the equation. List it in terms of P_0 and 0.08.
b) Suppose that $20,000 is invested. What is the balance after 1 yr? after 2 yr?
c) When will an investment of $20,000 double itself?

7. *Annual Interest Rate.* A bank advertises that it compounds interest continuously and that it will double your money in 10 yr. What is its annual interest rate?

8. *Annual Interest Rate.* A bank advertises that it compounds interest continuously and that it will double your money in 12 yr. What is its annual interest rate?

9. *Oil Demand.* The growth rate of the demand for oil in the United States is 10% per year. When will the demand be double that of 2000?

10. *Coal Demand.* The growth rate of the demand for coal in the world is 4% per year. When will the demand be double that of 2000?

Interest Compounded Continuously. For Exercises 11–14, complete the following.

Initial Investment at $t = 0$, P_0	Interest Rate, k	Doubling Time, T (in years)	Amount After 5 yr
11. $75,000	6.2%		
12. $5,000			$7,130.90
13.	8.4%		$11,414.71
14.		11 yr	$17,539.32

15. *Value of a Van Gogh Painting.* The Van Gogh painting "Irises," shown below, sold for $84,000 in 1947, but sold again for $53,900,000 in 1987. Assuming the exponential model:
a) Find the value k ($V_0 = \$84,000$), and write the function.
b) Estimate the value of the painting in 1997.
c) What is the doubling time for the value of the painting?
d) How long after 1947 will the value of the painting be $1 billion?

Van Gogh's "Irises," a 28-by-32-in. oil on canvas.

 16. *Sales of Digital Cameras.* Data in the following table show sales, in billions of dollars, of digital cameras. Photographs from these cameras allow consumers to take pictures and use them in Web pages or transmit them through e-mail.

Years, t, After 1996	Sales of Digital Cameras (in billions)
0	$0.386
1	0.518
2	0.573
3	0.819

Source: International Data Corporation

a) Use the **REGRESSION** feature to fit an exponential function $y = a \cdot b^x$ to the data. Then convert that formula to an exponential function, base e, where t is the number of years after 1996 and determine the exponential growth rate. (See the Technology Connection on p. 323.)
b) Estimate the sales of digital cameras in 2000 and in 2001.
c) After what amount of time will sales be $2.0 billion?
d) What is the doubling time of the sales of digital cameras?

17. *Sales of Digital Cameras.* Refer to the data in Exercise 16.

a) To find an exponential function that fits the data, find k using the data point $(2, 0.573)$ and write the function. Assume that $P_0 = 0.386$ and that t is the number of years after 1996.
b) Estimate the sales of digital cameras in 2000 and in 2001.
c) After what amount of time will sales be $2.0 billion?
d) What is the doubling time of the sales of digital cameras?
tw e) Compare your answers to Exercises 16 and 17. Decide which exponential function seems best to you and explain why.

18. *Consumer Price Index.* The *consumer price index* compares the costs of goods and services over various years, where 1967 is used as a base $(t = 0)$. The same goods and services that cost $100 in 1967 cost $184.50 in 1977. Assuming the exponential model:

a) Find the value k ($P_0 = \$100$), and write the function.
b) Estimate what the same goods and services will cost in 2005.
c) After what period of time did the same goods and services cost double that of 1967?

19. *Sales of Paper Shredders.* Data in the following bar graph show total sales of paper shredders in recent years.

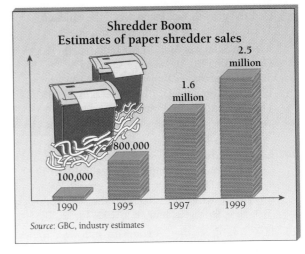

Shredder Boom
Estimates of paper shredder sales

2.5 million
1.6 million
800,000
100,000

1990   1995   1997   1999

Source: GBC, industry estimates

a) Use the **REGRESSION** feature to fit an exponential function $y = a \cdot b^x$ to the data. Then convert that formula to an exponential function, base e, where t is the number of years after 1990 and determine the exponential growth rate. (See the Technology Connection on p. 323.)
b) Estimate the total sales of paper shredders in 2004 and in 2010.
c) After what amount of time will sales be 200 million?
d) What is the doubling time of the total sales of paper shredders?

20. *Sales of Paper Shredders.* Refer to the data in Exercise 19.

a) To find an exponential function that fits the data, find k using the data point $(7, 1,600,000)$ and write the function. Assume $P_0 = 100,000$ and t is the number of years after 1990.
b) Estimate the total sales of paper shredders in 2004 and in 2010.
c) After what amount of time will total sales be 200 million?
d) What is the doubling time of the sales of paper shredders?
tw e) Compare your answers to Exercises 19 and 20. Decide which exponential function seems best to you and explain why.

21. *Value of Manhattan Island.* Peter Minuit of the Dutch West India Company purchased Manhattan Island from the Indians in 1626 for $24 worth of merchandise. Assuming an exponential rate of inflation of 8%, how much will Manhattan be worth in 2000?

22. *Total Revenue.* Intel, a computer chip manufacturer, reported $1265 million in total revenue in 1986. In 1994, the total revenue was $11,521 million. Assuming the exponential model, find the growth rate k and write the function. Then estimate the total revenue in 2006.

23. *Average Salary of Major-League Baseball Players.* In 1970, the average salary of major-league baseball players was $29,303. By 1998, the average salary had grown to $2,240,000. This was exponential growth. What was the growth rate? What will the average salary be in 2000? in 2008?

24. *Cost of a First-Class Postage Stamp.* The cost of a first-class postage stamp in 1962 was 4¢. In 1999, it was 33¢. This was exponential growth. What was the growth rate? What will the cost of a first-class postage stamp be in 2004? in 2010?

25. *Effect of Advertising.* A company introduces a new product on a trial run in a city. They advertised the product on television and found that the percentage P of people who bought the product after t ads had been run satisfied the function

$$P(t) = \frac{100\%}{1 + 49e^{-0.13t}}.$$

a) What percentage buy the product without seeing the ad ($t = 0$)?

b) What percentage buy the product after the ad is run 5 times? 10 times? 20 times? 30 times? 50 times? 60 times?

c) Find the rate of change $P'(t)$.

d) Sketch a graph of the function.

26. *Cost of a Hershey Bar.* The cost of a Hershey bar in 1962 was $0.05, and was increasing at an exponential growth rate of 9.7%. What will the cost of a Hershey bar be in 2000? in 2008?

LIFE AND PHYSICAL SCIENCES

Population Growth. For Exercises 27–32, complete the following.

	Population	Growth Rate, k	Doubling Time, T
27.	Mexico	3.5% per year	
28.	Europe		69.31 yr
29.	Oil reserves		6.931 yr
30.	Coal reserves		17.3 yr
31.	Alaska	2.794% per year	
32.	Central America		19.8 yr

33. *Population Growth.* The population of the United States was 241 million in 1986. At that time, it was estimated that the population P was growing exponentially at the rate of 0.9% per year, that is,

$$\frac{dP}{dt} = 0.009P,$$

where t is the time, in years.

a) Find the function that satisfies the equation. Assume that $P_0 = 241$ and $k = 0.009$.

b) Estimate the population of the United States in 2000 ($t = 14$).

c) After what period of time will the population be double that in 2000?

34. *Population Growth.* The population of western Europe was 430 million in 1961. At that time, it was estimated that the population was growing exponentially at the rate of 1% per year, that is,

$$\frac{dP}{dt} = 0.01P.$$

a) Find the function that satisfies the equation. Assume that $P_0 = 430$ and $k = 0.01$.

b) Estimate the population of Europe in 2000.

c) After what period of time will the population be double that of 2000?

35. *Blood Alcohol Level.* Refer to Example 7 (on alcohol absorption). At what blood alcohol level will the risk of an accident be 80%?

36. *Blood Alcohol Level.* Refer to Example 7. At what blood alcohol level will the risk of an accident be 90%?

37. *Bicentennial Growth of the United States.* The population of the United States in 1776 was about 2,508,000. In its bicentennial year, the population was about 216,000,000.

a) Assuming the exponential model, what was the growth rate of the United States through its bicentennial year?

tw b) Is this a reasonable assumption? Explain.

38. *Limited Population Growth.* A ship carrying 1000 passengers has the misfortune to be shipwrecked on a small island from which the passengers are never rescued. The natural resources of the island limit the growth of the population to a *limiting value* of 5780, to which the population gets closer and closer but which it never reaches. The population of the island after time t, in years, is given by the logistic equation

$$P(t) = \frac{5780}{1 + 4.78e^{-0.4t}}.$$

a) Find the population after 0 years; 1 year; 2 years; 5 years; 10 years; 20 years.
b) Find the rate of change $P'(t)$.
c) Sketch a graph of the function.

This island may have achieved its limited population growth.

39. *Limited Population Growth.* A lake is stocked with 400 fish of a new variety. The size of the lake, the availability of food, and the number of other fish restrict growth in the lake to a *limiting value* of 2500. (See Exercise 38.) The population of fish in the lake after time t, in months, is given by

$$P(t) = \frac{2500}{1 + 5.25e^{-0.32t}}.$$

a) Find the population after 0 months; 1 month; 5 months; 10 months; 15 months; 20 months.
b) Find the rate of change $P'(t)$.
c) Sketch a graph of the function.

SOCIAL SCIENCES

40. *Hullian Learning Model.* The Hullian learning model asserts that the probability p of mastering a task after t learning trials is given by

$$p(t) = 1 - e^{-kt},$$

where k is a constant that depends on the task to be learned. Suppose that a new task is introduced in a factory on an assembly line. For that particular task, the constant $k = 0.28$.

a) What is the probability of learning the task after 1 trial? 2 trials? 5 trials? 11 trials? 16 trials? 20 trials?
b) Find the rate of change $p'(t)$.
c) Sketch a graph of the function.

41. *Diffusion of Information.* Pharmaceutical firms spend an immense amount of money in order to test a new medication. After the drug is approved by the Federal Drug Administration, it still takes time for physicians to fully accept and make use of the medication. The use approaches a *limiting value* of 100%, or 1, after time t, in months. Suppose that for a new cancer medication the percentage P of physicians using the product after t months is given by

$$P(t) = 100\%(1 - e^{-0.4t}).$$

a) What percentage of doctors have accepted the medication after 0 months? 1 month? 2 months? 3 months? 5 months? 12 months? 16 months?
b) Find the rate of change $P'(t)$.
c) Sketch a graph of the function.

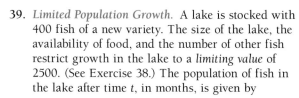 42. *Spread of a Rumor.* The rumor "People who study math are people you can count on" spreads across a college campus. Data in the table below show the number of students N who have heard the rumor after time t, in days.

Time, t (in days)	Number Who Have Heard the Rumor
1	1
2	2
3	4
4	7
5	12
6	18
7	24
8	26
9	28
10	28
11	29
12	30

a) Use the **REGRESSION** feature on a grapher to fit a logistic equation

$$N(t) = \frac{c}{1 + ae^{-bt}}$$

to the data.

    b) Graph the equation and estimate the limiting value of the function. At most, how many students will hear the rumor?

    c) Graph the function.

    d) Find the rate of change $N'(t)$.

tw   **e)** Find $\lim\limits_{t \to \infty} N'(t)$ and explain its meaning.

Synthesis

Business: Effective Annual Yield. Suppose that $100 is invested at 7%, compounded continuously, for 1 yr. We know from Example 4 that the balance will be $107.25. This is the same as if $100 were invested at 7.25% and compounded once a year (simple interest). The 7.25% is called the *effective annual yield*. In general, if P_0 is invested at k% compounded continuously, then the effective annual yield is that number i satisfying $P_0(1 + i) = P_0 e^k$. Then $1 + i = e^k$, or

$$\text{Effective annual yield} = i = e^k - 1.$$

43. An amount is invested at 7.3% per year compounded continuously. What is the effective annual yield?

44. An amount is invested at 8% per year compounded continuously. What is the effective annual yield?

45. The effective annual yield on an investment compounded continuously is 9.42%. At what rate was it invested?

46. The effective annual yield on an investment compounded continuously is 6.61%. At what rate was it invested?

47. Find an expression relating the growth rate k and the *tripling time T_3*.

48. Find an expression relating the growth rate k and the *quadrupling time T_4*.

49. Gather data concerning population growth in your city. Estimate the population in 2004; in 2010.

50. A quantity Q_1 grows exponentially with a doubling time of 1 yr. A quantity Q_2 grows exponentially with a doubling time of 2 yr. If the initial amounts of Q_1 and Q_2 are the same, when will Q_1 be twice the size of Q_2?

51. A growth rate of 100% per day corresponds to what exponential growth rate per hour?

52. Show that any two measurements of an exponentially growing population will determine k. That is, show that if y has the values y_1 at t_1 and y_2 at t_2, then

$$k = \frac{\ln (y_2/y_1)}{t_2 - t_1}.$$

tw **53.** Complete the table below, which relates growth rate k and doubling time T.

Growth Rate, k (in percent per year)	1%	2%			14%
Doubling Time, T (in years)			15	10	

Graph $T = \ln 2/k$. Is this a linear relationship? Explain.

tw **54.** Explain the differences in the graphs of an exponential function and a logistic function.

tw **55.** Explain the Rule of 70 to a fellow student.

4.4

Applications: Decay

In the equation of population growth $dP/dt = kP$, the constant k is actually given by

$$k = (\text{Birth rate}) - (\text{Death rate}).$$

Thus a population "grows" only when the *birth rate* is greater than the *death rate*. When the birth rate is less than the death rate, k will be negative so the population will be decreasing, or "decaying," at a rate proportional to its size. For convenience in our computations, we will express such a negative value as $-k$, where $k > 0$. The equation

$$\frac{dP}{dt} = -kP, \quad \text{where } k > 0,$$

shows P to be *decreasing* as a function of time, and the solution

$$P(t) = P_0 e^{-kt}$$

shows it to be decreasing exponentially. This is called **exponential decay.** The amount present initially at $t = 0$ is again P_0.

E x p l o r a t o r y .
Using the same set of axes, graph $y_1 = e^{2x}$ and $y_2 = e^{-2x}$.
Compare the graphs.
   Then, using the same set of axes, graph $y_1 = 100e^{-0.06x}$ and $y_2 = 100e^{0.06x}$.
Compare these graphs.

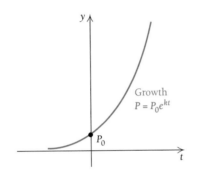

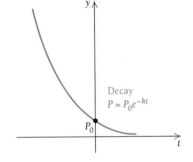

Radioactive Decay

Radioactive elements decay exponentially; that is, they disintegrate at a rate that is proportional to the amount present.

Example 1   *Life Science: Decay.* Strontium-90 has a decay rate of 2.8% per year. The rate of change of an amount N is given by

$$\frac{dN}{dt} = -0.028N.$$

a) Find the function that satisfies the equation in terms of N_0 (the amount present at $t = 0$).

b) Suppose that 1000 grams (g) of strontium-90 is present at $t = 0$. How much will remain after 100 yr?

c) After how long will half of the 1000 g remain?

Solution

a) $N(t) = N_0 e^{-0.028t}$

b) $N(100) = 1000e^{-0.028(100)}$
$\qquad\qquad = 1000e^{-2.8}$
$\qquad\qquad = 1000(0.060810)$
$\qquad\qquad \approx 60.8 \text{ g}$

c) We are asking, "At what time T will $N(T) = 500$?" The number T is called the **half-life.** To find T, we solve the equation

$$500 = 1000e^{-0.028T}$$

$$\frac{1}{2} = e^{-0.028T}$$

$$\ln \frac{1}{2} = \ln e^{-0.028T}$$

$$\ln 1 - \ln 2 = -0.028T$$

$$0 - \ln 2 = -0.028T$$

$$\frac{-\ln 2}{-0.028} = T$$

$$\frac{\ln 2}{0.028} = T$$

$$\frac{0.693147}{0.028} = T$$

$$25 \approx T.$$

Thus the half-life of strontium-90 is 25 yr.

We can find a general expression relating the decay rate k and the half-life T by solving the equation

$$\tfrac{1}{2}P_0 = P_0 e^{-kT}$$

$$\tfrac{1}{2} = e^{-kT}$$

$$\ln \tfrac{1}{2} = \ln e^{-kT}$$

$$\ln 1 - \ln 2 = -kT$$

$$0 - \ln 2 = -kT$$

$$-\ln 2 = -kT$$

$$\ln 2 = kT.$$

Again, we have the following.

> ### Theorem 11
>
> The *decay rate* k and the *half-life* T are related by
>
> $$kT = \ln 2 = 0.693147,$$
>
> or
>
> $$k = \frac{\ln 2}{T} \quad \text{and} \quad T = \frac{\ln 2}{k}.$$

Thus the half-life T depends only on the decay rate k. In particular, it is independent of the initial population size.

The effect of half-life is shown in the radioactive decay curve that follows.

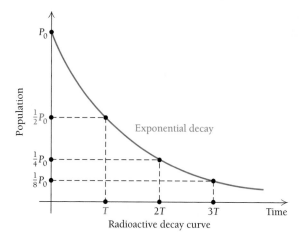

Radioactive decay curve

How can scientists determine that an animal bone has lost 30% of its carbon-14? The assumption is that the percentage of carbon-14 in the atmosphere and in living plants and animals is the same. When a plant or animal dies, the amount of carbon-14 decays exponentially. The scientist burns the animal bone and uses a Geiger counter to determine the percentage of the smoke that is carbon-14. The amount by which this varies from the percentage in the atmosphere indicates how much carbon-14 has been lost.

The process of carbon-14 dating was developed by the American chemist Willard F. Libby in 1952. It is known that the radioactivity in a living plant is 16 disintegrations per gram per minute. Since the half-life of carbon-14 is 5750 years, a dead plant with an activity of 8 disintegrations per gram per minute is 5750 years old, one with an activity of 4 disintegrations per gram per minute is 11,500 years old, and so on. Carbon-14 dating can be used to measure the age of objects from 30,000 to 40,000 years old. Beyond such an age, it is too difficult to measure the radioactivity, and some other method would have to be used.

Note that the exponential function gets close to, but never reaches, 0 as t gets larger. Thus, in theory, a radioactive substance never completely decays.

Example 2   *Life Science: Half-life.* Plutonium, a common product and ingredient of nuclear reactors, is of great concern to those who are against the building of nuclear reactors. Its decay rate is 0.003% per year. What is its half-life?

Solution   We have

$$T = \frac{\ln 2}{k}$$

$$= \frac{0.693147}{0.00003}$$

$$\approx 23{,}100 \text{ yr.}$$

Thus the half-life of plutonium is 23,100 yr.   ◁

Example 3   *Life Science: Carbon Dating.* The radioactive element carbon-14 has a half-life of 5750 yr. The percentage of carbon-14 present in the remains of plants and animals can be used to determine age. Archaeologists found that the linen wrapping from one of the Dead Sea Scrolls had lost 22.3% of its carbon-14. How old was the linen wrapping?

Solution

a) Find the decay rate k:

$$k = \frac{\ln 2}{T} = \frac{0.693147}{5750} \approx 0.0001205, \quad \text{or} \quad 0.01205\% \text{ per year.}$$

b) Find the exponential equation for the amount $N(t)$ that remains from an initial amount N_0 after t years:

$$N(t) = N_0 e^{-0.0001205t}.$$

(*Note*: This equation can be used for any subsequent carbon-dating problem.)

In 1947, a Bedouin youth looking for a stray goat climbed into a cave at Kirbet Qumran on the shores of the Dead Sea near Jericho and came upon earthenware jars containing an incalculable treasure of ancient manuscripts. Shown here are fragments of those so-called Dead Sea Scrolls, a portion of some 600 or so texts found so far and which concern the Jewish books of the Bible. Officials date them before 70 A.D., making them the oldest Biblical manuscripts by 1000 years.

c) If the Dead Sea Scrolls lost 22.3% of their carbon-14 from an initial amount P_0, then 77.7% (P_0) is the amount present. To find the age t of the scrolls, we solve the following equation for t:

$$77.7\%\ N_0 = N_0 e^{-0.0001205t}$$
$$0.777 = e^{-0.0001205t}$$
$$\ln 0.777 = \ln e^{-0.0001205t}$$
$$\ln 0.777 = -0.0001205t$$
$$-0.2523149 = -0.0001205t$$
$$\frac{0.2523149}{0.0001205} = t$$
$$2094 \approx t.$$

Thus the linen wrapping of the Dead Sea Scrolls is about 2094 yr old.

A Business Application: Present Value

A representative of a financial institution is often asked to solve a problem like the following.

Example 4 *Business: Present Value.* Following the birth of a child, a parent wants to make an initial investment of P_0 that will grow to $10,000 by the child's 20th birthday. Interest is compounded continuously at 8%. What should the initial investment be?

Solution Using the equation $P = P_0 e^{kt}$, we find P_0 such that

$$10,000 = P_0 e^{0.08 \cdot 20},$$

or

$$10,000 = P_0 e^{1.6}.$$

Now

$$\frac{10,000}{e^{1.6}} = P_0,$$

or

$$10,000 e^{-1.6} = P_0,$$

and, using a calculator, we have

$$P_0 = 10,000 e^{-1.6}$$
$$= 10,000(0.201897)$$
$$= \$2018.97.$$

Thus the parent must deposit $2018.97, which will grow to $10,000 by the child's 20th birthday.

Economists call $2018.97 the **present value** of $10,000 due 20 yr from now at 8%, compounded continuously. The process of computing present value is called **discounting.** Another way to think of this problem is "What do I have to invest now at 8%, compounded continuously, in order to have $10,000 in 20 years?" The answer is $2018.97 and it is the present value of $10,000.

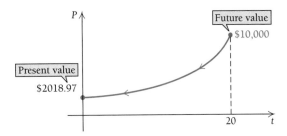

Computing present value can be interpreted as exponential decay from the future back to the present.

In general, the present value P_0 of an amount P due t years later is found by solving the following equation for P_0:

$$P_0 e^{kt} = P$$

$$P_0 = \frac{P}{e^{kt}} = P e^{-kt}.$$

Theorem 12

The **present value** P_0 of an amount P due t years later at interest rate k, compounded continuously, is given by

$$P_0 = P e^{-kt}.$$

Newton's Law of Cooling

Consider the following situation. A hot cup of soup, at a temperature of 200°, is placed in a room whose temperature is 70°. The temperature of the soup cools over time t, in minutes, according to the mathematical model, or equation, called **Newton's Law of Cooling.**

Newton's Law of Cooling

The temperature T of a cooling object drops at a rate that is proportional to the difference $T - C$, where C is the constant temperature of the surrounding medium. Thus,

$$\frac{dT}{dt} = -k(T - C). \tag{1}$$

The function that satisfies equation (1) is

$$T = T(t) = a e^{-kt} + C. \tag{2}$$

We can check this by differentiating:

$$\frac{dT}{dt} = -kae^{-kt}$$

$$= -k(ae^{-kt})$$

$$= -k(T - C). \qquad T = ae^{-kt} + C, \text{ or } T - C = ae^{-kt}$$

Example 5 *Life Science: Scalding Coffee.* McDivett's Pie Shoppes, a national restaurant firm, finds that the temperature of its freshly brewed coffee is 130°. They are naturally concerned that if a customer spills hot coffee on themselves, a lawsuit might result. Room temperature in the restaurants is generally 72°. The temperature of the coffee cools to 120° after 4.3 min. The company determines that it is safer to serve the coffee at a temperature of 105°. How long does it take a cup of coffee to cool to 105°?

Solution

a) We first find the value of a in equation (2) in Newton's Law of Cooling. At $t = 0$, $T = 130°$. We solve for a as follows:

$$130 = ae^{-k \cdot 0} + 72$$

$$130 = a + 72$$

$$58 = a.$$

Next, we find k using the fact that at $t = 4.3$, $T = 120°$. We solve the following equation for k:

$$120 = 58e^{-k \cdot (4.3)} + 72$$

$$48 = 58e^{-4.3k}$$

$$\frac{48}{58} = e^{-4.3k}$$

$$\ln \frac{48}{58} = \ln e^{-4.3k}$$

$$-4.3k = -0.1892420 \qquad \text{Dividing 48 by 58 and taking the natural logarithm}$$

$$k \approx 0.044.$$

b) To see how long it will take the coffee to cool to 105°, we solve for t:

$$105 = 58e^{-0.044t} + 72$$

$$33 = 58e^{-0.044t}$$

$$\frac{33}{58} = e^{-0.044t}$$

$$\ln \frac{33}{58} = \ln e^{-0.044t}$$

$$-0.5639354 = -0.044t$$

$$t \approx 12.8 \text{ min.}$$

Thus, if the coffee is allowed to cool for about 13 min, then it will be "safe" to serve.

The graph of $T(t) = ae^{-kt} + C$ shows that $\lim_{t \to \infty} T(t) = C$. The temperature of the object decreases toward the temperature of the surrounding medium.

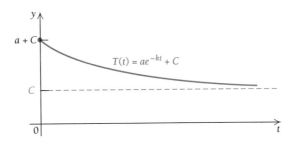

Mathematically, this model tells us that the temperature never reaches C, but in practice this happens eventually. At least, the temperature of the cooling object gets so close to that of the surrounding medium that no device could detect a difference. Let's now see how Newton's Law of Cooling can be used in solving a crime.

Example 6 *Life Science: When Was the Murder Committed?* The police discover the body of a calculus professor. Critical to solving the crime is determining when the murder was committed. The police call the coroner, who arrives at 12 noon. She immediately takes the temperature of the body and finds it to be 94.6°. She waits 1 hour, takes the temperature again, and finds it to be 93.4°. She also notes that the temperature of the room is 70°. When was the murder committed?

Solution We first find a in the equation $T(t) = ae^{-kt} + C$. Assuming that the temperature of the body was normal when the murder occurred, we have $T = 98.6°$ at $t = 0$. Thus,

$$98.6 = ae^{-k \cdot 0} + 70,$$

so $\qquad a = 28.6.$

Thus, T is given by $T(t) = 28.6e^{-kt} + 70$.

We want to find the number of hours N since the murder was committed. To do so, we must first determine k. From the two temperature readings the coroner made, we have

$$94.6 = 28.6e^{-kN} + 70, \quad \text{or} \quad 24.6 = 28.6e^{-kN}; \qquad (3)$$
$$93.4 = 28.6e^{-k(N+1)} + 70, \quad \text{or} \quad 23.4 = 28.6e^{-k(N+1)}. \qquad (4)$$

Dividing equation (3) by equation (4), we get

$$\frac{24.6}{23.4} = \frac{28.6e^{-kN}}{28.6e^{-k(N+1)}}$$
$$= e^{-kN+k(N+1)}$$
$$= e^{-kN+kN+k} = e^{k}.$$

We solve this equation for k:

$$\ln \frac{24.6}{23.4} = \ln e^{k} \qquad \text{Taking the natural logarithm on both sides}$$

$$0.05 \approx k.$$

Next, we substitute back into equation (3) and solve for N:

$$24.6 = 28.6e^{-0.05N}$$

$$\frac{24.6}{28.6} = e^{-0.05N}$$

$$\ln \frac{24.6}{28.6} = \ln e^{-0.05N}$$

$$-0.150660 = -0.05N$$

$$3 \approx N.$$

Since the coroner arrived at 12 noon, the murder was committed at about 9:00 A.M.

In summary, we have added several functions to our candidates for curve fitting. Let's review them.

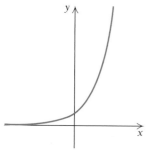

Exponential:
$f(x) = ab^x$, or ae^{kx}
$a, b > 0, k > 0$

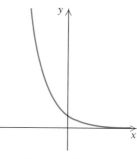

Exponential:
$f(x) = ab^{-x}$, or ae^{-kx}
$a, b > 0, k > 0$

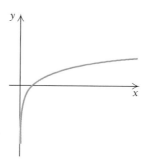

Logarithmic:
$f(x) = a + b \ln x$

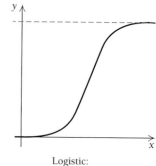

Logistic:
$$f(x) = \frac{a}{1 + be^{-kx}}$$

Now, when we analyze a set of data, we can consider these models, as well as linear, quadratic, polynomial, and rational functions, for curve fitting.

EXERCISE SET 4.4

Applications

BUSINESS AND ECONOMICS

1. *Present Value.* Following the birth of a child, a parent wants to make an initial investment P_0 that will grow to $50,000 by the child's 20th birthday. Interest is compounded continuously at 6%. What should the initial investment be?

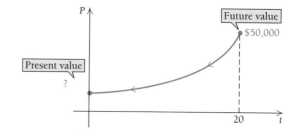

2. *Present Value.* Following the birth of a child, a parent wants to make an initial investment P_0 that will grow to $40,000 by the child's 20th birthday. Interest is compounded continuously at 7.2%. What should the initial investment be?

3. *Present Value.* Find the present value of $60,000 due 8 yr later at 6.4%, compounded continuously.

4. *Present Value.* Find the present value of $50,000 due 16 yr later at 8.2%, compounded continuously.

5. *Supply and Demand.* The supply and demand for the sale of stereos by a sound company are given by

$$x = S(p) = \ln p \quad \text{and} \quad x = D(p) = \ln \frac{163,000}{p},$$

where $S(p)$ is the number of stereos that the company is willing to sell at price p and $D(p)$ is the quantity that the public is willing to buy at price p. Find the equilibrium point. (For reference, see Section 1.5.)

6. *Salvage Value.* A business estimates that the salvage value V of a piece of machinery after t years is given by

$$V(t) = \$40,000e^{-t}.$$

a) What did the machinery cost initially?
b) What is the salvage value after 2 yr?
tw c) Find the rate of change of the salvage value and explain its meaning.

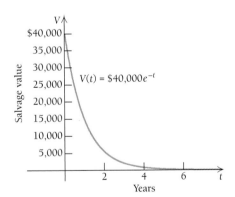

7. *Present Value.* A person knows that a trust fund will yield $80,000 in 13 yr. A CPA is preparing a financial statement for this client and wants to take into account the present value of the trust fund in computing the client's net worth. Interest is compounded continuously at 8.3%. What is the present value of the trust fund?

8. *Consumer Price Index.* The consumer price index compares the costs of goods and services over

various years, where 1967 is used as a base. The same goods and services that cost $100 in 1967 cost $42 in 1940. Assuming the exponential-decay model:

a) Find the value k, and write the equation. Round to the nearest hundredth.
b) Estimate what the same goods and services cost in 1900.

9. *Salvage Value.* A company tracks the value of a particular photocopier over a period of years. The data in the table below show the value of the copier at time t, in years after the date of purchase.

Time, t (in years)	Salvage Value
0	$34,000
1	22,791
2	15,277
3	10,241
4	6,865
5	4,600
6	3,084

Source: International Data Corporation

a) Use the REGRESSION feature on a grapher to fit an exponential function $y = a \cdot b^x$ to the data. Then convert that formula to $V(t) = V_0 e^{-kt}$, where V_0 is the value when the copier is purchased and t is the time, in years, from the date of purchase. (See the Technology Connection on p. 323.)
b) Estimate the salvage value of the copier after 7 yr; 10 yr.
c) After what amount of time will the salvage value be $1000?
d) After how long will the copier be worth half of its original value?
tw e) Find the rate of change of the salvage value and interpret its meaning.

Sales Modeling. For each of the following scatterplots, determine which, if any, of these functions might be used as a model for the data.

a) Quadratic: $f(x) = ax^2 + bx + c$
b) Polynomial, not quadratic
c) Exponential: $f(x) = ab^x$, or ae^{kx}, $k > 0$
d) Exponential: $f(x) = ab^x$, or ae^{-kx}, $k > 0$
e) Logarithmic: $f(x) = a + b \ln x$
f) Logistic: $f(x) = \dfrac{a}{1 + be^{-kx}}$

10.

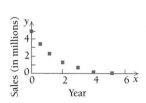

11.

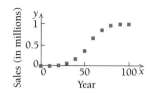

12.

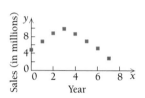

13.

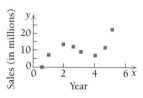

14.

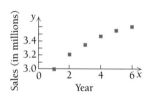

15.

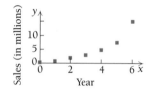

LIFE AND PHYSICAL SCIENCES

Radioactive Decay. For Exercises 16–21, complete the following.

Radioactive Substance	Decay Rate, k	Half-life, T
16. Polonium		3 min
17. Lead		22 yr
18. Iodine-131	9.6% per day	
19. Strontium-90		25 yr
20. Uranium-238		4,560 yr
21. Plutonium		23,105 yr

22. *Half-Life.* Of an initial amount of 1000 g of lead, how much will remain after 100 yr? See Exercise 17 for the value of k.

23. *Half-Life.* Of an initial amount of 1000 g of polonium, how much will remain after 20 min? See Exercise 16 for the value of k.

24. *Carbon Dating.* How old is an ivory tusk that has lost 40% of its carbon-14?

25. *Carbon Dating.* How old is a piece of wood that has lost 90% of its carbon-14?

26. *Carbon Dating.* How old is a skeleton that has lost 50% of its carbon-14?

27. *Carbon Dating.* How old is a Chinese artifact that has lost 60% of its carbon-14?

28. In a *chemical reaction,* substance A decomposes at a rate proportional to the amount of A present.

a) Write an equation relating A to the amount left of an initial amount A_0 after time t.

b) It is found that 10 lb of A will reduce to 5 lb in 3.3 hr. After how long will there be only 1 lb left?

29. In a *chemical reaction,* substance A decomposes at a rate proportional to the amount of A present.

a) Write an equation relating A to the amount left of an initial amount A_0 after time t.

b) It is found that 8 g of A will reduce to 4 g in 3 hr. After how long will there be only 1 g left?

30. *Weight Loss.* The initial weight of a starving animal is W_0. Its weight W after t days is given by

$$W = W_0 e^{-0.009t}.$$

a) What percentage of its weight does it lose each day?

b) What percentage of its initial weight remains after 30 days?

31. *Weight Loss.* The initial weight of a starving animal is W_0. Its weight W after t days is given by

$$W = W_0 e^{-0.008t}.$$

a) What percentage of its weight does it lose each day?

b) What percentage of its initial weight remains after 30 days?

The Beer–Lambert Law

A beam of light enters a medium such as water or smoky air with initial intensity I_0. Its intensity is decreased depending on the thickness (or concentration) of the medium. The intensity I at a depth (or concentration) of x units is given by

$$I = I_0 e^{-\mu x}.$$

The constant μ ("mu"), called the *coefficient of absorption,* varies with the medium.

32. *Light Through Smog.* Particulate concentrations of pollution reduce sunlight. In a smoggy area, $\mu = 0.01$ and x is the concentration of particulates measured in micrograms per cubic meter. What percentage of an initial amount I_0 of sunlight passes through smog that has a concentration of 100 micrograms per cubic meter?

33. **Light through sea water** has $\mu = 1.4$ when x is measured in meters.

   a) What percentage of I_0 remains at a depth of sea water that is 1 m? 2 m? 3 m?
   b) Plant life cannot exist below 10 m. What percentage of I_0 remains at 10 m?

34. *Cooling.* The temperature of a hot liquid is 100°. The liquid is placed in a refrigerator where the temperature is 40°, and it cools to 90° in 5 min.

   a) Find the value of the constant a in Newton's Law of Cooling.
   b) Find the value of the constant k. Round to the nearest hundredth.
   c) What is the temperature after 10 min?
   d) How long does it take the liquid to cool to 41°?
   tw e) Find the rate of change of the temperature and interpret its meaning.

35. *Cooling.* The temperature of a hot liquid is 100° and the room temperature is 75°. The liquid cools to 90° in 10 min.

   a) Find the value of the constant a in Newton's Law of Cooling.
   b) Find the value of the constant k. Round to the nearest hundredth.
   c) What is the temperature after 20 min?
   d) How long does it take the liquid to cool to 80°?
   tw e) Find the rate of change of the temperature and interpret its meaning.

36. *Cooling Body.* The coroner arrives at the scene of a murder at 2 A.M. He takes the temperature of the body and finds it to be 61.6°. He waits 1 hour,

takes the temperature again, and finds it to be 57.2°. The body is in a meat freezer, where the temperature is 10°. When was the murder committed?

37. *Cooling Body.* The coroner arrives at the scene of a murder at 11 P.M. She takes the temperature of the body and finds it to be 85.9°. She waits 1 hour, takes the temperature again, and finds it to be 83.4°. She notes that the room temperature is 60°. When was the murder committed?

38. *Population Decrease of Panama.* The population of Panama was 1,464,000 in 1970 and 1,260,000 in 1980. Assuming the population is decreasing according to the exponential-decay model:

   a) Find the value k, and write the equation.
   b) Estimate the population of Panama in 2002.
   c) After how long will the population of Panama be 100,000?

39. *Population Decrease of Cincinnati.* The population of Cincinnati was 453,000 in 1970 and 385,000 in 1980. Assuming the population is decreasing according to the exponential-decay model:

   a) Find the value k, and write the equation.
   b) Estimate the population of Cincinnati in 2008.
   c) After how long (theoretically) will Cincinnati have just 1 person?

40. *Atmospheric Pressure.* Atmospheric pressure P at altitude a is given by

$$P = P_0 e^{-0.00005a},$$

where P_0 is the pressure at sea level. Assume that $P_0 = 14.7$ lb/in^2 (pounds per square inch).

   a) Find the pressure at an altitude of 1000 ft.
   b) Find the pressure at an altitude of 20,000 ft.
   c) At what altitude is the pressure 1.47 lb/in^2?
   tw d) Find the rate of change of the pressure and interpret its meaning.

41. *Satellite Power.* The power supply of a satellite is a radioisotope. The power output P, in watts, decreases at a rate proportional to the amount present; P is given by

$$P = 50e^{-0.004t},$$

where t is the time, in days.

   a) How much power will be available after 375 days?
   b) What is the half-life of the power supply?
   c) The satellite's equipment cannot operate on fewer than 10 watts of power. How long can the satellite stay in operation?

d) How much power did the satellite have to begin with?

tw e) Find the rate of change of the power and interpret its meaning.

42. *Carbon Dating.* Recently, while digging in Chaco Canyon, New Mexico, archeologists found corn pollen that was 4000 yr old. This was evidence that Indians had begun cultivating crops in the Southwest centuries earlier than scientists had thought. What percent of the carbon-14 had been lost from the pollen?

 43. *Forgetting.* In an art class, students were tested at the end of the course on a final exam. Then they were retested with an equivalent test at subsequent time intervals. Their scores after time t, in months, are given in the following table.

Time, t (in months)	Score, y
1	84.9%
2	84.6%
3	84.4%
4	84.2%
5	84.1%
6	83.9%

a) Use the REGRESSION feature on a grapher to fit a logarithmic function $y = a + b \ln x$ to the data.
b) Use the function to predict test scores after 8 months; 10 months; 24 months; 36 months.
c) After how long will the test scores fall below 82%?
tw d) Find the rate of change of the scores and interpret its meaning.

44. *Decline in Beef Consumption.* The annual consumption of beef B per person was about 80 lb in 1985 and about 67 lb in 1996. Assuming consumption is decreasing according to the exponential-decay model:

a) Find the value k, and write the equation.
b) Estimate the consumption of beef in 2000.
c) In what year (theoretically) will the consumption of beef be 20 lb per person?

Synthesis

45. *Economics: Supply and Demand.* The demand and supply functions for a certain type of VCR are given by

$$x = D(p) = 480e^{-0.003p}$$

and

$$x = S(p) = 150e^{0.004p}.$$

Find the equilibrium point.

46. *Newton's Law of Cooling.* Consider the following exploratory situation. Draw a glass of hot tap water. Place a thermometer in the glass and check the temperature. Check the temperature every 30 min thereafter. Plot your data on this graph, and connect the points with a smooth curve.

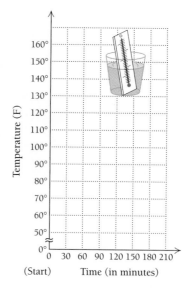

a) What was the temperature when you began?
b) At what temperature does there seem to be a leveling off of the graph?
c) What is the difference between your answers to parts (a) and (b)?
d) How does the temperature in part (b) compare with the room temperature?
e) Find an equation that "fits" the data. Use this equation to check values of other data points. How do they compare?
f) Is it ever "theoretically" possible for the temperature of the water to be the same as the room temperature? Explain.
tw g) Find the rate of change of the temperature and interpret its meaning.

tw 47. An interest rate decreases from 8% to 7.2%. Explain why this increases the present value of an amount due 10 yr later.

4.5

The Derivatives of a^x and $\log_a x$

The Derivative of a^x

To find the derivative of a^x, for any base a, we first express a^x as a power of e. In order to do this, we first prove the following additional property of a logarithm.

P7. $b^{\log_b x} = x$

To prove this, let

$$y = \log_b x.$$

Then, by the definition of a logarithm,

$$b^y = x.$$

Substituting $\log_b x$ for y, we have

$$b^{\log_b x} = x.$$

We can now express a^x as a power of e. Using Property 7, where $b = e$ and $x = a$, we have

$$a = e^{\ln a}. \qquad \text{Remember: } \ln a = \log_e a.$$

Raising both sides to the power x, we get

$$a^x = (e^{\ln a})^x$$
$$= e^{x \cdot \ln a}. \qquad \text{Multiplying exponents}$$

Thus we have the following.

Theorem 13

$$a^x = e^{x \cdot \ln a}$$

Example 1 Express as a power of e.

a) 3^2

$$3^2 = e^{2 \cdot \ln 3}$$
$$\approx e^{2(1.098612)}$$
$$\approx e^{2.1972}$$

b) 10^x

$$10^x = e^{x \cdot \ln 10}$$
$$\approx e^{x(2.302585)}$$
$$\approx e^{2.3026x}$$

Now we can differentiate.

Example 2

$$\frac{d}{dx}2^x = \frac{d}{dx}e^{x \cdot \ln 2} \qquad \text{Theorem 13}$$

$$= \left[\frac{d}{dx}(x \cdot \ln 2)\right] \cdot e^{x \cdot \ln 2}$$

$$= (\ln 2)(e^{\ln 2})^x$$

$$= (\ln 2)2^x$$

We completed this by taking the derivative of $x \ln 2$ and replacing $e^{x \ln 2}$ with 2^x. Note that $\ln 2 \approx 0.7$, so the above verifies our earlier approximation of the derivative of 2^x as $(0.7)2^x$ (see Section 4.1). ◄

In general,

$$\frac{d}{dx}a^x = \frac{d}{dx}e^{x \cdot \ln a} \qquad \text{Theorem 13}$$

$$= \left[\frac{d}{dx}(x \cdot \ln a)\right] \cdot e^{x \cdot \ln a}$$

$$= (\ln a)\,a^x.$$

Thus we have the following.

> **Theorem 14**
>
> $$\frac{d}{dx}a^x = (\ln a)\,a^x$$

Example 3

$$\frac{d}{dx}3^x = (\ln 3)\,3^x$$

◄

Example 4

$$\frac{d}{dx}(1.4)^x = (\ln 1.4)(1.4)^x$$

◄

Compare these formulas:

$$\frac{d}{dx}a^x = (\ln a)\,a^x \quad \text{and} \quad \frac{d}{dx}e^x = e^x.$$

It is the simplicity of the latter formula that is a reason for the use of base e in calculus. The many applications of e in natural phenomena provide other reasons.

One other result also follows from what we have done. If

$$f(x) = a^x,$$

we know that

$$f'(x) = a^x (\ln a).$$

In Section 4.1, we also showed that

$$f'(x) = a^x \left(\lim_{h \to 0} \frac{a^h - 1}{h} \right).$$

Thus,

$$a^x (\ln a) = a^x \left(\lim_{h \to 0} \frac{a^h - 1}{h} \right).$$

Since $a^x > 0$, we have the following.

> **Theorem 15**
>
> $$\ln a = \lim_{h \to 0} \frac{a^h - 1}{h}$$

The Derivative of $\log_a x$

Just as the derivative of a^x is expressed in terms of $\ln a$, so too is the derivative of $\log_a x$. To find this derivative, we first express $\log_a x$ in terms of $\ln a$ using Property 7:

$$a^{\log_a x} = x.$$

Then

$$\ln (a^{\log_a x}) = \ln x$$

$$(\log_a x) \cdot \ln a = \ln x \qquad \text{By P3, treating } \log_a x \text{ as an exponent}$$

and

$$\log_a x = \boxed{\frac{1}{\ln a}} \cdot \ln x.$$

$$\underline{\qquad} \text{ constant}$$

The derivative of $\log_a x$ follows.

> **Theorem 16**
>
> $$\frac{d}{dx} \log_a x = \frac{1}{\ln a} \cdot \frac{1}{x}$$

Comparing this with

$$\frac{d}{dx} \ln x = \frac{1}{x},$$

we again see a reason for the use of base e in calculus.

TECHNOLOGY CONNECTION

Check the results of Examples 3–7 using a grapher. Then differentiate $y = \log_2 x$ and check the result on the grapher.

Example 5

$$\frac{d}{dx} \log_3 x = \frac{1}{\ln 3} \cdot \frac{1}{x}$$

◁

Example 6

$$\frac{d}{dx} \log x = \frac{1}{\ln 10} \cdot \frac{1}{x} \qquad \log x = \log_{10} x$$

◁

Example 7

$$\frac{d}{dx} x^2 \log x = x^2 \frac{1}{\ln 10} \cdot \frac{1}{x} + 2x \log x \qquad \text{By the Product Rule}$$

$$= \frac{x}{\ln 10} + 2x \log x, \quad \text{or} \quad x\left(\frac{1}{\ln 10} + 2 \log x\right)$$

◁

EXERCISE SET 4.5

Express as a power of e.

1. 5^4
2. 2^3
3. $(3.4)^{10}$
4. $(5.3)^{20}$
5. 4^k
6. 5^R
7. 8^{kT}
8. 10^{kR}

Differentiate.

9. $y = 6^x$
10. $y = 7^x$
11. $f(x) = 10^x$
12. $f(x) = 100^x$
13. $f(x) = x(6.2)^x$
14. $f(x) = x(5.4)^x$
15. $y = x^3 10^x$
16. $y = x^4 5^x$
17. $y = \log_4 x$
18. $y = \log_5 x$
19. $f(x) = 2 \log x$
20. $f(x) = 5 \log x$
21. $f(x) = \log \frac{x}{3}$
22. $f(x) = \log \frac{x}{5}$
23. $y = x^3 \log_8 x$
24. $y = x \log_6 x$

Applications

BUSINESS AND ECONOMICS

25. *Recycling Aluminum Cans.* It is known that one fourth of all aluminum cans distributed will be recycled each year. A beverage company distributes 250,000 cans. The number still in use after time t, in years, is given by

$$N(t) = 250,000\left(\tfrac{1}{4}\right)^t.$$

a) Find $N'(t)$.

ⱶw b) Interpret the meaning of $N'(t)$.

26. *Double Declining-Balance Depreciation.* An office machine is purchased for \$5200. Under certain assumptions, its salvage value V depreciates according to a method called double declining balance, basically 80% each year, and is given by

$$V(t) = \$5200(0.80)^t,$$

where t is the time, in years.

  a) Find $V'(t)$.

  tw b) Interpret the meaning of $V'(t)$.

LIFE AND PHYSICAL SCIENCES

Earthquake Magnitude. The magnitude R (measured on the Richter scale) of an earthquake of intensity I is defined as

$$R = \log \frac{I}{I_0},$$

where I_0 is a minimum intensity used for comparison. When one earthquake is 10 times as intense as another, its magnitude on the Richter scale is 1 higher. If one earthquake is 100 times as intense as another, its magnitude on the Richter scale is 2 higher, and so on. Thus an earthquake whose magnitude is 7 on the Richter scale is 10 times as intense as an earthquake whose magnitude is 6. Earthquakes can be interpreted as multiples of the minimum intensity I_0.

27. In 1986, there was an earthquake near Cleveland, Ohio. It had an intensity of $10^5 \cdot I_0$. What was its magnitude on the Richter scale?

28. On October 17, 1989, there was an earthquake in San Francisco, California, during the World Series. It had an intensity of $10^{6.9} \cdot I_0$. What was its magnitude on the Richter scale?

This photograph shows part of the damage in the San Francisco, California, area earthquake in 1989.

29. *Earthquake Intensity.*   The intensity I of an earthquake is given by

$$I = I_0 10^R,$$

where R is the magnitude on the Richter scale and I_0 is the minimum intensity, where $R = 0$, used for comparison.

  a) Find I, in terms of I_0, for an earthquake of magnitude 7 on the Richter scale.

  b) Find I, in terms of I_0, for an earthquake of magnitude 8 on the Richter scale.

  c) Compare your answers to parts (a) and (b).

  d) Find the rate of change dI/dR.

  tw e) Interpret the meaning of dI/dR.

30. *Intensity of Sound.* The intensity of a sound is given by

$$I = I_0 10^{0.1L},$$

where L is the loudness of the sound as measured in decibels and I_0 is the minimum intensity detectable by the human ear.

  a) Find I, in terms of I_0, for the loudness of a power mower, which is 100 decibels.

  b) Find I, in terms of I_0, for the loudness of just audible sound, which is 10 decibels.

  c) Compare your answers to parts (a) and (b).

  d) Find the rate of change dI/dL.

  tw e) Interpret the meaning of dI/dL.

31. *Earthquake Magnitude.* The magnitude R (measured on the Richter scale) of an earthquake of intensity I is defined as

$$R = \log \frac{I}{I_0},$$

where I_0 is the minimum intensity (used for comparison). (The exponential form of this definition is given in Exercise 29.)

  a) Find the rate of change dR/dI.

  tw b) Interpret the meaning of dR/dI.

32. *Loudness of Sound.* The loudness L of a sound of intensity I is defined as

$$L = 10 \log \frac{I}{I_0},$$

where I_0 is the minimum intensity detectable by the human ear and L is the loudness measured in decibels. (The exponential form of this definition is given in Exercise 30.)

  a) Find the rate of change dL/dI.

  tw b) Interpret the meaning of dL/dI.

33. *Response to Drug Dosage.* The response y to a dosage x of a drug is given by

$$y = m \log x + b.$$

The response may be hard to measure with a number. The patient might perspire more, have an increase in temperature, or faint.

    a) Find the rate of change dy/dx.

tw **b)** Interpret the meaning of dy/dx.

Synthesis

34. Find $\lim\limits_{h \to 0} \dfrac{3^h - 1}{h}$.

Use the Chain Rule and other formulas given in this section to differentiate each of the following. In some cases, it may help to take the logarithm on both sides of the equation before differentiating.

35. $f(x) = 3^{2x}$

36. $y = 2^{x^4}$

37. $y = x^x, x > 0$

38. $y = \log_3 (x^2 + 1)$

39. $f(x) = x^{e^x}, x > 0$

40. $y = a^{f(x)}$

41. $y = \log_a f(x), f(x)$ positive

42. $y = [f(x)]^{g(x)}, f(x)$ positive

tw **43.** Explain in your own words how to justify the formula for finding the derivative of $f(x) = a^x$.

tw **44.** Explain in your own words how to justify the formula for finding the derivative of $f(x) = \log_a x$.

4.6

OBJECTIVES

➤ Find the elasticity of a demand function.
➤ Find the maximum of a total-revenue function.
➤ Characterize demand in terms of elasticity.

An Economics Application: Elasticity of Demand

Suppose that x represents a quantity of goods sold and p is the price per unit of the goods. Recall that x and p are related by the demand function

$$x = D(p).$$

Suppose that there is a change Δp in the price per unit of a product. The percent change in price is

$$\frac{\Delta p}{p} = \frac{\Delta p}{p} \cdot \frac{100}{100} = \frac{\Delta p \cdot 100}{p}\%.$$

A change in the price of a product produces a change Δx in the quantity sold. The percent change in quantity is

$$\frac{\Delta x}{x} = \frac{\Delta x \cdot 100}{x}\%.$$

The ratio of the percent change in quantity to the percent change in price is

$$\frac{(\Delta x/x)}{(\Delta p/p)},$$

which can be expressed as

$$\frac{p}{x} \cdot \frac{\Delta x}{\Delta p}. \tag{1}$$

For differentiable functions,

$$\lim_{\Delta p \to 0} \frac{\Delta x}{\Delta p} = \frac{dx}{dp},$$

and the limit as Δp approaches 0 of the expression in equation (1) becomes

$$\frac{p}{x} \cdot \frac{dx}{dp} = \frac{p}{D(p)} \cdot D'(p) = \frac{p\,D'(p)}{D(p)}.$$

Definition

The **elasticity of demand** E is given as a function of price p by

$$E(p) = -\frac{p\,D'(p)}{D(p)}.$$

The numbers x, or $D(p)$, and p are always nonnegative. The slope of the demand curve $dx/dp = D'(p)$ is always negative, since the demand curve is decreasing, and $D(p)$ is always nonnegative and must be positive in order for the elasticity to exist. Economists have used the $-$ sign in the definition of elasticity to make $E(p)$ nonnegative and easier to work with.

Example 1   *Economics: Demand for Videotape Rentals.* A videotape store works out a demand function for its videotape rentals and finds it to be

$$x = D(p) = 120 - 20p,$$

where x is the number of videotapes rented per day when p is the price per rental. Find each of the following.

a) The quantity demanded when the price is \$2 per rental
b) The elasticity as a function of p
c) The elasticity at $p = \$2$ and at $p = \$5$. Interpret the meaning of these values of the elasticity.
d) The value of p for which $E(p) = 1$. Interpret the meaning of this price.
e) The total-revenue function $R(p) = p\,D(p)$
f) The price p at which total revenue is a maximum

Solution

a) At $p = \$2$, $x = D(2) = 120 - 20(2) = 80$. Thus, 80 videotapes will be rented per day when the price per rental is \$2.
b) To find the elasticity, we first find the derivative $D'(p)$:

$$D'(p) = -20.$$

Then we substitute -20 for $D'(p)$ and $120 - 20p$ for $D(p)$ in the expression for elasticity:

$$E(p) = -\frac{p\,D'(p)}{D(p)} = -\frac{p \cdot (-20)}{120 - 20p} = \frac{20p}{120 - 20p} = \frac{p}{6 - p}.$$

c) $E(\$2) = \dfrac{2}{6 - 2} = \dfrac{1}{2}$

**Economics: Demand
for Hand-Held
Radios**

A company determines
that the demand
function for hand-held
radios is

$$x = D(p) = 300 - p,$$

where x is the number
sold per day when the
price is p dollars per
radio.

Exercises

1. Find the elasticity
    E and the total
    revenue R.
2. Using only the
    first quadrant,
    graph the
    demand,
    elasticity, and
    total-revenue
    functions on the
    same set of axes.
3. Find the price p
    for which the
    total revenue is a
    maximum. Use
    the method of
    Example 1.
    Check the answer
    on a grapher.

At $p = \$2$, the elasticity is $\frac{1}{2}$, which is less than 1. Thus the ratio of the percent change in quantity to the percent change in price is less than 1. A small increase in price will cause a percentage decrease in the quantity that is smaller than the percentage change in price.

$$E(\$5) = \frac{5}{6 - 5} = 5$$

At $p = \$5$, the elasticity is 5, which is greater than 1. Thus the ratio of the percent change in quantity to the percent change in price is greater than 1. A small increase in price will cause a percentage decrease in the quantity that is greater than the percentage change in price.

d) We set $E(p) = 1$ and solve for p:

$$\frac{p}{6 - p} = 1$$

$$p = 6 - p \qquad \text{We multiply by } 6 - p, \text{ assuming } p \neq 6.$$

$$2p = 6$$

$$p = 3.$$

Thus when the price is $3 per rental, the ratio of the percent change in quantity to the percent change in price is 1.

e) Recall that the total revenue $R(p)$ is given by $pD(p)$. Then

$$R(p) = pD(p) = p(120 - 20p) = 120p - 20p^2.$$

f) To find the price p that maximizes total revenue, we find $R'(p)$:

$$R'(p) = 120 - 40p.$$

Now $R'(p)$ exists for all values of p over the interval $[0, \infty)$. Thus we solve:

$$R'(p) = 120 - 40p = 0$$

$$-40p = -120$$

$$p = 3.$$

Since there is only one critical point, we can try to use the second derivative to see if we have a maximum:

$$R''(p) = -40, \quad \text{a constant.}$$

Thus, $R''(3)$ is negative, so $R(3)$ is a maximum. That is, total revenue is a maximum at $p = \$3$ per rental.   ◄

Note in Example 1 that the value of p for which $E(p) = 1$ is the same as the value of p for which the total revenue is a maximum. This is always true.

Theorem 17

Total revenue is a maximum at the value(s) of p for which $E(p) = 1$.

We can prove this as follows. We know that

$$R(p) = pD(p),$$

so $$R'(p) = p \cdot D'(p) + 1 \cdot D(p)$$

$$= D(p)\left[\frac{pD'(p)}{D(p)} + 1\right] \qquad \text{Check this by multiplying.}$$

$$= D(p)[-E(p) + 1]$$

$$= D(p)[1 - E(p)] \tag{2}$$

$$R'(p) = 0 \quad \text{when} \quad 1 - E(p) = 0, \quad \text{or} \quad E(p) = 1.$$

We have shown that total revenue is a maximum when $E(p) = 1$. What do we know when $E(p) < 1$ and $E(p) > 1$? We answer this by looking at equation (2) in the preceding proof.

For prices p for which $D(p) > 0$ and for which $E(p) < 1$, it follows that $1 - E(p) > 0$, so by equation (2), $R'(p) > 0$. This tells us that the total-revenue function is increasing.

For values of p for which $D(p) > 0$ and for which $E(p) > 1$, it follows that $1 - E(p) < 0$, so by equation (2), $R'(p) < 0$. This tells us that the total-revenue function is decreasing.

The following statement and the graphs shown below summarize our results.

For a particular value of the price p:

1. The demand is *inelastic* if $E(p) < 1$. An increase in price will bring an increase in revenue. If demand is inelastic, then revenue is increasing.

2. The demand has *unit elasticity* if $E(p) = 1$. The demand has unit elasticity when revenue is at a maximum.

3. The demand is *elastic* if $E(p) > 1$. An increase in price will bring a decrease in revenue. If demand is elastic, then revenue is decreasing.

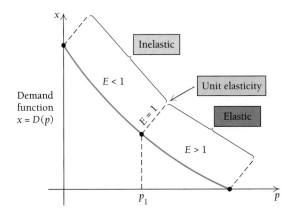

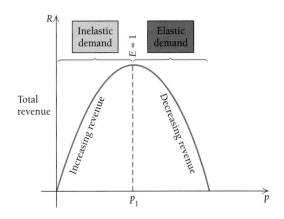

In summary, suppose that the videotape store in Example 1 raises the price per rental and that the total revenue increases. Then we say the demand is *inelastic*. If the total revenue decreases, we say the demand is *elastic*. Some price elasticities in the U.S. economy are listed in the following table.

PRICE ELASTICITIES IN THE U.S. ECONOMY

Industry	Elasticity
Elastic Demands	
Metals	1.52
Electrical engineering products	1.39
Mechanical engineering products	1.30
Furniture	1.26
Motor vehicles	1.14
Instrument engineering products	1.10
Professional services	1.09
Transportation services	1.03
Inelastic Demands	
Gas, electricity, and water	0.92
Oil	0.91
Chemicals	0.89
Beverages (all types)	0.78
Tobacco	0.61
Food	0.58
Banking and insurance services	0.56
Housing services	0.55
Clothing	0.49
Agricultural and fish products	0.42
Books, magazines, and newspapers	0.34
Coal	0.32

Source: Ahsan Mansur and John Whalley, "Numerical specification of applied general equilibrium models: Estimation, calibration, and data." In Scarf, H. E. and J. B. Shoven (eds.), *Applied General Equilibrium Analysis.* (New York: Cambridge University Press, 1984), p. 109.

EXERCISE SET 4.6

For the demand function in each of Exercises 1–12, find the following.

a) The elasticity
b) The elasticity at the given price, stating whether the demand is elastic or inelastic

c) The value(s) of p for which total revenue is a maximum

1. $x = D(p) = 400 - p$; $p = \$125$
2. $x = D(p) = 500 - p$; $p = \$38$

3. $x = D(p) = 200 - 4p; \quad p = \$46$

4. $x = D(p) = 500 - 2p; \quad p = \$57$

5. $x = D(p) = \dfrac{400}{p}; \quad p = \$50$

6. $x = D(p) = \dfrac{3000}{p}; \quad p = \$60$

7. $x = D(p) = \sqrt{500 - p}; \quad p = \$400$

8. $x = D(p) = \sqrt{300 - p}; \quad p = \$250$

9. $x = D(p) = 100e^{-0.25p}; \quad p = \$10$

10. $x = D(p) = 200e^{-0.05p}; \quad p = \$80$

11. $x = D(p) = \dfrac{100}{(p + 3)^2}; \quad p = \$1$

12. $x = D(p) = \dfrac{300}{(p + 8)^2}; \quad p = \$4.50$

Applications ....
BUSINESS AND ECONOMICS

13. *Demand for Chocolate Chip Cookies.* A bakery works out a demand function for its sale of chocolate chip cookies and finds it to be

    $$x = D(p) = 967 - 25p,$$

    where x is the quantity of cookies sold when the price per cookie, in cents, is p.
    a) Find the elasticity.
    b) At what price is the elasticity of demand equal to 1?
    c) At what prices is the elasticity of demand elastic?
    d) At what prices is the elasticity of demand inelastic?
    e) At what price is the revenue a maximum?
    f) At a price of 20¢ per cookie, will a small increase in price cause the total revenue to increase or decrease?

14. *Demand for Oil.* Suppose that you have been hired by OPEC as an economic consultant for the world demand for oil. The demand function is

    $$x = D(p) = 63,000 + 50p - 25p^2, \quad 0 \le p \le 50,$$

    where x is measured in millions of barrels of oil per day at a price of p dollars per barrel.
    a) Find the elasticity.
    b) Find the elasticity at a price of $10 per barrel, stating whether the demand is elastic or inelastic.
    c) Find the elasticity at a price of $20 per barrel, stating whether the demand is elastic or inelastic.

d) Find the elasticity at a price of $30 per barrel, stating whether the demand is elastic or inelastic.
e) At what price is the revenue a maximum?
f) What quantity of oil will be sold at the price that maximizes revenue? Check current world prices to see how this compares with your answer.
g) At a price of $30 per barrel, will a small increase in price cause the total revenue to increase or decrease?

15. *Demand for Videogames.* A computer software store determines the following demand function for a new videogame:

    $$x = D(p) = \sqrt{200 - p^3},$$

    where x is the number of videogames sold per day when the price is p dollars per game.

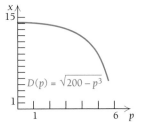

a) Find the elasticity.
b) Find the elasticity when $p = \$3$ per game.
c) At $p = \$3$, will a small increase in price cause the total revenue to increase or decrease?

16. *Demand for Tomato Plants.* A garden store determines the following demand function during early summer for a certain size of tomato plant:

    $$x = D(p) = \dfrac{2p + 300}{10p + 11},$$

    where x is the number of plants sold per day when the price is p dollars per plant.

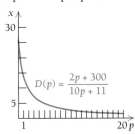

a) Find the elasticity.
b) Find the elasticity when $p = \$3$ per plant.
c) At $p = \$3$, will a small increase in price cause the total revenue to increase or decrease?

Synthesis

17. *Economics: Constant Elasticity Curve.*
    a) Find the elasticity of the demand function

    $$x = D(p) = \frac{k}{p^n},$$

    where k is a positive constant and n is an integer greater than 0.
    b) Is the value of the elasticity dependent on the price per unit?
    c) Does the total revenue have a maximum? When?

18. *Economics: Exponential Demand Curve.*
    a) Find the elasticity of the demand function

    $$x = D(p) = Ae^{-kp},$$

    where A and k are positive constants.

    b) Is the value of the elasticity dependent on the price per unit?
    c) Does the total revenue have a maximum? At what value of p?

19. Let

    $$L(p) = \ln D(p).$$

    Describe the elasticity in terms of $L'(p)$.

tw 20. Explain in your own words the concept of elasticity and its usefulness to economists. Do some library research or consult an economist in order to determine when and how this concept was first developed.

tw 21. Explain how the elasticity of demand for a product can be affected by the availability of substitutes for the product.

CHAPTER 4 SUMMARY AND REVIEW

Terms to Know

Exponential function, p. 287
Base, p. 287
The number e, pp. 291, 292
Logarithm, p. 300
Logarithmic base, p. 300
Common logarithm, p. 304
Natural logarithm, p. 304
Exponential equation, p. 306

Exponential growth, p. 319
Exponential growth rate, p. 319
Interest compounded continuously, p. 320
Doubling time, p. 321
Models of limited growth, p. 325
Exponential decay, p. 333
Half-life, p. 334

Decay rate, p. 334
Carbon dating, p. 335
Present value, p. 337
Newton's Law of Cooling, p. 337
Elasticity of demand, p. 351
Elastic demand, p. 353
Inelastic demand, p. 353
Unit elasticity, p. 353

Review Exercises

These review exercises are for test preparation. They can also be used as a lengthened practice test. Answers are at the back of the book. The answers also contain bracketed section references, which tell you where to restudy if your answer is incorrect.

Differentiate.

1. $y = \ln x$

2. $y = e^x$

3. $y = \ln(x^4 + 5)$

4. $y = e^{2\sqrt{x}}$

5. $f(x) = \ln x^6$

6. $f(x) = e^{4x} + x^4$

7. $f(x) = \dfrac{\ln x}{x^3}$

8. $f(x) = e^{x^2} \cdot \ln 4x$

9. $f(x) = e^{4x} - \ln \dfrac{x}{4}$

10. $g(x) = x^8 - 8\ln x$

11. $y = \dfrac{\ln e^x}{e^x}$

Given $\log_a 2 = 1.8301$ and $\log_a 7 = 5.0999$, find each of the following.

12. $\log_a 14$

13. $\log_a \frac{2}{7}$

14. $\log_a 28$

15. $\log_a 3.5$

16. $\log_a \sqrt{7}$

17. $\log_a \frac{1}{4}$

18. Find the function that satisfies $dQ/dt = kQ$. List the answer in terms of Q_0.

19. *Life Science: Population Growth.* The population of Boomtown doubled in 16 yr. What was the growth rate of the city? Round to the nearest tenth of a percent.

20. *Business: Interest Compounded Continuously.* Suppose that $8300 is deposited in a savings and loan association in which the interest rate is 6.8%, compounded continuously. How long will it take for the $8300 to double itself? Round to the nearest tenth of a year.

21. *Business: Cost of a Prime-Rib Dinner.* The average cost C of a prime-rib dinner was $4.65 in 1962. In 1986, it was $15.81. Assuming that the exponential-growth model applies:

    a) Find the exponential-growth rate, and write the equation.
    b) What will the cost of such a dinner be in 2000? in 2007?

22. *Business: Franchise Growth.* A clothing firm is selling franchises throughout the United States and Canada. It is estimated that the number of franchises N will increase at the rate of 12% per year, that is,

    $$\frac{dN}{dt} = 0.12N,$$

    where t is the time, in years.

    a) Find the function that satisfies the equation, assuming that the number of franchises in 1998 ($t = 0$) is 60.
    b) How many franchises will there be in 2004?
    c) After how long will the number of franchises be 120? Round to the nearest tenth of a year.

23. *Life Science: Decay Rate.* The decay rate of a certain radioactive isotope is 13% per year. What is its half-life? Round to the nearest tenth of a year.

24. *Life Science: Half-Life.* The half-life of radon-222 is 3.8 days. What is its decay rate? Round to the nearest tenth of a percent.

25. *Life Science: Decay Rate.* A certain radioactive element has a decay rate of 7% per day, that is,

    $$\frac{dA}{dt} = -0.07A,$$

where A is the amount of the element present at time t, in days.

   a) Find a function that satisfies the equation if the amount of the element present at $t = 0$ is 800 g.
   b) After 20 days, how much of the 800 g will remain? Round to the nearest gram.
   c) After how long does half the original amount remain?

26. *Social Science: The Hullian Learning Model.* The probability p of mastering a certain assembly line task after t learning trials is given by

    $$p(t) = 1 - e^{-0.7t}.$$

    a) What is the probability of learning the task after 1 trial? 2 trials? 5 trials? 10 trials? 14 trials?
    b) Find the rate of change $p'(t)$.
    tw c) Interpret the meaning of $p'(t)$.
    d) Sketch a graph of the function.

27. *Business: Present Value.* Find the present value of $1,000,000 due 40 yr later at 8.4%, compounded continuously.

Differentiate.

28. $y = 3^x$

29. $f(x) = \log_{15} x$

30. *Economics: Elasticity of Demand.* Consider the demand function

    $$x = D(p) = \frac{600}{(p + 4)^2}.$$

    a) Find the elasticity.
    b) Find the elasticity at $p = $1, stating whether the demand is elastic or inelastic.
    c) Find the elasticity at $p = $12, stating whether the demand is elastic or inelastic.
    d) At a price of $12, will a small increase in price cause the total revenue to increase or decrease?
    e) Find the value of p for which the total revenue is a maximum.

Synthesis ....

31. Differentiate: $y = \dfrac{e^{2x} + e^{-2x}}{e^{2x} - e^{-2x}}$.

32. Find the minimum value of $f(x) = x^4 \ln 4x$.

TECHNOLOGY CONNECTION

33. Graph: $f(x) = \dfrac{e^{1/x}}{(1 + e^{1/x})^2}$.

34. Find $\displaystyle\lim_{x \to 0} \dfrac{e^{1/x}}{(1 + e^{1/x})^2}$.

35. *Business: Shopping on the Internet.* Online (internet) sales of all types of consumer products has been increasing at exponential rates. Data in the table at right show online sales, in billions of dollars.

a) Use the **REGRESSION** feature on a grapher to fit an exponential function $y = a \cdot b^x$ to the data. Then convert that formula to an exponential function, base e, where t is the number of years after 1996, and determine the exponential growth rate.

Years, t, After 1996	Online Sales (in billions)
0	$ 0.71
1	2.6
2	5.8
3	9.9
4	15.6

Source: Jupiter Communications

b) Estimate online sales in 2010; in 2020.
c) After what amount of time will online sales be $40 billion?
d) What is the doubling time of online sales?

CHAPTER 4 TEST

Differentiate.

1. $y = e^x$

2. $y = \ln x$

3. $f(x) = e^{-x^2}$

4. $f(x) = \ln \dfrac{x}{7}$

5. $f(x) = e^x - 5x^3$

6. $f(x) = 3e^x \ln x$

7. $y = \ln(e^x - x^3)$

8. $y = \dfrac{\ln x}{e^x}$

Given $\log_b 2 = 0.2560$ and $\log_b 9 = 0.8114$, find each of the following.

9. $\log_b 18$

10. $\log_b 4.5$

11. $\log_b 3$

12. Find the function that satisfies $dM/dt = kM$. List the answer in terms of M_0.

13. The doubling time of a certain bacteria culture is 4 hr. What is the growth rate? Round to the nearest tenth of a percent.

14. *Business: Interest Compounded Continuously.* An investment is made at 6.931% per year, compounded continuously. What is the doubling time? Round to the nearest year.

15. *Business: Cost of Milk.* The cost C of a gallon of milk was $0.54 in 1941. In 1985, it was $2.31.

Assuming the exponential-growth model applies:

a) Find the exponential-growth rate, and write the equation.
b) Find the cost of a gallon of milk in 2000; in 2010.

16. *Life Science: Drug Dosage.* A dose of a drug is injected into the body of a patient. The drug amount in the body decreases at the rate of 10% per hour, that is,

$$\frac{dA}{dt} = -0.1A,$$

where A is the amount in the body and t is the time, in hours.

a) A dose of 3 cubic centimeters (cc) is administered. Assuming $A_0 = 3$ and $k = 0.1$, find the function that satisfies the equation.
b) How much of the initial dose of 3 cc will remain after 10 hr?
c) After how long does half the original dose remain?

17. *Life Science: Decay Rate.* The decay rate of zirconium is 1.1% per day. What is its half-life?

18. *Life Science: Half-Life.* The half-life of tellurium is 1,000,000 yr. What is its decay rate? As a percent, round to six decimal places.

19. *Business: Effect of Advertising.* A company introduces a new product on a trial run in a city. They advertised the product on television and found that the percentage P of people who bought the product after t ads had been run satisfied the function

$$P(t) = \frac{100\%}{1 + 24e^{-0.28t}}.$$

   a) What percentage buys the product without having seen the ad ($t = 0$)?
   b) What percentage buys the product after the ad has been run 1 time? 5 times? 10 times? 15 times? 20 times? 30 times? 35 times?
   c) Find the rate of change $P'(t)$.
   d) Interpret the meaning of $P'(t)$.
   e) Sketch a graph of the function.

20. *Business: Present Value.* Find the present value of $80,000 due 12 yr later at 8.6%, compounded continuously.

Differentiate.

21. $f(x) = 20^x$

22. $y = \log_{20} x$

23. *Economics: Elasticity of Demand.* Consider the demand function

$$x = D(p) = 400e^{-0.2p}.$$

   a) Find the elasticity.
   b) Find the elasticity at $p = \$3$, stating whether the demand is elastic or inelastic.
   c) Find the elasticity at $p = \$18$, stating whether the demand is elastic or inelastic.
   d) At a price of $3, will a small increase in price cause the total revenue to increase or decrease?
   e) Find the value of p for which the total revenue is a maximum.

Synthesis

24. Differentiate: $y = x\,(\ln x)^2 - 2x \ln x + 2x$.

25. Find the maximum and minimum values of $f(x) = x^4 e^{-x}$ over $[0, 10]$.

TECHNOLOGY CONNECTION

26. Graph: $f(x) = \dfrac{e^x - e^{-x}}{e^x + e^{-x}}$.

27. Find $\displaystyle\lim_{x \to 0} \dfrac{e^x - e^{-x}}{e^x + e^{-x}}$.

28. *Business: Average Price of a Television Commercial.* The cost of a 30-sec television commercial that runs during the Super Bowl has been increasing exponentially over the past several years. Data in the table below show costs for years after 1990.

Years, t, After 1990	Cost
1	$ 800,000
3	850,000
4	900,000
5	1,000,000
8	1,300,000

Source: National Football League

   a) Use the **REGRESSION** feature on a grapher to fit an exponential function $y = a \cdot b^x$ to the data. Then convert that formula to an exponential function, base e, where t is the number of years after 1990.
   b) Estimate the cost of a commercial run during the Super Bowl in 2000; in 2020.
   c) After what amount of time will the cost be $1 billion?
   d) What is the doubling time of the cost of a commercial run during the Super Bowl?

The Business of Motion-Picture Box-Office Revenue

In Example 8 of Section 4.3, we used a logistic equation to predict the total U.S. box-office revenue of the movie *Titanic*. The data used for that function are listed in the table below.

Movie Revenue: *Titanic*
Week 1 = December 19, 1997

Week in Release, t	Gross Revenue, G (in millions, current week)	Total Revenue, R (in millions)
1	$28.6	$ 28.6
2	59.8	88.4
3	97.7	157.5
4	40.5	198.0
5	44.3	242.3
6	32.3	274.6
8	33.5	308.1
11	29.3	337.4
12	38.9	376.3
13	26.3	402.6
15	24.4	427.0
16	22.2	449.2
17	22.3	471.5
18	23.0	494.5
19	20.8	515.3
20	15.1	530.4
21	12.5	542.9
22	11.2	554.1
23	6.5	560.6
24	5.1	565.7
25	4.1	569.8
26	2.9	572.7
27	4.4	577.1
28	2.3	579.4
29	2.5	581.9
30	2.0	583.9

Source: Exhibitor Relations Co., Inc.

Exercises

1. Make a scatterplot of the data points (t, R) on a grapher. Does it seem to fit a logistic equation? Explain.

2. Use the **REGRESSION** feature on a grapher to fit a logistic function

$$R(t) = \frac{c}{1 + ae^{-bt}}$$

to the data. Compare your answer to that in Example 8 of Section 4.3. (You may need to reduce the number of data points for your grapher.)

3. Find the rate of change $R'(t)$ and explain its meaning. Find $\lim_{t \to \infty} R'(t)$ and explain its meaning.

4. Next, make a scatterplot of the data points (t, G). Consider any function you have available for regression on your grapher. Choose the one that you think best fits the data and make a case for why you have chosen it. Find its domain.

5. Typically, the distributors of a movie wait until its weekly U.S. box-office revenue has become 0 before it begins retail videotape sales. Use the function found in Exercise 4 to predict when the distributor should begin videotape sales. Why, if at all, might you wait longer than the predicted time to begin such sales?

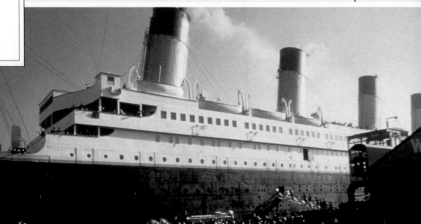

Now let's consider another movie.

Movie Revenue: *Jurassic Park: The Lost World*
Week 1 = May 21, 1997

Week in Release, t	Gross Revenue, G (in millions, current week)	Total Revenue, R (in millions)
1	$107.4	$107.4
2	34.1	141.5
3	29.6	171.1
4	19.6	190.7
5	14.3	205.0
6	8.2	213.2
7	5.1	218.3
8	2.8	221.1

Source: Exhibitor Relations Co., Inc.

Exercises

6. Make a scatterplot of the data points (t, R) on a grapher. Does it seem to fit a logistic equation? Explain.

7. Use the **REGRESSION** feature on a grapher to fit a logistic function

$$R(t) = \frac{c}{1 + ae^{-bt}}$$

to the data.

8. Find the rate of change $R'(t)$ and explain its meaning. Find $\lim_{t \to \infty} R'(t)$ and explain its meaning.

9. Next, make a scatterplot of the data points (t, G). Consider any function you have available for regression on your grapher. Choose the one that you think best fits the data and make a case for why you have chosen it. Find its domain.

10. Use the function found in Exercise 9 to predict when the distributor should begin videotape sales. (See Exercise 5.)

Let's consider a third movie.

Movie Revenue: *Air Force One*
Week 1 = August 4, 1997

Week in Release, t	Gross Revenue, G (in millions, current week)	Total Revenue, R (in millions)
1	$37.1	$ 37.1
2	43.6	80.7
3	29.9	110.6
4	19.8	130.4
5	12.7	143.1
6	11.1	154.2
7	5.1	159.3
8	3.9	163.2
9	2.9	166.1

Source: Exhibitor Relations Co., Inc.

Exercises

11. Make a scatterplot of the data points (t, R) on a grapher. Does a logistic equation appear to fit the data? Explain.

12. Use the **REGRESSION** feature on a grapher to fit a logistic function

$$R(t) = \frac{c}{1 + ae^{-bt}}$$

to the data.

13. Find the rate of change $R'(t)$ and explain its meaning. Find $\lim_{t \to \infty} R'(t)$ and explain its meaning.

14. Next, make a scatterplot of the data points (t, G). Consider any function you have available for regression on your grapher. Choose the one that you think best fits the data and make a case for why you have chosen it. Find its domain.

15. Use the function found in Exercise 14 to predict when the distributor should begin videotape sales. (See Exercise 5.)

Let's consider a fourth movie.

Movie Revenue: *Scream 2*
Week 1 = December 12, 1997 (ran for only 5 weeks)

Week in Release, t	Gross Revenue, G (in millions, current week)	Total Revenue, R (in millions)
1	$39.2	$39.2
2	15.9	55.1
3	22	71.1
4	14.4	85.5
5	5.3	90.8

Source: Exhibitor Relations Co., Inc.

Exercises

16. Make a scatterplot of the data points (t, R) on a grapher. Does it seem to fit a logistic equation? Explain.

17. Use the **REGRESSION** feature on a grapher to fit a logistic function

$$R(t) = \frac{c}{1 + ae^{-bt}}$$

to the data.

18. Find the rate of change $R'(t)$ and explain its meaning. Find $\lim_{t \to \infty} R'(t)$ and explain its meaning.

19. Next, make a scatterplot of the data points (t, G). Consider any function you have available for regression on your grapher. Choose the one that you think best fits the data and

make a case for why you have chosen it. Find its domain.

20. Use the function found in Exercise 19 to predict when the distributor should begin videotape sales. (See Exercise 5.)

Now let's consider some questions regarding all the preceding data.

Exercises

21. Does a logistic function seem to fit all graphs of the data points (t, R)? Why do you think this happens?

22. What type of function is used to fit the data points (t, G)? Explain why.

5

Integration

INTRODUCTION

Suppose that we do the reverse of differentiating, that is, suppose we try to find a function whose derivative is a given function. This process is called *antidifferentiation*, or *integration*; it is the main topic of this chapter and the second main branch of calculus, the first being differentiation. We will see that we can use integration to find the area under a curve over a closed interval, to find the accumulation of a certain quantity over an interval, and to calculate the average value of a function.

We first consider the meaning of integration and then learn several techniques for integrating.

AN APPLICATION

The rate of memorizing information initially increases with respect to time. Eventually, however, a maximum rate is reached, after which it begins to decrease. In a particular memory experiment, the rate of memorizing is given by

$$M'(t) = -0.009t^2 + 0.2t,$$

where $M'(t)$ is the memory rate, in words per minute. How many words are memorized in the first 10 min (from $t = 0$ to $t = 10$)?

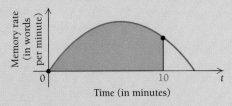

The total number of words memorized is given by

$$\int_0^{10} (-0.009t^2 + 0.2t)\, dt.$$

This is an integral.

This expression also gives the area under the graph of $M'(t)$ over the interval $[0, 10]$.

This problem appears as Exercise 36 in Exercise Set 5.3.

5.1

O B J E C T I V E S

➤ Find indefinite integrals and antiderivatives.
➤ Solve applied problems involving antiderivatives.

Integration

In Chapters 2, 3, and 4, we have considered several interpretations of the derivative, some of which are listed below.

Function	Derivative
Distance	Velocity
Revenue	Marginal revenue
Cost	Marginal cost
Population	Rate of growth of population

For population, we actually considered the derivative first and then the function. Many applications of mathematics involve the reverse of differentiation, called *antidifferentiation*. For example, if we know the marginal-revenue function in a particular situation, we can use antidifferentiation to find the total-revenue function.

The Antiderivative

Suppose that y is a function of x and that the derivative is the constant 8. Can we find y? It is easy to see that one such function is $8x$. That is, $8x$ is a function whose derivative is 8. Is there another function whose derivative is 8? In fact, there are many. Here are some examples:

$$8x + 3, \quad 8x - 10, \quad 8x + 7.4, \quad 8x + \sqrt{2}.$$

All these functions are $8x$ plus some constant. There are no other functions having a derivative of 8 other than those of the form $8x + C$. Another way of saying this is that any two functions having a derivative of 8 must differ by a constant. In general, any two functions having the same derivative differ by a constant.

> **Theorem 1**
>
> If two functions F and G have the same derivative over an interval, then
>
> $$F(x) = G(x) + C, \quad \text{where } C \text{ is a constant.}$$

The reverse of differentiating is *antidifferentiating,* and the result of antidifferentiating is called an **antiderivative.** Above, we found antiderivatives of the function 8. There are several of them, but they are all $8x$ plus some constant.

Example 1 Antidifferentiate (find the antiderivatives of) x^2.

Solution   One antiderivative is $x^3/3$. All other antiderivatives differ from this by a constant, so we can denote them as follows:

$$\frac{x^3}{3} + C.$$

To check this, differentiate $x^3/3 + C$:

$$\frac{d}{dx}\left(\frac{x^3}{3} + C\right) = 3\left(\frac{x^2}{3}\right) = x^2.$$

So $x^3/3 + C$ is an antiderivative of x^2.   ◄

The solution to Example 1 is the *general form* of the antiderivative of x^2.

Integrals and Integration

The process of antidifferentiating is often called **integration.** To indicate that the antiderivative of x^2 is $x^3/3 + C$, we write

$$\int x^2 \, dx = \frac{x^3}{3} + C, \tag{1}$$

and we note that

$$\int f(x) \, dx$$

is the symbolism we will use from now on to call for the antiderivative of a function $f(x)$. More generally, we can write

$$\int f(x) \, dx = F(x) + C, \tag{2}$$

where $F(x) + C$ is the general form of the antiderivative of $f(x)$. Equation (2) is read "the *indefinite integral* of $f(x)$ is $F(x) + C$." The constant C is called the *constant of integration* and can have any fixed value; hence the use of the word "indefinite."

The symbolism $\int f(x) \, dx$, from Leibniz, is called an *integral,* or more precisely, an **indefinite integral.** The symbol $\int$ is called an *integral sign.* The left side of equation (2) is often read "the integral of $f(x)$, dx" or more briefly, "the integral of $f(x)$." In this context, $f(x)$ is called the *integrand.* For now, think of "dx" as indicating that the variable involved in the integration is x. More will be said about dx later.

Example 2   Evaluate: $\int x^9 \, dx$.

Solution   We have

$$\int x^9 \, dx = \frac{x^{10}}{10} + C.$$

The symbol on the left is read "the integral of x^9, dx." (The "dx" is often omitted in the reading.) In this case, the integrand is x^9. The constant C is called the constant of integration.

Be careful to note that

$$\int e^x \, dx \neq \frac{e^{x+1}}{x+1} + C.$$

Example 3   Evaluate: $\int 5e^{4x} \, dx$.

Solution   We have

$$\int 5e^{4x} \, dx = \frac{5}{4} e^{4x} + C.$$

We can check by differentiating the antiderivative. The derivative of $\frac{5}{4}e^{4x} + C$ is $4\left(\frac{5}{4}\right)e^{4x}$, or $5e^{4x}$, which is the integrand.

To integrate (or antidifferentiate), we make use of differentiation formulas, in effect, reading them in reverse. Below are some of these, stated in reverse, as integration formulas. These can be checked by differentiating the right-hand side and noting that the result is, in each case, the integrand.

Theorem 2

Basic Integration Formulas

1. $\int k \, dx = kx + C$ (k is a constant)

2. $\int x^r \, dx = \frac{x^{r+1}}{r+1} + C,$ provided $r \neq -1$

   (To integrate a power of x other than -1, increase the power by 1 and divide by the increased power.)

3. $\int x^{-1} \, dx = \int \frac{1}{x} \, dx = \int \frac{dx}{x} = \ln x + C,$ $x > 0$

   $\int x^{-1} \, dx = \ln |x| + C,$ $x < 0$

   (We will generally assume that $x > 0$.)

4. $\int be^{ax} \, dx = \frac{b}{a} e^{ax} + C$

The following rules combined with the preceding formulas allow us to find many integrals. They can be derived by reversing two familiar differentiation rules.

> **Theorem 3**
>
> Rule A. $\displaystyle\int kf(x)\,dx = k\int f(x)\,dx$
>
> (The integral of a constant times a function is the constant times the integral of the function.)
>
> Rule B. $\displaystyle\int [f(x) \pm g(x)]\,dx = \int f(x)\,dx \pm \int g(x)\,dx$
>
> (The integral of a sum or a difference is the sum or the difference of the integrals.)

Example 4

$$\int (5x + 4x^3)\,dx = \int 5x\,dx + \int 4x^3\,dx \qquad \text{Rule B}$$

$$= 5\int x\,dx + 4\int x^3\,dx \qquad \text{Rule A}$$

Then

$$\int (5x + 4x^3)\,dx = 5 \cdot \frac{x^2}{2} + 4 \cdot \frac{x^4}{4} + C = \frac{5}{2}x^2 + x^4 + C.$$

Don't forget the constant of integration. It is not necessary to write two constants of integration. If we did, we could add them and consider C as the sum.

We can check this as follows:

$$\frac{d}{dx}\left(\frac{5}{2}x^2 + x^4 + C\right) = 2 \cdot \frac{5}{2} \cdot x + 4x^3 = 5x + 4x^3. \qquad \triangleleft$$

We can always check an integration by differentiating.

Example 5

$$\int (7e^{6x} - \sqrt{x})\,dx = \int 7e^{6x}\,dx - \int \sqrt{x}\,dx$$

$$= \int 7e^{6x}\,dx - \int x^{1/2}\,dx$$

$$= \frac{7}{6}e^{6x} - \frac{x^{(1/2)+1}}{\frac{1}{2}+1} + C$$

$$= \frac{7}{6}e^{6x} - \frac{x^{3/2}}{\frac{3}{2}} + C$$

$$= \frac{7}{6}e^{6x} - \frac{2}{3}x^{3/2} + C \qquad \triangleleft$$

Example 6

$$\int \left(1 - \frac{3}{x} + \frac{1}{x^4}\right) dx = \int 1 \, dx - 3\int \frac{dx}{x} + \int x^{-4} \, dx$$

$$= x - 3 \ln x + \frac{x^{-4+1}}{-4 + 1} + C$$

$$= x - 3 \ln x - \frac{x^{-3}}{3} + C$$

Exploratory.
Using the same set of
axes, graph

$$y = x^4 - 4x^3$$

and

$$y = x^4 - 4x^3 + 7.$$

Compare the graphs.
Draw a tangent line to
each graph at $x = -1$,
$x = 0$, and $x = 1$ and
find the slope of each
line. Compare the
slopes. Then find the
derivative of each
function. Compare the
derivatives. Is either
function an
antiderivative of
$y = 4x^3 - 12x^2$?

Another Look at Antiderivatives

The graphs of the antiderivatives of x^2 are the graphs of the functions

$$y = \int x^2 \, dx = \frac{x^3}{3} + C$$

for the various values of the constant C.

As we can see in the figures below, x^2 is the derivative of each function; that is,
the tangent line at the point

$$\left(a, \frac{a^3}{3} + C\right)$$

has slope a^2. The curves $(x^3/3) + C$ fill up the plane, with exactly one curve going
through any given point (x_0, y_0).

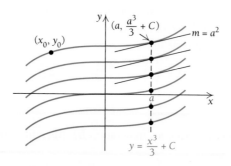
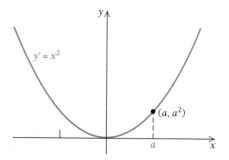

Suppose that we look for an antiderivative of x^2 with a specified value at a certain point—say, $f(-1) = 2$. We find that there is only one such function.

Example 7   Find the function f such that

$$f'(x) = x^2$$

and

$$f(-1) = 2.$$

Solution

a) We find $f(x)$ by integrating:

$$f(x) = \int x^2 \, dx = \frac{x^3}{3} + C.$$

b) The condition $f(-1) = 2$ allows us to find C,

$$f(-1) = \frac{(-1)^3}{3} + C = 2,$$

and solving for C, we get

$$-\tfrac{1}{3} + C = 2$$
$$C = 2 + \tfrac{1}{3}, \text{ or } \tfrac{7}{3}.$$

Thus, $f(x) = \dfrac{x^3}{3} + \dfrac{7}{3}$.

We are often given further information about a function, such as, in Example 7, $f(-1) = 2$ or $(-1, 2)$. The fact that $f(-1) = 2$ is called a *boundary condition*, or an *initial condition*. Sometimes the terms *boundary value*, or *initial value*, are used.

Applications

DETERMINING TOTAL COST FROM MARGINAL COST

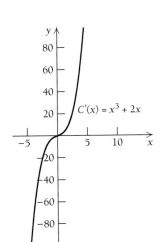

Example 8 *Business: Total Cost from Marginal Cost.* A company determines that the marginal cost C' of producing the xth unit of a certain product is given by

$$C'(x) = x^3 + 2x.$$

Find the total-cost function C, assuming that fixed costs (costs when 0 units are produced) are \$45. The boundary condition, or initial condition, is that $C = \$45$ when $x = 0$; that is, $C(0) = \$45$.

Solution

a) We integrate to find $C(x)$, using K for the integration constant to avoid confusion with the cost function C:

$$C(x) = \int C'(x)\, dx = \int (x^3 + 2x)\, dx = \frac{x^4}{4} + x^2 + K.$$

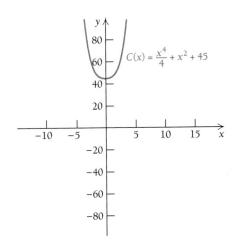

b) Fixed costs are \$45; that is, $C(0) = 45$. This allows us to determine the value of K:

$$C(0) = \frac{0^4}{4} + 0^2 + K = 45$$

$$K = 45.$$

Thus, $C(x) = \frac{x^4}{4} + x^2 + 45$.

FINDING VELOCITY AND DISTANCE FROM ACCELERATION

Recall that the position coordinate at time t of an object moving along a number line is $s(t)$. Then

$$s'(t) = v(t) = \text{the \textbf{velocity} at time } t,$$

$$v'(t) = a(t) = \text{the \textbf{acceleration} at time } t.$$

Example 9   *Physical Science: Distance.* Suppose that $v(t) = 5t^4$ and $s(0) = 9$. Find $s(t)$.

Solution

a) We find $s(t)$ by integrating:

$$s(t) = \int v(t)\, dt = \int 5t^4\, dt = t^5 + C.$$

b) We determine C by using the boundary condition $s(0) = 9$, which is the initial value, or position, of s at time $t = 0$:

$$s(0) = 0^5 + C = 9$$

$$C = 9.$$

Thus, $s(t) = t^5 + 9$.

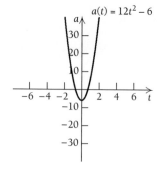
$a(t) = 12t^2 - 6$

Example 10   *Physical Science: Distance.* Suppose that $a(t) = 12t^2 - 6$, $v(0) = $ the initial velocity $= 5$, and $s(0) = $ the initial position $= 10$. Find $s(t)$.

Solution

a) We find $v(t)$ by integrating $a(t)$:

$$v(t) = \int a(t)\, dt$$

$$= \int (12t^2 - 6)\, dt$$

$$= 4t^3 - 6t + C_1.$$

b) The condition $v(0) = 5$ allows us to find C_1:

$$v(0) = 4 \cdot 0^3 - 6 \cdot 0 + C_1 = 5$$

$$C_1 = 5.$$

Thus, $v(t) = 4t^3 - 6t + 5$.

$v(t) = 4t^3 - 6t + 5$

$s(t) = t^4 - 3t^2 + 5t + 10$

c) We find $s(t)$ by integrating $v(t)$:

$$s(t) = \int v(t)\, dt$$

$$= \int (4t^3 - 6t + 5)\, dt$$

$$= t^4 - 3t^2 + 5t + C_2.$$

d) The condition $s(0) = 10$ allows us to find C_2:

$$s(0) = 0^4 - 3 \cdot 0^2 + 5 \cdot 0 + C_2 = 10$$

$$C_2 = 10.$$

Thus, $s(t) = t^4 - 3t^2 + 5t + 10$.

EXERCISE SET 5.1

Evaluate.

1. $\displaystyle\int x^6\, dx$

2. $\displaystyle\int x^7\, dx$

3. $\displaystyle\int 2\, dx$

4. $\displaystyle\int 4\, dx$

5. $\displaystyle\int x^{1/4}\, dx$

6. $\displaystyle\int x^{1/3}\, dx$

7. $\displaystyle\int (x^2 + x - 1)\, dx$

8. $\displaystyle\int (x^2 - x + 2)\, dx$

9. $\displaystyle\int (t^2 - 2t + 3)\, dt$

10. $\displaystyle\int (3t^2 - 4t + 7)\, dt$

11. $\displaystyle\int 5e^{8x}\, dx$

12. $\displaystyle\int 3e^{5x}\, dx$

13. $\displaystyle\int (x^3 - x^{8/7})\, dx$

14. $\displaystyle\int (x^4 - x^{6/5})\, dx$

15. $\displaystyle\int \frac{1000}{x}\, dx$

16. $\displaystyle\int \frac{500}{x}\, dx$

17. $\displaystyle\int \frac{dx}{x^2}\left(\text{or }\int \frac{1}{x^2}\, dx\right)$

18. $\displaystyle\int \frac{dx}{x^3}$

19. $\displaystyle\int \sqrt{x}\, dx$

20. $\displaystyle\int \sqrt[3]{x^2}\, dx$

21. $\displaystyle\int \frac{-6}{\sqrt[3]{x^2}}\, dx$

22. $\displaystyle\int \frac{20}{\sqrt[5]{x^4}}\, dx$

23. $\displaystyle\int 8e^{-2x}\, dx$

24. $\displaystyle\int 7e^{-0.25x}\, dx$

25. $\displaystyle\int \left(x^2 - \frac{3}{2}\sqrt{x} + x^{-4/3}\right) dx$

26. $\displaystyle\int \left(x^4 + \frac{1}{8\sqrt{x}} - \frac{4}{5}x^{-2/5}\right) dx$

Find f such that:

27. $f'(x) = x - 3, \quad f(2) = 9$

28. $f'(x) = x - 5, \quad f(1) = 6$

29. $f'(x) = x^2 - 4, \quad f(0) = 7$

30. $f'(x) = x^2 + 1, \quad f(0) = 8$

Applications ....

BUSINESS AND ECONOMICS

31. *Total Cost from Marginal Cost.* A company determines that the marginal cost C' of producing the xth unit of a certain product is given by

$$C'(x) = x^3 - 2x.$$

Find the total-cost function C, assuming fixed costs to be $100.

32. *Total Cost from Marginal Cost.* A company determines that the marginal cost C' of producing the xth unit of a certain product is given by

$$C'(x) = x^3 - x.$$

Find the total-cost function C, assuming fixed costs to be $200.

33. *Total Revenue from Marginal Revenue.* A company determines that the marginal revenue R' from selling the xth unit of a certain product is given by

$$R'(x) = x^2 - 3.$$

a) Find the total-revenue function R, assuming that $R(0) = 0$.

tw b) Why is $R(0) = 0$ a reasonable assumption?

34. *Total Revenue from Marginal Revenue.* A company determines that the marginal revenue R' from selling the xth unit of a certain product is given by

$$R'(x) = x^2 - 1.$$

a) Find the total-revenue function R, assuming that $R(0) = 0$.

tw b) Why is $R(0) = 0$ a reasonable assumption?

35. *Demand from Marginal Demand.* A company finds that the rate at which consumer-demand quantity changes with respect to price is given by the marginal-demand function

$$D'(p) = -\frac{4000}{p^2}.$$

Find the demand function if it is known that 1003 units of the product are demanded by consumers when the price is \$4 per unit.

36. *Supply from Marginal Supply.* A company finds that the rate at which a seller's quantity changes with respect to price is given by the marginal supply function

$$S'(p) = 0.24p^2 + 4p + 10.$$

Find the supply function if it is known that the seller will sell 121 units of the product when the price is \$5 per unit.

37. *Efficiency of a Machine Operator.* The rate at which a machine operator's efficiency E (expressed as a percentage) changes with respect to time t is given by

$$\frac{dE}{dt} = 30 - 10t,$$

where t is the number of hours that the operator has been at work.

a) Find $E(t)$, given that the operator's efficiency after working 2 hr is 72%; that is, $E(2) = 72$.

b) Use the answer to part (a) to find the operator's efficiency after 3 hr; after 5 hr.

38. *Efficiency of a Machine Operator.* The rate at which a machine operator's efficiency E (expressed

as a percentage) changes with respect to time t is given by

$$\frac{dE}{dt} = 40 - 10t,$$

where t is the number of hours that the operator has been at work.

A machine operator's efficiency changes with respect to time.

a) Find $E(t)$, given that the operator's efficiency after working 2 hr is 72%; that is, $E(2) = 72$.

b) Use the answer to part (a) to find the operator's efficiency after 4 hr; after 8 hr.

LIFE AND PHYSICAL SCIENCES

Find $s(t)$.

39. $v(t) = 3t^2$, $s(0) = 4$

40. $v(t) = 2t$, $s(0) = 10$

Find $v(t)$.

41. $a(t) = 4t$, $v(0) = 20$

42. $a(t) = 6t$, $v(0) = 30$

Find $s(t)$.

43. $a(t) = -2t + 6$, $v(0) = 6$, and $s(0) = 10$

44. $a(t) = -6t + 7$, $v(0) = 10$, and $s(0) = 20$

45. *Distance.* For a freely falling object, $a(t) = -32$ ft/sec^2, $v(0) = $ initial velocity $= v_0$, and $s(0) = $ initial height $= s_0$. Find a general expression for $s(t)$ in terms of v_0 and s_0.

46. *Time.* A ball is thrown from a height of 10 ft, where $s(0) = 10$, at an initial velocity of 80 ft/sec, where $v(0) = 80$. How long will it take before the ball hits the ground? (See Exercise 45.)

47. *Distance.* A car with constant acceleration goes from 0 to 60 mph in $\frac{1}{2}$ min. How far does the car travel during that time?

48. *Area of a Healing Wound.* The area A of a healing wound is decreasing at a rate given by

$$A'(t) = -43.4t^{-2}, \quad 1 \leq t \leq 7,$$

where t is the time, in days, and A is in square centimeters.

a) Find $A(t)$ if $A(1) = 39.7$.
b) Find the area of the wound after 7 days.

SOCIAL SCIENCES

49. *Memory.* In a certain memory experiment, the rate of memorizing is given by

$$M'(t) = 0.2t - 0.003t^2,$$

where $M(t)$ is the number of Spanish words memorized in t minutes.

a) Find $M(t)$ if it is known that $M(0) = 0$.
b) How many words are memorized in 8 min?

Synthesis

Find f.

50. $f'(t) = \sqrt{t} + \dfrac{1}{\sqrt{t}}, \quad f(4) = 0$

51. $f'(t) = t^{\sqrt{3}}, \quad f(0) = 8$

Evaluate. Each of the following can be integrated using the rules developed in this section, but some algebra may be required beforehand.

52. $\displaystyle\int (5t + 4)^2 \, dt$

53. $\displaystyle\int (x - 1)^2 x^3 \, dx$

54. $\displaystyle\int (1 - t)\sqrt{t} \, dt$

55. $\displaystyle\int \dfrac{(t + 3)^2}{\sqrt{t}} \, dt$

56. $\displaystyle\int \dfrac{x^4 - 6x^2 - 7}{x^3} \, dx$

57. $\displaystyle\int (t + 1)^3 \, dt$

58. $\displaystyle\int \dfrac{1}{\ln 10} \dfrac{dx}{x}$

59. $\displaystyle\int be^{ax} \, dx$

60. $\displaystyle\int (3x - 5)(2x + 1) \, dx$

61. $\displaystyle\int \sqrt[3]{64x^4} \, dx$

62. $\displaystyle\int \dfrac{x^2 - 1}{x + 1} \, dx$

63. $\displaystyle\int \dfrac{t^3 + 8}{t + 2} \, dt$

tw 64. On a test, a student makes the statement, "The $f(x) = x^2$ has a unique integral." Discuss.

tw 65. Describe the graphical interpretation of an antiderivative.

5.2

OBJECTIVES

➤ Find the area under a curve over a given closed interval.
➤ Interpret the area under a curve in two other ways.
➤ Evaluate a definite integral.

Area and Definite Integrals

We now use integration to find areas of certain regions. Consider a function whose outputs are nonnegative over an interval. We wish to find the area of the region between the graph of the function and the x-axis over that interval, as illustrated in the following figures.

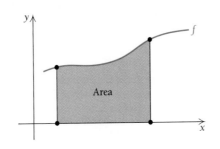

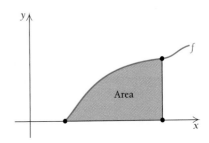

Let's first consider a constant function $f(x) = m$ over the interval from 0 to x, $[0, x]$.

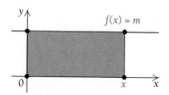

Refer to the figure above, which shows a rectangle whose area is the base times the height, $x \cdot m$, or mx. Suppose that we allow x to vary, giving us rectangles of different areas. The area of each rectangle is still mx. We now have an area function:

$A(x) = mx.$

Its graph is shown below.

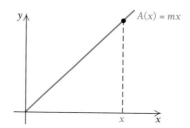

Let's consider next the linear function $f(x) = mx$ over the interval from 0 to x, $[0, x]$, as shown below.

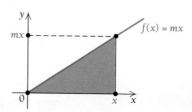

The figure formed this time is a triangle, and its area is one-half the base times the height, $\frac{1}{2} \cdot x \cdot (mx)$, or $\frac{1}{2}mx^2$. If we allow x to vary, we again get an area function:

$A(x) = \frac{1}{2}mx^2.$

Its graph is shown below.

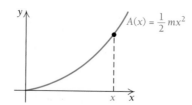

Next, let's consider the linear function $f(x) = mx + b$ over the interval from 0 to x, $[0, x]$, as shown at the top of the following page.

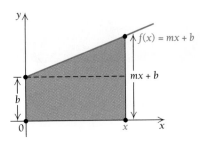

The figure formed this time is a trapezoid, and its area is one-half the height times the sum of the lengths of its parallel sides (or, noting the dashed line, the area of the triangle plus the area of the rectangle):

$$\tfrac{1}{2} \cdot x \cdot [b + (mx + b)],$$

or $\tfrac{1}{2} \cdot x \cdot (mx + 2b)$,

or $\tfrac{1}{2} mx^2 + bx$.

If we allow x to vary, we again get an area function:

$$A(x) = \tfrac{1}{2} mx^2 + bx.$$

Its graph is shown below.

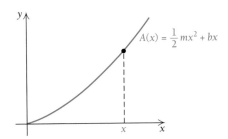

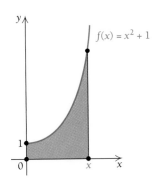

Now we consider the function $f(x) = x^2 + 1$ over the interval from 0 to x, $[0, x]$. The graph of the region in question is shown at left, but it is not so easy this time to find the area function because the graph of $f(x)$ is not a straight line. Let's look for a pattern in the following table.

$f(x)$	$A(x)$
$f(x) = 3$	$A(x) = 3x$
$f(x) = m$	$A(x) = mx$
$f(x) = 3x$	$A(x) = \tfrac{3}{2}x^2$
$f(x) = mx$	$A(x) = \tfrac{1}{2}mx^2$
$f(x) = mx + b$	$A(x) = \tfrac{1}{2}mx^2 + bx$

You may have conjectured that the area function $A(x)$ is an antiderivative of $f(x)$. In the following exploratory exercises, you will investigate further.

EXPLORATORY EXERCISES

Finding Areas

1. Consider the constant function $f(x) = 3$.

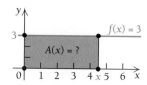

a) Find $A(x)$.
b) Find $A(1)$, $A(2)$, and $A(5)$.
c) Graph $A(x)$.
d) How are $f(x)$ and $A(x)$ related?

2. Consider the function $f(x) = 3x$.

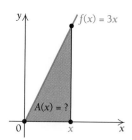

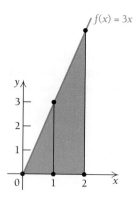

a) Find $A(x)$.
b) Find $A(1)$, $A(2)$, and $A(3.5)$.
c) Graph $A(x)$.
d) How are $f(x)$ and $A(x)$ related?

3. The region under the graph of $f(x) = x^2 + 1$, over the interval $[0, 2]$, is shown below.

a) On a sheet of thin paper, make a copy of the shaded region.
b) Cut up the shaded region in any way you wish in order to fill up squares in the grid. Make an estimate of the total area.

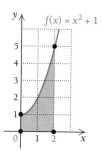

c) Using the antiderivative

$$F(x) = \frac{x^3}{3} + x,$$

find $F(2)$.

d) Compare your answers in parts (b) and (c).

4. Repeat Exercises 3(a) and (b) for the shaded region of the following graph, using the grid shown below.

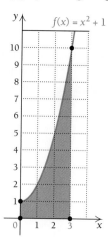

(*continued*)

c) Using the antiderivative

$$F(x) = \frac{x^3}{3} + x,$$

find $F(3)$.

d) Compare your answers in parts (b) and (c).

The conjecture concerning areas and antiderivatives (or integrals) is true. It is expressed as follows.

> **Theorem 4**
>
> Let f be a nonnegative, continuous function on an interval $[a, b]$, and let $A(x)$ be the area of the region between the graph of f and the x-axis on the interval $[a, x]$, with $x \le b$. Then $A(x)$ is a differentiable function of x and
>
> $$A'(x) = f(x).$$

The situation described in the theorem is shown in the following figure.

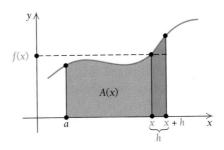

The derivative of $A(x)$ is, by the definition of a derivative,

$$A'(x) = \lim_{h \to 0} \frac{A(x + h) - A(x)}{h}.$$

Note from the figure that $A(x + h) - A(x)$ is the area of the small, orange, vertical strip. The area of this small strip is approximately that of a rectangle of base h and height $f(x)$, especially for small values of h. Thus we have

$$A(x + h) - A(x) \approx f(x) \cdot h.$$

Now

$$A'(x) = \lim_{h \to 0} \frac{A(x + h) - A(x)}{h}$$

$$= \lim_{h \to 0} \frac{f(x) \cdot h}{h}$$

$$= \lim_{h \to 0} f(x) = f(x),$$

since $f(x)$ does not involve h. Therefore,

$$A(x) = \int f(x)\,dx = F(x) + C,$$

where $F(x)$ is any antiderivative of f.

Definite Integrals

Let's see how we can use the remarkable conclusion of Theorem 4 to find the area under the graph of a nonnegative, continuous function f from a to x. We know that the area function A is an antiderivative.

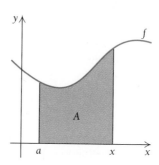

Let F be any other antiderivative of f. Then

$$A(x) - A(a) = F(x) - F(a).$$

To understand this, recall from Section 5.1 that we know that A and F differ by a constant since they are both antiderivatives; that is, $A(x) = F(x) + C$. Then

$$A(x) - A(a) = [F(x) + C] - [F(a) + C] = F(x) - F(a).$$

Thus the difference $A(x) - A(a)$ has the same value for all antiderivatives of f.

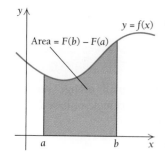

To find the area under the graph of a nonnegative, continuous function f over the interval $[a, b]$:

1. Find any antiderivative $F(x)$ of $f(x)$. (The simplest is the one for which the constant of integration is 0.)
2. Substitute b and a and find the difference $F(b) - F(a)$. The result is the area under the graph over the interval $[a, b]$.

Example 1   Find the area under the graph of $y = x^2 + 1$ over the interval $[-1, 2]$.

Solution   In this case, $f(x) = x^2 + 1$, $a = -1$, and $b = 2$.

1) Find any antiderivative $F(x)$ of $f(x)$. We choose the simplest one:

$$F(x) = \frac{x^3}{3} + x.$$

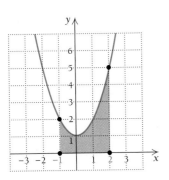

2) Substitute 2 and -1 and find the difference $F(2) - F(-1)$:

$$F(2) - F(-1) = \left[\frac{2^3}{3} + 2\right] - \left[\frac{(-1)^3}{3} + (-1)\right]$$

$$= \frac{8}{3} + 2 - \left[\frac{-1}{3} - 1\right]$$

$$= \frac{8}{3} + 2 + \frac{1}{3} + 1$$

$$= 6.$$

We can make a partial check by counting the squares or parts of squares as in the preceding exploratory exercises.

Example 2   Find the area under the graph of $y = x^3$ over the interval $[0, 5]$.

Solution   In this case, $f(x) = x^3$, $a = 0$, and $b = 5$.

1) Find any antiderivative $F(x)$ of $f(x)$. We choose the simplest one:

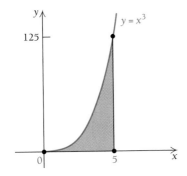

$$F(x) = \frac{x^4}{4}.$$

2) Substitute 5 and 0 and find the difference $F(5) - F(0)$:

$$F(5) - F(0) = \frac{5^4}{4} - \frac{0^4}{4} = \frac{625}{4} = 156\tfrac{1}{4}.$$

The difference $F(b) - F(a)$ has the same value for all antiderivatives of a function f whether the function is nonnegative or not. It is called the **definite integral** of f from a to b.

Definition

Let f be any continuous function over the interval $[a, b]$ and F be any antiderivative of f. Then the **definite integral** of f from a to b is

$$\int_a^b f(x)\,dx = F(b) - F(a).$$

Evaluating definite integrals is called *integrating*. The numbers a and b are known as the **limits of integration.**

Example 3   Evaluate: $\int_a^b x^2\,dx$.

Solution   Using the antiderivative $F(x) = x^3/3$, we have

$$\int_a^b x^2\,dx = \frac{b^3}{3} - \frac{a^3}{3}.$$

It is convenient to use an intermediate notation:

$$\int_a^b f(x)\ dx = [F(x)]_a^b = F(b) - F(a).$$

We now evaluate several definite integrals.

Example 4

$$\int_{-1}^{2} x^2\ dx = \left[\frac{x^3}{3}\right]_{-1}^{2}$$

$$= \frac{2^3}{3} - \frac{(-1)^3}{3}$$

$$= \frac{8}{3} - \left(-\frac{1}{3}\right) = \frac{8}{3} + \frac{1}{3} = 3$$

Example 5

$$\int_0^3 e^x\ dx = [e^x]_0^3 = e^3 - e^0 = e^3 - 1$$

Example 6

$$\int_1^4 (x^2 - x)\ dx = \left[\frac{x^3}{3} - \frac{x^2}{2}\right]_1^4 = \left(\frac{4^3}{3} - \frac{4^2}{2}\right) - \left(\frac{1^3}{3} - \frac{1^2}{2}\right)$$

$$= \left(\frac{64}{3} - \frac{16}{2}\right) - \left(\frac{1}{3} - \frac{1}{2}\right)$$

$$= \frac{64}{3} - 8 - \frac{1}{3} + \frac{1}{2} = 13\frac{1}{2}$$

Example 7

$$\int_1^e \left(1 + 2x - \frac{1}{x}\right) dx = [x + x^2 - \ln x]_1^e$$

$$= (e + e^2 - \ln e) - (1 + 1^2 - \ln 1)$$

$$= (e + e^2 - 1) - (1 + 1 - 0)$$

$$= e + e^2 - 1 - 1 - 1$$

$$= e + e^2 - 3$$

More on Area

When we evaluate the definite integral of a nonnegative function, we get the area under the graph over an interval.

Example 8   Find the area under $y = 1/x$ over the interval $[1, 4]$.

Solution

$$\int_1^4 \frac{dx}{x} = [\ln x]_1^4 = \ln 4 - \ln 1$$

$$= \ln 4 \approx 1.3863$$

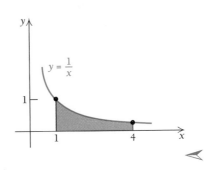

Example 9   Find the area under $y = 1/x^2$ over the interval $[1, b]$.

Solution

$$\int_1^b \frac{dx}{x^2} = \int_1^b x^{-2} \, dx$$

$$= \left[\frac{x^{-2+1}}{-2+1} \right]_1^b$$

$$= \left[\frac{x^{-1}}{-1} \right]_1^b = \left[-\frac{1}{x} \right]_1^b$$

$$= \left(-\frac{1}{b} \right) - \left(-\frac{1}{1} \right)$$

$$= 1 - \frac{1}{b}$$

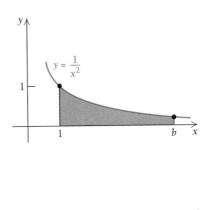

Now let's compare the definite integrals of the functions $y = x^2$ and $y = -x^2$. Note that one is nonnegative and one is nonpositive.

$$\int_0^2 x^2 \, dx = \left[\frac{x^3}{3} \right]_0^2$$

$$= \frac{2^3}{3} - \frac{0^3}{3} = \frac{8}{3}$$

$$\int_0^2 -x^2 \, dx = \left[-\frac{x^3}{3} \right]_0^2$$

$$= -\frac{2^3}{3} + \frac{0^3}{3} = -\frac{8}{3}$$

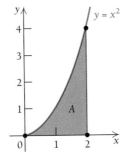

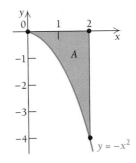

The graphs of the functions $y = x^2$ and $y = -x^2$ are reflections of each other across

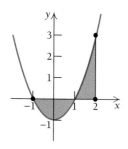

the x-axis. Thus the areas of the shaded regions are the same—that is, $\frac{8}{3}$. The evaluation procedure in the second case gave us $-\frac{8}{3}$. This illustrates that for negative-valued functions, the definite integral gives us the opposite, or additive inverse, of the area between the curve and the x-axis.

Now let's consider the function $x^2 - 1$ over the interval $[-1, 2]$. It has both positive and negative values. We will apply the preceding evaluation procedure, even though function values are not all nonnegative. We do so in two ways.

First, let's use the fact that for any a, b, c, if $a < c < b$, then

$$\int_a^b f(x)\, dx = \int_a^c f(x)\, dx + \int_c^b f(x)\, dx.$$

We will consider this property of integrals in Section 5.4. We have

$$\int_{-1}^{2} (x^2 - 1)\, dx = \int_{-1}^{1} (x^2 - 1)\, dx + \int_{1}^{2} (x^2 - 1)\, dx$$

$$= \left[\frac{x^3}{3} - x\right]_{-1}^{1} + \left[\frac{x^3}{3} - x\right]_{1}^{2}$$

$$= \left[\left(\frac{1^3}{3} - 1\right) - \left(\frac{(-1)^3}{3} - (-1)\right)\right]$$

$$+ \left[\left(\frac{2^3}{3} - 2\right) - \left(\frac{1^3}{3} - 1\right)\right]$$

$$= \left[-\frac{4}{3}\right] + \left[\frac{4}{3}\right]$$

$$= 0.$$

This shows that the area of the region under the x-axis is the same as the area of the region above the x-axis.

Now let's evaluate in another way:

$$\int_{-1}^{2} (x^2 - 1)\, dx = \left[\frac{x^3}{3} - x\right]_{-1}^{2}$$

$$= \left(\frac{2^3}{3} - 2\right) - \left[\frac{(-1)^3}{3} - (-1)\right]$$

$$= \left(\frac{8}{3} - 2\right) - \left(-\frac{1}{3} + 1\right)$$

$$= \frac{8}{3} - 2 + \frac{1}{3} - 1$$

$$= 0.$$

The definite integral of a continuous function over an interval is the sum of the areas above the x-axis minus the sum of the areas below the x-axis.

Example 10   In each of the following figures, decide visually whether

$$\int_a^b f(x)\, dx$$

is positive, negative, or zero.

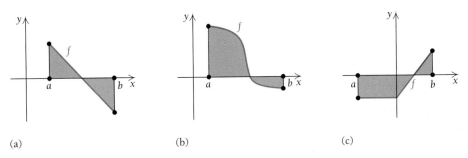

(a)   (b)   (c)

Solution

a) In this figure, there is the same area above the *x*-axis as below. Thus,

$$\int_a^b f(x)\, dx = 0.$$

b) In this figure, there is more area above the *x*-axis than below. Thus,

$$\int_a^b f(x)\, dx > 0.$$

c) In this figure, there is more area below the *x*-axis than above. Thus,

$$\int_a^b f(x)\, dx < 0.$$

Example 11   Evaluate: $\int_{-1}^{2} (x^3 - 3x + 1)\, dx$. Interpret the results in terms of area.

Solution   We have

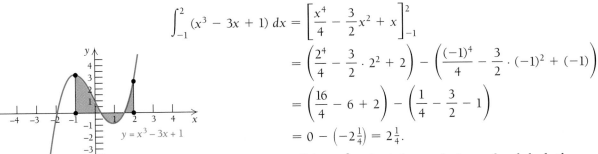

$$\int_{-1}^{2} (x^3 - 3x + 1)\, dx = \left[\frac{x^4}{4} - \frac{3}{2}x^2 + x \right]_{-1}^{2}$$

$$= \left(\frac{2^4}{4} - \frac{3}{2} \cdot 2^2 + 2 \right) - \left(\frac{(-1)^4}{4} - \frac{3}{2} \cdot (-1)^2 + (-1) \right)$$

$$= \left(\frac{16}{4} - 6 + 2 \right) - \left(\frac{1}{4} - \frac{3}{2} - 1 \right)$$

$$= 0 - \left(-2\tfrac{1}{4} \right) = 2\tfrac{1}{4}.$$

$y = x^3 - 3x + 1$

We can graph the function $f(x) = x^3 - 3x + 1$ over the interval and shade the area between the curve and the *x*-axis. The sum of the areas above the axis minus the area below is $2\tfrac{1}{4}$.

In this section, we have considered definite integrals in relation to area. In the section that follows and in Chapter 6, we will see many other interpretations and applications of the idea of a definite integral.

TECHNOLOGY CONNECTION

Approximating Definite Integrals

There are two methods for evaluating definite integrals using a grapher. Let's consider the function of Example 11,

$$f(x) = x^3 - 3x + 1.$$

Method 1: fnInt

We first enter $y_1 = x^3 - 3x + 1$. Then we select the fnInt feature from the **MATH** menu. Next, we enter the function, the variable, and the endpoints of the interval over which we are integrating. The grapher returns the value of the definite integral found in Example 11.

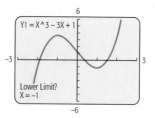

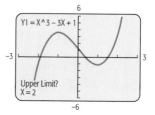

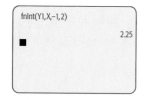

Method 2: $\int f(x)\, dx$

We first graph $y_1 = x^3 - 3x + 1$. Then we select the $\int f(x)\, dx$ feature from the **CALC** menu and enter the lower and upper limits of integration. The grapher shades the area and returns the value of the definite integral in Example 11.

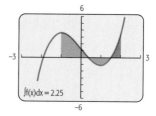

Exercises
Evaluate the definite integral.

1. $\displaystyle\int_{-1}^{2} (x^2 - 1)\, dx$ 2. $\displaystyle\int_{-2}^{3} (x^3 - 3x + 1)\, dx$

3. $\displaystyle\int_{1}^{6} \frac{\ln x}{x^2}\, dx$ 4. $\displaystyle\int_{-8}^{2} \frac{4}{(1 + e^x)^2}\, dx$

5. $\displaystyle\int_{-10}^{10} (0.002x^4 - 0.3x^2 + 4x - 7)\, dx$

EXERCISE SET 5.2

Find the area under the given curve over the indicated interval.

1. $y = 4$; $[1, 3]$

2. $y = 5$; $[1, 3]$

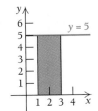

3. $y = 2x$; $[1, 3]$

4. $y = x^2$; $[0, 3]$

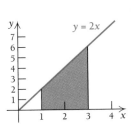

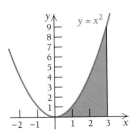

5. $y = x^2$; $[0, 5]$

6. $y = x^3$; $[0, 2]$

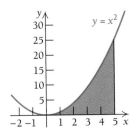

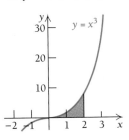

7. $y = x^3$; $[0, 1]$

8. $y = 1 - x^2$; $[-1, 1]$

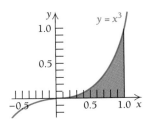

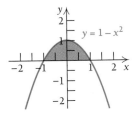

9. $y = 4 - x^2$; $[-2, 2]$ **10.** $y = e^x$; $[0, 2]$

11. $y = e^x$; $[0, 3]$

12. $y = \dfrac{1}{x}$; $[1, 2]$

13. $y = \dfrac{1}{x}$; $[1, 3]$

14. $y = x^2 - 4x$; $[-4, -2]$

Evaluate. Then interpret the results.

15. $\displaystyle\int_0^{1.5} (x - x^2)\,dx$

16. $\displaystyle\int_0^2 (x^2 - x)\,dx$

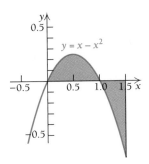

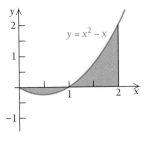

17. $\displaystyle\int_{-1}^1 (x^4 - x^2)\,dx$

18. $\displaystyle\int_0^b -2e^{3x}\,dx$

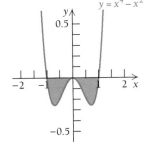

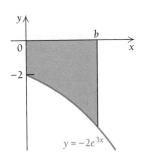

 19.–36. Check the results of each of Exercises 1–18 using a grapher.

Evaluate.

37. $\displaystyle\int_a^b e^t\,dt$

38. $\displaystyle\int_0^a (ax - x^2)\,dx$

39. $\displaystyle\int_a^b 3t^2\,dt$

40. $\displaystyle\int_{-5}^2 4t^3\,dt$

41. $\displaystyle\int_1^e \left(x + \frac{1}{x}\right) dx$

42. $\displaystyle\int_1^e \left(x - \frac{1}{x}\right) dx$

43. $\int_0^1 \sqrt{x} \, dx$

44. $\int_0^1 3\sqrt{x} \, dx$

45. $\int_{-4}^1 \frac{10}{17} t^3 \, dt$

46. $\int_0^1 \frac{12}{13} t^2 \, dt$

Find the area under the graph over the indicated interval.

47. $y = x^3$; $[0, 2]$

48. $y = x^4$; $[0, 1]$

49. $y = x^2 + x + 1$; $[2, 3]$

50. $y = 2 - x - x^2$; $[-2, 1]$

51. $y = 5 - x^2$; $[-1, 2]$

52. $y = e^x$; $[-2, 3]$

53. $y = e^x$; $[-1, 5]$

54. $y = 2x + \frac{1}{x^2}$; $[1, 4]$

In each exercise, determine visually whether $\int_a^b f(x) \, dx$ is positive, negative, or zero, and express $\int_a^b f(x) \, dx$ in terms of A. Explain your result.

tw 55. a)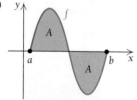

   b)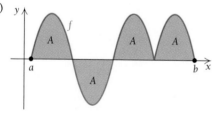

tw 56. a)

b)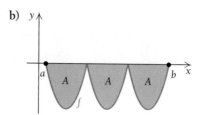

Synthesis

Evaluate.

57. $\int_2^3 \frac{x^2 - 1}{x - 1} \, dx$

58. $\int_1^5 \frac{x^5 - x^{-1}}{x^2} \, dx$

59. $\int_4^{16} (x - 1) \sqrt{x} \, dx$

60. $\int_0^1 (x + 2)^3 \, dx$

61. $\int_1^8 \frac{\sqrt[3]{x^2} - 1}{\sqrt[3]{x}} \, dx$

62. $\int_0^1 \frac{x^3 + 8}{x + 2} \, dx$

63. $\int_1^2 (4x + 3)(5x - 2) \, dx$

64. $\int_2^5 (t + \sqrt{3})(t - \sqrt{3}) \, dt$

65. $\int_0^1 (t + 1)^3 \, dt$

66. $\int_1^3 \left(x - \frac{1}{x} \right)^2 \, dx$

67. $\int_1^3 \frac{t^5 - t}{t^3} \, dt$

68. $\int_4^9 \frac{t + 1}{\sqrt{t}} \, dt$

69. $\int_3^5 \frac{x^2 - 4}{x - 2} \, dx$

70. $\int_0^1 \frac{t^3 + 1}{t + 1} \, dt$

Find the error in each of the following. Explain.

tw 71. $\int_1^2 (x^2 + x + 1) \, dx = \left[\frac{1}{3} x^3 + \frac{1}{2} x^2 + x \right]_1^2$

$$= \left(\frac{1}{3} \cdot 2^3 + \frac{1}{2} \cdot 2^2 + 2 \right)$$

$$= \frac{10}{3}$$

tw 72. $\int_1^2 (\ln x - e^x) \, dx = \left[\frac{1}{x} - e^x \right]_1^2$

$$= \left(\frac{1}{2} - e^2 \right) - (1 - e^1)$$

$$= e - e^2 - \frac{1}{2}$$

TECHNOLOGY   CONNECTION

Evaluate.

73. $\int_{-2}^{3} (x^3 - 4x)\, dx$ **74.** $\int_{-2}^{3} (x - x^3)\, dx$

75. $\int_{-1.2}^{6.3} (x^3 - 9x^2 + 27x + 50)\, dx$

76. $\int_{-8}^{1.4} (x^4 + 4x^3 - 36x^2 - 160x + 300)\, dx$

77. $\int_{-1}^{20} \dfrac{4x}{x^2 + 1}\, dx$

78. $\int_{-1}^{1} (1 - \sqrt{1 - x^2})\, dx$

79. $\int_{-2}^{2} \sqrt{4 - x^2}\, dx$

80. $\int_{0}^{10} (x^3 - 6x^2 + 15x)\, dx$

81. $\int_{0}^{8} x(x - 5)^4\, dx$

82. $\int_{-2}^{2} x^{2/3}\left(\dfrac{5}{2} - x\right) dx$

83. $\int_{2}^{4} \dfrac{x^2 - 4}{x^2 - 3}\, dx$

84. $\int_{-10}^{10} \dfrac{8}{x^2 + 4}\, dx$

5.3

OBJECTIVES

➤ Interpret definite integrals as limits of sums and approximate definite integrals by adding areas of rectangles.
➤ Use definite integrals to find accumulations.
➤ Find the average value of a function.

Limits of Sums and Accumulations

In Section 5.2, we developed the idea of a definite integral and used it to find areas. Here we look at the definite integral as a limit of a sum. Then we consider applications involving accumulations of quantities such as cost.

Limits of Sums

Let's consider the area under a curve over an interval $[a, b]$. We divide $[a, b]$ into subintervals of equal length and construct rectangles, the sum of whose areas approximates the area under the curve.

In the following figure, $[a, b]$ has been divided into 4 subintervals, each having width $\Delta x = (b - a)/4$.

The area under a curve can be approximated by a sum of rectangular areas.

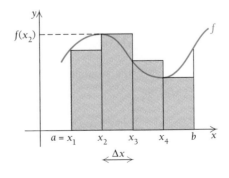

The heights of the rectangles shown are

$$f(x_1), \qquad f(x_2), \qquad f(x_3), \quad \text{and} \quad f(x_4).$$

The area of the region under the curve is approximately the sum of the areas of the four rectangles:

$$f(x_1)\,\Delta x + f(x_2)\,\Delta x + f(x_3)\,\Delta x + f(x_4)\,\Delta x.$$

We can denote this sum with **summation notation,** which uses the Greek capital letter sigma, Σ:

$$\sum_{i=1}^{4} f(x_i)\,\Delta x, \quad \text{or} \quad \sum_{i=1}^{4} f(x_i)\,\Delta x.$$

This is read "the sum of the numbers $f(x_i)\,\Delta x$ from $i = 1$ to $i = 4$." To recover the original expression, we substitute the numbers 1 through 4 successively into $f(x_i)\,\Delta x$ and write plus signs between the results.

Before we continue, let's consider some examples involving summation notation.

Example 1   Write summation notation for $2 + 4 + 6 + 8 + 10$.

Solution

$$2 + 4 + 6 + 8 + 10 = \sum_{i=1}^{5} 2i$$

Example 2   Write summation notation for

$$g(x_1)\,\Delta x + g(x_2)\,\Delta x + \cdots + g(x_{19})\,\Delta x.$$

Solution

$$g(x_1)\,\Delta x + g(x_2)\,\Delta x + \cdots + g(x_{19})\,\Delta x = \sum_{i=1}^{19} g(x_i)\,\Delta x$$

Example 3   Express

$$\sum_{i=1}^{4} 3^i$$

without using summation notation.

Solution

$$\sum_{i=1}^{4} 3^i = 3^1 + 3^2 + 3^3 + 3^4, \quad \text{or} \quad 120$$

Example 4   Express

$$\sum_{i=1}^{30} h(x_i)\,\Delta x$$

without using summation notation.

Solution

$$\sum_{i=1}^{30} h(x_i)\,\Delta x = h(x_1)\,\Delta x + h(x_2)\,\Delta x + \cdots + h(x_{30})\,\Delta x$$

Approximation of area by rectangles becomes more accurate as we use smaller subintervals and hence more rectangles, as shown in the figures at the top of the following page.

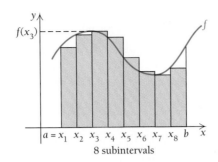
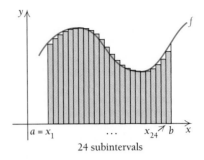

8 subintervals   24 subintervals

In general, suppose that the interval $[a, b]$ is divided into n equal subintervals, each of width $\Delta x = (b - a)/n$. We construct rectangles with heights

$$f(x_1), f(x_2), \ldots, f(x_n).$$

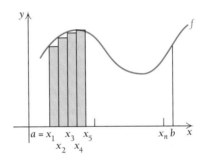

The width of each rectangle is Δx, so the first rectangle has area

$$f(x_1)\ \Delta x,$$

the second rectangle has area

$$f(x_2)\ \Delta x,$$

and so on. The area of the region under the curve is approximated by the sum of the areas of the rectangles:

$$\sum_{i=1}^{n} f(x_i)\ \Delta x.$$

We obtain the actual area by finding the sum of the areas of the rectangles as n increases without bound. The exact area is thus given by

$$A = \lim_{n \to \infty} \sum_{i=1}^{n} f(x_i)\ \Delta x.$$

The area is also given by a definite integral:

$$\int_a^b f(x)\ dx = \lim_{n \to \infty} \sum_{i=1}^{n} f(x_i)\ \Delta x.$$

The fact that we can express the integral of a function (positive or otherwise) as a limit of a sum or in terms of an antiderivative is so important that it has a name: *The Fundamental Theorem of Integral Calculus.*

Historical Note: Sums, or limits of sums, used in this context are generally called *Riemann sums* in honor of the German mathematician G. F. Bernhard Riemann, 1826–1866. Riemann received his doctorate at the University of Göttingen and did much of his research in geometry.

The Fundamental Theorem of Integral Calculus

If a function f has an antiderivative F over $[a, b]$, then

$$\lim_{n \to \infty} \sum_{i=1}^{n} f(x_i)\, \Delta x = \int_{a}^{b} f(x)\, dx = F(b) - F(a).$$

It is interesting to envision that, as we take the limit, the summation sign stretches into something reminiscent of an S (the integral sign) and Δx is defined to be dx. This is why dx appears in the integral notation.

In the following example, we find the exact area using integration. Then, to provide a better understanding of a limit of sums, we approximate the value of the definite integral by a sum, making it as good as we please by taking n sufficiently large.

Example 5 Consider the function

$$f(x) = 600x - x^2$$

over the interval $[0, 600]$.

a) Find the exact area using integration.

b) Approximate the area by dividing the interval into 6 subintervals.

c) Approximate the area by dividing the interval into 12 subintervals.

Solution

a) We find the exact area by integrating as follows:

$$\int_{0}^{600} (600x - x^2)\, dx = \left[300x^2 - \frac{x^3}{3} \right]_{0}^{600}$$

$$= \left(300 \cdot 600^2 - \frac{600^3}{3} \right) - \left(300 \cdot 0^2 - \frac{0^3}{3} \right)$$

$$= 36{,}000{,}000.$$

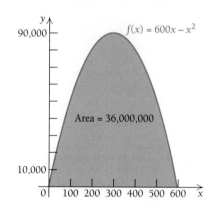

b) The interval $[0, 600]$ is divided into 6 subintervals, each of length $\Delta x = (600 - 0)/6 = 100$, and x_i ranges from $x_1 = 0$ to $x_6 = 500$.

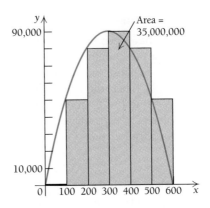

Thus we have

$$\sum_{i=1}^{6} f(x_i)\,\Delta x = f(0) \cdot 100 + f(100) \cdot 100 + f(200) \cdot 100$$
$$+ f(300) \cdot 100 + f(400) \cdot 100 + f(500) \cdot 100$$
$$= 0 \cdot 100 + 50{,}000 \cdot 100 + 80{,}000 \cdot 100$$
$$+ 90{,}000 \cdot 100 + 80{,}000 \cdot 100 + 50{,}000 \cdot 100$$
$$= 35{,}000{,}000.$$

c) The interval $[0, 600]$ is divided into 12 subintervals, each of length $\Delta x = (600 - 0)/12 = 50$, and x_i ranges from $x_1 = 0$ to $x_{12} = 550$.

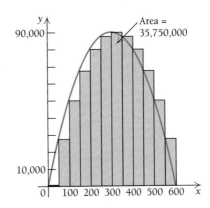

Thus we have

$$\sum_{i=1}^{12} f(x_i)\,\Delta x = f(0) \cdot 50 + f(50) \cdot 50 + f(100) \cdot 50 + f(150) \cdot 50$$
$$+ f(200) \cdot 50 + f(250) \cdot 50 + f(300) \cdot 50 + f(350) \cdot 50$$
$$+ f(400) \cdot 50 + f(450) \cdot 50 + f(500) \cdot 50 + f(550) \cdot 50$$
$$= 0 \cdot 50 + 27{,}500 \cdot 50 + 50{,}000 \cdot 50 + 67{,}500 \cdot 50$$
$$+ 80{,}000 \cdot 50 + 87{,}500 \cdot 50 + 90{,}000 \cdot 50 + 87{,}500 \cdot 50$$
$$+ 80{,}000 \cdot 50 + 67{,}500 \cdot 50 + 50{,}000 \cdot 50 + 27{,}500 \cdot 50$$
$$= 35{,}750{,}000.$$

Note that the approximation using $n = 12$ is closer to the exact value than the one found using $n = 6$.   ◁

Accumulations

We have seen that the definite integral of $f(x)$ can be interpreted as an accumulating area that eventually gives the total area.

Suppose that we have a velocity function $v(t)$, where $v(t) > 0$ over the interval $[0, b]$. Then the area under the curve $y = v(t)$ is the total distance traveled from $t = 0$ to $t = b$. In Example 2 of Section 5.2, we found that the area under the graph of $y = x^3$ over the interval $[0, 5]$ is $156\frac{1}{4}$. Suppose that we rewrite $y = x^3$ as the velocity function

$$v(t) = t^3,$$

where t is in hours and distance is in miles. Then the total distance traveled in the 5-hr period from $t = 0$ to $t = 5$ is the area under the curve $y = v(t)$ over $[0, 5]$, or $156\frac{1}{4}$ mi.

For a marginal-cost function $C'(x)$, the area under the curve $y = C'(x)$ over the interval $[0, x]$ is the total cost, or the accumulated cost, of producing x units.

Example 6   *Business: Cost from Marginal Cost.* Raggs, Ltd., determines that the marginal cost per suit is given by

$$C'(x) = 0.0003x^2 - 0.2x + 50.$$

a) Approximate the total cost of producing 400 suits by computing the sum $\sum_{i=1}^{4} C'(x_i)\, \Delta x$.

b) Find the exact total cost of producing 400 suits by integrating.

Solution

a) The interval $[0, 400]$ is divided into 4 subintervals, each of length $\Delta x = (400 - 0)/4 = 100$, and x_i ranges from $x_1 = 0$ to $x_4 = 300$.

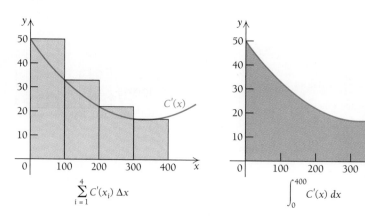

$$\sum_{i=1}^{4} C'(x_i)\, \Delta x \qquad\qquad \int_0^{400} C'(x)\, dx$$

Thus we have

$$\sum_{i=1}^{4} C'(x_i)\, \Delta x = C'(0) \cdot 100 + C'(100) \cdot 100 + C'(200) \cdot 100$$
$$+ C'(300) \cdot 100$$
$$= 50 \cdot 100 + 33 \cdot 100 + 22 \cdot 100 + 17 \cdot 100$$
$$= \$12,200.$$

.   **b)** The exact total cost is

$$\int_0^{400} C'(x)\,dx = [0.0001x^3 - 0.1x^2 + 50x]_0^{400} = \$10{,}400.$$

Thus the approximation found in part (a) is not too far off, considering that the number of subintervals is small.   ◁

The fact that an integral can be approximated by a sum is useful when the antiderivative of a function does not have an elementary formula. For example, for the function $f(x) = e^{-x^2/2}$, important in probability, there is no formula for the antiderivative. Thus tables of approximate values of its integral have been computed using summation methods.

The Average Value of a Function

Suppose that

$$T = f(t)$$

is the temperature at time t recorded at a weather station on a certain day. The station uses a 24-hr clock, so the domain of the temperature function is the interval $[0, 24]$. The function is continuous, as shown below.

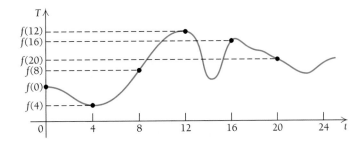

To find the average temperature for the given day, we might take six temperature readings at 4-hr intervals, starting at midnight:

$$T_0 = f(0), \qquad T_1 = f(4),$$
$$T_2 = f(8), \qquad T_3 = f(12),$$
$$T_4 = f(16), \qquad T_5 = f(20).$$

The average reading would then be the sum of these six readings divided by 6:

$$T_{av} = \frac{T_0 + T_1 + T_2 + T_3 + T_4 + T_5}{6}.$$

This computation of the average temperature may not give the most useful answer. For example, suppose it is a hot summer day, and at 2:00 in the afternoon (hour 14 on the 24-hr clock), there is a short thunderstorm that cools the air for an hour between our readings. This temporary dip would not show up in the average computed above.

What can we do? We could take 48 readings at half-hour intervals. This should give us a better result. In fact, the shorter the time between readings, the better the result should be. It seems reasonable that we might define the **average value** of T over the interval $[0, 24]$ to be the limit, as n approaches ∞, of the average of n values:

$$\text{Average value of } T = \lim_{n \to \infty} \frac{1}{n} \sum_{i=1}^{n} T_i$$

$$= \lim_{n \to \infty} \frac{1}{n} \sum_{i=1}^{n} f(t_i).$$

Note that this is not too far from our definition of an integral. All we would need is to get Δt, which is $(24 - 0)/n$, or $24/n$, into the summation:

$$\text{Average value of } T = \lim_{n \to \infty} \frac{1}{n} \sum_{i=1}^{n} f(t_i)$$

$$= \lim_{n \to \infty} \frac{1}{\Delta t} \cdot \frac{1}{n} \sum_{i=1}^{n} f(t_i)\, \Delta t$$

$$= \lim_{n \to \infty} \frac{n}{24} \cdot \frac{1}{n} \sum_{i=1}^{n} f(t_i)\, \Delta t \qquad \Delta t = \frac{24}{n}, \text{ or } \frac{1}{\Delta t} = \frac{n}{24}$$

$$= \frac{1}{24} \lim_{n \to \infty} \sum_{i=1}^{n} f(t_i)\, \Delta t$$

$$= \frac{1}{24} \int_{0}^{24} f(t)\, dt.$$

Definition

Let f be a continuous function over a closed interval $[a, b]$. Its **average value, y_{av},** over $[a, b]$ is given by

$$y_{av} = \frac{1}{b - a} \int_{a}^{b} f(x)\, dx.$$

Let's consider average value in another way. If we multiply on both sides of

$$y_{av} = \frac{1}{b - a} \int_{a}^{b} f(x)\, dx$$

by $b - a$, we get

$$(b - a) y_{av} = \int_{a}^{b} f(x)\, dx.$$

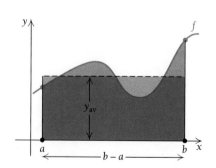

Now the expression on the left will give the area of a rectangle of length $b - a$ and height y_{av}. The area of such a rectangle is the same as the area under the graph of $y = f(x)$ over the interval $[a, b]$, as shown in the figure.

Example 7   Find the average value of $f(x) = x^2$ over the interval $[0, 2]$.

Graph $f(x) = x^4$. Compute the average value of the function over the interval $[0, 2]$, using the method of Example 7. Then use that value y_{av} and draw a graph of it as a horizontal line using the same set of axes. What does this line represent in comparison to the graph of $f(x) = x^4$?

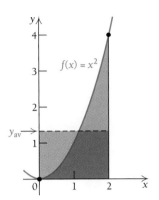

Solution   The average value is

$$\frac{1}{2 - 0}\int_0^2 x^2\, dx = \frac{1}{2}\left[\frac{x^3}{3}\right]_0^2$$

$$= \frac{1}{2}\left(\frac{2^3}{3} - \frac{0^3}{3}\right)$$

$$= \frac{1}{2}\cdot\frac{8}{3} = \frac{4}{3}, \quad \text{or} \quad 1\frac{1}{3}.$$

Note that although the values of $f(x)$ increase from 0 to 4 over $[0, 2]$, we would not expect the average value to be 2, because we see from the graph that $f(x)$ is less than 2 over more than half the interval.   ◁

Example 8   *Life Science: Engine Emissions.* The emissions of an engine are given by

$$E(t) = 2t^2,$$

where $E(t)$ is the engine's rate of emission, in billions of pollution particulates per year, at time t, in years. Find the average emissions from $t = 1$ to $t = 5$.

Solution   The average emissions are

$$\frac{1}{5 - 1}\int_1^5 2t^2\, dt = \frac{1}{4}\left[\frac{2}{3}t^3\right]_1^5$$

$$= \frac{1}{4}\cdot\frac{2}{3}(5^3 - 1^3)$$

$$= \frac{1}{6}(125 - 1)$$

$$= 20\frac{2}{3} \text{ billion pollution particulates per year.}$$ ◁

EXERCISE SET 5.3

1. a) Approximate

$$\int_1^7 \frac{dx}{x^2}$$

by computing the area of each rectangle to four decimal places and then adding.

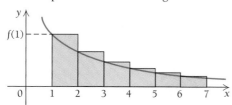

b) Approximate

$$\int_1^7 \frac{dx}{x^2}$$

by computing the area of each rectangle to four decimal places and then adding.

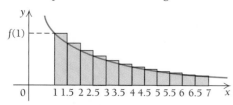

c) Evaluate

$$\int_1^7 \frac{dx}{x^2}.$$

Compare your answers to parts (a) and (b).

2. a) Approximate

$$\int_0^5 (x^2 + 1)\,dx$$

by computing the area of each rectangle and then adding.

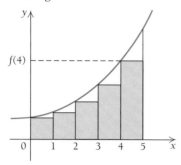

b) Approximate

$$\int_0^5 (x^2 + 1)\,dx$$

by computing the area of each rectangle and then adding.

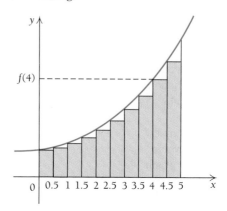

c) Evaluate

$$\int_0^5 (x^2 + 1)\,dx.$$

Compare your answers to parts (a) and (b).

In each case, give two interpretations of the shaded region other than as area.

3.

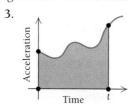

4.

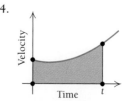

5.

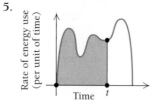

6.

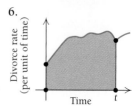

7.

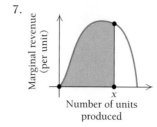

8.

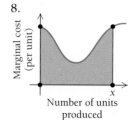

9.

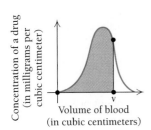

10.

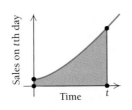

11.

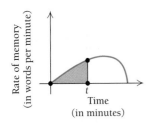

12.

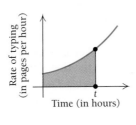

Find the average value over the given interval.

13. $y = 2x^3$; $[-1, 1]$

14. $y = 4 - x^2$; $[-2, 2]$

15. $y = e^x$; $[0, 1]$

16. $y = e^{-x}$; $[0, 1]$

17. $y = x^2 - x + 1$; $[0, 2]$

18. $y = x^2 + x - 2$; $[0, 4]$

19. $y = 3x + 1$; $[2, 6]$

20. $y = 4x + 1$; $[3, 7]$

21. $y = x^n$; $[0, 1]$

22. $y = x^n$; $[1, 2]$

Applications
BUSINESS AND ECONOMICS

23. *Accumulated Sales.* A company estimates that its sales will grow continuously at a rate given by the function

$$S'(t) = 20e^t,$$

where $S'(t)$ is the sales rate, in dollars per day, at time t, in days.

a) Find the accumulated sales for the first 5 days.
b) Find the sales from the 2nd day through the 5th day. (This is the integral from 1 to 5.)
c) On what day will accumulated sales exceed $20,000?

24. *Accumulated Sales.* Raggs, Ltd., estimates that its

sales will grow continuously at a rate given by the function

$$S'(t) = 10e^t,$$

where $S'(t)$ is the sales rate, in dollars per day, at time t, in days.

a) Find the accumulated sales for the first 5 days.
b) Find the sales from the 2nd day through the 5th day. (This is the integral from 1 to 5.)
c) On what day will accumulated sales exceed $40,000?

25. *Total Cost from Marginal Cost.* Raggs, Ltd., determines that the marginal cost per suit is given by

$$C'(x) = 0.0003x^2 - 0.2x + 50.$$

Ignoring fixed costs, find the total cost of producing the 101st suit through the 400th suit (that is, integrate from $x = 100$ to $x = 400$).

26. *Total Cost from Marginal Cost.* Using the information in Exercise 25, find the cost of producing the 201st suit through the 400th suit (that is, integrate from $x = 200$ to $x = 400$).

27. *Total Revenue from Marginal Revenue.* A fight promoter sells x tickets and has a marginal-revenue function given by

$$R'(x) = 200x - 1080.$$

This means that the rate of change of total revenue with respect to the number of tickets sold x is $R'(x)$. Find the total revenue from the sale of the 1001st ticket through the 1300th ticket.

28. *Total Profit from Marginal Profit.* A company has the marginal-profit function given by

$$P'(x) = -2x + 980.$$

This means that the rate of change of total profit with respect to the number of units x produced is $P'(x)$. Find the total profit from the production and sale of the 101st unit through the 800th unit of the product.

29. *Results of Practice.* A typist's speed over a 5-min interval is given by

$$W(t) = -6t^2 + 12t + 90,$$

t in $[0, 5]$,

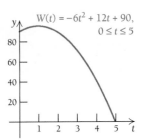

where $W(t)$ is the speed, in words per minute, at time t.

a) Find the speed at the beginning of the interval.
b) Find the maximum speed and when it occurs.
c) Find the average speed over the 5-min interval.

LIFE AND PHYSICAL SCIENCES

30. A particle starts out from the origin. Its velocity at time t is given by

$$v(t) = 3t^2 + 2t.$$

How far does it travel from the 2nd hour through the 5th hour (from $t = 1$ to $t = 5$)?

31. A particle starts out from the origin. Its velocity at time t is given by

$$v(t) = 4t^3 + 2t.$$

How far does it travel from the start through the 3rd hour (from $t = 0$ to $t = 3$)?

32. *Total Pollution.* A factory is polluting a lake in such a way that the rate of pollutants entering the lake at time t, in months, is given by

$$N'(t) = 280t^{3/2},$$

where N is the total number of pounds of pollutants in the lake at time t.

a) How many pounds of pollutants enter the lake in 16 months?
b) An environmental expert tells the factory that it will have to begin cleanup procedures after 50,000 lb of pollutants have entered the lake. After what amount of time will this occur?

33. *Bacteria Growth.* A population of bacteria in a biological experiment grows at the rate of

$$P'(t) = 1200e^{0.32t},$$

where $P(t)$ is the total population at time t, in days.

a) Find the increase in the bacteria population in the first 20 days.
b) The experiment will cease when the population reaches 4 million. After what amount of time will this occur?

34. *Average Drug Dosage.* The amount of a drug in the body at time t is given by

$$A(t) = 3e^{-0.1t},$$

where A is in cubic centimeters and t is the time, in hours.

a) What is the initial dosage of the drug?
b) What is the average amount in the body over the first 4 hr?

35. *Outside Temperature.* The temperature over a 10-hr period is given by

$$f(t) = -t^2 + 5t + 40, \quad 0 \le t \le 10.$$

a) Find the average temperature.
b) Find the minimum temperature.
c) Find the maximum temperature.

SOCIAL SCIENCES

Memorizing. The rate of memorizing information (say, in words per minute) initially increases with respect to time. Eventually, however, a maximum rate is reached, after which it begins to decrease.

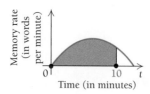

36. Suppose that in a certain memory experiment the rate of memorizing is given by

$$M'(t) = -0.009t^2 + 0.2t,$$

where $M'(t)$ is the memory rate, in words per minute. How many words are memorized in the first 10 min (from $t = 0$ to $t = 10$)?

37. Suppose that in a certain memory experiment the rate of memorizing is given by

$$M'(t) = -0.003t^2 + 0.2t,$$

where $M'(t)$ is the memory rate, in words per minute. How many words are memorized in the first 10 min (from $t = 0$ to $t = 10$)?

38. *Industrial Learning Curve.* A company is producing a new product. Due to the nature of the product, it is felt that the time required to produce each unit will decrease as the workers become more familiar with production procedure. It is determined that the function for the learning process is

$$T(x) = ax^b,$$

where $T(x)$ is the average time to produce x units, x is the number of units produced, a is the number of hours required to produce the first unit, and b is the slope of the learning curve.

a) Find an expression for the total time required to produce 100 units.
b) Suppose that $a = 100$ hr and $b = -0.322$. Find the total time required to produce 100 units.

39. *Average Population.* The population of the United States is given by

$$P(t) = 241e^{0.009t},$$

where P is in millions and t is the number of years since 1986. Find the average value of the population from 1986 to 2000.

40. *Average Population of a City.* The population of a city increased and then decreased over an 8-yr period according to the function

$$P(t) = -0.1t^2 + t + 3, \quad 0 \le t \le 8,$$

where P is in millions and t is the time.

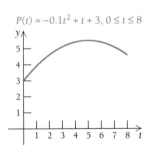

$P(t) = -0.1t^2 + t + 3, \ 0 \le t \le 8$

a) Find the average population.
b) Find the minimum population.
c) Find the maximum population.

41. *Results of Studying.* A student's score on a test is given by the function

$$S(t) = t^2, \quad t \text{ in } [0, 10],$$

where $S(t)$ is the score after t hours of study.

a) Find the maximum score that the student can achieve and the number of hours of study required to attain it.

b) Find the average score over the 10-hr interval.

A student's test score is a function of the time spent studying.

Synthesis

The Trapezoidal Rule. Another way to approximate an integral is to replace each rectangle in the sum (see Fig. 1) with a trapezoid (see Fig. 2). The area of a trapezoid is $h(c_1 + c_2)/2$, where c_1 and c_2 are the lengths of the parallel sides. Thus, in Fig. 2,

$$\int_a^b f(x)\,dx = \int_a^m f(x)\,dx + \int_m^b f(x)\,dx$$

$$\approx \Delta x \frac{f(a) + f(m)}{2} + \Delta x \frac{f(m) + f(b)}{2}$$

$$\approx \Delta x \left[\frac{f(a)}{2} + f(m) + \frac{f(b)}{2} \right].$$

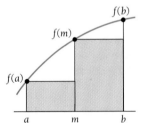

FIGURE 1

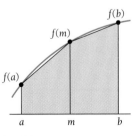

FIGURE 2

For an interval $[a, b]$ subdivided into n equal sub-intervals of length $\Delta x = (b - a)/n$, we get the approximation

$$\int_a^b f(x)\, dx \approx$$

$$\Delta x \left[\frac{f(a)}{2} + f(x_2) + f(x_3) + \cdots + f(x_n) + \frac{f(b)}{2} \right],$$

where $x_1 = a$ and

$$x_n = x_{n-1} + \Delta x \quad \text{or} \quad x_n = a + (n - 1)\, \Delta x.$$

This is called the **Trapezoidal Rule.**

42. Use the Trapezoidal Rule and the interval subdivision of Exercise 1(a) to approximate

$$\int_1^7 \frac{dx}{x^2}.$$

43. Use the Trapezoidal Rule and the interval subdivision of Exercise 2(a) to approximate

$$\int_0^5 (x^2 + 1)\, dx.$$

 TECHNOLOGY CONNECTION

Graph the function and evaluate the integral using a grapher.

44. $\displaystyle\int_0^2 (x^4 + 1)\, dx$ **45.** $\displaystyle\int_0^4 \sqrt{x}\, dx$

46. $\displaystyle\int_0^3 \frac{2}{\sqrt{x + 1}}\, dx$ **47.** $\displaystyle\int_2^4 \ln x\, dx$

5.4

Properties of Definite Integrals

Properties of Definite Integrals

The following properties of definite integrals can be derived rather easily from the definition of a definite integral and from the properties of the indefinite integral.

Property 1

$$\int_a^b k \cdot f(x)\, dx = k \cdot \int_a^b f(x)\, dx$$

The integral of a constant times a function is the constant times the integral of the function. That is, a constant can be "factored out" of the integrand.

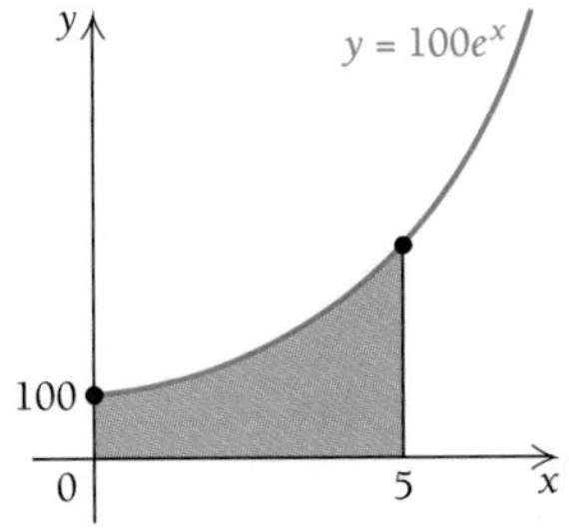

Example 1

$$\begin{aligned}\int_0^5 100e^x\,dx &= 100\int_0^5 e^x\,dx\\ &= 100[e^x]_0^5\\ &= 100(e^5 - e^0)\\ &= 100(e^5 - 1)\\ &\approx 14{,}741.32\end{aligned}$$

Property 2

$$\int_a^b [f(x) + g(x)]\,dx = \int_a^b f(x)\,dx + \int_a^b g(x)\,dx$$

The integral of a sum is the sum of the integrals.

Property 3

For $a < c < b$,

$$\int_a^b f(x)\,dx = \int_a^c f(x)\,dx + \int_c^b f(x)\,dx.$$

For any number c between a and b, the integral from a to b is the integral from a to c plus the integral from c to b.

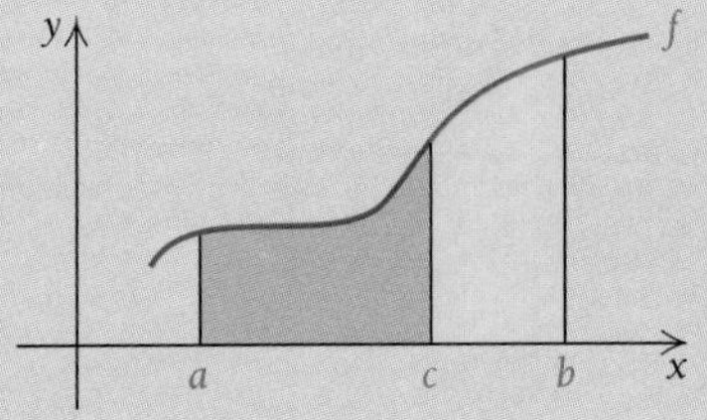

Property 3 has particular application when a function is defined piecewise in different ways over subintervals.

Example 2 Find the area under the graph of $y = f(x)$ from -4 to 5, where

$$f(x) = \begin{cases} 9, & \text{if } x < 3,\\ x^2, & \text{if } x \ge 3.\end{cases}$$

Solution

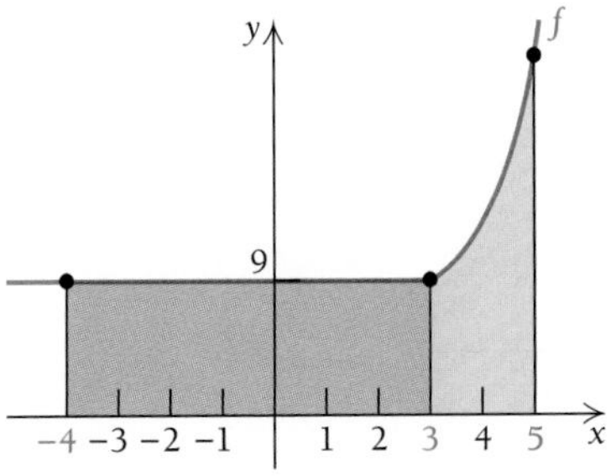

$$\begin{aligned}\int_{-4}^5 f(x)\,dx &= \int_{-4}^3 f(x)\,dx + \int_3^5 f(x)\,dx = \int_{-4}^3 9\,dx + \int_3^5 x^2\,dx\\ &= 9\int_{-4}^3 dx + \int_3^5 x^2\,dx = 9[x]_{-4}^3 + \left[\frac{x^3}{3}\right]_3^5\\ &= 9[3 - (-4)] + \left(\frac{5^3}{3} - \frac{3^3}{3}\right)\\ &= 95\tfrac{2}{3}\end{aligned}$$

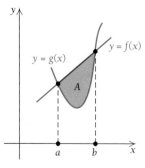

FIGURE 1

The Area of a Region Bounded by Two Graphs

Suppose that we want to find the area of a region bounded by the graphs of two functions, $y = f(x)$ and $y = g(x)$, as shown in Fig. 1.

Note that the area of the desired region A in Fig. 2(a) is that of A_2 in Fig. 2(b) minus that of A_1 in Fig. 2(c).

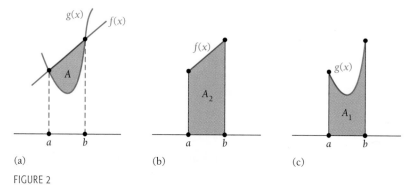

(a)   (b)   (c)

FIGURE 2

Thus,

$$A = \int_a^b f(x) \, dx - \int_a^b g(x) \, dx,$$

or

$$A = \int_a^b [f(x) - g(x)] \, dx.$$

In general, we have the following.

Theorem 5

Let f and g be continuous functions and suppose that $f(x) \geq g(x)$ over the interval $[a, b]$. Then the area of the region between the two curves, from $x = a$ to $x = b$, is

$$\int_a^b [f(x) - g(x)] \, dx.$$

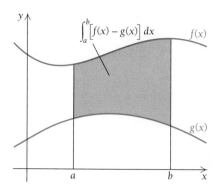

Example 3   Find the area of the region bounded by the graphs of $f(x) = 2x + 1$ and $g(x) = x^2 + 1$.

Solution

a) First, we make a reasonably accurate sketch, as in the following figure, to ensure that we have the right configuration. Note which is the *upper* graph. In this case, $2x + 1 \geq x^2 + 1$ over the interval $[0, 2]$.

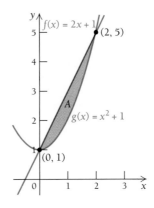

b) Second, if boundaries are not stated, we determine the first coordinates of possible points of intersection. Occasionally we can do this just by looking at the graph. If not, we can solve the system of equations as follows. At the points of intersection, $f(x) = g(x)$, so

$$x^2 + 1 = 2x + 1$$
$$x^2 - 2x = 0$$
$$x(x - 2) = 0$$
$$x = 0 \quad or \quad x = 2.$$

Thus the interval with which we are concerned is $[0, 2]$.

c) We compute the area as follows:

$$\int_0^2 [(2x + 1) - (x^2 + 1)] \, dx = \int_0^2 (2x - x^2) \, dx$$
$$= \left[x^2 - \frac{x^3}{3} \right]_0^2$$
$$= \left(2^2 - \frac{2^3}{3} \right) - \left(0^2 - \frac{0^3}{3} \right)$$
$$= 4 - \frac{8}{3}$$
$$= \frac{4}{3}.$$

TECHNOLOGY CONNECTION

To find the area bounded by two graphs, such as $y_1 = -2x - 7$ and $y_2 = -x^2 - 4$, we graph each function and use the **INTERSECT** feature from the **CALC** menu to determine the points of intersection.

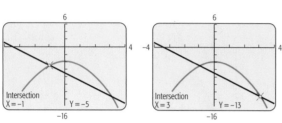

Next, we use the fnInt feature from the **MATH** menu.

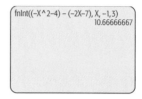

The area bounded by the two curves is 10.6.

Find the area of the region in Example 3 using a grapher.

Example 4   Find the area of the region bounded by

$$y = x^4 - 3x^3 - 4x^2 + 10, \qquad y = 40 - x^2, \qquad x = 1, \quad \text{and} \quad x = 3.$$

Solution

a) First, we make a reasonably accurate sketch, as in the following figure, to ensure that we have the correct configuration. Note that over $[1, 3]$, the upper graph is $y = 40 - x^2$. Thus, $40 - x^2 \geq x^4 - 3x^3 - 4x^2 + 10$ over $[1, 3]$.

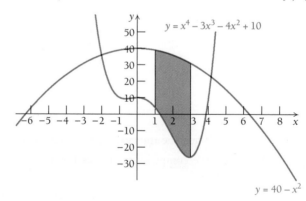

b) The limits of integration are stated, so we can compute the area as follows:

$$\int_1^3 [(40 - x^2) - (x^4 - 3x^3 - 4x^2 + 10)]\, dx = \int_1^3 (-x^4 + 3x^3 + 3x^2 + 30)\, dx$$

$$= \left[-\frac{x^5}{5} + \frac{3}{4}x^4 + x^3 + 30x \right]_1^3$$

$$= \left(-\frac{3^5}{5} + \frac{3}{4} \cdot 3^4 + 3^3 + 30 \cdot 3 \right)$$

$$- \left(-\frac{1^5}{5} + \frac{3}{4} \cdot 1^4 + 1^3 + 30 \cdot 1 \right)$$

$$= 97.6.$$

An Application

Example 5 *Life Science: Emission Control.* A clever college student develops an engine that is believed to meet federal standards for emission control. The engine's rate of emission is given by

$$E(t) = 2t^2,$$

where $E(t)$ is the emissions, in billions of pollution particulates per year, at time t, in years. The emission rate of a conventional engine is given by

$$C(t) = 9 + t^2.$$

The graphs of both curves are shown below.

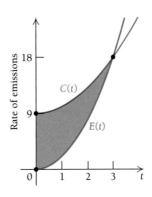

a) At what point in time will the emission rates be the same?

b) What is the reduction in emissions resulting from using the student's engine?

Solution

a) The rate of emission will be the same when $E(t) = C(t)$, or

$$2t^2 = 9 + t^2$$
$$t^2 - 9 = 0$$
$$(t - 3)(t + 3) = 0$$
$$t = 3 \quad or \quad t = -3.$$

Since negative time has no meaning in this problem, the emission rates will be the same when $t = 3$ yr.

b) The reduction in emissions is represented by the area of the shaded region in the figure above. It is the area between $C(t) = 9 + t^2$ and $E(t) = 2t^2$, from $t = 0$ to $t = 3$, and is computed as follows:

$$\int_0^3 [(9 + t^2) - 2t^2]\, dt = \int_0^3 (9 - t^2)\, dt = \left[9t - \frac{t^3}{3} \right]_0^3$$

$$= \left(9 \cdot 3 - \frac{3^3}{3} \right) - \left(9 \cdot 0 - \frac{0^3}{3} \right)$$

$$= 27 - 9$$

$$= 18 \text{ billion pollution particulates per year.}$$

EXERCISE SET 5.4

Find the area of the region bounded by the given graphs.

1. $y = x, y = x^3, x = 0, x = 1$
2. $y = x, y = x^4$
3. $y = x + 2, y = x^2$
4. $y = x^2 - 2x, y = x$
5. $y = 6x - x^2, y = x$
6. $y = x^2 - 6x, y = -x$
7. $y = 2x - x^2, y = -x$
8. $y = x^2, y = \sqrt{x}$
9. $y = x, y = \sqrt{x}$
10. $y = 3, y = x, x = 0$
11. $y = 5, y = \sqrt{x}, x = 0$
12. $y = x^2, y = x^3$
13. $y = 4 - x^2, y = 4 - 4x$
14. $y = x^2 + 1, y = x^2, x = 1, x = 3$
15. $y = x^2 + 3, y = x^2, x = 1, x = 2$

Find the area of the shaded region.

16. $f(x) = 2x + x^2 - x^3, \quad g(x) = 0$

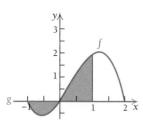

17. $f(x) = x^3 + 3x^2 - 9x - 12, \quad g(x) = 4x + 3$

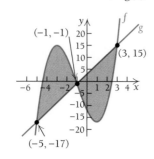

18. $f(x) = x^4 - 8x^3 + 18x^2, \quad g(x) = x + 28$

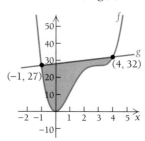

19. $f(x) = 4x - x^2, \quad g(x) = x^2 - 6x + 8$

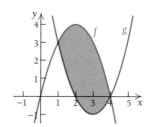

Find the area under the graph over the interval $[-2, 3]$, where:

20. $f(x) = \begin{cases} x^2, & \text{if } x < 1, \\ 1, & \text{if } x \geq 1. \end{cases}$

21. $f(x) = \begin{cases} 4 - x^2, & \text{if } x < 0, \\ 4, & \text{if } x \geq 0. \end{cases}$

Applications

BUSINESS AND ECONOMICS

22. *Total Profit.* A company determines that its marginal revenue per day is given by

$$R'(t) = 100e^t, \quad R(0) = 0,$$

where $R(t)$ is the revenue, in dollars, on the tth day. The company's marginal cost per day is given by

$$C'(t) = 100 - 0.2t, \quad C(0) = 0,$$

where $C(t)$ is the cost, in dollars, on the tth day. Find the total profit from $t = 0$ to $t = 10$ (the first 10 days). *Note*:

$$P(T) = R(T) - C(T) = \int_0^T [R'(t) - C'(t)] \, dt.$$

SOCIAL SCIENCES

23. *Memorizing.* In a certain memory experiment, subject A is able to memorize words at the rate given by

$$m'(t) = -0.009t^2 + 0.2t \quad \text{(words per minute)}.$$

In the same memory experiment, subject B is able to memorize at the rate given by

$$M'(t) = -0.003t^2 + 0.2t \quad \text{(words per minute)}.$$

a) Which subject has the higher rate of memorization?

b) How many more words does that subject memorize from $t = 0$ to $t = 10$ (during the first 10 min)?

Synthesis

Find the area of the region bounded by the given graphs.

24. $y = x^2$, $y = x^{-2}$, $x = 1$, $x = 5$

25. $y = e^x$, $y = e^{-x}$, $x = 0$, $x = 1$

26. $y = x^2$, $y = \sqrt[3]{x^2}$, $x = 1$, $x = 8$

27. $y = x^2$, $y = x^3$, $x = -1$, $x = 1$

28. $x + 2y = 2$, $y - x = 1$, $2x + y = 7$

29. Find the area of the region bounded by $y = 3x^5 - 20x^3$, the x-axis, and the first coordinates of the relative maximum and minimum values of the function.

30. Find the area of the region bounded by $y = x^3 - 3x + 2$, the x-axis, and the first coordinates of the relative maximum and minimum values of the function.

31. *Life Science: Poiseuille's Law.* The flow of blood in a blood vessel is faster toward the center of the vessel and slower toward the outside. The speed of the blood is given by

$$V = \frac{p}{4Lv}(R^2 - r^2),$$

where R is the radius of the blood vessel, r is the distance of the blood from the center of the vessel, and p, v, and L are physical constants related to the

pressure and viscosity of the blood and the length of the blood vessel. If R is constant, we can think of V as a function of r:

$$V(r) = \frac{p}{4Lv}(R^2 - r^2).$$

The *total blood flow* Q is given by

$$Q = \int_0^R 2\pi \cdot V(r) \cdot r \cdot dr.$$

Find Q.

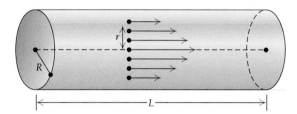

 TECHNOLOGY CONNECTION

Find the area of the region bounded by the given graphs.

32. $y = x + 6$, $y = -2x$, $y = x^3$

33. $y = x^2 + 4x$, $y = \sqrt{16 - x^2}$

34. $y = x\sqrt{4 - x^2}$, $y = \dfrac{-4x}{x^2 + 1}$, $x = 0$, $x = 2$

35. $y = 2x^2 + x - 4$, $y = 1 - x + 8x^2 - 4x^4$

36. $y = \sqrt{1 - x^2}$, $y = 1 - x^2$, $x = -1$, $x = 1$

37. Consider the following functions:

$$f(x) = 3.8x^5 - 18.6x^3,$$
$$g(x) = 19x^4 - 55.8x^2.$$

a) Graph these functions in the window $[-4, 4, -70, 70]$, with $\text{Yscl} = 10$.

b) Estimate the first coordinates a, b, and c of the three points of intersection of the two graphs.

c) Find the area between the curves on the interval $[a, b]$.

d) Find the area between the curves on the interval $[b, c]$.

5.5

➤ Evaluate integrals using substitution.
➤ Solve applied problems involving integration by substitution.

Integration Techniques: Substitution

The following formulas provide a basis for an integration technique called *substitution*.

A. $\int u^r \, du = \dfrac{u^{r+1}}{r+1} + C,$ provided $r \neq -1$

B. $\int e^u \, du = e^u + C$

C. $\int \dfrac{1}{u} \, du = \ln |u| + C;$ or

$\int \dfrac{1}{u} \, du = \ln u + C, \quad u > 0$

(We will generally consider $u > 0$.)

Recall the Leibniz notation, dy/dx, for a derivative. We gave specific definitions of the differentials dy and dx in Section 3.6. Recall that

$$\frac{dy}{dx} = f'(x)$$

and

$$dy = f'(x) \, dx.$$

We will make extensive use of this notation in this section.

Example 1 For $y = f(x) = x^3$, find dy.

Solution We have

$$\frac{dy}{dx} = f'(x) = 3x^2,$$

so

$$dy = f'(x) \, dx = 3x^2 \, dx.$$

Example 2 For $u = g(x) = \ln x$, find du.

Solution We have

$$\frac{du}{dx} = g'(x) = \frac{1}{x},$$

so

$$du = g'(x) \, dx = \frac{1}{x} \, dx, \quad \text{or} \quad \frac{dx}{x}.$$

Example 3   For $y = f(x) = e^{x^2}$, find dy.

Solution   Using the Chain Rule, we have

$$\frac{dy}{dx} = f'(x) = 2xe^{x^2},$$

so

$$dy = f'(x)\ dx = 2xe^{x^2}\ dx. \qquad \blacktriangleleft$$

So far the dx in

$$\int f(x)\ dx$$

has played no role in integrating other than to indicate the variable of integration. Now it will be convenient to make use of dx. Consider the integral

$$\int 2xe^{x^2}\ dx.$$

At first look, it would seem that the integral could not be evaluated at this point using the rules we know so far. But, after an appropriate substitution, we can evaluate the integral. If we set

$$u = x^2,$$

then

$$\frac{du}{dx} = 2x,$$

so

$$du = 2x\ dx.$$

If we substitute u for x^2 and du for $2x\ dx$, the integral takes on the form

$$\int e^u\ du.$$

Since

$$\int e^u\ du = e^u + C,$$

it follows that

$$\int 2xe^{x^2}\ dx = \int e^u\ du$$
$$= e^u + C$$
$$= e^{x^2} + C.$$

In effect, we have used the Chain Rule in reverse. We can check the result by differentiating. The procedure is referred to as *substitution*, or *change of variable*. It is a *trial-and-error* procedure that you will become more proficient with after much practice. If you try a substitution that doesn't result in an integrand that can be easily integrated, try another. There are many integrations that cannot be carried out using substitution. We do know that any integrations that fit rules A, B, or C on p. 408 can be done with substitution.

Let's consider some additional examples.

Example 4 Evaluate: $\displaystyle\int \frac{2x\ dx}{1 + x^2}$.

Solution

$$\int \frac{2x\ dx}{1 + x^2} = \int \frac{du}{u} \qquad \underline{\text{Substitution}} \qquad \boxed{\text{Let } u = 1 + x^2; \\ \text{then } du = 2x\ dx.}$$

$$= \ln u + C$$

$$= \ln (1 + x^2) + C$$

Example 5 Evaluate: $\displaystyle\int \frac{2x\ dx}{(1 + x^2)^2}$.

Solution

$$\int \frac{2x\ dx}{(1 + x^2)^2} = \int \frac{du}{u^2} \qquad \underline{\text{Substitution}} \qquad \boxed{u = 1 + x^2, \\ du = 2x\ dx}$$

$$= \int u^{-2}\ du$$

$$= -u^{-1} + C$$

$$= -\frac{1}{u} + C$$

$$= \frac{-1}{1 + x^2} + C$$

Example 6 Evaluate: $\displaystyle\int \frac{\ln 3x\ dx}{x}$.

Solution

$$\int \frac{\ln 3x\ dx}{x} = \int u\ du \qquad \underline{\text{Substitution}} \qquad \boxed{u = \ln 3x, \\ du = \frac{1}{x}\ dx}$$

$$= \frac{u^2}{2} + C$$

$$= \frac{(\ln 3x)^2}{2} + C$$

Example 7 Evaluate: $\displaystyle\int xe^{x^2}\ dx$.

Solution   Suppose we try

$$u = x^2;$$

then we have

$$du = 2x\, dx.$$

We don't have $2x\, dx$ in $\int xe^{x^2}\, dx$. We have $x\, dx$ and need to supply a 2. We do so by multiplying by 1, using $\frac{1}{2} \cdot 2$, to obtain

$$\frac{1}{2} \cdot 2 \cdot \int xe^{x^2}\, dx = \frac{1}{2} \int 2xe^{x^2}\, dx$$

$$= \frac{1}{2} \int e^{x^2}(2x\, dx) = \frac{1}{2} \int e^u\, du$$

$$= \frac{1}{2}e^u + C = \frac{1}{2}e^{x^2} + C.$$

Example 8   Evaluate: $\displaystyle\int \frac{dx}{x + 3}$.

Solution

$$\int \frac{dx}{x + 3} = \int \frac{du}{u} \qquad \underline{\text{Substitution}} \quad \boxed{\begin{array}{l} u = x + 3, \\ du = dx \end{array}}$$

$$= \ln u + C$$

$$= \ln (x + 3) + C$$

With practice, you will be able to make certain substitutions mentally and just write down the answer. Example 8 is a good illustration of this.

Example 9   Evaluate: $\displaystyle\int x^2(x^3 + 1)^{10}\, dx$.

Solution

$$\int x^2(x^3 + 1)^{10}\, dx = \frac{1}{3} \int (x^3 + 1)^{10}(3x^2\, dx) \qquad \underline{\text{Substitution}} \quad \boxed{\begin{array}{l} u = x^3 + 1, \\ du = 3x^2\, dx \end{array}}$$

$$= \frac{1}{3} \int u^{10}\, du$$

$$= \frac{1}{3} \cdot \frac{u^{11}}{11} + C$$

$$= \frac{1}{33}(x^3 + 1)^{11} + C$$

Example 10   Evaluate: $\displaystyle\int_0^1 x^2(x^3 + 1)^{10}\, dx$.

Solution

a) First we find the indefinite integral (shown in Example 9).

Exercise

1. Use a grapher to evaluate

$$\int_0^1 x^2(x^3 + 1)^{10}\, dx.$$

b) Then we evaluate the definite integral over $[0, 1]$:

$$\int_0^1 x^2(x^3 + 1)^{10}\, dx = \left[\frac{1}{33}(x^3 + 1)^{11} \right]_0^1$$

$$= \frac{1}{33}[(1^3 + 1)^{11} - (0^3 + 1)^{11}]$$

$$= \frac{1}{33}(2^{11} - 1^{11})$$

$$= \frac{2^{11} - 1}{33}.$$

EXERCISE SET 5.5

Evaluate. (Be sure to check by differentiating!)

1. $\displaystyle\int \frac{3x^2\, dx}{7 + x^3}$

2. $\displaystyle\int \frac{3x^2\, dx}{1 + x^3}$

3. $\displaystyle\int e^{4x}\, dx$

4. $\displaystyle\int e^{3x}\, dx$

5. $\displaystyle\int e^{x/2}\, dx$

6. $\displaystyle\int e^{x/3}\, dx$

7. $\displaystyle\int x^3 e^{x^4}\, dx$

8. $\displaystyle\int x^4 e^{x^5}\, dx$

9. $\displaystyle\int t^2 e^{-t^3}\, dt$

10. $\displaystyle\int t e^{-t^2}\, dt$

11. $\displaystyle\int \frac{\ln 4x\, dx}{x}$

12. $\displaystyle\int \frac{\ln 5x\, dx}{x}$

13. $\displaystyle\int \frac{dx}{1 + x}$

14. $\displaystyle\int \frac{dx}{5 + x}$

15. $\displaystyle\int \frac{dx}{4 - x}$

16. $\displaystyle\int \frac{dx}{1 - x}$

17. $\displaystyle\int t^2(t^3 - 1)^7\, dt$

18. $\displaystyle\int t(t^2 - 1)^5\, dt$

19. $\displaystyle\int (x^4 + x^3 + x^2)^7(4x^3 + 3x^2 + 2x)\, dx$

20. $\displaystyle\int (x^3 - x^2 - x)^9(3x^2 - 2x - 1)\, dx$

21. $\displaystyle\int \frac{e^x\, dx}{4 + e^x}$

22. $\displaystyle\int \frac{e^t\, dt}{3 + e^t}$

23. $\displaystyle\int \frac{\ln x^2}{x}\, dx$

24. $\displaystyle\int \frac{(\ln x)^2}{x}\, dx$

25. $\displaystyle\int \frac{dx}{x \ln x}$

26. $\displaystyle\int \frac{dx}{x \ln x^2}$

27. $\displaystyle\int \sqrt{ax + b}\, dx$

28. $\displaystyle\int x\sqrt{ax^2 + b}\, dx$

29. $\displaystyle\int be^{ax}\, dx$

30. $\displaystyle\int P_0 e^{kt}\, dt$

31. $\displaystyle\int \frac{3x^2\, dx}{(1 + x^3)^5}$

32. $\displaystyle\int \frac{x^3\, dx}{(2 - x^4)^7}$

33. $\displaystyle\int 7x \sqrt[3]{4 - x^2}\, dx$

34. $\displaystyle\int 12x \sqrt[5]{1 + 6x^2}\, dx$

Evaluate.

35. $\displaystyle\int_0^1 2x e^{x^2}\, dx$

36. $\displaystyle\int_0^1 3x^2 e^{x^3}\, dx$

37. $\displaystyle\int_0^1 x(x^2 + 1)^5\, dx$

38. $\displaystyle\int_1^2 x(x^2 - 1)^7\, dx$

39. $\displaystyle\int_1^3 \frac{dt}{1 + t}$

40. $\displaystyle\int_1^3 e^{2x}\, dx$

41. $\displaystyle\int_1^4 \frac{2x + 1}{x^2 + x - 1}\, dx$

42. $\displaystyle\int_1^3 \frac{2x + 3}{x^2 + 3x}\, dx$

43. $\displaystyle\int_0^b e^{-x}\, dx$

44. $\displaystyle\int_0^b 2e^{-2x}\, dx$

45. $\displaystyle\int_0^b me^{-mx}\, dx$

46. $\displaystyle\int_0^b ke^{-kx}\, dx$

47. $\displaystyle\int_0^4 (x-6)^2\, dx$

48. $\displaystyle\int_0^3 (x-5)^2\, dx$

49. $\displaystyle\int_0^2 \frac{3x^2\, dx}{(1+x^3)^5}$

50. $\displaystyle\int_{-1}^0 \frac{x^3\, dx}{(2-x^4)^7}$

51. $\displaystyle\int_0^{\sqrt{7}} 7x\sqrt[3]{1+x^2}\, dx$

52. $\displaystyle\int_0^1 12x\sqrt[5]{1-x^2}\, dx$

 53. Use a grapher to check the results of any of Exercises 35–52.

Applications ....

BUSINESS AND ECONOMICS

54. *Demand from Marginal Demand.* A firm has the marginal-demand function

$$D'(p) = \frac{-2000p}{\sqrt{25-p^2}}.$$

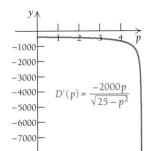

Find the demand function given that $D = 13,000$ when $p = \$3$ per unit.

55. *Value of an Investment.* A company buys a new machine for \$250,000. The marginal revenue from the sale of products produced by the machine is projected to be

$$R'(t) = 4000t.$$

The salvage value of the machine decreases at the rate of

$$V'(t) = 25,000e^{-0.1t}.$$

The total profit from the machine after T years is

given by

$$P(T) = \begin{pmatrix} \text{Revenue} \\ \text{from} \\ \text{sale of} \\ \text{product} \end{pmatrix} + \begin{pmatrix} \text{Revenue} \\ \text{from} \\ \text{sale of} \\ \text{machine} \end{pmatrix} - \begin{pmatrix} \text{Cost} \\ \text{of} \\ \text{machine} \end{pmatrix}$$

$$= \int_0^T R'(t)\, dt + \int_0^T V'(t)\, dt - \$250,000.$$

a) Find $P(T)$.
b) Find $P(10)$.

56. *Profit from Marginal Profit.* A firm has the marginal-profit function

$$\frac{dP}{dx} = \frac{9000-3000x}{(x^2-6x+10)^2}.$$

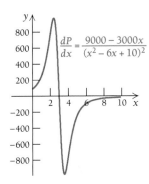

Find the total-profit function given that $P = \$1500$ at $x = 3$.

SOCIAL SCIENCES

57. *Divorce.* The divorce rate in the United States is approximated by

$$D(t) = 100,000e^{0.025t},$$

where $D(t)$ is the number of divorces occurring at time t and t is the number of years measured from 1900. That is, $t = 0$ corresponds to 1900, $t = 98\frac{9}{365}$ corresponds to January 9, 1998, and so on.

a) Find the total number of divorces from 1900 to 1999. Note that this is

$$\int_0^{99} D(t)\, dt.$$

b) Find the total number of divorces from 1980 to 1999. Note that this is

$$\int_{80}^{99} D(t)\, dt.$$

Synthesis

Find the area of the shaded region.

58.

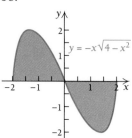

$y = -x\sqrt{4 - x^2}$

59.

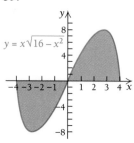

$y = x\sqrt{16 - x^2}$

Evaluate.

60. $\int \dfrac{dx}{ax + b}$

61. $\int 5x\sqrt{1 - 4x^2}\, dx$

62. $\int \dfrac{e^{\sqrt{t}}}{\sqrt{t}}\, dt$

63. $\int \dfrac{x^2}{e^{x^3}}\, dx$

64. $\int \dfrac{(\ln x)^{99}}{x}\, dx$

65. $\int \dfrac{e^{1/t}}{t^2}\, dt$

66. $\int (e^t + 2)e^t\, dt$

67. $\int \dfrac{dx}{x(\ln x)^4}$

68. $\int \dfrac{t^2}{\sqrt[4]{2 + t^3}}\, dt$

69. $\int x^2\sqrt{x^3 + 1}\, dx$

70. $\int \dfrac{[(\ln x)^2 + 3(\ln x) + 4]}{x}\, dx$

71. $\int \dfrac{x - 3}{(x^2 - 6x)^{1/3}}\, dx$

72. $\int \dfrac{t^3 \ln (t^4 + 8)}{t^4 + 8}\, dt$

73. $\int \dfrac{t^2 + 2t}{(t + 1)^2}\, dt$

$$\left(\text{Hint: } \dfrac{t^2 + 2t}{(t + 1)^2} = \dfrac{t^2 + 2t + 1 - 1}{t^2 + 2t + 1}\right.$$
$$\left. = 1 - \dfrac{1}{(t + 1)^2}.\right)$$

74. $\int \dfrac{x^2 + 6x}{(x + 3)^2}\, dx$

75. $\int \dfrac{x + 3}{x + 1}\, dx$

$$\left(\text{Hint: } \dfrac{x + 3}{x + 1} = 1 + \dfrac{2}{x + 1}.\right)$$

76. $\int \dfrac{t - 5}{t - 4}\, dt$

77. $\int \dfrac{dx}{x(\ln x)^n}$

78. $\int \dfrac{dx}{e^x + 1}$

$$\left(\text{Hint: } \dfrac{1}{e^x + 1} = \dfrac{e^{-x}}{1 + e^{-x}}.\right)$$

79. $\int \dfrac{e^x - e^{-x}}{e^x + e^{-x}}\, dx$

80. $\int \dfrac{(\ln x)^n}{x}\, dx$

81. $\int \dfrac{dx}{x \ln x\, [\ln (\ln x)]}$

82. $\int \dfrac{e^{-mx}}{1 + ae^{-mx}}\, dx$

83. $\int 9x(7x^2 + 9)^n\, dx$

84. $\int 5x^2(2x^3 - 7)^n\, dx$

tw 85. Determine whether the following is a theorem:

$$\int [f(x)]^2\, dx = 2\int f(x)\, dx.$$

5.6

Integration Techniques: Integration by Parts

Recall the Product Rule for derivatives:

$$\dfrac{d}{dx}uv = u\dfrac{dv}{dx} + v\dfrac{du}{dx}.$$

Integrating both sides with respect to x, we get

$$uv = \int u \frac{dv}{dx}\, dx + \int v \frac{du}{dx}\, dx$$

$$= \int u\, dv + \int v\, du.$$

Solving for $\int u\, dv$, we get the following.

Theorem 6

The Integration-by-Parts Formula

$$\int u\, dv = uv - \int v\, du$$

This equation can be used as a formula for integrating in certain situations—that is, situations in which an integrand is a product of two functions, and one of the functions can be integrated using the techniques we have already developed. For example,

$$\int xe^x\, dx$$

can be considered as

$$\int x(e^x\, dx) = \int u\, dv,$$

where we let

$$u = x \quad \text{and} \quad dv = e^x\, dx.$$

If so, we have

$$du = dx, \quad \text{by differentiating}$$

and

$$v = e^x, \quad \text{by integrating and using the simplest antiderivative.}$$

Then the Integration-by-Parts Formula gives us

$$\int \overset{u}{(x)}\overset{dv}{(e^x\, dx)} = \overset{u}{(x)}\overset{v}{(e^x)} - \int \overset{v}{(e^x)}\overset{du}{(dx)}$$

$$= xe^x - e^x + C.$$

This method of integrating is called **integration by parts.**

Note that integration by parts, like substitution, is a trial-and-error process. In the preceding example, suppose that we had reversed the roles of x and e^x. We would have obtained

$$u = e^x, \qquad dv = x\, dx,$$

$$du = e^x\, dx, \qquad v = \frac{x^2}{2},$$

and

$$\int (e^x)(x\, dx) = (e^x)\left(\frac{x^2}{2}\right) - \int \left(\frac{x^2}{2}\right)(e^x\, dx).$$

Now the integrand on the right is more difficult to integrate than the one with which we began. When we can integrate *both* factors of an integrand, and thus have a choice as to how to apply the Integration-by-Parts Formula, it can happen that only one (and maybe none) of the possibilities will work.

Tips on Using Integration by Parts

1. If you have tried substitution, and have had no success, then try integration by parts.
2. Use integration by parts when an integral is of the form

$$\int f(x)\, g(x)\, dx.$$

Then match it with an integral of the form

$$\int u\, dv$$

by choosing a function to be $u = f(x)$, where $f(x)$ can be differentiated, and the remaining factor to be $dv = g(x)\, dx$, where $g(x)$ can be integrated.
3. Find du by differentiating and v by integrating.
4. If the resulting integral is more complicated than the original, make some other choice for $u = f(x)$ and $dv = g(x)\, dx$.

Let's consider some additional examples.

Example 1   Evaluate: $\int \ln x\, dx$.

Solution   Note that $\int (dx/x) = \ln x + C$, but we do not yet know how to find $\int \ln x\, dx$ since we do not know how to find an antiderivative of $\ln x$. Since we can differentiate $\ln x$, we let

$$u = \ln x \quad \text{and} \quad dv = dx.$$

Then

$$du = \frac{1}{x}\, dx \quad \text{and} \quad v = x.$$

Using the Integration-by-Parts Formula gives

$$\underset{u}{} \quad \underset{dv}{} \qquad \underset{u}{} \; \underset{v}{} \qquad \underset{v}{} \quad \underset{du}{}$$

$$\int (\ln x)(dx) = (\ln x)x - \int x\left(\frac{1}{x}\,dx\right)$$

$$= x \ln x - \int dx$$

$$= x \ln x - x + C.$$

Example 2   Evaluate: $\int x \ln x \, dx$.

Solution   Let's examine several choices, as follows.

   Choice 1:   We let

   $$u = 1 \quad \text{and} \quad dv = x \ln x \, dx.$$

This will not work because we are back to our original integral, in which we do not know how to integrate $dv = x \ln x \, dx$.

   Choice 2:   We let

   $$u = x \ln x \qquad\qquad\qquad \text{and} \quad dv = dx.$$

Then

$$du = \left[x\left(\frac{1}{x}\right) + 1(\ln x)\right] dx \quad \text{and} \quad v = x.$$
$$= (1 + \ln x) \, dx$$

Using the Integration-by-Parts Formula, we have

$$\int u \, dv = uv - \int v \, du$$

$$= (x \ln x)x - \int x(1 + \ln x) \, dx$$

$$= x^2 \ln x - \int (x + x \ln x) \, dx.$$

This integral seems more complicated than the original.

   Choice 3:   We let

   $$u = \ln x \quad \text{and} \quad dv = x \, dx.$$

Then

$$du = \frac{1}{x} \, dx \quad \text{and} \quad v = \frac{x^2}{2}.$$

Using the Integration-by-Parts Formula, we have

$$\int u \, dv = uv - \int v \, du = (\ln x)\frac{x^2}{2} - \int \frac{x^2}{2} \cdot \frac{1}{x} \, dx$$

$$= \frac{x^2}{2} \ln x - \int \frac{x}{2} \, dx$$

$$= \frac{x^2}{2} \ln x - \frac{x^2}{4} + C.$$

This choice allows us to evaluate the integral.

Example 3   Evaluate: $\int x\sqrt{x + 1} \, dx$.

Solution   We let

$$u = x \quad \text{and} \quad dv = (x + 1)^{1/2} \, dx.$$

Then

$$du = dx \quad \text{and} \quad v = \tfrac{2}{3}(x + 1)^{3/2}.$$

Note that we had to use substitution in order to integrate dv. We see this as follows:

$$\int (x + 1)^{1/2} \, dx = \int w^{1/2} \, dw \quad \underline{\text{Substitution}} \quad \boxed{\begin{array}{l} w = x + 1, \\ dw = dx \end{array}}$$

$$= \frac{w^{1/2+1}}{\tfrac{1}{2} + 1} = \tfrac{2}{3}w^{3/2} = \tfrac{2}{3}(x + 1)^{3/2}.$$

Using the Integration-by-Parts Formula gives us

$$\int x\sqrt{x + 1} \, dx = x \cdot \tfrac{2}{3}(x + 1)^{3/2} - \int \tfrac{2}{3}(x + 1)^{3/2} \, dx$$

$$= \tfrac{2}{3}x(x + 1)^{3/2} - \tfrac{2}{3} \cdot \tfrac{2}{5}(x + 1)^{5/2} + C$$

$$= \tfrac{2}{3}x(x + 1)^{3/2} - \tfrac{4}{15}(x + 1)^{5/2} + C.$$

Example 4   Evaluate: $\int_1^2 \ln x \, dx$.

Solution

a) First we find the indefinite integral (Example 1).

b) Then we evaluate the definite integral:

$$\int_1^2 \ln x \, dx = [x \ln x - x]_1^2$$

$$= (2 \ln 2 - 2) - (1 \cdot \ln 1 - 1)$$

$$= 2 \ln 2 - 2 + 1$$

$$= 2 \ln 2 - 1.$$

Repeated Integration by Parts

In some cases, we may need to apply the Integration-by-Parts Formula more than once.

**TECHNOLOGY
CONNECTION**

Exercise

1. Use a grapher to
    evaluate

$$\int_0^7 x^2 e^{-x}\, dx.$$

Example 5   Evaluate $\int_0^7 x^2 e^{-x}\, dx$ to find the area of the shaded region below.

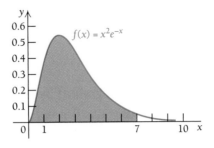

Solution

a) We let

$$u = x^2 \qquad \text{and} \quad dv = e^{-x}\, dx.$$

Then

$$du = 2x\, dx \quad \text{and} \quad v = -e^{-x}.$$

Using the Integration-by-Parts Formula gives

$$\int u\, dv = uv - \int v\, du$$

$$= x^2(-e^{-x}) - \int -e^{-x}(2x\, dx)$$

$$= -x^2 e^{-x} + \int 2x e^{-x}\, dx.$$

To evaluate the integral on the right, we can apply integration by parts again, as
follows. We let

$$u = 2x \qquad \text{and} \quad dv = e^{-x}\, dx.$$

Then

$$du = 2\, dx \quad \text{and} \quad v = -e^{-x}.$$

Using the Integration-by-Parts Formula once again, we get

$$\int u\, dv = uv - \int v\, du$$

$$= 2x(-e^{-x}) - \int -e^{-x}(2\, dx)$$

$$= -2x e^{-x} - 2e^{-x} + C.$$

Thus the original integral becomes

$$\int x^2 e^{-x}\, dx = -x^2 e^{-x} + \int 2x e^{-x}\, dx$$

$$= -x^2 e^{-x} - 2x e^{-x} - 2e^{-x} + C$$

$$= -e^{-x}(x^2 + 2x + 2) + C.$$

b) We now evaluate the definite integral:

$$\int_0^7 x^2 e^{-x}\, dx = [-e^{-x}(x^2 + 2x + 2)]_0^7$$

$$= [-e^{-7}(7^2 + 2(7) + 2)] - [-e^{-0}(0^2 + 2(0) + 2)]$$

$$= -65e^{-7} + 2 \approx 1.94.$$

Tabular Integration by Parts

In situations like that in Example 5, we have an integral

$$\int f(x)\, g(x)\, dx,$$

where $f(x)$ can be repeatedly differentiated easily to a derivative that is eventually 0. The function $g(x)$ can also be repeatedly integrated easily. In such cases, we can use integration by parts more than once to evaluate the integral. As we saw in Example 5, the procedures can get complicated. When this happens, we can use **tabular integration,** which is illustrated in Example 6.

Example 6   Evaluate: $\int x^3 e^x\, dx$.

Solution   We let $f(x) = x^3$ and $g(x) = e^x$. Then we make a tabulation as follows.

$f(x)$ and Repeated Derivatives		$g(x)$ and Repeated Integrals
x^3	$(+)$	e^x
$3x^2$	$(-)$	e^x
$6x$	$(+)$	e^x
6	$(-)$	e^x
0		e^x

We then add products along the arrows, making the alternating sign changes to obtain

$$\int x^3 e^x\, dx = x^3 e^x - 3x^2 e^x + 6xe^x - 6e^x + C.$$

EXERCISE SET 5.6

Evaluate using integration by parts. Check by differentiating.

1. $\displaystyle\int 5xe^{5x}\,dx$

2. $\displaystyle\int 2xe^{2x}\,dx$

3. $\displaystyle\int x^3(3x^2\,dx)$

4. $\displaystyle\int x^2(2x\,dx)$

5. $\displaystyle\int xe^{2x}\,dx$

6. $\displaystyle\int xe^{3x}\,dx$

7. $\displaystyle\int xe^{-2x}\,dx$

8. $\displaystyle\int xe^{-x}\,dx$

9. $\displaystyle\int x^2\ln x\,dx$

10. $\displaystyle\int x^3\ln x\,dx$

11. $\displaystyle\int x\ln x^2\,dx$

12. $\displaystyle\int x^2\ln x^3\,dx$

13. $\displaystyle\int \ln(x+3)\,dx$

14. $\displaystyle\int \ln(x+1)\,dx$

15. $\displaystyle\int (x+2)\ln x\,dx$

16. $\displaystyle\int (x+1)\ln x\,dx$

17. $\displaystyle\int (x-1)\ln x\,dx$

18. $\displaystyle\int (x-2)\ln x\,dx$

19. $\displaystyle\int x\sqrt{x+2}\,dx$

20. $\displaystyle\int x\sqrt{x+4}\,dx$

21. $\displaystyle\int x^3\ln 2x\,dx$

22. $\displaystyle\int x^2\ln 5x\,dx$

23. $\displaystyle\int x^2e^x\,dx$

24. $\displaystyle\int (\ln x)^2\,dx$

25. $\displaystyle\int x^2e^{2x}\,dx$

26. $\displaystyle\int x^{-5}\ln x\,dx$

27. $\displaystyle\int x^3e^{-2x}\,dx$

28. $\displaystyle\int x^5e^{4x}\,dx$

29. $\displaystyle\int (x^4+1)e^{3x}\,dx$

30. $\displaystyle\int (x^3-x+1)e^{-x}\,dx$

Evaluate using integration by parts.

31. $\displaystyle\int_1^2 x^2\ln x\,dx$

32. $\displaystyle\int_1^2 x^3\ln x\,dx$

33. $\displaystyle\int_2^6 \ln(x+3)\,dx$

34. $\displaystyle\int_0^5 \ln(x+1)\,dx$

35. $\displaystyle\int_0^1 xe^x\,dx$

36. $\displaystyle\int_0^1 xe^{-x}\,dx$

Applications ....

BUSINESS AND ECONOMICS

37. *Cost from Marginal Cost.* A company determines that its marginal-cost function is given by
$$C'(x) = 4x\sqrt{x+3}.$$
Find the total cost given that $C(13) = \$1126.40$.

38. *Profit from Marginal Profit.* A firm determines that its marginal-profit function is given by
$$P'(x) = 1000x^2e^{-0.2x}.$$

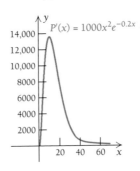

Find the total profit given that $P = -\$2000$ when $x = 0$.

LIFE AND PHYSICAL SCIENCES

39. *Electrical Energy Use.* The rate of electrical energy used by a family, in kilowatt hours per day, is given by
$$K(t) = 10te^{-t},$$
where t is the time, in hours. That is, t is in the interval $[0, 24]$.

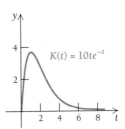

a) How many kilowatt hours does the family use in the first T hours of a day ($t = 0$ to $t = T$)?

b) How many kilowatt hours does the family use in the first 4 hours of the day?

40. *Drug Dosage.* Suppose that an oral dose of a drug is taken. From that time, the drug is assimilated in the body and excreted through the urine. The total amount of the drug that has passed through the body in time T is given by

$$\int_0^T E(t)\, dt,$$

where E is the rate of excretion of the drug. A typical rate-of-excretion function is

$$E(t) = te^{-kt},$$

where $k > 0$ and t is the time, in hours.

a) Use integration by parts to find a formula for

$$\int_0^T E(t)\, dt.$$

b) Find

$$\int_0^{10} E(t)\, dt, \quad \text{when } k = 0.2 \text{ mg/hr}.$$

47. $\displaystyle \int \frac{13t^2 - 48}{\sqrt[5]{4t + 7}}\, dt$

48. $\displaystyle \int (27x^3 + 83x - 2)\, \sqrt[6]{3x + 8}\, dx$

49. Verify that for any positive integer n,

$$\int x^n e^x\, dx = x^n e^x - n \int x^{n-1} e^x\, dx.$$

50. Verify that for any positive integer n,

$$\int (\ln x)^n\, dx = x(\ln x)^n - n \int (\ln x)^{n-1}\, dx.$$

tw 51. Determine whether the following is a theorem:

$$\int f(x)\, g(x)\, dx = \int f(x)\, dx \cdot \int g(x)\, dx.$$

Explain.

tw 52. Compare the procedures of differentiation and integration. Which seems to be the most complicated or difficult and why?

Synthesis

Evaluate using integration by parts.

41. $\displaystyle \int \sqrt{x}\, \ln x\, dx$

42. $\displaystyle \int x^n \ln x\, dx$

43. $\displaystyle \int \frac{te^t}{(t + 1)^2}\, dt$

44. $\displaystyle \int x^2\, (\ln x)^2\, dx$

45. $\displaystyle \int \frac{\ln x}{\sqrt{x}}\, dx$

46. $\displaystyle \int x^n\, (\ln x)^2\, dx$

 TECHNOLOGY CONNECTION

53. Use a grapher to evaluate

$$\int_1^{10} x^5 \ln x\, dx.$$

5.7

OBJECTIVES

➤ Evaluate integrals using a table of integration formulas.

Integration Techniques: Tables

Tables of Integration Formulas

You have probably noticed that, generally speaking, integration is more difficult and "tricky" than differentiation. Because of this, integral formulas that are reasonable and/or important have been gathered into tables. Table 1, listed below as well as at the back of the book on p. 567, though quite brief, is such an example. Entire books of integration formulas are available in libraries, and lengthy tables are also available in mathematics handbooks. Such tables are usually classified by the form of the integrand. The idea is to properly match the integral in question with a formula in the table. Sometimes some algebra or a technique such as integration by substitution or parts may be needed as well as a table.

TABLE 1
INTEGRATION FORMULAS

(Whenever $\ln X$ is used, it is assumed that $X > 0$.)

1. $\displaystyle\int x^n\, dx = \frac{x^{n+1}}{n+1} + C,\ n \neq -1$

2. $\displaystyle\int \frac{dx}{x} = \ln x + C$

3. $\displaystyle\int u\, dv = uv - \int v\, du$

4. $\displaystyle\int e^x\, dx = e^x + C$

5. $\displaystyle\int e^{ax}\, dx = \frac{1}{a} \cdot e^{ax} + C$

6. $\displaystyle\int x e^{ax}\, dx = \frac{1}{a^2} \cdot e^{ax}(ax - 1) + C$

7. $\displaystyle\int x^n e^{ax}\, dx = \frac{x^n e^{ax}}{a} - \frac{n}{a} \int x^{n-1} e^{ax}\, dx + C$

8. $\displaystyle\int \ln x\, dx = x \ln x - x + C$

9. $\displaystyle\int (\ln x)^n\, dx = x(\ln x)^n - n \int (\ln x)^{n-1}\, dx + C,\ n \neq -1$

10. $\displaystyle\int x^n \ln x\, dx = x^{n+1}\left[\frac{\ln x}{n+1} - \frac{1}{(n+1)^2}\right] + C,\ n \neq -1$

11. $\displaystyle\int a^x\, dx = \frac{a^x}{\ln a} + C,\ a > 0,\ a \neq 1$

12. $\displaystyle\int \frac{1}{\sqrt{x^2 + a^2}}\, dx = \ln\left(x + \sqrt{x^2 + a^2}\right) + C$

13. $\displaystyle\int \frac{1}{\sqrt{x^2 - a^2}}\, dx = \ln\left(x + \sqrt{x^2 - a^2}\right) + C$

14. $\displaystyle\int \frac{1}{x^2 - a^2}\, dx = \frac{1}{2a} \ln\left(\frac{x - a}{x + a}\right) + C$

15. $\displaystyle\int \frac{1}{a^2 - x^2}\, dx = \frac{1}{2a} \ln\left(\frac{a + x}{a - x}\right) + C$

16. $\displaystyle\int \frac{1}{x\sqrt{a^2 + x^2}}\, dx = -\frac{1}{a} \ln\left(\frac{a + \sqrt{a^2 + x^2}}{x}\right) + C$

(continued)

TABLE 1
INTEGRATION FORMULAS (*continued*)

17. $\displaystyle\int \frac{1}{x\sqrt{a^2 - x^2}}\, dx = -\frac{1}{a}\ln\!\left(\frac{a + \sqrt{a^2 - x^2}}{x}\right) + C,\ 0 < x < a$

18. $\displaystyle\int \frac{x}{ax + b}\, dx = \frac{b}{a^2} + \frac{x}{a} - \frac{b}{a^2}\ln\,(ax + b) + C$

19. $\displaystyle\int \frac{x}{(ax + b)^2}\, dx = \frac{b}{a^2(ax + b)} + \frac{1}{a^2}\ln\,(ax + b) + C$

20. $\displaystyle\int \frac{1}{x(ax + b)}\, dx = \frac{1}{b}\ln\!\left(\frac{x}{ax + b}\right) + C$

21. $\displaystyle\int \frac{1}{x(ax + b)^2}\, dx = \frac{1}{b(ax + b)} + \frac{1}{b^2}\ln\!\left(\frac{x}{ax + b}\right) + C$

22. $\displaystyle\int \sqrt{x^2 \pm a^2}\, dx = \tfrac{1}{2}\left[x\sqrt{x^2 \pm a^2} \pm a^2\ln\,(x + \sqrt{x^2 \pm a^2})\right] + C$

23. $\displaystyle\int x\sqrt{a + bx}\, dx = \frac{2}{15b^2}(3bx - 2a)(a + bx)^{3/2} + C$

24. $\displaystyle\int x^2\sqrt{a + bx}\, dx = \frac{2}{105b^3}(15b^2x^2 - 12abx + 8a^2)(a + bx)^{3/2} + C$

25. $\displaystyle\int \frac{x\, dx}{\sqrt{a + bx}} = \frac{2}{3b^2}(bx - 2a)\sqrt{a + bx} + C$

26. $\displaystyle\int \frac{x^2\, dx}{\sqrt{a + bx}} = \frac{2}{15b^3}(3b^2x^2 - 4abx + 8a^2)\sqrt{a + bx} + C$

Example 1   Evaluate:

$$\int \frac{dx}{x(3 - x)}.$$

Solution   This integral fits *Formula 20* in Table 1:

$$\int \frac{1}{x(ax + b)}\, dx = \frac{1}{b}\ln\left(\frac{x}{ax + b}\right) + C.$$

In our integral, $a = -1$ and $b = 3$, so we have, by the formula,

$$\int \frac{1}{x(3 - x)}\, dx = \int \frac{dx}{x(-1 \cdot x + 3)}$$

$$= \frac{1}{3}\ln\left(\frac{x}{-1 \cdot x + 3}\right) + C$$

$$= \frac{1}{3}\ln\left(\frac{x}{3 - x}\right) + C.$$

Example 2   Evaluate:

$$\int \frac{5x}{7x - 8} \, dx.$$

Solution   If we first factor 5 out of the integral, then the integral fits *Formula 18* in Table 1:

$$\int \frac{x}{ax + b} \, dx = \frac{b}{a^2} + \frac{x}{a} - \frac{b}{a^2} \ln (ax + b) + C.$$

In our integral, $a = 7$ and $b = -8$, so we have, by the formula,

$$\int \frac{5x}{7x - 8} \, dx = 5 \int \frac{x}{7x - 8} \, dx$$

$$= 5 \left[\frac{-8}{7^2} + \frac{x}{7} - \frac{-8}{7^2} \ln (7x - 8) \right] + C$$

$$= 5 \left[\frac{-8}{49} + \frac{x}{7} + \frac{8}{49} \ln (7x - 8) \right] + C$$

$$= -\frac{40}{49} + \frac{5x}{7} + \frac{40}{49} \ln (7x - 8) + C.$$

Example 3   Evaluate:

$$\int \sqrt{16x^2 + 3} \, dx.$$

Solution   This integral almost fits *Formula 22* in Table 1:

$$\int \sqrt{x^2 \pm a^2} \, dx = \frac{1}{2} \left[x\sqrt{x^2 \pm a^2} \pm a^2 \ln (x + \sqrt{x^2 \pm a^2}) \right] + C.$$

But the x^2-coefficient needs to be 1. To achieve this, we first factor out 16. Then we apply *Formula 22* in Table 1:

$$\int \sqrt{16x^2 + 3} \, dx = \int \sqrt{16\left(x^2 + \tfrac{3}{16}\right)} \, dx$$

$$= \int 4\sqrt{x^2 + \tfrac{3}{16}} \, dx$$

$$= 4 \int \sqrt{x^2 + \tfrac{3}{16}} \, dx$$

$$= 4 \cdot \frac{1}{2} \left[x \sqrt{x^2 + \tfrac{3}{16}} + \tfrac{3}{16} \ln \left(x + \sqrt{x^2 + \tfrac{3}{16}} \right) \right] + C$$

$$= 2 \left[x \sqrt{x^2 + \tfrac{3}{16}} + \tfrac{3}{16} \ln \left(x + \sqrt{x^2 + \tfrac{3}{16}} \right) \right] + C.$$

In our integral, $a^2 = 3/16$ and $a = \sqrt{3}/4$, though we did not need to use a in this form when applying the formula.

Example 4 Evaluate:

$$\int \frac{dx}{x^2 - 25}.$$

Solution This integral fits *Formula 14* in Table 1:

$$\int \frac{1}{x^2 - a^2}\, dx = \frac{1}{2a} \ln \left(\frac{x - a}{x + a} \right) + C.$$

In our integral, $a = 5$, so we have, by the formula,

$$\int \frac{dx}{x^2 - 25} = \frac{1}{10} \ln \left(\frac{x - 5}{x + 5} \right) + C.$$

Example 5 Evaluate: $\int (\ln x)^3\, dx.$

Solution This integral fits *Formula 9* in Table 1:

$$\int (\ln x)^n\, dx = x(\ln x)^n - n \int (\ln x)^{n-1}\, dx + C, \quad n \neq -1.$$

We must apply the formula three times:

$$\int (\ln x)^3\, dx$$

$$= x(\ln x)^3 - 3 \int (\ln x)^2\, dx + C \qquad \text{Formula 9}$$

$$= x(\ln x)^3 - 3\left[x(\ln x)^2 - 2 \int \ln x\, dx \right] + C \qquad \begin{array}{l}\text{Applying Formula 9}\\ \text{again}\end{array}$$

$$= x(\ln x)^3 - 3\left[x(\ln x)^2 - 2\left(x \ln x - \int dx \right) \right] + C \qquad \begin{array}{l}\text{Applying}\\ \text{Formula 9 for}\\ \text{the third time}\end{array}$$

$$= x(\ln x)^3 - 3x(\ln x)^2 + 6x \ln x - 6x + C.$$

EXERCISE SET 5.7

Evaluate using Table 1.

1. $\int xe^{-3x}\, dx$

2. $\int xe^{4x}\, dx$

3. $\int 5^x\, dx$

4. $\int \frac{1}{\sqrt{x^2 - 9}}\, dx$

5. $\int \frac{1}{16 - x^2}\, dx$

6. $\int \frac{1}{x\sqrt{4 + x^2}}\, dx$

7. $\int \frac{x}{5 - x}\, dx$

8. $\int \frac{x}{(1 - x)^2}\, dx$

9. $\int \frac{1}{x(5 - x)^2}\, dx$

10. $\int \sqrt{x^2 + 9}\, dx$

11. $\int \ln 3x\, dx$

12. $\int \ln \frac{4}{5}x\, dx$

13. $\int x^4 e^{5x}\, dx$

14. $\int x^3 e^{-2x}\, dx$

15. $\int x^3 \ln x\, dx$

16. $\int 5x^4 \ln x\, dx$

17. $\int \dfrac{dx}{\sqrt{x^2 + 7}}$

18. $\int \dfrac{3\,dx}{x\sqrt{1 - x^2}}$

19. $\int \dfrac{10\,dx}{x(5 - 7x)^2}$

20. $\int \dfrac{2}{5x(7x + 2)}\,dx$

21. $\int \dfrac{-5}{4x^2 - 1}\,dx$

22. $\int \sqrt{9t^2 - 1}\,dt$

23. $\int \sqrt{4m^2 + 16}\,dm$

24. $\int \dfrac{3\ln x}{x^2}\,dx$

25. $\int \dfrac{-5\ln x}{x^3}\,dx$

26. $\int (\ln x)^4\,dx$

27. $\int \dfrac{e^x}{x^{-3}}\,dx$

28. $\int \dfrac{3}{\sqrt{4x^2 + 100}}\,dx$

Applications

BUSINESS AND ECONOMICS

29. *Supply from Marginal Supply.* A lawn machinery company introduces a new kind of lawn seeder. It finds that its marginal supply for the seeder satisfies the function

$$S'(p) = \dfrac{100p}{(20 - p)^2}, \quad 0 \le p \le 19,$$

where S is the quantity purchased when the price is p thousand dollars per seeder. Find the supply function $S(p)$ given that the company will sell 2000 seeders when the price is \$19 thousand.

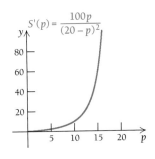

SOCIAL SCIENCES

30. *Learning Rate.* The rate of change of the probability that an employee learns a task on a new assembly line is given by

$$p'(t) = \dfrac{1}{t(2 + t)^2},$$

where p is the probability of learning the task after time t, in months. Find the function $p(t)$ given that $p = 0.8267$ when $t = 2$.

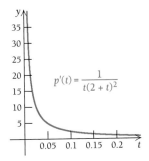

Synthesis

Evaluate using Table 1.

31. $\int \dfrac{8}{3x^2 - 2x}\,dx$

32. $\int \dfrac{x\,dx}{4x^2 - 12x + 9}$

33. $\int \dfrac{dx}{x^3 - 4x^2 + 4x}$

34. $\int e^x \sqrt{e^{2x} + 1}\,dx$

35. $\int \dfrac{-e^{-2x}\,dx}{9 - 6e^{-x} + e^{-2x}}$

36. $\int \dfrac{\sqrt{(\ln x)^2 + 49}}{2x}\,dx$

CHAPTER 5 SUMMARY AND REVIEW

Terms to Know

Antiderivative, p. 364
Integration, p. 365
Indefinite integral, p. 365
Area, pp. 374, 375
Definite integral, p. 379

Limits of integration, p. 379
Summation notation, p. 388
Fundamental Theorem of Integral
  Calculus, p. 390
Average value, p. 394

Trapezoidal Rule, p. 399
Integration by substitution, p. 408
Integration by parts, p. 415
Tabular integration by parts, p. 420
Integration using tables, p. 422

Review Exercises

These review exercises are for test preparation. They can also be used as a lengthened practice test. Answers are at the back of the book. The answers also contain bracketed section references, which tell you where to restudy if your answer is incorrect.

Evaluate.

1. $\int 8x^4 \, dx$

2. $\int (3e^x + 2) \, dx$

3. $\int \left(3t^2 + 7t + \frac{1}{t} \right) dt$

Find the area under the curve over the indicated interval.

4. $y = 4 - x^2$; $[-2, 1]$
5. $y = x^2 + 3x + 6$; $[0, 2]$

In each case, give two interpretations of the shaded region.

6.

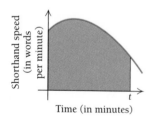

7.

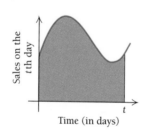

Evaluate.

8. $\int_a^b x^5 \, dx$

9. $\int_{-1}^{1} (x^3 - x^4) \, dx$

10. $\int_0^1 (e^x + x) \, dx$

11. $\int_1^3 \frac{3}{x} \, dx$

Decide whether $\int_a^b f(x) \, dx$ is positive, negative, or zero.

12.

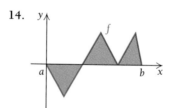

13.

14.

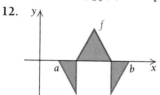

15. Find the area of the region bounded by $y = 3x^2$ and $y = 9x$.

Evaluate using substitution. Do not use Table 1.

16. $\int x^3 e^{x^4} \, dx$

17. $\int \frac{24t^5}{4t^6 + 3} \, dt$

18. $\int \frac{\ln 4x}{2x} \, dx$

19. $\int 2e^{-3x} \, dx$

Evaluate using integration by parts. Do not use Table 1.

20. $\int 3x\, e^{3x}\, dx$

21. $\int \ln x^7\, dx$

22. $\int 3x^2 \ln x\, dx$

Evaluate using Table 1.

23. $\int \dfrac{1}{49 - x^2}\, dx$

24. $\int x^2 e^{5x}\, dx$

25. $\int \dfrac{x}{7x + 1}\, dx$

26. $\int \dfrac{dx}{\sqrt{x^2 - 36}}$

27. $\int x^6 \ln x\, dx$

28. $\int xe^{8x}\, dx$

29. Approximate $\int_1^4 (2/x)\, dx$ by computing the area of each rectangle to three decimal places and adding.

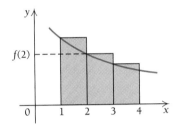

30. Find the average value of $y = e^{-x} + 5$ over $[0, 2]$.

31. A particle starts out from the origin. Its velocity at any time t, $t \geq 0$, is given by $v(t) = 3t^2 + 2t$. Find the distance that the particle travels during the first 4 hr (from $t = 0$ to $t = 4$).

32. *Business: Accumulated Sales.* A company estimates that its sales will grow continuously according to the function $S'(t) = 3e^{3t}$, where $S'(t)$ is the sales,

in dollars, on the tth day. Find the accumulated sales for the first 4 days.

Integrate using any method.

33. $\int x^3 e^{0.1x}\, dx$

34. $\int \dfrac{12t^2}{4t^3 + 7}\, dt$

35. $\int \dfrac{x\, dx}{\sqrt{4 + 5x}}$

36. $\int 5x^4 e^{x^5}\, dx$

37. $\int \dfrac{dx}{x + 9}$

38. $\int t^7 (t^8 + 3)^{11}\, dt$

39. $\int \ln 7x\, dx$

40. $\int x \ln 8x\, dx$

Synthesis

Evaluate.

41. $\int \dfrac{t^4 \ln (t^5 + 3)}{t^5 + 3}\, dt$

42. $\int \dfrac{dx}{e^x + 2}$

43. $\int \dfrac{\ln \sqrt{x}}{x}\, dx$

44. $\int x^{91} \ln x\, dx$

45. $\int \ln \left(\dfrac{x - 3}{x - 4}\right)\, dx$

46. $\int \dfrac{dx}{x\, (\ln x)^4}$

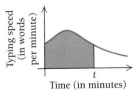 TECHNOLOGY CONNECTION

47. Use a grapher to approximate the area between the following curves:

$$y = 2x^2 - 2x,$$
$$y = 12x^2 - 12x^3,$$
$$x = 0, \quad x = 1.$$

CHAPTER 5 TEST

Evaluate.

1. $\int dx$

2. $\int 1000x^4\, dx$

3. $\int \left(e^x + \dfrac{1}{x} + x^{3/8}\right) dx$

Find the area under the curve over the indicated interval.

4. $y = x - x^2$; $[0, 1]$

5. $y = \dfrac{4}{x}$; $[1, 3]$

6. Give two interpretations of the shaded area.

Evaluate.

7. $\displaystyle\int_{-1}^{2} (2x + 3x^2)\, dx$ **8.** $\displaystyle\int_{0}^{1} e^{-2x}\, dx$

9. $\displaystyle\int_{a}^{b} \frac{dx}{x}$

10. Decide whether $\int_{a}^{b} f(x)\, dx$ is positive, negative, or zero.

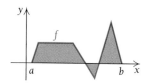

Evaluate using substitution. Do not use Table 1.

11. $\displaystyle\int \frac{dx}{x + 8}$ **12.** $\displaystyle\int e^{-0.5x}\, dx$

13. $\displaystyle\int t^3 (t^4 + 1)^9\, dt$

Evaluate using integration by parts. Do not use Table 1.

14. $\displaystyle\int xe^{5x}\, dx$ **15.** $\displaystyle\int x^3 \ln x^4\, dx$

Evaluate using Table 1.

16. $\displaystyle\int 2^x\, dx$ **17.** $\displaystyle\int \frac{dx}{x(7 - x)}$

18. Find the average value of $y = 4t^3 + 2t$ over $[-1, 2]$.

19. Find the area of the region bounded by $y = x$, $y = x^5$, $x = 0$, and $x = 1$.

20. Approximate

$$\int_{0}^{5} (25 - x^2)\, dx$$

by computing the area of each rectangle and adding.

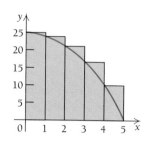

21. *Business: Cost from Marginal Cost.* An air conditioning company determines that the marginal cost of the xth air conditioner is given by

$$C'(x) = -0.2x + 500, \quad C(0) = 0.$$

Find the total cost of producing 100 air conditioners.

22. *Social Science: Learning Curve.* A typist's speed over a 4-min interval is given by

$$W(t) = -6t^2 + 12t + 90, \quad t \text{ in } [0, 4],$$

where $W(t)$ is the speed, in words per minute, at time t. How many words are typed during the second minute (from $t = 1$ to $t = 2$)?

Integrate using any method.

23. $\displaystyle\int \frac{dx}{x(10 - x)}$ **24.** $\displaystyle\int x^5 e^x\, dx$

25. $\displaystyle\int x^5 e^{x^6}\, dx$ **26.** $\displaystyle\int \sqrt{x}\, \ln x\, dx$

27. $\displaystyle\int x^3 \sqrt{x^2 + 4}\, dx$ **28.** $\displaystyle\int \frac{dx}{64 - x^2}$

29. $\displaystyle\int x^4 e^{0.1x}\, dx$ **30.** $\displaystyle\int x \ln 13x\, dx$

Synthesis

Evaluate using any method.

31. $\displaystyle\int \frac{[(\ln x)^3 - 4(\ln x)^2 + 5]}{x}\, dx$

32. $\displaystyle\int \ln\left(\frac{x + 3}{x + 5}\right) dx$

33. $\displaystyle\int \frac{8x^3 + 10}{\sqrt[3]{5x - 4}}\, dx$

TECHNOLOGY CONNECTION

34. Use a grapher to approximate the area between the following curves:

$$y = 3x - x^2,$$
$$y = 2x^3 - x^2 - 5x,$$
$$x = -2, \quad x = 0.$$

E X T E N D E D   T E C H N O L O G Y   A P P L I C A T I O N

Total Sales of Sherwin-Williams, Intel, Sprint, and The Gap

One of the many tasks of corporate leaders is the analysis of all kinds of data regarding the company they represent. The data might be total revenue, total costs, operating income, interest expense, costs of sales and marketing, and so on.

One very important use of the data can be the prediction, or forecasting, of the future. In this application, we will ask you to analyze factual data of several actual companies. You will be asked to do curve fitting to find models and make predictions.

Exercises

Sherwin-Williams® Company specializes in many types of paint products and coatings. The following table lists the total sales of Sherwin-Williams Company for several years.

Year		Total Sales (in millions)
0.	1990	$2267
1.	1991	2541
2.	1992	2748
3.	1993	2949
4.	1994	3100
5.	1995	3274
6.	1996	4133
7.	1997	4881

1. Make a scatterplot of the data. What kind of function—linear, quadratic, cubic polynomial, quartic polynomial, exponential, or logarithmic—seems to best fit the data?

2. Use the **REGRESSION** feature on a grapher to fit the chosen function to the data.

3. Use the function to predict the total sales of Sherwin-Williams Company in 2001, 2005, and 2010.

4. Use integration to predict the total sales of the company from 1990 through 2020.

 Intel® is a company that specializes in microprocessors and other semiconductor products, such as computer chips. The growth of personal computer sales is a top priority of the company. The following table lists the total sales of Intel® for several years.

Year		Total Sales (in millions)
0.	1986	$ 1,265
1.	1987	1,907
2.	1988	2,875
3.	1989	3,127
4.	1990	3,921
5.	1991	4,779
6.	1992	5,844
7.	1993	8,782
8.	1994	11,521
9.	1995	13,888
10.	1996	14,873

5. Make a scatterplot of the data. What kind of function—linear, quadratic, cubic polynomial, quartic polynomial, exponential, or logarithmic—seems to best fit the data?

6. Use the **REGRESSION** feature on a grapher to fit the chosen function to the data.

7. Use the function to predict the total sales of Intel® in 2005, 2008, and 2010.

8. Use integration to predict the total sales of the company from 1986 through 2020.

 Sprint® is a diversified telecommunications company that uses fiberoptic networks for long-distance voice, video, and data transmission. The following table lists the total sales of Sprint® for several years.

Year		Total Sales (in billions)
0.	1990	$ 8.6
1.	1991	9.5
2.	1992	9.9
3.	1993	10.4
4.	1994	11.4
5.	1995	12.7
6.	1996	13.9
7.	1997	14.9

9. Make a scatterplot of the data. What kind of function—linear, quadratic, cubic polynomial, quartic polynomial, exponential, or logarithmic—seems to best fit the data?

10. Use the **REGRESSION** feature on a grapher to fit the chosen function to the data.

11. Use the function to predict the total sales of Sprint® in 2004, 2009, and 2014.

12. Use integration to find the total sales of the company from 1990 through 2020.

The GAP

The Gap® is a specialty retailer that sells casual apparel under private-label brand names. The following table lists the total sales of The Gap® for several years.

Year		Total Sales (in billions)
0.	1992	$2.5
1.	1993	3.0
2.	1994	3.3
3.	1995	3.7
4.	1996	4.4
5.	1997	5.3
6.	1998	6.5

13. Make a scatterplot of the data. What kind of function—linear, quadratic, cubic polynomial, quartic polynomial, exponential, or logarithmic—seems to best fit the data?

14. Use the **REGRESSION** feature on a grapher to fit the chosen function to the data.

15. Use the function to predict the total sales of The Gap® in 2003, 2008, and 2011.

16. Use integration to find the total sales of the company from 1992 through 2020.

6

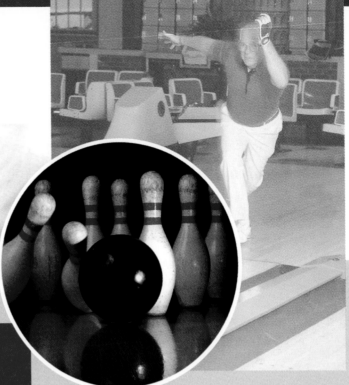

Applications of Integration

INTRODUCTION

In this chapter, we study a wide variety of applications of integration. We first consider an economics application to finding consumer's and producer's surplus. Then we study the integration of functions involved in exponential growth and decay. We will discover applications to not only business but also such environmental concerns as the depletion of resources and the buildup of radioactivity in the atmosphere. Integration has extensive application to probability and statistics, to finding volume, and to many situations involving the solution of differential equations.

Topics in this chapter can be chosen to fit the needs of the student and the course.

AN APPLICATION

At the time this book was written, the bowling scores S of the author were normally distributed with mean 201 and standard deviation 23. Find the probability that a score is from 190 to 213.

To find the probability, we evaluate the integral

$$P(190 \leq x \leq 213) = \int_{190}^{213} \frac{1}{23\sqrt{2\pi}} e^{-1/2[(x - 201)/23]^2} \, dx.$$

We do so by using Table 2 (at the back of the book) and find it to be about 0.383.

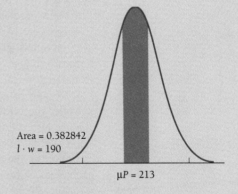

Area = 0.382842
$l \cdot w = 190$

$\mu P = 213$

This problem appears as Exercise 50 in Exercise Set 6.5.

6.1

➢ Given a demand and supply function, find the consumer's surplus and producer's surplus at the equilibrium point.

An Economics Application: Consumer's Surplus and Producer's Surplus

It has been convenient to think of demand and supply as quantities that are functions of price. For purposes of this section, it will be convenient to think of them as prices that are functions of quantity: $p = D(x)$ and $p = S(x)$. Indeed, such interpretation is common in a study of economics.

The consumer's demand curve, $p = D(x)$, gives the demand price per unit that the consumer is willing to pay for x units. It is usually a decreasing function. The producer's supply curve, $p = S(x)$, gives the price per unit at which the seller is willing to supply x units. It is usually an increasing function. The equilibrium point (x_E, p_E) is the intersection of the two curves.

Exercise

1. Graph the demand and supply functions

   $D(x) = (x - 5)^2$ and
   $S(x) = x^2 + x + 3$

   using the viewing window $[0, 5, 0, 30]$, with Yscl $= 5$. Find the equilibrium point using the INTERSECT feature.

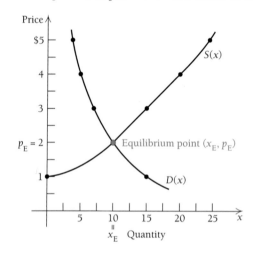

Utility is a function often considered in economics. When a consumer receives x units of a product, a certain amount of pleasure, or utility, U, is derived from them (see Exercise 13 in Exercise Set 2.3). For example, the number of movies that you see in a month gives you a certain utility. If you see 4 movies (unless they are unentertaining), you get more utility than if you see no movies. The same notion applies to having a meal in a restaurant or paying your heating bill to warm your home.

To help to explain the concepts of *consumer's* and *producer's surplus*, we will consider the utility of seeing movies over a fixed amount of time, say, one month. We are also going to make the assumption that the movies seen are of about the same quality.

Suppose the graphs in Figs. 1 and 2 show the demand curve of college students for movies. Suppose we consider this curve for just one such student, Samantha, and we want to examine the utility that she receives from going to the movies. Samantha goes to 0 movies per month if the price is $10. She will go to 1 movie per month if the price is $8.80. Suppose she goes to 1 movie. Then the total expenditure to her is $8.80 · 1, or $8.80, as shown in Fig. 1.

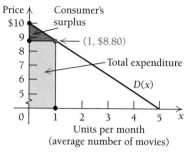

FIGURE 1

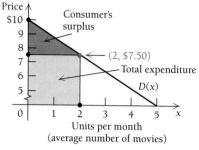

FIGURE 2

The area of the blue region represents her total expenditure, $8.80. Look at the area of the orange region. It is $\frac{1}{2}(1)(\$1.20)$, or $0.60. The total area under the curve is a measure of the *total utility* of going to 1 movie, and is $9.40. The area of the orange region, $0.60, is a measure of the satisfaction Samantha gets but for which she does not have to pay. Economists define this amount as **consumer's surplus.** It is the utility that consumers derive from living in a society in which the price consumers are willing to pay decreases when more units are purchased. It is the extra utility that is gained from this purchase. The consumer's surplus is the source of revenue that becomes available to a firm that discriminates in its prices, say, by offering discounts to college students, senior citizens, and children, or by selling discount coupons.

Suppose that Samantha goes to 2 movies per month if the price is $7.50 (Fig. 2). The total amount that Samantha actually spends is $7.50(2), or $15, and is the area of the blue region. It is her total expenditure. The total area under the curve is $17.50 and represents the total satisfaction to Samantha of going to 2 movies. Look at the area of the orange region. It (the consumer's surplus) is $\frac{1}{2}(2)(\$2.50)$, or $2.50. It is a measure of the utility Samantha received but for which she did not have to pay.

Suppose that the graph of the demand function is a curve, as shown below.

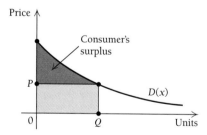

If Samantha goes to Q movies when the price is P, then Samantha's total expenditure is QP. The total area under the curve is the total utility, or the total satisfaction received, and is

$$\int_0^Q D(x)\ dx.$$

The *consumer's surplus* is the total area under the curve minus the total expenditure, quantity times price, or QP, and it is the total utility minus the total cost, which is

given by

$$\int_0^Q D(x)\,dx - QP.$$

Definition

Suppose that $p = D(x)$ describes the demand function for a commodity. Then the **consumer's surplus** is defined for the point (Q, P) as

$$\int_0^Q D(x)\,dx - QP.$$

Example 1   Find the consumer's surplus for the demand function $D(x) = (x - 5)^2$ when $x = 3$.

TECHNOLOGY
CONNECTION

E x p l o r a t o r y .   **Graph**
$D(x) = (x - 5)^2$, the demand function
in Example 1, using the viewing
window $[0, 5, 0, 30]$, with Yscl $= 5$. To
find the consumer's surplus at $x = 3$,
we first find $D(3)$. Then we graph
$y = D(3)$. What is the point of
intersection of $y = D(x)$ and $y = D(3)$?
    From the intersection, draw a
vertical line down to the x-axis. What
does the area of the resulting rectangle
represent? What does the area above
the horizontal line and below the curve
represent?

Solution   When $x = 3$, $D(3) = (3 - 5)^2 = (-2)^2 = 4$. Then

$$\text{Consumer's surplus} = \int_0^3 (x - 5)^2\,dx - 3 \cdot 4$$

$$= \int_0^3 (x^2 - 10x + 25)\,dx - 12$$

$$= \left[\frac{x^3}{3} - 5x^2 + 25x\right]_0^3 - 12$$

$$= \left[\left(\frac{3^3}{3} - 5(3)^2 + 25(3)\right)\right.$$

$$\left. - \left(\frac{0^3}{3} - 5(0)^2 + 25(0)\right)\right] - 12$$

$$= (9 - 45 + 75) - 0 - 12$$

$$= \$27.$$

Suppose that we now look at the supply curve for the movies, as shown in Figs. 3 and 4. At a price of \$0 per movie, the producer is willing to supply 0 movies.

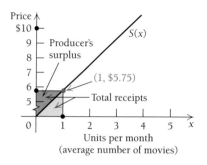

FIGURE 3

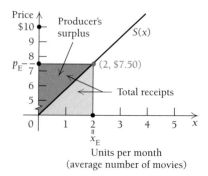

FIGURE 4

At a price of $5.75 per movie, the producer is willing to supply 1 movie and makes total receipts of 1($5.75), or $5.75. The area of the beige triangle in Fig. 3 represents the total cost to the producer of producing 1 movie, and is $\frac{1}{2}(1)(\$5.75)$, or $2.875. The area of the green triangle is also $2.875, and represents the surplus over cost. It is a contribution to profit. Economists call this number the **producer's surplus.** It is the utility, or satisfaction, to the producer of living in a society in which a greater amount of a commodity will be supplied when the price increases. It is the extra revenue that the producer receives when the consumer cannot purchase the individual units along the producer's supply curve.

At a price of $7.50, the producer is willing to supply 2 movies and makes total receipts of 2($7.50), or $15. The area of the beige triangle in Fig. 4 represents the total cost to the producer of actually making the 2 movies, and is $\frac{1}{2}(2)(\$7.50)$, or $7.50. The area of the green triangle is also $7.50 and is the producer's surplus. It is a contribution to the profit of the producer.

Suppose that the graph of the supply function is a curve, as shown at left. The producer will supply Q movies if the price is P. The total receipts are QP. The *producer's surplus* is the total receipts minus the area under the curve and is given by

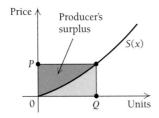

$$QP - \int_0^Q S(x)\, dx.$$

Example 2   Find the producer's surplus for $S(x) = x^2 + x + 3$ when $x = 3$.

Solution   When $x = 3$, $S(3) = 3^2 + 3 + 3 = 15$. Then

$$\text{Producer's surplus} = 3 \cdot 15 - \int_0^3 (x^2 + x + 3)\, dx$$

$$= 45 - \left[\frac{x^3}{3} + \frac{x^2}{2} + 3x \right]_0^3$$

$$= 45 - \left[\left(\frac{3^3}{3} + \frac{3^2}{2} + 3(3) \right) - \left(\frac{0^3}{3} + \frac{0^2}{2} + 3(0) \right) \right]$$

$$= 45 - \left(9 + \frac{9}{2} + 9 - 0 \right)$$

$$= \$22.50.$$

The **equilibrium point** (x_E, p_E) in Fig. 5 is the point at which the supply and demand curves intersect. It is that point at which the sellers and buyers come together and purchases and sales actually occur. In Fig. 6, we see the equilibrium point and the consumer's and producer's surpluses for the movie curves.

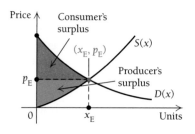

FIGURE 5

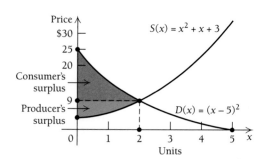

FIGURE 6

Example 3   Given

$$D(x) = (x - 5)^2 \quad \text{and} \quad S(x) = x^2 + x + 3,$$

find each of the following.

a) The equilibrium point

b) The consumer's surplus at the equilibrium point

c) The producer's surplus at the equilibrium point

Solution

a) To find the equilibrium point, we set $D(x) = S(x)$ and solve:

$$(x - 5)^2 = x^2 + x + 3$$
$$x^2 - 10x + 25 = x^2 + x + 3$$
$$-10x + 25 = x + 3$$
$$22 = 11x$$
$$2 = x.$$

Thus, $x_E = 2$. To find p_E, we substitute x_E into either $D(x)$ or $S(x)$. If we choose $D(x)$, then

$$p_E = D(x_E) = D(2)$$
$$= (2 - 5)^2$$
$$= (-3)^2$$
$$= \$9 \text{ per unit.}$$

Thus the equilibrium point is (2, $9).

b) The consumer's surplus at the equilibrium point is

$$\int_0^{x_E} D(x) \, dx - x_E \, p_E,$$

or

$$\int_0^2 (x - 5)^2 \, dx - 2 \cdot 9 = \left[\frac{(x-5)^3}{3} \right]_0^2 - 18$$

$$= \left[\frac{(2-5)^3}{3} - \frac{(0-5)^3}{3} \right] - 18$$

$$= \frac{(-3)^3}{3} - \frac{(-5)^3}{3} - 18 = -\frac{27}{3} + \frac{125}{3} - \frac{54}{3}$$

$$= \frac{44}{3} \approx \$14.67.$$

c) The producer's surplus at the equilibrium point is

$$x_E \, p_E - \int_0^{x_E} S(x) \, dx,$$

or

$$2 \cdot 9 - \int_0^2 (x^2 + x + 3) \, dx$$

$$= 2 \cdot 9 - \left[\frac{x^3}{3} + \frac{x^2}{2} + 3x \right]_0^2$$

$$= 18 - \left[\left(\frac{2^3}{3} + \frac{2^2}{2} + 3 \cdot 2 \right) - \left(\frac{0^3}{3} + \frac{0^2}{2} + 3 \cdot 0 \right) \right]$$

$$= 18 - \left(\frac{8}{3} + 2 + 6 \right)$$

$$= \frac{22}{3} \approx \$7.33.$$

EXERCISE SET 6.1

In each of Exercises 1–12, find (a) the equilibrium point, (b) the consumer's surplus at the equilibrium point, and (c) the producer's surplus at the equilibrium point.

1. $D(x) = -\frac{5}{6}x + 10, \quad S(x) = \frac{1}{2}x + 2$

2. $D(x) = -2x + 8, \quad S(x) = x + 2$

3. $D(x) = (x - 4)^2, \quad S(x) = x^2 + 2x + 6$

4. $D(x) = (x - 3)^2, \quad S(x) = x^2 + 2x + 1$

5. $D(x) = (x - 6)^2, \quad S(x) = x^2$

6. $D(x) = (x - 8)^2, \quad S(x) = x^2$

7. $D(x) = 1000 - 10x, \quad S(x) = 250 + 5x$

8. $D(x) = 8800 - 30x, \quad S(x) = 7000 + 15x$

9. $D(x) = 5 - x, 0 \le x \le 5; \quad S(x) = \sqrt{x + 7}$

10. $D(x) = 7 - x, 0 \le x \le 7; \quad S(x) = 2\sqrt{x} + 1$

Synthesis ....

11. $D(x) = e^{-x+4.5}, \quad S(x) = e^{x-5.5}$

12. $D(x) = \sqrt{56 - x}, \quad S(x) = x$

 13. Do some research on consumer's surplus in an economics book. Write a brief description.

 14. Do some research on producer's surplus in an economics book. Write a brief description.

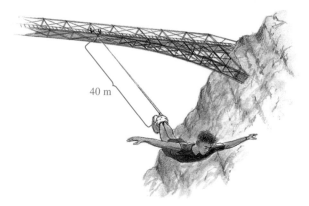

Time Spent (in half hours per month)	Price (per half hour)
8	$ 2.50
7	5.00
6	7.50
5	10.00
4	12.50
3	15.00
2	17.50
1	20.00

40 m

~~ TECHNOLOGY CONNECTION

Graph each pair of demand and supply functions. Then:

a) Find the equilibrium point using the INTERSECT feature or another feature that will allow you to find this point of intersection.

b) Then graph $y = D(x_E)$ and shade the regions of both consumer's and producer's surpluses.

c) Find the consumer's surplus.

d) Find the producer's surplus.

15. $D(x) = (x - 4)^2, \quad S(x) = x^2 + 2x + 6$

16. $D(x) = 5 - x, \quad S(x) = \sqrt{x} + 7$

17. *Bungee Jumping.* Reggie loves to go bungee jumping. The table at right above shows the number of half hours that Reggie is willing to go bungee jumping at various prices.

   a) Make a scatterplot of the data and determine the type of function that you think best fits the data.

b) Fit that function to these data using the REGRESSION feature.

c) If Reggie goes bungee jumping for 6 half hours per month, what is his consumer's surplus?

d) At a price of $11.50 per half hour, what is Reggie's consumer's surplus?

6.2

OBJECTIVES

➤ Do computations involving interest compounded continuously and continuous money flow.

➤ Find the total use of a natural resource.

➤ Find the present value of an investment.

➤ Find the accumulated present value of an investment.

Applications of the Models $\int_0^T P_0 e^{kt}\, dt$ and $\int_0^T P_0 e^{-kt}\, dt$

Recall the basic model of exponential growth (Section 4.3):

$$P'(t) = 2k \cdot P(t), \quad \text{or} \quad \frac{dP}{dt} = kP, \quad \text{where } P = P_0 \text{ when } t = 0.$$

The function that satisfies the equation is

$$P(t) = P_0 e^{kt}. \tag{1}$$

One application of equation (1) is to compute the balance of a savings account after t years from an initial investment of P_0 at interest rate k, compounded continuously.

Exercise

1. Graph the function
    $f(x) = 1000e^{0.08x}$
    of Example 1
    using the view-
    ing window [0, 3,
    0, 1500], with
    Xscl = 1 and
    Yscl = 100. Then
    find function
    values for $x = 0$,
    1, 2, and 3.

Example 1   *Business: Growth in an Investment.* Find the balance in a savings account after 3 yr from an initial investment of $1000 at interest rate 8%, compounded continuously.

Solution   Using equation (1) with $k = 0.08$, $t = 3$, and $P_0 = \$1000$, we get

$$P(3) = 1000e^{0.08(3)}$$
$$= 1000e^{0.24}$$
$$= 1000(1.271249)$$
$$\approx \$1271.25.$$

The balance in the account after 3 yr will be approximately $1271.25.   ◁

The Integral $\int_0^T P_0 e^{kt} \, dt$

Suppose that we consider the integral of $P_0 e^{kt}$ over the interval $[0, T]$:

$$\int_0^T P_0 e^{kt} \, dt = \left[\frac{P_0}{k} \cdot e^{kt} \right]_0^T = \frac{P_0}{k}(e^{kT} - e^{k \cdot 0}) = \frac{P_0}{k}(e^{kT} - 1).$$

We have a formula for this integral, given as

$$\int_0^T P(t) \, dt = \int_0^T P_0 e^{kt} \, dt = \frac{P_0}{k}(e^{kT} - 1). \tag{2}$$

We know that this formula represents the area under the graph of $P(t) = P_0 e^{kt}$ over the interval $[0, T]$.

In the remainder of this section, we consider two other applications or interpretations of this integral.

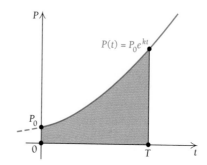

Continuous Money Flow

Suppose that money is flowing continuously into a savings account at the rate of $1000 per year at interest rate 8%, compounded continuously. This means that over a small amount of time dt, the bank pays interest on all money accumulated in the account up to this time (since $t = 0$). The amount that is paid in over time dt is

$$\$1000e^{0.08t} \, dt.$$

Suppose that we want to find the accumulation of all these amounts over a 5-yr

Exercise

1. Evaluate the integral

$$\int_0^3 \$1000e^{0.08x}\, dx.$$

    Explain what the answer represents.

period. That accumulation is given by the integral

$$\int_0^5 \$1000e^{0.08t}\, dt = \left[\frac{1000}{0.08}e^{0.08t}\right]_0^5$$

$$= 12{,}500(e^{0.08(5)} - e^{0.08(0)})$$

$$= 12{,}500(e^{0.4} - 1)$$

$$= 12{,}500(1.491825 - 1)$$

$$\approx \$6147.81.$$

Economists call $6147.81 the *amount of a continuous money flow*. In this case, the money is flowing according to a constant function $R(t) = \$1000$. Money could also be flowing according to some variable function, say, $R(t) = 2t - 7$ or $R(t) = t^2$.

> **Theorem 1**
>
> If the rate of flow of money into an investment is given by some constant function $R(t)$, then the **amount of continuous money flow** at interest rate k, compounded continuously, over time T is given by
>
> $$\int_0^T R(t)e^{kt}\, dt.$$

Exploratory.   Graph $f(x) = 1000e^{0.08x}$ using the viewing window [1, 15, 0, 5000], with Xscl = 2 and Yscl = 1000. Then graph the line $y = f(0)$. Consider the rectangle formed by the lines $y = f(0)$, $x = 15$, the y-axis, and the x-axis. Graph it if you can. What does the area of the rectangle represent? What does the area above the rectangle and below the curve represent? What does the entire area under the curve represent?

Example 2   *Business: Amount of a Continuous Money Flow.* Find the amount of a continuous money flow where $1000 per year is being invested at 8%, compounded continuously, for 15 yr.

Solution

$$\int_0^{15} \$1000e^{0.08t}\, dt = \frac{1000}{0.08}(e^{0.08(15)} - 1) \quad \text{Using equation (2)}$$

$$= 12{,}500(e^{1.2} - 1)$$

$$= 12{,}500(3.320117 - 1)$$

$$\approx \$29{,}001.46$$

Sometimes we might want to know how money should be flowing into an investment so that we end up with a specified amount.

Example 3   *Business: Continuous Money Flow.* Consider a continuous flow of money into an investment at the constant rate of P_0 dollars per year. What should P_0 be so that the amount of a continuous money flow over 20 yr at interest rate 8%, compounded continuously, will be $10,000?

Solution   We find P_0 such that

$$10{,}000 = \int_0^{20} P_0 e^{0.08t}\, dt.$$

We solve the following equation:

$$10,000 = \frac{P_0}{0.08}(e^{0.08(20)} - 1) \qquad \text{Using equation (2)}$$

$$800 = P_0(e^{1.6} - 1)$$

$$800 = P_0(4.953032 - 1)$$

$$800 = P_0(3.953032)$$

$$\$202.38 \approx P_0.$$

Life and Physical Sciences: Depletion of Natural Resources

Another application of the integral of exponential growth concerns

$$P(t) = P_0 e^{kt}$$

as a model of the demand for natural resources. Suppose that P_0 represents the amount of a natural resource (such as coal or oil) used at time $t = 0$ and that the growth rate for the use of this resource is k. Then, assuming exponential growth (which is the case for the use of many resources), the amount to be used at time t is $P(t)$, given by

$$P(t) = P_0 e^{kt}.$$

The total amount used during an interval $[0, T]$ is given again by equation (2):

$$\int_0^T P(t)\, dt = \int_0^T P_0 e^{kt}\, dt$$

$$= \frac{P_0}{k}(e^{kT} - 1). \tag{2}$$

Example 4 *Physical Science: Demand for Copper.* In 1997 ($t = 0$), the world use of copper was 11,300,000 metric tons, and the demand for copper was growing exponentially at the rate of 15% per year. If the growth continues at this rate, how many tons of copper will the world use from 1997 to 2010?

Solution   Using equation (2), we have

$$\int_0^{13} 11,300,000 e^{0.15t}\, dt = \frac{11,300,000}{0.15}(e^{0.15(13)} - 1) \qquad \text{Using equation (2)}$$

$$\approx 75,333,333(e^{1.95} - 1)$$

$$\approx 75,333,333(7.028687581)$$

$$\approx 75,333,333(6.028687581)$$

$$\approx 454,161,000. \qquad \text{Rounded to the nearest thousand}$$

Thus from 1997 to 2010, the world will use approximately 454,161,000 metric tons of copper.

Example 5 *Physical Science: Depletion of Copper Ore.* The world reserves of copper ore in 1997 were estimated to be 2,300,000,000 metric tons. Assuming that the

Bingham Canyon mine in Utah has produced more copper than any other mine in history. The grade of ore, however, has dropped from 1.93 percent copper in 1906 to 0.6 percent today. At present, it is planned that mining will cease when the percentage of copper reaches 0.4.

growth rate in Example 4 continues and that no new reserves are discovered, when will the world reserves of copper ore be exhausted?

Solution   Using equation (2), we want to find T such that

$$2,300,000,000 = \frac{11,300,000}{0.15}(e^{0.15T} - 1).$$

We solve for T as follows:

$$2,300,000,000 \approx 75,333,333(e^{0.15T} - 1)$$

$$\frac{2,300,000,000}{75,333,333} \approx e^{0.15T} - 1$$

$$30.530974 \approx e^{0.15T} - 1$$

$$31.530974 = e^{0.15T}$$

$\ln 31.530974 \approx \ln e^{0.15T}$ Taking the natural logarithm on both sides

$\ln 31.530974 = 0.15T$ Recall that $\ln e^k = k$.

$$\frac{\ln 31.530974}{0.15} = T$$

$$\frac{31.530974}{0.15} \approx T$$

$$23 \approx T.$$ Rounding to the nearest one

Thus 23 yr from 1997 (or by 2020), the world reserves of copper ore will be exhausted.

Applications of the Model $\int_0^T Pe^{-kt}\,dt$

Recall our study of present value in Section 4.4. If you did not study that topic there, you should do so now. The **present value** P_0 of an amount P due t years later is found by solving the following equation for P_0:

$$P_0 e^{kt} = P$$

$$P_0 = \frac{P}{e^{kt}} = Pe^{-kt}.$$

Theorem 2

The **present value P_0** of an amount P due t years later at interest rate k, compounded continuously, is given by

$$P_0 = Pe^{-kt}.$$

Note that this can be interpreted as exponential decay from the future back to the present.

Example 6   *Business: Present Value.* Find the present value of $200,000 due 25 yr from now at 8.7%, compounded continuously.

Solution   We substitute $200,000 for P, 0.087 for k, and 25 for t in the equation for present value:

$$P_0 = 200{,}000e^{-0.087(25)} \approx \$22{,}721.63.$$

Thus the present value is $22,721.63.   ◁

Suppose that we know that a continuous flow of money will go into an investment at the constant rate of P dollars per year, from now until some time T in the future. If an infinitesimal amount of time dt passes,

$$P \cdot dt$$

dollars will have accumulated. The present value of that amount is

$$(P \cdot dt)e^{-kt},$$

where k is the current interest rate, compounded continuously. The accumulation of all the present values is given by the integral

$$\int_0^T Pe^{-kt}\,dt$$

and is called the **accumulated present value.** Evaluating this integral, we get

$$\int_0^T Pe^{-kt}\,dt = \frac{P}{-k}(e^{-kT} - e^{-k\cdot 0})$$

$$= \frac{P}{k}(1 - e^{-kT}). \tag{3}$$

In the preceding case, money is flowing according to a constant function $R(t) = P$. Money could also be flowing according to some variable function, such as $R(t) = 2t + 8$ or $R(t) = t^3$.

Definition

The **accumulated present value** of a continuous money flow into an investment at a constant rate of $R(t)$ dollars per year from now until some time T in the future is given by

$$\int_0^T R(t)e^{-kt}\, dt,$$

where k is the current interest rate, compounded continuously.

Example 7 *Business: Accumulated Present Value.* Find the accumulated present value of an investment over a 5-yr period if there is a continuous money flow of $2400 per year and the current interest rate is 14%, compounded continuously.

Solution   The accumulated present value is

$$\int_0^5 \$2400e^{-0.14t}\, dt = \frac{2400}{0.14}(1 - e^{-0.14 \cdot 5}) \qquad \text{Using equation (3)}$$

$$\approx 17{,}142.86(1 - e^{-0.7})$$

$$= 17{,}142.86(1 - 0.496585)$$

$$\approx \$8629.97.$$

The preceding example is an application of the model

$$\int_0^T Pe^{-kt}\, dt = \frac{P}{k}(1 - e^{-kT}).$$

This model can be applied to a calculation of the buildup of a specific amount of radioactive material released into the atmosphere annually. Some of the material decays, but more continues to be released. The amount present at time T is given by the integral above.

EXERCISE SET 6.2

Applications
BUSINESS AND ECONOMICS

1. Find the amount in a savings account after 3 yr from an initial investment of $100 at interest rate 9%, compounded continuously.

2. Find the amount in a savings account after 4 yr from an initial investment of $100 at interest rate 10%, compounded continuously.

3. Find the amount of a continuous money flow in which $100 per year is being invested at 9%, compounded continuously, for 20 yr.

4. Find the amount of a continuous money flow in which $100 per year is being invested at 10%, compounded continuously, for 20 yr.

5. Find the amount of a continuous money flow in which $1000 per year is being invested at 8.5%, compounded continuously, for 40 yr.

6. Find the amount of a continuous money flow in which $1000 per year is being invested at 7.5%, compounded continuously, for 40 yr.

7. What should P_0 be so that the amount of a continuous money flow over 20 yr at interest rate 8.5%, compounded continuously, will be $50,000?

8. What should P_0 be so that the amount of a continuous money flow over 20 yr at interest rate 7.5%, compounded continuously, will be $50,000?

9. What should P_0 be so that the amount of a continuous money flow over 30 yr at interest rate 9%, compounded continuously, will be $40,000?

10. What should P_0 be so that the amount of a continuous money flow over 30 yr at interest rate 10%, compounded continuously, will be $40,000?

11. Following the birth of a child, a parent wants to make an initial investment P_0 that will grow to $50,000 by the child's 20th birthday. Interest is compounded continuously at 9%. What should the initial investment be?

12. Following the birth of a child, a parent wants to make an initial investment P_0 that will grow to $60,000 by the child's 20th birthday. Interest is compounded continuously at 10%. What should the initial investment be?

13. Find the present value of $60,000 due 8 yr later at 8.8%, compounded continuously.

14. Find the present value of $50,000 due 16 yr later at 7.4%, compounded continuously.

15. Find the accumulated present value of an investment over a 10-yr period if there is a continuous money flow of $2700 per year and the current interest rate is 9%, compounded continuously.

16. Find the accumulated present value of an investment over a 10-yr period if there is a continuous money flow of $2700 per year and the current interest rate is 10%, compounded continuously.

17. A woman with an MBA accepts a position as president of a company at the age of 35. Assuming retirement at age 65 and an annual salary of $85,000 that is paid in a continuous money flow, what is the president's accumulated present value? The current interest rate is 8%, compounded continuously.

18. A college dropout takes a job as a truck driver at the age of 25. Assuming retirement at age 65 and an annual salary of $34,000 that is paid in a continuous money flow, what is the truck driver's accumulated present value? The current interest rate is 7%, compounded continuously.

LIFE AND PHYSICAL SCIENCES

19. *The Demand for Aluminum Ore (Bauxite).* In 1997 ($t = 0$), the world use of aluminum ore was 115,000,000 tons, and the demand for aluminum ore was growing exponentially at the rate of 12% per year. If the demand continues to grow at this rate, how many tons of aluminum ore will the world use from 1997 to 2004?

20. *The Demand for Natural Gas.* In 1990 ($t = 0$), the world use of natural gas was 72,137 billion cubic feet, and the demand for natural gas was growing exponentially at the rate of 4% per year. If the demand continues to grow at this rate, how many cubic feet of natural gas will the world use from 1990 to 2020?

21. *The Depletion of Aluminum Ore.* The world reserves of aluminum ore are 23,000,000,000 tons. Assuming that the growth rate of Exercise 19 continues and that no new reserves are discovered, when will the world reserves of aluminum ore be exhausted?

22. *The Depletion of Natural Gas.* The world reserves of natural gas are 3,926,000 billion cubic feet. Assuming that the growth rate of Exercise 20 continues and that no new reserves are discovered, when will the world reserves of natural gas be exhausted?

23. *Radioactive Buildup.* Plutonium has a decay rate of 0.003% per year. Suppose plutonium is released into the atmosphere each year for 20 yr at the rate of 1 lb per year. What is the total amount of radioactive buildup?

24. *Radioactive Buildup.* Cesium-137 has a decay rate of 2.3% per year. Suppose cesium-137 is released into the atmosphere each year for 20 yr at the rate of 1 lb per year. What is the total amount of radioactive buildup?

25. *Demand for Oil.* In 1990 ($t = 0$), the world use of oil was 6570 million barrels, and the demand for oil was growing exponentially at the rate of 10% per year. If the demand continues at this rate, how many barrels of oil will the world use from 1990 to 2000?

26. *Depletion of Oil.* The world reserves of oil are 920,700 million barrels. In 1990 ($t = 0$), the world use of oil was 6570 million barrels, and the growth rate for the use of oil was 10%. Assuming that this growth rate continues and that no new reserves are discovered, when will the world reserves of oil be exhausted?

Synthesis

For a nonconstant function $R(t)$, the amount of a continuous money flow is

$$\int_0^T R(t)e^{k(T-t)}\, dt.$$

Find the amount of a continuous money flow for each of the following.

27. $R(t) = 2000t + 7$, $k = 8\%$, and $T = 30$ yr

28. $R(t) = t^2$, $k = 7\%$, and $T = 40$ yr

For a nonconstant function $R(t)$, the accumulated present value is

$$\int_0^T R(t)e^{-k(T-t)}\, dt.$$

Find the accumulated present value for each of the following.

29. $R(t) = t$, $k = 8\%$, and $T = 20$ yr

30. $R(t) = e^t$, $k = 7\%$, and $T = 10$ yr

 31. Look up some data on rate of use and world reserves of a natural resource not considered in this section. Predict when the world reserves for that resource will be depleted.

 32. Describe the idea of present value to a friend who is not a business major. Then describe accumulated present value.

TECHNOLOGY CONNECTION

33. Graph the area under the curve represented by the continuous money flow in Exercise 3. Then compute the area and compare the result with the answer to the exercise.

34. Graph the area under the curve represented by the demand for aluminum ore in Exercise 19. Then compute the area and compare the result with the answer to the exercise.

6.3

OBJECTIVES

➤ Determine whether an improper integral is convergent or divergent.

➤ Solve applied problems involving improper integrals.

Improper Integrals

Let's try to find the area of the region under the graph of $y = 1/x^2$ over the interval $[1, \infty)$.

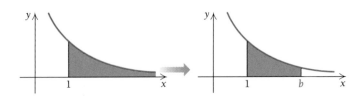

Note that this region is of infinite extent. We have not yet considered how to find the area of such a region. Let's find the area under the curve over the interval from 1 to b, and then see what happens as b gets very large. The area under the graph over $[1, b]$ is

$$\int_1^b \frac{dx}{x^2} = \left[-\frac{1}{x} \right]_1^b$$

$$= \left(-\frac{1}{b} \right) - \left(-\frac{1}{1} \right)$$

$$= -\frac{1}{b} + 1$$

$$= 1 - \frac{1}{b}.$$

Then

$$\lim_{b \to \infty} [\text{area from 1 to } b] = \lim_{b \to \infty} \left(1 - \frac{1}{b} \right) = 1.$$

We *define* the area from 1 to infinity to be this limit. Here we have an example of an infinitely long region with a finite area.

Such areas may not always be finite. Let's try to find the area of the region under the graph of $y = 1/x$ over the interval $[1, \infty)$.

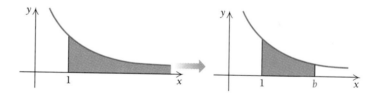

By definition, the area A from 1 to infinity is the limit as b approaches ∞ of the area from 1 to b, so

$$A = \lim_{b \to \infty} \int_1^b \frac{dx}{x} = \lim_{b \to \infty} [\ln x]_1^b$$

$$= \lim_{b \to \infty} (\ln b - \ln 1)$$

$$= \lim_{b \to \infty} \ln b.$$

In Section 4.2, we learned that $\ln b$ increases indefinitely as b increases. Therefore, the limit does not exist.

Thus we have an infinitely long region with an infinite area. Note that the graphs of $y = 1/x^2$ and $y = 1/x$ have similar shapes, but the region under one of them has a finite area and the other does not.

An integral such as

$$\int_a^\infty f(x)\, dx,$$

with an upper limit of infinity, is called an **improper integral.** Its value is defined to be the following limit.

> ### Definition
>
> $$\int_a^\infty f(x)\, dx = \lim_{b \to \infty} \int_a^b f(x)\, dx$$

If the limit exists, then we say that the improper integral **converges,** or is **convergent.** If the limit does not exist, then we say that the improper integral **diverges,** or is **divergent.** Thus,

$$\int_1^\infty \frac{dx}{x^2} = 1 \; \text{converges} \quad \text{and} \quad \int_1^\infty \frac{dx}{x} \; \text{diverges}.$$

Example 1   Determine whether the following integral is convergent or divergent, and calculate its value if it is convergent:

$$\int_0^\infty 2e^{-2x}\, dx.$$

Solution   We have

$$\int_0^\infty 2e^{-2x}\, dx = \lim_{b \to \infty} \int_0^b 2e^{-2x}\, dx$$

$$= \lim_{b \to \infty} \left[\frac{2}{-2} e^{-2x} \right]_0^b$$

$$= \lim_{b \to \infty} \left[-e^{-2x} \right]_0^b$$

$$= \lim_{b \to \infty} \left[-e^{-2b} - (-e^{-2\cdot 0}) \right]$$

$$= \lim_{b \to \infty} \left(-e^{-2b} + 1 \right)$$

$$= \lim_{b \to \infty} \left(1 - \frac{1}{e^{2b}} \right).$$

Now as b approaches ∞, we know that e^{2b} approaches ∞ (from Chapter 4), so

$$\frac{1}{e^{2b}} \to 0 \quad \text{and} \quad \left(1 - \frac{1}{e^{2b}} \right) \to 1.$$

Thus,

$$\int_0^\infty 2e^{-2x}\, dx = \lim_{b \to \infty} \left(1 - \frac{1}{e^{2b}} \right) = 1.$$

The integral is convergent.

Following are definitions of two other types of improper integrals.

Definitions

1. $\displaystyle\int_{-\infty}^{b} f(x)\ dx = \lim_{a\to-\infty} \int_{a}^{b} f(x)\ dx$

2. $\displaystyle\int_{-\infty}^{\infty} f(x)\ dx = \int_{-\infty}^{c} f(x)\ dx + \int_{c}^{\infty} f(x)\ dx$

In order for $\int_{-\infty}^{\infty} f(x)\ dx$ to converge, both integrals on the right in Definition 2 above must converge.

Applications

In Section 6.2, we learned that the accumulated present value of a continuous money flow of P dollars per year from now until time T in the future is given by

$$\int_{0}^{T} Pe^{-kt}\ dt = \frac{P}{k}(1 - e^{-kT}),$$

where k is the current interest rate. Suppose that the money flow is to continue perpetually. Under this assumption, the accumulated present value over this infinite time period would be

$$\int_{0}^{\infty} Pe^{-kt}\ dt = \lim_{T\to\infty} \int_{0}^{T} Pe^{-kt}\ dt$$

$$= \lim_{T\to\infty} \frac{P}{k}(1 - e^{-kT})$$

$$= \lim_{T\to\infty} \frac{P}{k}\left(1 - \frac{1}{e^{kT}}\right) = \frac{P}{k}.$$

 **TECHNOLOGY CONNECTION**

Exploratory. Consider Example 2 with a grapher. To evaluate the integral $\int_{0}^{\infty} 2000e^{-0.08x}\ dx$, we would first consider

$$\int_{0}^{t} 2000e^{-0.08x}\ dx = \frac{2000}{0.08}(1 - e^{-0.08t}).$$

Let's examine what happens as t gets large. Graph

$$f(x) = \frac{2000}{0.08}(1 - e^{-0.08x})$$

using the viewing window [0, 10, 0, 30000], with Xscl = 1 and Yscl = 5000. Using the same set of axes, graph $y = 25,000$. Then change the viewing window to [0, 50, 0, 30000], with Xscl = 10 and Yscl = 5000, and finally to [0, 100, 0, 30000]. What happens as x gets larger? What is the significance of 25,000?

Theorem 3

The **accumulated present value** of a continuous money flow into an investment at the rate of P dollars per year perpetually is given by

$$\frac{P}{k},$$

where k is the current interest rate, compounded continuously.

Example 2   *Business: Accumulated Present Value.* Find the accumulated present value of an investment for which there is a perpetual continuous money flow of $2000 per year. The current interest rate is 8%, compounded continuously.

Solution The accumulated present value is 2000/0.08, or $25,000.

When an amount P of radioactive material is being released into the atmosphere annually, the amount present at time T is given by

$$\int_0^T Pe^{-kt}\,dt = \frac{P}{k}(1 - e^{-kT}).$$

As T approaches ∞ (the radioactive material is to be released forever), the buildup of radioactive material approaches a limiting value P/k. It is no wonder that scientists and environmentalists are so concerned about continued nuclear detonations. Eventually the buildup is "here to stay."

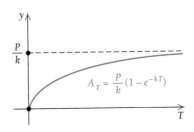

EXERCISE SET 6.3

Determine whether the improper integral is convergent or divergent, and calculate its value if it is convergent.

1. $\int_3^\infty \dfrac{dx}{x^2}$

2. $\int_4^\infty \dfrac{dx}{x^2}$

3. $\int_3^\infty \dfrac{dx}{x}$

4. $\int_4^\infty \dfrac{dx}{x}$

5. $\int_0^\infty 3e^{-3x}\,dx$

6. $\int_0^\infty 4e^{-4x}\,dx$

7. $\int_1^\infty \dfrac{dx}{x^3}$

8. $\int_1^\infty \dfrac{dx}{x^4}$

9. $\int_0^\infty \dfrac{dx}{1+x}$

10. $\int_0^\infty \dfrac{4\,dx}{1+x}$

11. $\int_1^\infty 5x^{-2}\,dx$

12. $\int_1^\infty 7x^{-2}\,dx$

13. $\int_0^\infty e^x\,dx$

14. $\int_0^\infty e^{2x}\,dx$

15. $\int_3^\infty x^2\,dx$

16. $\int_5^\infty x^4\,dx$

17. $\int_0^\infty xe^x\,dx$

18. $\int_1^\infty \ln x\,dx$

19. $\int_0^\infty me^{-mx}\,dx,\ m > 0$

20. $\int_0^\infty Qe^{-kt}\,dt,\ k > 0$

21. Find the area, if it exists, of the region under the graph of $y = 1/x^2$ over the interval $[2, \infty)$.

22. Find the area, if it exists, of the region under the graph of $y = 1/x$ over the interval $[2, \infty)$.

23. Find the area, if it exists, of the region bounded by $y = 2xe^{-x^2}$ and the lines $x = 0$ and $y = 0$.

24. Find the area, if it exists, of the region bounded by $y = 1/\sqrt{(3x - 2)^3}$ and the lines $x = 6$ and $y = 0$.

Applications ....
BUSINESS AND ECONOMICS

25. *Accumulated Present Value.* Find the accumulated present value of an investment for which there is a perpetual continuous money flow of $3600 per year. The current interest rate is 8%.

26. *Accumulated Present Value.* Find the accumulated present value of an investment for which there is a perpetual continuous money flow of $3500 per year. The current interest rate is 7%.

27. *Total Profit from Marginal Profit.* A firm is able to determine that its marginal profit is given by

$$P'(x) = 200e^{-0.032x}.$$

Suppose that it were possible for the firm to make infinitely many units of this product. What would its total profit be?

28. *Total Profit from Marginal Profit.* In Exercise 27, find the total profit if

$$P'(x) = 200x^{-1.032}, \quad \text{where } x \geq 1.$$

29. *Total Cost from Marginal Cost.* A company determines that its marginal cost for the production of x units of a product is given by

$$C'(x) = 3600x^{-1.8}, \quad \text{where } x \geq 1.$$

Suppose that it were possible for the firm to make infinitely many units of this product. What would the total cost be?

30. *Total Production.* A firm determines that it can produce tires at a rate of

$$r(t) = 2000e^{-0.42t},$$

where t is the time, in years. Assuming the firm endures forever, how many tires can it make?

LIFE AND PHYSICAL SCIENCES

31. *Radioactive Buildup.* Plutonium has a decay rate of 0.003% per year. Suppose that a nuclear accident causes plutonium to be released into the atmosphere each year perpetually at the rate of 1 lb per year. What is the limiting value of the radioactive buildup?

32. *Radioactive Buildup.* Cesium-137 has a decay rate of 2.3% per year. Suppose that a nuclear accident causes cesium-137 to be released into the atmosphere each year perpetually at the rate of 1 lb per year. What is the limiting value of the radioactive buildup?

Synthesis ....

Determine whether the improper integral is convergent or divergent, and calculate its value if it is convergent.

33. $\displaystyle\int_0^\infty \frac{dx}{x^{2/3}}$

34. $\displaystyle\int_1^\infty \frac{dx}{\sqrt{x}}$

35. $\displaystyle\int_0^\infty \frac{dx}{(x+1)^{3/2}}$

36. $\displaystyle\int_{-\infty}^0 e^{2x}\, dx$

37. $\displaystyle\int_0^\infty xe^{-x^2}\, dx$

38. $\displaystyle\int_{-\infty}^\infty xe^{-x^2}\, dx$

Life Science: Drug Dosage. Suppose that an oral dose of a drug is taken. From that time, the drug is assimilated in the body and excreted through the urine. The total amount of the drug that has passed through the body in time T is given by

$$\int_0^T E(t)\, dt,$$

where E is the rate of excretion of the drug. A typical rate-of-excretion function is $E(t) = te^{-kt}$, where $k > 0$ and t is the time, in hours. (Now do Exercises 39 and 40.)

39. Find $\int_0^\infty E(t)\, dt$ and interpret the answer. That is, what does the integral represent?

40. A physician prescribes a dosage of 100 mg. Find k.

tw 41. Consider the functions

$$y = \frac{1}{x^2} \quad \text{and} \quad y = \frac{1}{x}.$$

Suppose that you go to a paint store to buy paint to cover the region under each graph over the interval $[1, \infty)$. Discuss whether you could be successful and why or why not.

tw 42. Suppose that you are the owner of a building that yields a continuous series of rental payments and you decide to sell the building. Explain how you would use the concept of the accumulated present value of a perpetual continuous money flow to determine a fair selling price.

TECHNOLOGY CONNECTION

43. Graph the function E and shade the area under the curve for the situation in Exercises 39 and 40.

Approximate the integral.

44. $\displaystyle\int_1^\infty \frac{4}{1+x^2}\, dx$

45. $\displaystyle\int_1^\infty \frac{6}{5+e^x}\, dx$

6.4

Probability

A number between 0 and 1 that represents the chances that an event will occur is referred to as the event's **probability.** There are two types of probability, **experimental** and **theoretical.**

Experimental and Theoretical Probability

If we toss a coin a great number of times—say, 1000—and count the number of times it falls heads, we can determine the probability of it falling heads. If it falls heads 503 times, we would calculate the probability of it falling heads to be

$$\frac{503}{1000}, \quad \text{or} \quad 0.503.$$

This is an **experimental** determination of probability. Such a determination of probability is discovered by the observation and study of data and is quite common and very useful. Here, for example, are some probabilities that have been determined *experimentally*:

1. If a person has a heart attack, the probability he or she will die is $\frac{1}{3}$.
2. If you kiss someone who has a cold, the probability of your catching a cold is 0.07.
3. A person who has just been released from prison has an 80% probability of returning.

If we consider a coin and reason that it is just as likely to fall heads as tails when tossed, we would calculate the probability of it falling heads to be $\frac{1}{2}$. This is a **theoretical** determination of probability. Here, for example, are some probabilities that have been determined *theoretically*, using mathematics:

1. If there are 30 people in a room, the probability that two of them have the same birthday (excluding year) is 0.706.
2. While on a trip, you meet someone, and after a period of conversation, discover that you have a common acquaintance. The typical reaction, "It's a small world!", is actually not appropriate, because the probability of such an occurrence is quite high—just over 22%.

In summary, experimental probabilities are determined by making observations and gathering data. Theoretical probabilities are determined by reasoning mathematically. Examples of experimental and theoretical probabilities like those above, especially those we do not expect, lead us to see the value of a study of probability. You might ask, "What is *true* probability?" In fact, there is none. Experimentally, we can determine probabilities within certain limits. These may or may not agree with the probabilities that we obtain theoretically. There are situations in which it is much easier to determine one of these types of probabilities than the other. For example, it would be quite difficult to arrive at the probability of catching a cold using theoretical probability.

In the discussion that follows, we will consider primarily theoretical probability. Eventually, calculus will come to bear on our considerations.

A desire to calculate odds in games of chance gave rise to the theory of probability.

Example 1   What is the probability of drawing an ace from a well-shuffled deck of cards?

Solution   Since there are 52 possible outcomes and each card has the same chance of being drawn, and since there are 4 aces, the probability of drawing an ace is $\frac{4}{52}$ or $\frac{1}{13}$, or about 7.7%.

In practice, we may not draw an ace 7.7% of the time, but in a large number of trials, after shuffling the cards and drawing one, replacing the card, and shuffling the cards and drawing one, we would expect to draw an ace about 7.7% of the time. That is, the more draws we make, the closer we expect to get to 7.7%.   ◅

Example 2   A jar contains 7 black balls, 6 yellow balls, 4 green balls, and 3 red balls. The jar is shaken well and you remove 1 ball without looking. What is the probability that the ball is red? that it is white?

Solution   There are 20 balls altogether and of these 3 are red, so the probability of drawing a red ball is $\frac{3}{20}$. There are no white balls, so the probability of drawing a white one is $\frac{0}{20}$, or 0.   ◅

Let's consider a table of probabilities from Example 2. Note that the sum of these probabilities is 1. We are certain that we will draw either a black, yellow, green, or red ball. The probability of that event is 1. Let's arrange these data from the table into what is called a **relative frequency graph,** or **histogram,** which shows the proportion of times that each event occurs (the probability of each event).

Color	Probability
Black (B)	$\frac{7}{20}$
Yellow (Y)	$\frac{6}{20}$
Green (G)	$\frac{4}{20}$
Red (R)	$\frac{3}{20}$

If we assign a width of 1 to each rectangle, then the sum of the areas of the rectangles is 1. That is, it is certain that you will get a ball of one of these colors.

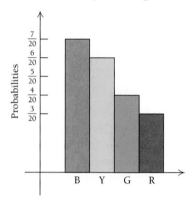

Continuous Random Variables

Suppose we throw a dart at a number line in such a way that it always lands in the interval [1, 3]. Let x be the number that the dart hits. There is an infinite number of possibilities for x. Note that x is a quantity that can be observed (or measured) repeatedly and whose possible values consist of an entire interval of real numbers. Such a variable is called a **continuous random variable.**

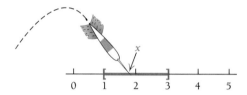

Suppose we throw the dart a large number of times and it lands 43% of the time in the subinterval [1.6, 2.8] of the main interval [1, 3]. The probability, then, that the dart lands in the interval [1.6, 2.8] is 0.43.

Let's consider some other examples of continuous random variables.

Example 3 Suppose that x is the arrival time of buses at a bus stop in a 3-hr period from 2:00 P.M. to 5:00 P.M. The interval is [2, 5]. Then x is a continuous random variable distributed over the interval [2, 5]. ◀

Example 4 Suppose that x is the corn acreage of any farm in the United States and Canada. The interval is [0, a], where a is the highest acreage. (Not knowing what the highest acreage might be, we could say that the interval is [0, ∞) to allow for all possibilities. *Note*: It might be argued that there is a value in [0, a] or [0, ∞) for which no farm has that acreage, but for practical convenience, all values are included in our consideration.)

(a)   (b)

Then x is a continuous random variable distributed over the interval $[0, a]$ or $[0, \infty)$.   ◄

Considering Example 3 on the arrival times of buses, suppose we want to know the probability that a bus will arrive between 4:00 P.M. and 5:00 P.M., as represented by

$$P([4, 5]), \quad \text{or} \quad P(4 \le x \le 5).$$

There may be a function $y = f(x)$ such that the area under the graph over subintervals of $[2, 5]$ will give the probabilities that a bus will arrive during these subintervals. For example, suppose we have a constant function $f(x) = \frac{1}{3}$ that will give us these probabilities. Look at its graph.

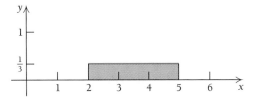

The area under the curve is $3 \cdot \frac{1}{3}$, or 1. The probability that a bus will arrive between 4:00 P.M. and 5:00 P.M. is that fraction of the large area that lies over the interval $[4, 5]$. That is,

$$P([4, 5]) = \frac{1}{3} = 33\tfrac{1}{3}\%.$$

This is the area of the rectangle over $[4, 5]$.

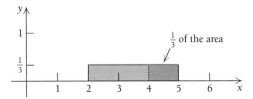

The probability that a bus will arrive between 2:00 P.M. and 4:30 P.M. is $\frac{5}{6}$, or $83\tfrac{1}{3}\%$. This is the area of the rectangle over $[2, 4.5]$.

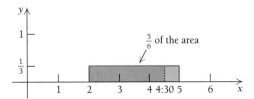

Note that in the above example, where $f(x) = \frac{1}{3}$, any interval between 2:00 P.M. and 5:00 P.M. of length 1 has probability $\frac{1}{3}$. This might not always happen. Suppose instead that

$$f(x) = \frac{3}{117}x^2.$$

As shown in the graph below, the area under the graph of $f(x)$ from 4 to 5 is given by the definite integral over the interval $[4, 5]$ and would yield the probability that a bus will arrive between 4:00 P.M. and 5:00 P.M.

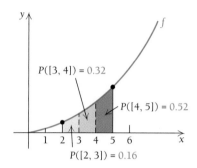

Exploratory.
Graph the function

$$f(x) = \tfrac{3}{117}x^2$$

using a viewing window of $[0, 5, 0, 1]$. Then successively evaluate each of the following integrals, shading the appropriate area if possible:

$$\int_2^3 \tfrac{3}{117}x^2 \, dx,$$

$$\int_3^4 \tfrac{3}{117}x^2 \, dx,$$

and

$$\int_4^5 \tfrac{3}{117}x^2 \, dx.$$

Add your results and explain the meaning of the answer.

Thus,

$$P([4, 5]) = \int_4^5 f(x) \, dx$$

$$= \int_4^5 \frac{3}{117}x^2 \, dx$$

$$= \frac{3}{117}\left[\frac{x^3}{3}\right]_4^5$$

$$= \frac{1}{117}\left[x^3\right]_4^5$$

$$= \frac{1}{117}(5^3 - 4^3) = \frac{61}{117} \approx 0.52.$$

Thus there is a probability of 0.52 that at least one bus will arrive between 4:00 P.M. and 5:00 P.M. The function f is called a **probability density function.** Its integral over *any* subinterval gives the probability that x "lands" in that subinterval.

Similar calculations are shown in the following table.

Time Interval	Probability That a Bus Arrives During Interval
2:00 P.M. and 3:00 P.M.	$\int_2^3 \tfrac{3}{117}x^2 \, dx = 0.16 = P([2, 3])$
3:00 P.M. and 4:00 P.M.	$\int_3^4 \tfrac{3}{117}x^2 \, dx = 0.32 = P([3, 4])$
4:00 P.M. and 5:00 P.M.	$\int_4^5 \tfrac{3}{117}x^2 \, dx = 0.52 = P([4, 5])$
2:00 P.M. and 5:00 P.M.	$\int_2^5 \tfrac{3}{117}x^2 \, dx = 1.00 = P([2, 5])$

The results in the table lead us to the following definition of a probability density function.

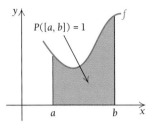

FIGURE 1

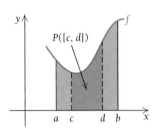

FIGURE 2

Definition

Let x be a continuous random variable distributed over some interval $[a, b]$. A function f is said to be a **probability density function** for x if:

1. f is nonnegative over $[a, b]$; that is, $f(x) \geq 0$ for all x in $[a, b]$.

2. $\displaystyle\int_a^b f(x)\, dx = 1.$

Since $P([a, b]) = \displaystyle\int_a^b f(x)\, dx = 1$, it is "certain" that x lands in the interval (see Figure 1).

3. For any subinterval $[c, d]$ of $[a, b]$ (see Figure 2), the probability $P([c, d])$, or $P(c \leq x \leq d)$, that x lands in that subinterval is given by

$$P([c, d]) = \int_c^d f(x)\, dx.$$

Example 5 Verify Property 2 of the definition above for

$$f(x) = \frac{3}{117}x^2, \quad \text{for } 2 \leq x \leq 5.$$

Solution

$$
\begin{aligned}
\int_2^5 \frac{3}{117}x^2\, dx &= \frac{3}{117}\left[\frac{1}{3}x^3\right]_2^5 \\
&= \frac{1}{117}\left[x^3\right]_2^5 \\
&= \frac{1}{117}(5^3 - 2^3) \\
&= \frac{117}{117} = 1
\end{aligned}
$$

Example 6 *Business: Life of a Product.* A company that produces transistors determines that the life t of a transistor is from 3 to 6 yr and that the probability density function for t is given by

$$f(t) = \frac{24}{t^3}, \quad \text{for } 3 \leq t \leq 6.$$

a) Verify Property 2.

b) Find the probability that a transistor will last no more than 4 yr.

c) Find the probability that a transistor will last at least 4 yr and at most 5 yr.

Solution

a) We want to show that $\int_3^6 f(t)\, dt = 1$. Now

$$\int_3^6 \frac{24}{t^3}\, dt = 24 \int_3^6 t^{-3}\, dt = 24 \left[\frac{t^{-2}}{-2} \right]_3^6$$

$$= -12 \left[\frac{1}{t^2} \right]_3^6 = -12 \left(\frac{1}{6^2} - \frac{1}{3^2} \right)$$

$$= -12 \left(\frac{1}{36} - \frac{1}{9} \right) = -12 \left(-\frac{3}{36} \right) = 1.$$

b) The probability that a transistor will last no more than 4 yr is

$$P(3 \le t \le 4) = \int_3^4 \frac{24}{t^3}\, dt = 24 \int_3^4 t^{-3}\, dt$$

$$= 24 \left[\frac{t^{-2}}{-2} \right]_3^4 = -12 \left[\frac{1}{t^2} \right]_3^4$$

$$= -12 \left(\frac{1}{4^2} - \frac{1}{3^2} \right) = -12 \left(\frac{1}{16} - \frac{1}{9} \right)$$

$$= -12 \left(-\frac{7}{144} \right) = \frac{7}{12} \approx 0.58.$$

c) The probability that a transistor will last at least 4 yr and at most 5 yr is

$$P(4 \le t \le 5) = \int_4^5 \frac{24}{t^3}\, dt = 24 \int_4^5 t^{-3}\, dt$$

$$= 24 \left[\frac{t^{-2}}{-2} \right]_4^5 = -12 \left[\frac{1}{t^2} \right]_4^5$$

$$= -12 \left(\frac{1}{5^2} - \frac{1}{4^2} \right) = -12 \left(\frac{1}{25} - \frac{1}{16} \right)$$

$$= -12 \left(-\frac{9}{400} \right) = \frac{27}{100} = 0.27.$$

Constructing Probability Density Functions

Suppose that you have an arbitrary nonnegative function $f(x)$ whose definite integral over some interval $[a, b]$ is K. Then

$$\int_a^b f(x)\, dx = K.$$

Multiplying on both sides by $1/K$ gives us

$$\frac{1}{K} \int_a^b f(x)\, dx = \frac{1}{K} \cdot K = 1, \quad \text{or} \quad \int_a^b \frac{1}{K} \cdot f(x)\, dx = 1.$$

Thus when we multiply the function $f(x)$ by $1/K$, we have a function whose area over the given interval is 1.

Example 7 Find k such that

$$f(x) = kx^2$$

is a probability density function over the interval $[2, 5]$. Then write the probability density function.

Solution We have

$$\int_2^5 x^2 \, dx = \left[\frac{x^3}{3}\right]_2^5$$

$$= \frac{5^3}{3} - \frac{2^3}{3} = \frac{125}{3} - \frac{8}{3} = \frac{117}{3}.$$

Thus,

$$k = \frac{1}{\frac{117}{3}} = \frac{3}{117}$$

and the probability density function is

$$f(x) = \frac{3}{117}x^2.$$

Uniform Distributions

Suppose that the probability density function of a continuous random variable is constant. How is it described? Consider the graph shown below.

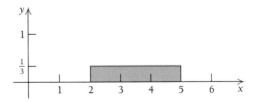

The length of the shaded rectangle is the length of the interval $[2, 5]$, which is 3. In order for the shaded area to be 1, the height of the rectangle must be $\frac{1}{3}$. Thus, $f(x) = \frac{1}{3}$.

The length of the shaded rectangle shown below is the length of the interval $[a, b]$, which is $b - a$. In order for the shaded area to be 1, the height of the rectangle must be $1/(b - a)$. Thus, $f(x) = 1/(b - a)$.

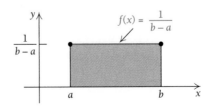

> **Definition**
>
> A continuous random variable x is said to be **uniformly distributed** over an interval $[a, b]$ if it has a probability density function f given by
>
> $$f(x) = \frac{1}{b - a}, \quad \text{for } a \leq x \leq b.$$

Example 8 A number x is selected at random from the interval $[40, 50]$. The probability density function for x is given by

$$f(x) = \frac{1}{10}, \quad \text{for } 40 \leq x \leq 50.$$

Find the probability that a number selected is in the subinterval $[42, 48]$.

Solution The probability is

$$P(42 \leq x \leq 48) = \int_{42}^{48} \tfrac{1}{10}\, dx = \tfrac{1}{10}[x]_{42}^{48}$$

$$= \tfrac{1}{10}(48 - 42) = \tfrac{6}{10} = 0.6.$$

Example 9 *Business: Quality Control.* A company produces sirens for tornado warnings. The maximum loudness L of the sirens ranges from 70 to 100 decibels. The probability density function for L is

$$f(L) = \tfrac{1}{30}, \quad \text{for } 70 \leq L \leq 100.$$

A siren is selected at random off the assembly line. Find the probability that its maximum loudness is from 70 to 92 decibels.

Solution The probability is

$$P(70 \leq L \leq 92) = \int_{70}^{92} \tfrac{1}{30}\, dL = \tfrac{1}{30}[L]_{70}^{92}$$

$$= \tfrac{1}{30}(92 - 70) = \tfrac{22}{30} = \tfrac{11}{15} \approx 0.73.$$

Exponential Distributions

The duration of a phone call, the distance between successive cars on a highway, and the amount of time required to learn a task are all examples of exponentially distributed random variables. That is, their probability density functions are exponential.

> **Definition**
>
> A continuous random variable is **exponentially distributed** if it has a probability density function given by
>
> $$f(x) = ke^{-kx}, \quad \text{over the interval } [0, \infty).$$

The function $f(x) = 2e^{-2x}$ is such a probability density function. That

$$\int_0^\infty 2e^{-2x}\, dx = 1$$

is shown in Section 6.3. The general case

$$\int_0^\infty ke^{-kx}\, dx = 1$$

can be verified in a similar way.

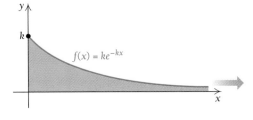

Why is it reasonable to assume that the distance between cars is exponentially distributed? Part of the reason is that there are many more cases in which distances are small, though we can find other distributions that are "skewed" in this manner. The same argument holds for the duration of a phone call. That is, there are more short calls than long ones. The rest of the reason might lie in an analysis of the data involving such distances or phone calls.

A transportation planner can determine probabilities that cars are certain distances apart.

Example 10   *Business: Transportation Planning.* The distance x, in feet, between successive cars on a certain stretch of highway has a probability density function

$$f(x) = ke^{-kx}, \quad \text{for } 0 \le x < \infty,$$

where $k = 1/a$ and a is the average distance between successive cars over some period of time.

A transportation planner determines that the average distance between cars on a certain stretch of highway is 166 ft. What is the probability that if we choose two successive cars at random, the distance between those cars is 50 ft or less?

Solution   We first determine k:

$$k = \tfrac{1}{166}$$
$$\approx 0.006024.$$

The probability density function for x is

$$f(x) = 0.006024e^{-0.006024x}, \quad \text{for } 0 \le x < \infty.$$

The probability that the distance between the cars is 50 ft or less is

$$P(0 \le x \le 50) = \int_0^{50} 0.006024e^{-0.006024x}\, dx$$

$$= \left[\frac{0.006024}{-0.006024} e^{-0.006024x} \right]_0^{50}$$

$$= \left[-e^{-0.006024x} \right]_0^{50}$$

$$= \left(-e^{-0.006024 \cdot 50} \right) - \left(-e^{-0.006024 \cdot 0} \right)$$

$$= -e^{-0.301200} + 1$$

$$= 1 - e^{-0.301200}$$

$$= 1 - 0.739930 \approx 0.260.$$

EXERCISE SET 6.4

Verify Property 2 of the definition of a probability density function over the given interval.

1. $f(x) = 2x$, $[0, 1]$

2. $f(x) = \frac{1}{4}x$, $[1, 3]$

3. $f(x) = \frac{1}{3}$, $[4, 7]$

4. $f(x) = \frac{1}{4}$, $[9, 13]$

5. $f(x) = \frac{3}{26}x^2$, $[1, 3]$

6. $f(x) = \frac{3}{64}x^2$, $[0, 4]$

7. $f(x) = \dfrac{1}{x}$, $[1, e]$

8. $f(x) = \dfrac{1}{e - 1}e^x$, $[0, 1]$

9. $f(x) = \frac{3}{2}x^2$, $[-1, 1]$

10. $f(x) = \frac{1}{3}x^2$, $[-2, 1]$

11. $f(x) = 3e^{-3x}$, $[0, \infty)$

12. $f(x) = 4e^{-4x}$, $[0, \infty)$

Find k such that the function is a probability density function over the given interval. Then write the probability density function.

13. $f(x) = kx$, $[1, 3]$

14. $f(x) = kx$, $[1, 4]$

15. $f(x) = kx^2$, $[-1, 1]$

16. $f(x) = kx^2$, $[-2, 2]$

17. $f(x) = k$, $[2, 7]$

18. $f(x) = k$, $[3, 9]$

19. $f(x) = k(2 - x)$, $[0, 2]$

20. $f(x) = k(4 - x)$, $[0, 4]$

21. $f(x) = \dfrac{k}{x}$, $[1, 3]$

22. $f(x) = \dfrac{k}{x}$, $[1, 2]$

23. $f(x) = ke^x$, $[0, 3]$

24. $f(x) = ke^x$, $[0, 2]$

25. A dart is thrown at a number line in such a way that it always lands in the interval $[0, 10]$. Let $x = $ the number that the dart hits. Suppose that the probability density function for x is given by

$$f(x) = \tfrac{1}{50}x, \quad \text{for } 0 \le x \le 10.$$

a) Find $P(2 \le x \le 6)$, the probability that it lands in $[2, 6]$.

b) Interpret your answer to part (a).

26. In Exercise 25, suppose that the dart always lands in the interval $[0, 5]$, and that the probability density function for x is given by

$$f(x) = \tfrac{3}{125}x^2, \quad \text{for } 0 \le x \le 5.$$

a) Find $P(1 \le x \le 4)$, the probability that it lands in $[1, 4]$.

b) Interpret your answer to part (a).

27. A number x is selected at random from the interval $[4, 20]$. The probability density function for x is given by

$$f(x) = \tfrac{1}{16}, \quad \text{for } 4 \le x \le 20.$$

Find the probability that a number selected is in the subinterval $[9, 17]$.

28. A number x is selected at random from the interval $[5, 29]$. The probability density function for x is given by

$$f(x) = \tfrac{1}{24}, \quad \text{for } 5 \le x \le 29.$$

Find the probability that a number selected is in the subinterval $[13, 29]$.

Applications
BUSINESS AND ECONOMICS

29. *Transportation Planning.* A transportation planner determines that the average distance between cars on a certain highway is 100 ft. What is the probability that if we choose two successive cars at random, the distance between those cars is 40 ft or less?

30. *Transportation Planning.* A transportation planner determines that the average distance between cars on a certain highway is 200 ft. What is the probability that if we choose two successive cars at random, the distance between those cars is 10 ft or less?

31. *Duration of a Phone Call.* A telephone company determines that the duration t of a phone call is an exponentially distributed random variable with probability density function

$$f(t) = 2e^{-2t}, \quad 0 \le t < \infty.$$

Find the probability that a phone call will last no more than 5 min.

32. *Duration of a Phone Call.* Referring to Exercise 31, find the probability that a phone call will last no more than 2 min.

33. *Time to Failure.* The *time to failure t*, in hours, of a certain machine can often be assumed to be exponentially distributed with probability density function

$$f(t) = ke^{-kt}, \quad 0 \le t < \infty,$$

where $k = 1/a$ and a is the average amount of time that will pass before a failure occurs. Suppose the average amount of time that will pass before a failure occurs is 100 hr. What is the probability that a failure will occur in 50 hr or less?

34. *Reliability of a Machine.* The *reliability* of the machine (the probability that it will work) in Exercise 33 is defined as

$$R(T) = 1 - \int_0^T 0.01e^{-0.01t}\, dt,$$

where $R(T)$ is the reliability at time T. Find $R(T)$.

SOCIAL SCIENCES

35. *Time in a Maze.* In a psychology experiment, the time t, in seconds, that it takes a rat to learn its way through a maze is an exponentially distributed random variable with probability density function

$$f(t) = 0.02e^{-0.02t}, \quad 0 \le t < \infty.$$

Find the probability that a rat will learn its way through a maze in 150 sec or less.

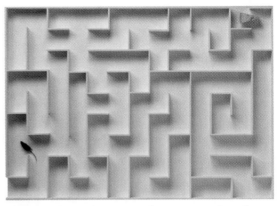

The time that it takes a rat to learn its way through a maze is an exponentially distributed random variable.

36. *Time in a Maze.* Using the situation and the equation in Exercise 35, find the probability that a rat will learn its way through the maze in 50 sec or less.

Synthesis

37. The function $f(x) = x^3$ is a probability density function over $[0, b]$. What is b?

38. The function $f(x) = 12x^2$ is a probability density function over $[-a, a]$. What is a?

tw 39. Explain the idea of a probability density function.

tw 40. Give as many examples as you can of the use of probability in daily life.

 TECHNOLOGY CONNECTION

41.–52. Verify Property 2 of the definition of a probability density function for each of the functions in Exercises 1–12.

6.5

Probability: Expected Value; The Normal Distribution

Expected Value

Let's again consider throwing a dart at a number line in such a way that it always lands in the interval $[1, 3]$. This time we will assume that it is *equally likely* for the dart to land anywhere in the interval.

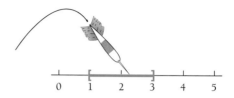

Suppose that we throw the dart at the line 100 times and keep track of the numbers it hits. Then suppose we calculate the arithmetic mean (or average) $\bar{x}$ of all these numbers:

$$\bar{x} = \frac{x_1 + x_2 + x_3 + \cdots + x_{100}}{100} = \frac{\sum\limits_{i=1}^{100} x_i}{100}$$

$$= \sum_{i=1}^{100} x_1 \cdot \frac{1}{100}.$$

The expression

$$\sum_{i=1}^{n} x_i \cdot \frac{1}{n}, \quad \text{or} \quad \sum_{i=1}^{n} x_i \cdot \frac{1}{2} \cdot \frac{2}{n}$$

is analogous to the integral

$$\int_1^3 x \cdot f(x)\, dx,$$

where f is called a *probability density function* for x and where, in this case, $f(x)$ is the constant function $\frac{1}{2}$. The probability density function gives a "weight" to x. Also, $2/n$ can be considered to be Δx. We add all the

$$x_i \cdot \frac{1}{2} \cdot \left(\frac{2}{n}\right)$$

values when we find $\sum_{i=1}^{n} x_i \cdot \frac{1}{2} \cdot (2/n)$; and, similarly, we add all the

$$x \cdot f(x) \cdot \Delta x$$

values when we find $\int_1^3 x \cdot f(x)\, dx$.

Suppose that we have the probability density function $f(x) = \frac{1}{4}x$ over the interval $[1, 3]$. As we can see from the graph below, this function gives more "weight" to the right side of the interval than to the left. Perhaps more points are awarded when the dart lands on the right. Then

$$\int_1^3 x \cdot f(x)\, dx = \int_1^3 x \cdot \frac{1}{4}x\, dx$$

$$= \frac{1}{4} \int_1^3 x^2\, dx$$

$$= \frac{1}{4}\left[\frac{x^3}{3}\right]_1^3$$

$$= \frac{1}{12}(3^3 - 1^3) = \frac{26}{12} \approx 2.17.$$

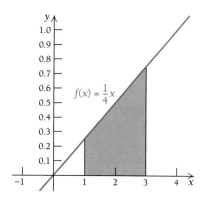

Suppose that we continue to throw the dart and compute averages. The more times we throw the dart, the closer we expect the averages to come to 2.17.

Let x be a continuous random variable over the interval $[a, b]$ with probability density function f.

Definition

The **expected value** of x is defined by

$$E(x) = \int_a^b x \cdot f(x)\, dx,$$

where f is a probability density function for x.

The concept of expected value of a random variable generalizes to other functions of a random variable. Suppose that $y = g(x)$ is a function of the random variable x. Then we have the following.

Definition

The **expected value** of $g(x)$ is defined by

$$E(g(x)) = \int_a^b g(x) \cdot f(x)\, dx,$$

where f is a probability density function for x.

For example,

$$E(x) = \int_a^b x f(x)\, dx,$$

$$E(x^2) = \int_a^b x^2 f(x)\, dx,$$

$$E(e^x) = \int_a^b e^x f(x)\, dx,$$

and

$$E(2x + 3) = \int_a^b (2x + 3)f(x)\, dx.$$

Example 1   Given the probability density function

$$f(x) = \tfrac{1}{2}x \quad \text{over } [0, 2],$$

find $E(x)$ and $E(x^2)$.

Solution

$$E(x) = \int_0^2 x \cdot \frac{1}{2}x\, dx = \int_0^2 \frac{1}{2}x^2\, dx$$

$$= \frac{1}{2}\left[\frac{x^3}{3}\right]_0^2 = \frac{1}{6}\left[x^3\right]_0^2$$

$$= \frac{1}{6}(2^3 - 0^3)$$

$$= \frac{1}{6} \cdot 8 = \frac{4}{3};$$

$$E(x^2) = \int_0^2 x^2 \cdot \frac{1}{2}x\, dx = \int_0^2 \frac{1}{2}x^3\, dx = \frac{1}{2}\left[\frac{x^4}{4}\right]_0^2$$

$$= \frac{1}{8}\left[x^4\right]_0^2 = \frac{1}{8}(2^4 - 0^4)$$

$$= \frac{1}{8} \cdot 16 = 2$$

Definition

The **mean μ** of a continuous random variable x is defined to be $E(x)$. That is,

$$\mu = E(x) = \int_a^b xf(x)\, dx,$$

where f is a probability density function for x. (The symbol μ is the lower-case Greek letter "mu.")

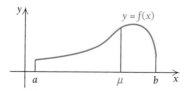

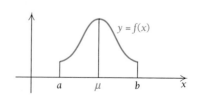

We can get a physical idea of a mean of a random variable by pasting the graph of the probability density function on cardboard and cutting out the area under the curve over the interval $[a, b]$. Then we try to find a fulcrum, or balance point, on the x-axis. That balance point is the mean.

Definition

The **variance σ^2** of a continuous random variable x is defined as

$$\sigma^2 = E(x^2) - \mu^2$$
$$= E(x^2) - [E(x)]^2$$
$$= \int_a^b x^2 f(x) \, dx - \left[\int_a^b x f(x) \, dx\right]^2.$$

The **standard deviation σ** of a continuous random variable is defined as

$$\sigma = \sqrt{\text{variance}}.$$

(The symbol σ is the lower-case Greek letter "sigma.")

Example 2   Given the probability density function

$$f(x) = \tfrac{1}{2}x \quad \text{over } [0, 2],$$

find the mean, the variance, and the standard deviation.

Solution   From Example 1, we have

$$E(x) = \tfrac{4}{3} \quad \text{and} \quad E(x^2) = 2.$$

Then

$$\textit{the mean} = \mu = E(x) = \tfrac{4}{3};$$
$$\textit{the variance} = \sigma^2 = E(x^2) - [E(x)]^2$$
$$= 2 - \left(\tfrac{4}{3}\right)^2 = 2 - \tfrac{16}{9}$$
$$= \tfrac{18}{9} - \tfrac{16}{9} = \tfrac{2}{9};$$
$$\textit{the standard deviation} = \sigma = \sqrt{\tfrac{2}{9}}$$
$$= \tfrac{1}{3}\sqrt{2} \approx 0.47.$$

Loosely speaking, we say that the standard deviation is a measure of how close the graph of f is to the mean, that is, the line $x = \mu$ as indicated below.

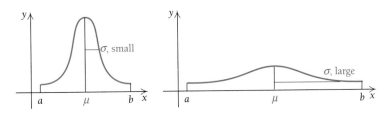

The Normal Distribution

Suppose that the average on a test is 70. Usually there are about as many scores above the average as there are below the average; and the further away from the average a particular score is, the fewer people there are who get that score. In this example, it is probable that more people would score in the 80s than in the 90s, and more people would score in the 60s than in the 50s. Test scores, heights of human beings, and weights of human beings are all examples of random variables that are often *normally* distributed.

Consider the function

$$g(x) = e^{-x^2/2} \quad \text{over the interval } (-\infty, \infty).$$

This function has the entire set of real numbers as its domain. Its graph is the bell-shaped curve shown below. We can find function values by using a calculator:

$$y = e^{-x^2/2}.$$

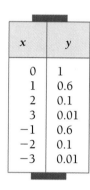

x	y
0	1
1	0.6
2	0.1
3	0.01
−1	0.6
−2	0.1
−3	0.01

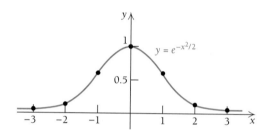

This function has an antiderivative, but that antiderivative has no elementary formula. Nevertheless, it has been shown that its improper integral converges over the integral $(-\infty, \infty)$ to a number given by

$$\int_{-\infty}^{\infty} e^{-x^2/2}\, dx = \sqrt{2\pi}.$$

That is, although an elementary expression for the antiderivative cannot be found, there is a numerical value for the improper integral evaluated over the set of real numbers. Note that since the area is not 1, the function g is not a probability density function, but the following is:

$$\frac{1}{\sqrt{2\pi}} e^{-x^2/2}.$$

> **Definition**
>
> A continuous random variable x has a **standard normal distribution** if its probability density function is
>
> $$f(x) = \frac{1}{\sqrt{2\pi}} e^{-x^2/2} \quad \text{over } (-\infty, \infty).$$

This distribution has a mean of 0 and a standard deviation of 1. Its graph follows.

Exploratory.   Use a grapher to approximate

$$\int_{-b}^{b} \frac{1}{\sqrt{2\pi}} e^{-x^2/2} \, dx$$

for $b = 10$, 100, and 1000. What does this suggest about

$$\int_{-\infty}^{\infty} \frac{1}{\sqrt{2\pi}} e^{-x^2/2} \, dx?$$

This is a way to verify part of the assertion that

$$f(x) = \frac{1}{\sqrt{2\pi}} e^{-x^2/2}$$

is a probability density function. Then use a similar approximation procedure to show that the mean is 0 and the standard deviation is 1.

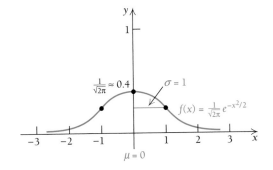

The general case is defined as follows.

Definition

A continuous random variable x is **normally distributed** with mean μ and standard deviation σ if its probability density function is given by

$$f(x) = \frac{1}{\sigma\sqrt{2\pi}} e^{-(1/2)[(x-\mu)/\sigma]^2} \quad \text{over } (-\infty, \infty).$$

The graph is a transformation of the graph of the standard density function. This can be shown by translating the graph along the x-axis and changing the way in which the graph is clustered about the mean. Some examples follow.

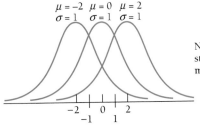

Normal distributions with same standard deviations but different means

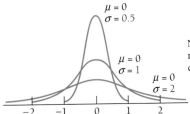

Normal distributions with same means but different standard deviations

The normal distribution is extremely important in statistics; it underlies much of the research in the behavioral and social sciences. Because of this, tables of approximate values of the definite integral of the standard density functions have been prepared using numerical approximation methods like the Trapezoidal Rule given in Exercise Set 5.3. Table 2 at the back of the book is such a table. It contains values of

$$P(0 \le x \le t) = \int_0^t \frac{1}{\sqrt{2\pi}} e^{-x^2/2} \, dx.$$

The symmetry of the graph about the mean allows many types of probabilities to be computed from the table.

Example 3   Let x be a continuous random variable with standard normal density. Using Table 2, find each of the following.

a) $P(0 \le x \le 1.68)$ b) $P(-0.97 \le x \le 0)$

c) $P(-2.43 \le x \le 1.01)$ d) $P(1.90 \le x \le 2.74)$

e) $P(-2.98 \le x \le -0.42)$ f) $P(x \ge 0.61)$

Solution

a) $P(0 \le x \le 1.68)$ is the area bounded by the standard normal curve and the lines $x = 0$ and $x = 1.68$. We look this up in Table 2 by going down the left column to 1.6, then moving to the right to the column headed 0.08. There we read 0.4535. Thus,

$$P(0 \le x \le 1.68) = 0.4535.$$

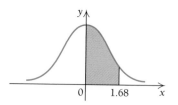

b) Because of the symmetry of the graph,

$$P(-0.97 \le x \le 0)$$
$$= P(0 \le x \le 0.97)$$
$$= 0.3340.$$

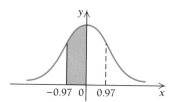

c) $P(-2.43 \le x \le 1.01)$
$$= P(-2.43 \le x \le 0) + P(0 \le x \le 1.01)$$
$$= P(0 \le x \le 2.43) + P(0 \le x \le 1.01)$$
$$= 0.4925 + 0.3438$$
$$= 0.8363$$

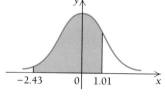

d) $P(1.90 \le x \le 2.74)$
$$= P(0 \le x \le 2.74) - P(0 \le x \le 1.90)$$
$$= 0.4969 - 0.4713$$
$$= 0.0256$$

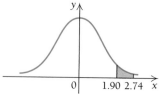

e) $P(-2.98 \leq x \leq -0.42)$
$= P(0.42 \leq x \leq 2.98)$
$= P(0 \leq x \leq 2.98) - P(0 \leq x \leq 0.42)$
$= 0.4986 - 0.1628$
$= 0.3358$

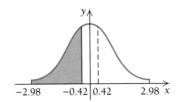

f) $P(x \geq 0.61)$
$= P(x \geq 0) - P(0 \leq x \leq 0.61)$
$= 0.5000 - 0.2291$
$= 0.2709$

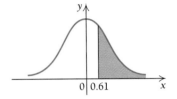

(Because of the symmetry about the line $x = 0$, half the area is on each side of the line, and since the entire area is 1, $P(x \geq 0) = 0.5000$.)

In many applications, a normal distribution is not standard. It would be a hopeless task to make tables for all values of the mean μ and the standard deviation σ. In such cases, the transformation

$$X = \frac{x - \mu}{\sigma}$$

standardizes the distribution, permitting the use of Table 2 at the back of the book. That is,

$$P(a \leq x \leq b) = P\left(\frac{a - \mu}{\sigma} \leq X \leq \frac{b - \mu}{\sigma}\right),$$

and the probability on the right can be found using Table 2. To see this, consider

$$P(a \leq x \leq b) = \int_a^b \frac{1}{\sigma\sqrt{2\pi}} e^{-(1/2)[(x-\mu)/\sigma]^2} \, dx,$$

and make the substitution

$$X = \frac{x - \mu}{\sigma}$$

$$= \frac{x}{\sigma} - \frac{\mu}{\sigma}.$$

Then

$$dX = \frac{1}{\sigma} dx.$$

When $x = a$, $X = (a - \mu)/\sigma$; and when $x = b$, $X = (b - \mu)/\sigma$. Then

$$P(a \leq x \leq b) = \int_a^b \frac{1}{\sigma\sqrt{2\pi}} e^{-(1/2)[(x-\mu)/\sigma]^2} \, dx$$

and

$$P(a \leq x \leq b) = \int_{(a-\mu)/\sigma}^{(b-\mu)/\sigma} \frac{1}{\sqrt{2\pi}} e^{-(1/2)X^2} \, dX$$

The integrand is now in the form of the standard density.

$$= P\left(\frac{a - \mu}{\sigma} \leq X \leq \frac{b - \mu}{\sigma} \right).$$

We can look this up in Table 2.

Example 4   The weights w of the students in a calculus class are normally distributed with mean 150 lb and standard deviation 25 lb. Find the probability that a student's weight is from 160 lb to 180 lb.

Solution   We first standardize the weights:

$$180 \text{ is standardized to } \frac{b - \mu}{\sigma} = \frac{180 - 150}{25} = 1.2;$$

$$160 \text{ is standardized to } \frac{a - \mu}{\sigma} = \frac{160 - 150}{25} = 0.4.$$

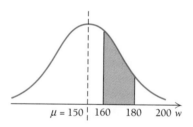

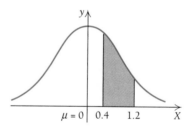

Then

$$\begin{aligned}
P(160 \leq w \leq 180) &= P(0.4 \leq X \leq 1.2) \qquad \text{Now we can use Table 2.}\\
&= P(0 \leq X \leq 1.2) - P(0 \leq X \leq 0.4)\\
&= 0.3849 - 0.1554\\
&= 0.2295.
\end{aligned}$$

Thus the probability that a student's weight is from 160 lb to 180 lb is 0.2295. That is, about 23% of the students have weights from 160 lb to 180 lb.   ◄

TECHNOLOGY CONNECTION

Statistics on a Grapher

Many new graphers have extensive statistical capabilities. While most of these will be advantageous to you in other courses or applications, we will use the TI-83 here to make an approximation in Example 4 without performing a standard conversion, using Table 2, or entering the normal probability density function. We first select an appropriate window, [0, 300, −0.002, 0.02], with Xscl = 50 and Yscl = 0.01. Next, we use the ShadeNorm command, which we find by pressing ⟨2nd⟩ ⟨DISTR⟩ ⟨▷⟩ ⟨ENTER⟩.

Left endpoint of interval   Right endpoint of interval

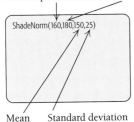
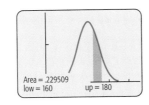

Mean   Standard deviation

The area is shaded and given as 0.229509, or about 23%. If we are entering intervals involving infinity without endpoints, using extremely large numbers like 1 E 99 or −1 E 99 will allow adequate approximations.

Exercises

1. The weights of the students in a calculus class are normally distributed with mean 150 lb and standard deviation 25 lb.
   a) What is the probability that a student's weight is from 125 lb to 170 lb?
   b) What is the probability that a student's weight is greater than 200 lb?

2. *SAT Scores.* SAT test scores are normally distributed with mean 800 and standard deviation 430.
   a) What percentage of the scores are between 270 and 500?
   b) What percentage of the scores are above 640?

EXERCISE SET 6.5

For each probability density function, over the given interval, find $E(x)$, $E(x^2)$, the mean, the variance, and the standard deviation.

1. $f(x) = \frac{1}{3}$, [2, 5]
2. $f(x) = \frac{1}{4}$, [3, 7]
3. $f(x) = \frac{2}{9}x$, [0, 3]
4. $f(x) = \frac{1}{8}x$, [0, 4]
5. $f(x) = \frac{2}{3}x$, [1, 2]
6. $f(x) = \frac{1}{4}x$, [1, 3]
7. $f(x) = \frac{1}{3}x^2$, [−2, 1]
8. $f(x) = \frac{3}{2}x^2$, [−1, 1]
9. $f(x) = \frac{1}{\ln 3} \cdot \frac{1}{x}$, [1, 3]
10. $f(x) = \frac{1}{\ln 2} \cdot \frac{1}{x}$, [1, 2]

Let x be a continuous random variable with standard normal density. Using Table 2, find each of the following.

11. $P(0 \le x \le 2.69)$
12. $P(0 \le x \le 0.04)$
13. $P(-1.11 \le x \le 0)$
14. $P(-2.61 \le x \le 0)$
15. $P(-1.89 \le x \le 0.45)$
16. $P(-2.94 \le x \le 2.00)$
17. $P(1.76 \le x \le 1.86)$
18. $P(0.76 \le x \le 1.45)$
19. $P(-1.45 \le x \le -0.69)$

20. $P(-2.45 \leq x \leq -1.69)$

21. $P(x \geq 3.01)$ 22. $P(x \geq 1.01)$

23. a) $P(-1 \leq x \leq 1)$
    b) What percentage of the area is from -1 to 1?

24. a) $P(-2 \leq x \leq 2)$
    b) What percentage of the area is from -2 to 2?

Let x be a continuous random variable that is normally distributed with mean $\mu = 22$ and standard deviation $\sigma = 5$. Using Table 2, find each of the following.

25. $P(24 \leq x \leq 30)$ 26. $P(22 \leq x \leq 27)$

27. $P(19 \leq x \leq 25)$ 28. $P(18 \leq x \leq 26)$

 29.–46. Use a grapher to do Exercises 11–28.

Applications

BUSINESS AND ECONOMICS

47. *Mail Orders.* The number of daily orders N received by a mail-order firm is normally distributed with mean 250 and standard deviation 20. The company has to hire extra help or pay overtime on those days when the number of orders received is 300 or higher. What percentage of the days will the company have to hire extra help or pay overtime?

48. *Stereo Production.* The daily production N of stereos by a recording company is normally distributed with mean 1000 and standard deviation 50. The company promises to pay bonuses to its employees on those days when the production of stereos is 1100 or more. What percentage of the days will the company have to pay a bonus?

SOCIAL SCIENCES

49. *Test Score Distribution.* The scores S on a psychology test are normally distributed with mean 65 and standard deviation 20. A score of 80 to 89 is a B. What is the probability of getting a B?

GENERAL INTEREST

tw 50. *Bowling Scores.* At the time this book was written, the bowling scores S of the author were normally distributed with mean 201 and standard deviation 23.
    a) Find the probability that a score is from 190 to 213 and interpret your results.
    b) Find the probability that a score is from 160 to 175 and interpret your results.

c) Find the probability that a score is greater than 200 and interpret your results.

Synthesis

For each probability density function, over the given interval, find $E(x)$, $E(x^2)$, the mean, the variance, and the standard deviation.

51. The uniform probability density
$$f(x) = \frac{1}{b - a} \quad \text{over } [a, b]$$

52. The probability density
$$f(x) = \frac{3a^3}{x^4} \quad \text{over } [a, \infty)$$

Median. Let x be a continuous random variable over $[a, b]$ with probability density function f. Then the *median* of x is that number m for which
$$\int_a^m f(x)\, dx = \tfrac{1}{2}.$$

Find the median.

53. $f(x) = \tfrac{1}{2}x$, $[0, 2]$

54. $f(x) = \tfrac{3}{2}x^2$, $[-1, 1]$

55. $f(x) = ke^{-kx}$, $[0, \infty)$

56. *Business: Sugar Production.* Suppose that the amount of sugar going in a sack has a mean μ, which can be adjusted on the filling machine. Suppose that the amount dispensed is normally distributed with $\sigma = 0.1$ oz. What should be the

setting of μ to ensure that only one bag in 20 will have less than 60 oz?

57. *Business: Does Thy Cup Overflow?* Suppose that the mean amount of coffee μ dispensed by a coffee machine can be set. If a cup holds 6.5 oz and the amount of coffee dispensed is normally distributed with $\sigma = 0.3$ oz, what should the setting of μ be to ensure that the cup will overflow only one time in a hundred?

tw 58. Explain the uses of integration in the study of probability.

tw 59. You are told that "The antiderivative of the function $f(x) = e^{-x^2/2}$ has no elementary

formula." Make some guesses of functions that might seem reasonable to you as antiderivatives and show why they are not.

TECHNOLOGY CONNECTION

60. Approximate the integral

$$\int_{-\infty}^{\infty} e^{-x^2} \, dx.$$

6.6

OBJECTIVES

➤ Find the volume of a solid of revolution.

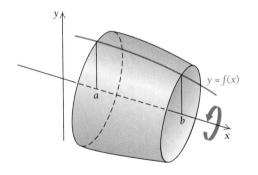

Volume

Consider the graph of $y = f(x)$. If the upper half-plane is rotated about the x-axis, then each point on the graph has a circular path, and the whole graph sweeps out a certain surface, called a *surface of revolution*.

The plane region bounded by the graph, the x-axis, and the interval $[a, b]$ sweeps out a *solid of revolution*. To calculate the volume of this solid, we first approximate it by a finite sum of thin right circular cylinders (Fig. 1). We divide the interval $[a, b]$ into equal subintervals, each of length Δx. Thus the height of each cylinder is Δx (Fig. 2). The radius of each cylinder is $f(x_i)$ if f is nonnegative, or $|f(x_i)|$ in general, where x_i is the righthand endpoint of the subinterval that determines that cylinder.

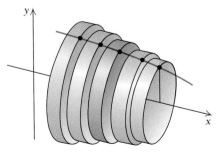

FIGURE 1

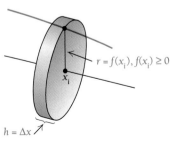

FIGURE 2

Since the volume of a right circular cylinder is given by

$$V = \pi r^2 h,$$

each of the approximating cylinders has volume

$$\pi |f(x_i)|^2 \, \Delta x = \pi [f(x_i)]^2 \, \Delta x.$$

The volume of the solid of revolution is approximated by the sum of the volumes of all the cylinders:

$$V \approx \sum_{i=1}^{n} \pi [f(x_i)]^2 \, \Delta x.$$

The actual volume is the limit as the thickness of the cylinders approaches zero or the number of them approaches infinity:

$$V = \lim_{n \to \infty} \sum_{i=1}^{n} \pi [f(x_i)]^2 \, \Delta x = \int_a^b \pi [f(x)]^2 \, dx.$$

(See Section 5.3.) That is, the volume is the value of the definite integral of the function $y = \pi [f(x)]^2$ from a to b.

Theorem 4

For a continuous function f defined on an interval $[a, b]$, the **volume V of the solid of revolution** obtained by rotating the area under the graph of f over $[a, b]$ is given by

$$V = \int_a^b \pi [f(x)]^2 \, dx.$$

Example 1 Find the volume of the solid of revolution generated by rotating the region under the graph of $y = \sqrt{x}$ from $x = 0$ to $x = 1$ about the x-axis.

Solution

$$V = \int_0^1 \pi [f(x)]^2 \, dx$$

$$= \int_0^1 \pi [\sqrt{x}]^2 \, dx$$

$$= \int_0^1 \pi x \, dx$$

$$= \pi \left[\frac{x^2}{2} \right]_0^1$$

$$= \frac{\pi}{2} \left[x^2 \right]_0^1$$

$$= \frac{\pi}{2}(1^2 - 0^2) = \frac{\pi}{2}$$

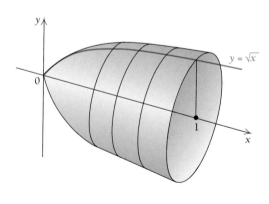

Example 2   Find the volume of the solid of revolution, "horn," generated by rotating the region under the graph of

$$y = e^x$$

from $x = -1$ to $x = 2$ about the x-axis.

Solution

$$V = \int_{-1}^{2} \pi [f(x)]^2 \, dx$$

$$= \int_{-1}^{2} \pi [e^x]^2 \, dx$$

$$= \int_{-1}^{2} \pi e^{2x} \, dx$$

$$= \left[\frac{\pi}{2} e^{2x} \right]_{-1}^{2}$$

$$= \frac{\pi}{2} \left[e^{2x} \right]_{-1}^{2}$$

$$= \frac{\pi}{2} (e^{2 \cdot 2} - e^{2(-1)})$$

$$= \frac{\pi}{2} (e^4 - e^{-2})$$

Explain how this could be interpreted as a solid of revolution.

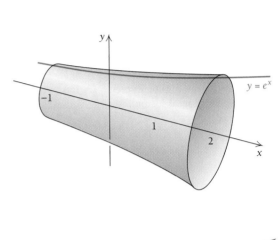

EXERCISE SET 6.6

Find the volume generated by revolving about the x-axis the regions bounded by the graphs of the following equations.

1. $y = x$, $x = 0$, $x = 1$

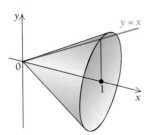

2. $y = \sqrt{x}$, $x = 0$, $x = 2$

3. $y = x$, $x = 1$, $x = 2$

4. $y = x$, $x = 1$, $x = 3$

5. $y = e^x$, $x = -2$, $x = 5$

6. $y = e^x$, $x = -3$, $x = 2$

7. $y = \dfrac{1}{x}$, $x = 1$, $x = 3$

8. $y = \dfrac{1}{x}$, $x = 1$, $x = 4$

9. $y = \dfrac{1}{\sqrt{x}}, x = 1, x = 3$

10. $y = \dfrac{1}{\sqrt{x}}, x = 1, x = 4$

11. $y = 4, x = 1, x = 3$

12. $y = 5, x = 1, x = 3$

13. $y = x^2, x = 0, x = 2$

14. $y = x + 1, x = -1, x = 2$

15. $y = \sqrt{1 + x}, x = 2, x = 10$

16. $y = 2\sqrt{x}, x = 1, x = 2$

17. $y = \sqrt{4 - x^2}, x = -2, x = 2$

18. $y = \sqrt{r^2 - x^2}, x = -r, x = r$

Make a drawing for Exercise 18. Here you will derive a general formula for the volume of a sphere.

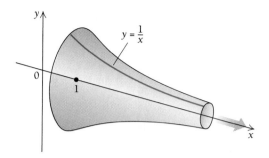

Find the volume of the solid of revolution formed by rotating the region under the graph of $y = 1/x$ about the x-axis over the interval $[1, \infty)$; that is, find

$$\int_1^\infty \pi \left[\frac{1}{x}\right]^2 dx.$$

This solid is sometimes referred to as *Gabriel's horn*.

Synthesis

Find the volume generated by revolving about the x-axis the regions bounded by the following graphs.

19. $y = \sqrt{\ln x}, x = e, x = e^3$

20. $y = \sqrt{xe^{-x}}, x = 1, x = 2$

21. Consider the function $y = 1/x$ over the interval $[1, \infty)$. We showed in Section 6.3 that the area under the curve does not exist; that is,

$$\int_1^\infty \frac{1}{x} dx$$

diverges.

 TECHNOLOGY CONNECTION

22. *Paradox of Gabriel's Horn or the Infinite Paint Can.* Though we cannot prove it here, the surface area of Gabriel's horn (see Exercise 21) is given by

$$S = \int_1^\infty \frac{2\pi}{x} \sqrt{1 + \frac{1}{x^4}} dx.$$

Show that the surface area of Gabriel's horn does not exist. The paradox occurs because the volume of the horn exists, but the surface area does not. This is like a can of paint that has a finite volume, but not enough paint to paint the outside of the can.

6.7

OBJECTIVES

➤ Solve differential equations.
➤ Verify that a given function is a solution of a differential equation.
➤ Solve differential equations using separation of variables.

Differential Equations

A **differential equation** is an equation that involves derivatives or differentials. In Chapter 4, we studied one very important differential equation,

$$\frac{dP}{dt} = kP,$$

where P, or $P(t)$, is the population at time t. This equation is a model of uninhibited population growth. Its solution is the function

$$P = P_0 e^{kt},$$

where the constant P_0 is the size of the initial population, that is, at $t = 0$. As this one equation illustrates, differential equations are rich in application.

Solving Certain Differential Equations

In this chapter, we will frequently use the notation y' for a derivative—mainly because it is simple. Thus, if $y = f(x)$, then

$$y' = \frac{dy}{dx} = f'(x).$$

We have already found solutions of certain differential equations when we found antiderivatives or indefinite integrals. The differential equation

$$\frac{dy}{dx} = g(x), \quad \text{or} \quad y' = g(x),$$

has the solution

$$y = \int g(x)\, dx.$$

Waves can be represented by differential equations.

Example 1 Solve: $y' = 2x$.

Solution

$$y = \int 2x\, dx$$

$$= x^2 + C$$

Look again at the solution to Example 1. Note the constant of integration. This solution is called a *general solution* because taking all values of C gives *all* the solutions. Taking specific values of C gives *particular solutions*. For example, the following are particular solutions to $y' = 2x$:

$$y = x^2 + 3,$$
$$y = x^2,$$
$$y = x^2 - 3.$$

T E C H N O L O G Y
C O N N E C T I O N

E x p l o r a t o r y .
For $y' = 3x^2$, write
the general solution.
Then graph the
particular solutions for
$C = -2$, $C = 0$, and
$C = 1$.

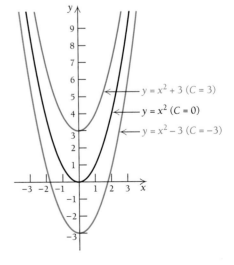

The graph on the preceding page shows the curves of a few particular solutions. The general solution can be regarded as the set of all particular solutions, a *family* of curves.

Knowing the value of a function at a particular point may allow us to select a particular solution from the general solution.

Example 2 Solve
$$f'(x) = e^x + 5x - x^{1/2},$$
given that $f(0) = 8$.

Solution

a) We first find the general solution:
$$f(x) = \int f'(x)\,dx$$
$$= e^x + \tfrac{5}{2}x^2 - \tfrac{2}{3}x^{3/2} + C.$$

b) Since $f(0) = 8$, we substitute to find C:
$$8 = e^0 + \tfrac{5}{2}\cdot 0^2 - \tfrac{2}{3}\cdot 0^{3/2} + C$$
$$8 = 1 + C$$
$$7 = C.$$

Thus the particular solution is
$$f(x) = e^x + \tfrac{5}{2}x^2 - \tfrac{2}{3}x^{3/2} + 7.$$

Verifying Solutions

To verify that a function is a solution to a differential equation, we find the necessary derivatives and substitute.

Example 3 Show that $y = 4e^x + 5e^{3x}$ is a solution to
$$y'' - 4y' + 3y = 0.$$

Solution

a) We first find y' and y'':
$$y' = 4e^x + 15e^{3x};$$
$$y'' = 4e^x + 45e^{3x}.$$

b) Then we substitute in the differential equation, as follows:

$$
\begin{array}{c|c}
y'' - 4y' + 3y & 0 \\
\hline
(4e^x + 45e^{3x}) - 4(4e^x + 15e^{3x}) + 3(4e^x + 5e^{3x}) & 0 \\
4e^x + 45e^{3x} - 16e^x - 60e^{3x} + 12e^x + 15e^{3x} & \\
0 &
\end{array}
$$

Separation of Variables

Consider the differential equation

$$\frac{dy}{dx} = 2xy. \tag{1}$$

We treat dy/dx as a quotient, as we did in Chapter 5. We multiply equation (1) by dx and then by $1/y$ to get

$$\frac{dy}{y} = 2x\, dx, \quad y \neq 0. \tag{2}$$

We say that we have **separated the variables,** meaning that all the expressions involving y are on one side and all those involving x are on the other. We then integrate both sides of equation (2):

$$\int \frac{dy}{y} = \int 2x\, dx$$

$$\ln y = x^2 + C, \quad y > 0.$$

We use only one constant because any two antiderivatives differ by a constant. Recall that the definition of logarithms says that if $\log_a b = t$, then $b = a^t$. Now, $\ln y = \log_e y = x^2 + C$, so by the definition of logarithms, we have

$$y = e^{x^2+C} = e^{x^2} \cdot e^C.$$

Thus the solution to differential equation (1) is

$$y = C_1 e^{x^2}, \quad \text{where } C_1 = e^C.$$

In fact, C_1 is still an arbitrary constant.

Example 4   Solve

$$3y^2 \frac{dy}{dx} + x = 0,$$

where $y = 5$ when $x = 0$.

Solution

a) We first separate the variables as follows:

$$3y^2 \frac{dy}{dx} = -x$$

$$3y^2\, dy = -x\, dx.$$

We then integrate both sides:

$$\int 3y^2\, dy = \int -x\, dx$$

$$y^3 = -\frac{x^2}{2} + C = C - \frac{x^2}{2}$$

$$y = \sqrt[3]{C - \frac{x^2}{2}}. \qquad \text{\small Taking the cube root}$$

At the 1968 Olympic Games in Mexico City, Bob Beamon made a miraculous long jump of 29 ft, $2\frac{1}{2}$ in. Many believed this was due to the altitude, which was 7400 ft. Using differential equations for analysis, M. N. Bearley refuted the altitude theory in "The Long Jump Miracle of Mexico City" (Mathematics Magazine, 45, November 1972, pp. 241–246). Bearley argues that the world record jump was a result of Beamon's exceptional speed (9.5 sec in the 100-yd dash) and the fact that he hit the take-off board in perfect position.

b) Since $y = 5$ when $x = 0$, we substitute to find C:

$$5 = \sqrt[3]{C - \frac{0^2}{2}} \qquad \text{Substituting 5 for } y \text{ and 0 for } x$$

$$5 = \sqrt[3]{C}$$

$$125 = C. \qquad \text{Cubing both sides}$$

The particular solution is

$$y = \sqrt[3]{125 - \frac{x^2}{2}}.$$

Example 5   Solve:

$$\frac{dy}{dx} = \frac{x}{y}.$$

Solution   We first separate the variables:

$$y\frac{dy}{dx} = x$$

$$y\,dy = x\,dx.$$

We then integrate both sides:

$$\int y\,dy = \int x\,dx$$

$$\frac{y^2}{2} = \frac{x^2}{2} + C$$

$$y^2 = x^2 + 2C$$

$$y^2 = x^2 + C_1,$$

where $C_1 = 2C$. We make this substitution in order to simplify the equation. We then obtain the solutions

$$y = \sqrt{x^2 + C_1}$$

and

$$y = -\sqrt{x^2 + C_1}.$$

Example 6   Solve: $y' = x - xy$.

Solution   Before we separate the variables, we replace y' with dy/dx:

$$\frac{dy}{dx} = x - xy.$$

Then we separate the variables:

$$dy = (x - xy)\, dx$$
$$dy = x(1 - y)\, dx$$
$$\frac{dy}{1 - y} = x\, dx.$$

Next, we integrate both sides:

$$\int \frac{dy}{1 - y} = \int x\, dx$$

$$-\ln(1 - y) = \frac{x^2}{2} + C \qquad 1 - y > 0$$

$$\ln(1 - y) = -\frac{x^2}{2} - C$$

$$1 - y = e^{-x^2/2 - C}$$

$$-y = e^{-x^2/2 - C} - 1$$

$$y = -e^{-x^2/2 - C} + 1$$

$$= -e^{-x^2/2} \cdot e^{-C} + 1.$$

Thus,

$$y = 1 + C_1 e^{-x^2/2}, \quad \text{where } C_1 = -e^{-C}.$$

TECHNOLOGY CONNECTION

Exploratory.
Solve $y' = 2x + xy$.
Graph the particular
solutions for
$C_1 = -2$, $C_1 = 0$,
and $C_1 = 1$.

An Application to Economics: Elasticity

Example 7 Suppose that for a certain product, the elasticity of demand is 1 for all $p > 0$. That is, $E(p) = 1$ for all $p > 0$. Find the demand function $x = D(p)$. (See Section 4.6.)

Solution Since $E(p) = 1$ for all $p > 0$,

$$1 = E(p) = -\frac{p\, D'(p)}{D(p)}$$

$$= -\frac{p}{x} \cdot \frac{dx}{dp}.$$

Then

$$\frac{dx}{dp} = -\frac{x}{p}. \tag{3}$$

Separating the variables, we get

$$\frac{dp}{p} = -\frac{dx}{x}.$$

Now we integrate both sides:

$$\int \frac{dp}{p} = -\int \frac{dx}{x}$$

$$\ln p = -\ln x + C. \qquad \text{\footnotesize $p > 0$ and $dx/dp < 0$, since}$$

p > 0 and *dx/dp* < 0, since
demand functions are decreasing.
So *x* > 0. See equation (3).

Then

$$\ln p + \ln x = C$$

$$\ln px = C$$

$$px = e^{C}.$$

We let $C_1 = e^{C} = px$. Then

$$p = \frac{C_1}{x} \quad \text{and} \quad x = \frac{C_1}{p}.$$

This characterizes those demand functions for which the elasticity is always 1.

An Application to Psychology: Reaction to a Stimulus

The Weber–Fechner Law

In psychology, one model of stimulus–response asserts that the rate of change dR/dS of the reaction R with respect to a stimulus S is inversely proportional to the stimulus; that is,

$$\frac{dR}{dS} = \frac{k}{S},$$

where k is some positive constant.

To solve this equation, we first separate the variables:

$$dR = k \cdot \frac{dS}{S}.$$

We then integrate both sides:

$$\int dR = \int k \cdot \frac{dS}{S}$$

$$R = k \ln S + C. \tag{4}$$

Now suppose that we let S_0 be the lowest level of the stimulus that can be detected consistently. This is the *threshold value,* or the *detection threshold.* For example, the lowest level of sound that can be consistently detected is the tick of a watch at 20 ft, under very quiet conditions. If S_0 is the lowest level of sound that can be detected, it seems reasonable that the reaction to it would be 0. That is, $R(S_0) = 0$. Substituting this condition into equation (4), we get

$$0 = k \ln S_0 + C,$$

or

$$-k \ln S_0 = C.$$

Replacing C in equation (4) with $-k \ln S_0$ gives us

$$R = k \cdot \ln S - k \cdot \ln S_0$$
$$= k(\ln S - \ln S_0).$$

Using a property of logarithms, we have

$$R = k \cdot \ln \frac{S}{S_0}.$$

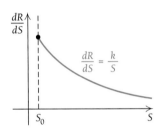

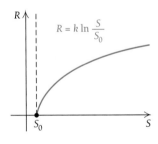

Look at the graphs of dR/dS and R shown at left. Note that as the stimulus gets larger, the rate of change decreases; that is, the change in reaction becomes smaller as the stimulation received becomes stronger. For example, suppose that you are in a room with one lamp and that lamp has a 50-watt bulb in it. If the bulb were suddenly changed to 100 watts, you would probably be very aware of the difference. That is, your reaction would be strong. If the bulb were then changed to 150 watts, your reaction would not be as great as it was to the change from 50 to 100 watts. A change from a 150- to a 200-watt bulb would cause even less reaction, and so on.

For your interest, here are some other detection thresholds.

Stimulus	Detection Threshold
Light	The flame of a candle 30 miles away on a dark night
Taste	Water diluted with sugar in the ratio of 1 teaspoon to 2 gallons
Smell	One drop of perfume diffused into the volume of three average-size rooms
Touch	The wing of a bee dropped on your cheek at a distance of 1 centimeter $\left(\text{about } \frac{3}{8} \text{ of an inch}\right)$

EXERCISE SET 6.7

Find the general solution and three particular solutions.

1. $y' = 4x^3$

2. $y' = 6x^5$

3. $y' = e^{2x} + x$

4. $y' = e^{3x} - x$

5. $y' = \dfrac{3}{x} - x^2 + x^5$

6. $y' = \dfrac{5}{x} + x^2 - x^4$

Find the particular solution determined by the given condition.

7. $y' = x^2 + 2x - 3;\quad y = 4$ when $x = 0$

8. $y' = 3x^2 - x + 5;\quad y = 6$ when $x = 0$

9. $f'(x) = x^{2/3} - x;\quad f(1) = -6$

10. $f'(x) = x^{2/5} + x;\quad f(1) = -7$

11. Show that $y = x \ln x + 3x - 2$ is a solution to
$$y'' - \frac{1}{x} = 0.$$

12. Show that $y = x \ln x - 5x + 7$ is a solution to
$$y'' - \frac{1}{x} = 0.$$

13. Show that $y = e^x + 3xe^x$ is a solution to
$$y'' - 2y' + y = 0.$$

14. Show that $y = -2e^x + xe^x$ is a solution to
$$y'' - 2y' + y = 0.$$

Solve.

15. $\dfrac{dy}{dx} = 4x^3 y$

16. $\dfrac{dy}{dx} = 5x^4 y$

17. $3y^2 \dfrac{dy}{dx} = 5x$

18. $3y^2 \dfrac{dy}{dx} = 7x$

19. $\dfrac{dy}{dx} = \dfrac{2x}{y}$

20. $\dfrac{dy}{dx} = \dfrac{x}{2y}$

21. $\dfrac{dy}{dx} = \dfrac{3}{y}$

22. $\dfrac{dy}{dx} = \dfrac{4}{y}$

23. $y' = 3x + xy;\quad y = 5$ when $x = 0$

24. $y' = 2x - xy;\quad y = 9$ when $x = 0$

25. $y' = 5y^{-2};\quad y = 3$ when $x = 2$

26. $y' = 7y^{-2};\quad y = 3$ when $x = 1$

27. $\dfrac{dy}{dx} = 3y$

28. $\dfrac{dy}{dx} = 4y$

29. $\dfrac{dP}{dt} = 2P$

30. $\dfrac{dP}{dt} = 4P$

31. Solve
$$f'(x) = \frac{1}{x} - 4x + \sqrt{x},$$
given that $f(1) = \frac{23}{3}$.

Applications
BUSINESS AND ECONOMICS

32. *Total Revenue from Marginal Revenue.* The marginal revenue for a certain product is given by $R'(x) = 300 - 2x$. Find the total-revenue function $R(x)$, assuming that $R(0) = 0$.

33. *Total Cost from Marginal Cost.* The marginal cost for a certain product is given by $C'(x) = 2.6 - 0.02x$. Find the total-cost function $C(x)$ and the average cost $A(x)$, assuming fixed costs are 120; that is, $C(0) = \$120$.

34. *Domar's capital expansion model* is
$$\frac{dI}{dt} = hkI,$$
where I is the investment, h is the investment productivity (constant), k is the marginal productivity to the consumer (constant), and t is the time.

a) Use separation of variables to solve the differential equation.

b) Rewrite the solution in terms of the condition $I_0 = I(0)$.

35. *Total Profit from Marginal Profit.* A firm's marginal profit P as a function of its total cost C is given by
$$\frac{dP}{dC} = \frac{-200}{(C + 3)^{3/2}}.$$

a) Find the profit function $P(C)$ if $P = \$10$ when $C = \$61$.

b) At what cost will the firm break even $(P = 0)$?

36. *Stock Growth.* The *growth rate of a certain stock* is modeled by

$$\frac{dV}{dt} = k(L - V), \quad V = \$20 \text{ when } t = 0,$$

where V is the value of the stock, per share, after time t (in months), $L = \$24.81$, the *limiting value* of the stock, and k is a constant. Find the solution to the differential equation in terms of t and k.

37. *Utility.* The reaction R in pleasure units by a consumer receiving S units of a product can be modeled by the differential equation

$$\frac{dR}{dS} = \frac{k}{S + 1},$$

where k is a positive constant.
a) Use separation of variables to solve the differential equation.
b) Rewrite the solution in terms of the initial condition $R(0) = 0$.
c) Explain why the condition $R(0) = 0$ is reasonable.

Elasticity. Find the demand function $p = D(x)$ given the following elasticity conditions.

38. $E(p) = \dfrac{4}{p}; \quad x = e$ when $p = 4$

39. $E(p) = \dfrac{p}{200 - p}; \quad x = 190$ when $p = 10$

40. $E(p) = 2$ for all $p > 0$

41. $E(p) = n$ for some constant n and all $p > 0$

LIFE AND PHYSICAL SCIENCES

42. *Exponential Growth.*
a) Use separation of variables to solve the differential-equation model of uninhibited growth,

$$\frac{dP}{dt} = kP.$$

b) Rewrite the solution in terms of the condition $P_0 = P(0)$.

SOCIAL SCIENCES

43. *The Brentano–Stevens Law.* The Weber–Fechner Law has been the subject of great debate among psychologists as to its validity. The model

$$\frac{dR}{dS} = k \cdot \frac{R}{S},$$

where k is a positive constant, has also been conjectured and experimented with. Find the general solution to this equation. (This has also been referred to as the *Power Law of Stimulus–Response.*)

Synthesis ....

 44. Discuss as many applications as you can of the use of integration in this chapter.

TECHNOLOGY CONNECTION

45. Solve $dy/dx = 5/y$. Graph the particular solutions for $C_1 = 5$, $C_1 = -200$, and $C_1 = 100$.

CHAPTER 6 SUMMARY AND REVIEW

Terms to Know

Review Exercises

These review exercises are for test preparation. They can also be used as a lengthened practice test. Answers are at the back of the book. The answers also contain bracketed section references, which tell you where to restudy if your answer is incorrect.

Given $p = D(x) = (x - 6)^2$ and $p = S(x) = x^2 + 12$, find each of the following.

1. The equilibrium point

2. The consumer's surplus at the equilibrium point

3. The producer's surplus at the equilibrium point

4. *Business: Amount of a Continuous Money Flow.* Find the amount of a continuous money flow in which $2400 per year is being invested at 10%, compounded continuously, for 15 yr.

5. *Business: Continuous Money Flow.* Consider a continuous money flow into an investment at the constant rate of P_0 dollars per year. What should P_0 be so that the amount of a continuous money flow over 25 yr at 12%, compounded continuously, will be $40,000?

6. *Physical Science: Demand for Potash.* In 1997 ($t = 0$), the world use of potash was 24,000 thousand tons, and the demand for potash was growing at the rate of 3% per year. If the demand continues to grow at this rate, how many tons of potash will the world use from 1997 to 2002?

7. *Physical Science: Depletion of Potash.* The world reserves of potash are 8,400,000 thousand tons. Assuming the growth rate in Exercise 6 continues and no new reserves are discovered, when will the world reserves of potash be exhausted?

8. *Business: Present Value.* Find the present value of $100,000 due 50 yr later at 12%, compounded continuously.

9. *Business: Accumulated Present Value.* Find the accumulated present value of an investment over a 20-yr period if there is a continuous money flow of $4800 per year and the current interest rate is 9%.

10. *Business: Accumulated Present Value.* Find the accumulated present value of an investment for which the continuous money flow in Exercise 9 is perpetual.

Determine whether the improper integral is convergent or divergent, and calculate its value if it is convergent.

11. $\int_1^\infty \frac{1}{x^2}\, dx$

12. $\int_1^\infty e^{4x}\, dx$

13. $\int_0^\infty e^{-2x}\, dx$

14. Find k such that $f(x) = k/x^3$ is a probability density function over the interval $[1, 2]$. Then write the probability density function.

15. *Business: Waiting Time.* A person randomly arrives at a bus stop where the waiting time t for a bus is no more than 25 min. The probability density function for t is $f(t) = \frac{1}{25}$, for $0 \le t \le 25$. Find the probability that a person will have to wait no more than 15 min for a bus.

Given the probability density function $f(x) = 3x^2$ over $[0, 1]$, find each of the following.

16. $E(x^2)$

17. $E(x)$

18. The mean

19. The variance

20. The standard deviation

Let x be a continuous random variable with standard normal density. Using Table 2, find each of the following.

21. $P(0 \le x \le 1.85)$

22. $P(-1.74 \le x \le 1.43)$

23. $P(-2.08 \le x \le -1.18)$

24. $P(x \ge 0)$

25. *Business: Pizza Sales.* The number of pizzas sold daily at Benito's Pizzeria is normally distributed with mean 400 and standard deviation 60. What is the probability that the number sold during one day is 480 or more?

Find the volume generated by revolving about the x-axis the region bounded by each of the following graphs.

26. $y = x^3,\ x = 1,\ x = 2$

27. $y = \dfrac{1}{x + 2},\ x = 0,\ x = 1$

Solve the differential equation.

28. $\dfrac{dy}{dx} = 11x^{10}y$

29. $\dfrac{dy}{dx} = \dfrac{2}{y}$

30. $\dfrac{dy}{dx} = 4y;\quad y = 5$ when $x = 0$

31. $\dfrac{dv}{dt} = 5v^{-2};\quad v = 4$ when $t = 3$

32. $y' = \dfrac{3x}{y}$

33. $y' = 8x - xy$

34. *Economics: Elasticity.* Find the demand function given the elasticity condition

$$E(p) = \frac{p}{p - 100}, \quad x = 70 \text{ when } p = 30.$$

35. *Business: Stock Growth.* The growth rate of a stock is modeled by

$$\frac{dV}{dt} = k(L - V), \quad V = \$30 \text{ when } t = 0,$$

where V is the value of the stock, per share, after time t (in months), $L = \$36.37$, the limiting value of the stock, and k is a constant. Find the solution to the differential equation in terms of t and k.

Synthesis

36. The function $f(x) = x^8$ is a probability density function over the interval $[-c, c]$. Find c.

Determine whether the improper integral is convergent or divergent, and calculate its value if it is convergent.

37. $\displaystyle\int_{-\infty}^{0} x^4 e^{-x^5}\, dx$

38. $\displaystyle\int_{0}^{\infty} \frac{dx}{(x + 1)^{4/3}}$

 TECHNOLOGY CONNECTION

39. Approximate the integral

$$\int_{1}^{\infty} \frac{\ln x}{x^2}\, dx.$$

CHAPTER 6 TEST

Given the demand and supply functions

$$p = D(x) = (x - 7)^2$$

and

$$p = S(x) = x^2 + x + 4,$$

find each of the following.

1. The equilibrium point

2. The consumer's surplus at the equilibrium point

3. The producer's surplus at the equilibrium point

4. *Business: Amount of a Continuous Money Flow.* Find the amount of a continuous money flow if $1200 per year is being invested at 6%, compounded continuously, for 15 yr.

5. *Business: Continuous Money Flow.* Consider a continuous money flow into an investment at the rate of P_0 dollars per year. What should P_0 be so that the amount of a continuous money flow over 25 yr at 6%, compounded continuously, will be $20,000?

6. *Physical Science: Demand for Iron Ore.* In 1997 ($t = 0$), the world use of iron ore was 1,030,000 thousand tons, and the demand for iron ore was

growing exponentially at the rate of 6% per year. If the demand continues to grow at this rate, how many tons of iron ore will the world use from 1997 to 2010?

7. *Physical Science: Depletion of Iron Ore.* The world reserves of iron ore are 270,000,000 thousand tons. Assuming the demand for iron ore continues to grow at the rate of 6% per year and no new reserves are discovered, when will the world reserves of iron ore be exhausted?

8. *Business: Present Value.* Following the birth of a child, a parent wants to make an initial investment P_0 that will grow to $10,000 by the child's 20th birthday. Interest is compounded continuously at 7%. What should the initial investment be?

9. *Business: Accumulated Present Value.* Find the accumulated present value of an investment over a 20-yr period if there is a continuous money flow of $3800 per year and the current interest rate is 11%.

10. *Business: Accumulated Present Value.* Find the accumulated present value of an investment for

which the continuous money flow in Question 9 is perpetual.

Determine whether the improper integral is convergent or divergent, and calculate its value if it is convergent.

11. $\displaystyle\int_1^\infty \frac{dx}{x^5}$ **12.** $\displaystyle\int_0^\infty \frac{3}{1+x}\, dx$

13. Find k such that $f(x) = kx^3$ is a probability density function over the interval $[0, 2]$. Then find the probability density function.

14. *Business: Times of Telephone Calls.* A telephone company determines that the length of time t of a phone call is an exponentially distributed random variable with probability density function

$$f(t) = 2e^{-2t}, \quad 0 \le t < \infty.$$

Find the probability that a phone call will last no more than 1 min.

Given the probability density function $f(x) = \frac{1}{4}x$ over $[1, 3]$, find each of the following.

15. $E(x)$ **16.** $E(x^2)$

17. The mean   **18.** The variance

19. The standard deviation

Let x be a continuous random variable with standard normal density. Using Table 2, find each of the following.

20. $P(0 \le x \le 1.5)$ **21.** $P(0.12 \le x \le 2.32)$

22. $P(-1.61 \le x \le 1.76)$

23. The price per pound p of a T-bone steak at various stores in a certain city is normally distributed with mean \$4.75 and standard deviation \$0.25. What is the probability that the price per pound is \$4.80 or more?

Find the volume generated by revolving about the x-axis the regions bounded by the following graphs.

24. $y = \dfrac{1}{\sqrt{x}}$, $x = 1$, $x = 5$

25. $y = \sqrt{2 + x}$, $x = 0$, $x = 1$

Solve the differential equation.

26. $\dfrac{dy}{dx} = 8x^7 y$ **27.** $\dfrac{dy}{dx} = \dfrac{9}{y}$

28. $\dfrac{dy}{dt} = 6y;\quad y = 11$ when $t = 0$

29. $y' = 5x^2 - x^2 y$ **30.** $\dfrac{dv}{dt} = 2v^{-3}$

31. $y' = 4y + xy$

32. *Economics: Elasticity.* Find the demand function given the elasticity condition

$$E(p) = 4 \quad \text{for all } p > 0.$$

33. *Business: Stock Growth.* The growth rate of stock for Glamour Industries is modeled by

$$\frac{dV}{dt} = k(L - V),$$

where V is the value of the stock per share, after time t (in months), $L = \$36$, the *limiting value* of the stock, k is a constant, and $V(0) = \$0$.

a) Write the solution $V(t)$ in terms of L and k.
b) If $V(6) = \$18$, determine k to the nearest hundredth.
c) Rewrite $V(t)$ in terms of t and k using the value of k found in part (b).
d) Use the equation in part (c) to find $V(12)$, the value of the stock after 12 months.
e) In how many months will the value be \$30?

Synthesis ....

34. The function $f(x) = x^5$ is a probability density function over the interval $[0, b]$. What is b?

35. Determine whether the following improper integral is convergent or divergent, and calculate its value if it is convergent:

$$\int_{-\infty}^0 x^3 e^{-x^4}\, dx.$$

 TECHNOLOGY CONNECTION

36. Approximate the integral

$$\int_{-\infty}^\infty \frac{1}{1+x^2}\, dx.$$

E X T E N D E D   T E C H N O L O G Y   A P P L I C A T I O N

Curve Fitting and the Volume of a Bottle of Soda

Consider the urn or vase below. How could we estimate the volume? One way would be to simply fill the container with a liquid and then pour the liquid into a measuring device.

Another way, using calculus and the curve-fitting or REGRESSION features of a grapher, would be to turn the urn on its side, as shown below, take a series of vertical measurements from the center to the top, do a curve-fitting or REGRESSION procedure with the grapher, and then estimate the volume using the integration technique of the grapher.

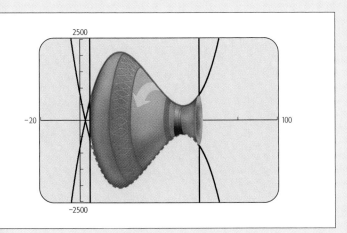

Suppose that the following is a table of values for the red curve.

x	y
5	547.5
10	1440.0
15	1932.5
20	2100.0
25	2017.5
30	1760.0
35	1402.5
40	1020.0
45	687.5
50	480.0
55	472.5
60	740.0
61	833.1

Exercises

1. Use the REGRESSION feature on a grapher to fit a cubic polynomial function to the data.

2. Using the function found in Exercise 1, integrate over the interval $[5, 61]$ to find the volume of the urn.

Now consider the soda bottle pictured to the right. To find its volume in a similar manner, we can turn the bottle on its side and use a measuring device to take vertical measurements and proceed as we did with the urn.

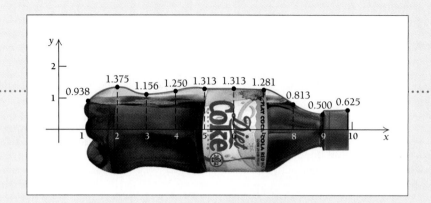

The table of measurements is as follows.

x	y
1	0.938
2	1.375
3	1.156
4	1.250
5	1.313
6	1.313
7	1.281
8	0.813
9	0.500
9.875	0.625

Exercises

3. Use the REGRESSION feature on a grapher to fit a quartic polynomial function to the data.

4. Using the function found in Exercise 3, integrate to find the volume of the bottle. Your answer will be in cubic inches. Convert it to fluid ounces using the fact that $1 \text{ in}^3 = 0.55424$ fluid ounce.

5. The bottle in question has a capacity of 20 oz. How good was our curve-fitting procedure for making this estimate?

6. Try to find a curve that gives a better estimate of the volume. What is the curve and what is the estimated volume?

7

Functions of Several Variables

INTRODUCTION

Functions that have more than one input are called *functions of several variables.* We introduce these functions in this chapter and learn to differentiate them to find *partial derivatives.* Then we use these functions and their partial derivatives to solve maximum–minimum problems. Finally, we consider the integration of such functions.

AN APPLICATION

A firm produces two kinds of golf ball, one that sells for $3 and one for $2. The total profit, in thousands of dollars, from the production and sale of x thousand balls at $3 each and y thousand at $2 each is given by

$$P(x, y) = -2x^2 + 2xy - y^2 + 12x - 4y - 7.$$

This is an example of a *function of several variables.* Find the amount of each type of ball that must be produced and sold in order to maximize profit.

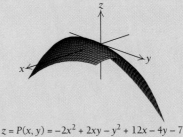

$$z = P(x, y) = -2x^2 + 2xy - y^2 + 12x - 4y - 7$$

To solve this problem, we must learn to find what are called *partial derivatives* and use a maximum–minimum test similar to one we used in Chapter 3. It turns out that profit is maximized by making and selling 4 thousand of the $3 balls and 2 thousand of the $2 balls.

This problem appears as Example 3 in Section 7.4.

7.1

O B J E C T I V E S

➤ Find a function value for a function of several variables.

Functions of Several Variables

Suppose that a one-product firm produces x items of its product at a profit of $4 per item. Then its total profit $P(x)$ is given by

$$P(x) = 4x.$$

This is a function of one variable.

Suppose that a two-product firm produces x items of one product at a profit of $4 per item and y items of a second at a profit of $6 per item. Then its total profit P is a function of the *two* variables x and y, and is given by

$$P(x, y) = 4x + 6y.$$

This function assigns to the input pair (x, y) a unique output number $4x + 6y$.

Definition

A **function of two variables** assigns to each input pair (x, y) exactly one output number $f(x, y)$.

We can also think of a function of two variables as a machine that has two inputs. The domain of a function of two variables is a set of pairs (x, y) in the plane. Unless otherwise restricted, when such a function is given by a formula, the domain consists of all ordered pairs (x, y) that are meaningful replacements in the formula.

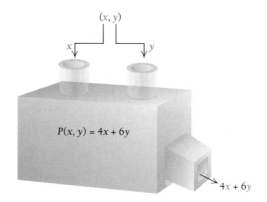

Example 1   For $P(x, y) = 4x + 6y$, find $P(25, 10)$.

Solution $P(25, 10)$ is defined to be the value of the function found by substituting 25 for x and 10 for y:

$$P(25, 10) = 4 \cdot 25 + 6 \cdot 10$$
$$= 100 + 60$$
$$= \$160.$$

This means that by selling 25 items of the first product and 10 of the second, the two-product firm will make a profit of $160.   ◁

   The following are examples of **functions of several variables,** that is, functions of two or more variables. If there are n variables, then there are n inputs for such a function.

Example 2   *Business: Total Cost.* The total cost of a company, in thousands of dollars, is given by

$$C(x, y, z, w) = 4x^2 + 5y + z - \ln(w + 1),$$

where x dollars is spent for labor, y dollars for raw materials, z dollars for advertising, and w dollars for machinery. This is a function of four variables (all in thousands of dollars). Find $C(3, 2, 0, 10)$.

Solution   We substitute 3 for x, 2 for y, 0 for z, and 10 for w:

$$C(3, 2, 0, 10) = 4 \cdot 3^2 + 5 \cdot 2 + 0 - \ln(10 + 1)$$
$$= 4 \cdot 9 + 10 + 0 - 2.397895$$
$$\approx \$43.6 \text{ thousand.}$$ ◁

Example 3   *Business: Cost of Storage Equipment.* A business purchases a piece of storage equipment that costs C_1 dollars and has capacity V_1. Later it wishes to replace the original with a new piece of equipment that costs C_2 dollars and has capacity V_2. The ratio of the new capacity to the original is

$$\frac{V_2}{V_1} = k,$$

so $V_2 = kV_1$.

It has been found in industrial economics that in this case, the cost of the new piece of equipment can be estimated by the function of three variables

$$C_2 = \left(\frac{V_2}{V_1}\right)^{0.6} C_1 = k^{0.6}C_1. \tag{1}$$

For $45,000, a beverage company buys a manufacturing tank that has a capacity of 10,000 gallons. Later it decides to buy a tank with double the capacity of the original. Estimate the cost of the new tank.

Solution   We substitute 20,000 for V_2, 10,000 for V_1, and 45,000 for C_1 in equation (1):

$$C_2 = \left(\frac{20,000}{10,000}\right)^{0.6}(45,000)$$
$$= 2^{0.6}(45,000)$$
$$\approx \$68,207.25.$$

Note that a 100% increase in capacity was achieved by about a 52% increase in cost. This is independent of any increase in the costs of labor, management, or other equipment resulting from the purchase of the tank.   ◁

498

Exercise

1. According to a recent census, the population of San Francisco was 1,149,000, and the population of Oakland was 373,000. The distance between the cities is 8 miles. What is the average number of telephone calls in a day between these two cities?

This NASA photograph shows an overhead view of San Francisco and Oakland. Sociologists say that as two cities merge, the communication between them increases.

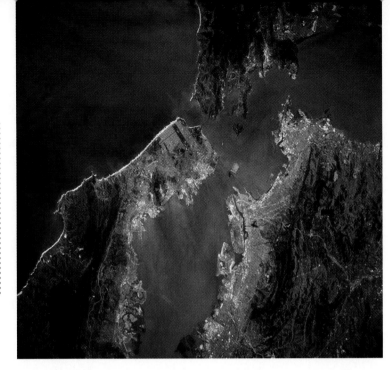

Example 4 *Social Science: The Gravity Model.* The average number of telephone calls in a day between two cities is given by

$$N(d, P_1, P_2) = \frac{2.8 P_1 P_2}{d^{2.4}},$$

where d is the distance, in miles, between the cities and P_1 and P_2 are their populations.

A constant can also be thought of as a function of several variables.

Example 5 The constant function f is given by

$$f(x, y) = -3 \quad \text{for all inputs } x \text{ and } y.$$

Find $f(5, 7)$ and $f(-2, 0)$.

Solution Since this is a constant function, it has the value -3 for any x and y. Thus,

$$f(5, 7) = -3 \quad \text{and} \quad f(-2, 0) = -3.$$

Geometric Interpretations

Consider a function of two variables

$$z = f(x, y).$$

As a mapping, a function of two variables can be thought of as mapping a point (x_1, y_1) in an xy-plane onto a point z_1 on a number line.

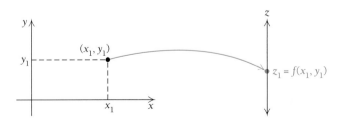

To graph a function of two variables, we need a three-dimensional coordinate system. The axes are generally placed as shown below. The line z, called the z-axis, is placed perpendicular to the xy-plane at the origin.

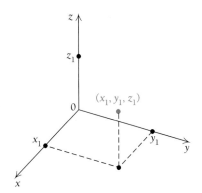

To help visualize this, think of looking into the corner of a room, where the floor is the xy-plane and the z-axis is the intersection of the two walls. To plot a point (x_1, y_1, z_1), we locate the point (x_1, y_1) in the xy-plane and move up or down in space according to the value of z_1.

Example 6   Plot these points:

$$P_1(2, 3, 5), \quad P_2(2, -2, -4), \quad P_3(0, 5, 2), \quad \text{and} \quad P_4(2, 3, 0).$$

Solution   The solution is shown below.

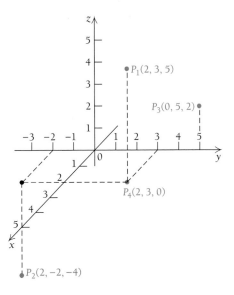

The *graph* of a function of two variables

$$z = f(x, y)$$

consists of ordered triples (x_1, y_1, z_1), where $z_1 = f(x_1, y_1)$. The domain of f is a region D in the xy-plane, and the graph of f is a surface S.

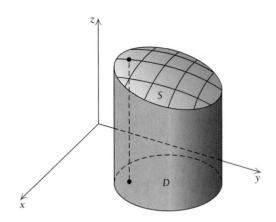

The following graphs have been generated from their equations by a computer graphics program called *Mathematica*. There are many excellent graphics programs for generating graphs of functions of two variables. The graphs can be generated with very few keystrokes. Seeing one of these graphs often reveals more insight to the behavior of the function than does its formula.

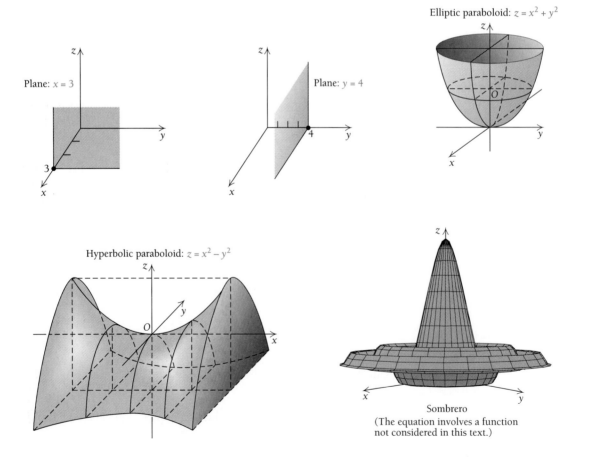

Plane: $x = 3$

Plane: $y = 4$

Elliptic paraboloid: $z = x^2 + y^2$

Hyperbolic paraboloid: $z = x^2 - y^2$

Sombrero
(The equation involves a function
not considered in this text.)

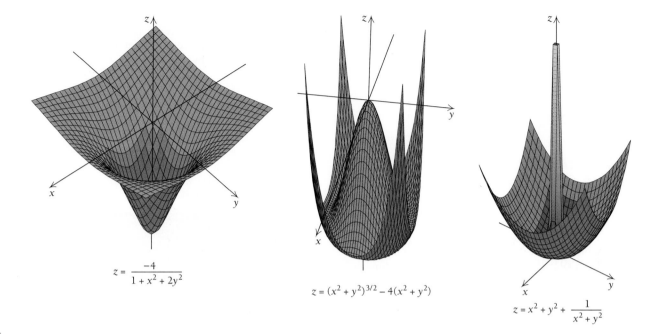

$$z = \frac{-4}{1 + x^2 + 2y^2}$$

$$z = (x^2 + y^2)^{3/2} - 4(x^2 + y^2)$$

$$z = x^2 + y^2 + \frac{1}{x^2 + y^2}$$

TECHNOLOGY CONNECTION

There are many 3D graphers capable of generating graphs of functions of two variables. Use one to verify the graphs shown here. What problems do you have with viewing windows?

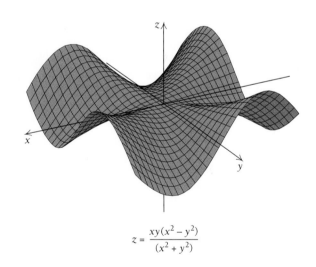

$$z = \frac{xy(x^2 - y^2)}{(x^2 + y^2)}$$

EXERCISE SET 7.1

1. For $f(x, y) = x^2 - 2xy$, find $f(0, -2)$, $f(2, 3)$, and $f(10, -5)$.

2. For $f(x, y) = (y^2 + 3xy)^3$, find $f(-2, 0)$, $f(3, 2)$, and $f(-5, 10)$.

3. For $f(x, y) = 3^x + 7xy$, find $f(0, -2)$, $f(-2, 1)$, and $f(2, 1)$.

4. For $f(x, y) = \log_{10} x - 5y^2$, find $f(10, 2)$, $f(1, -3)$, and $f(100, 4)$.

5. For $f(x, y) = \ln x + y^3$, find $f(e, 2)$, $f(e^2, 4)$, and $f(e^3, 5)$.

6. For $f(x, y) = 2^x - 3^y$, find $f(0, 0)$, $f(1, 1)$, and $f(2, 2)$.

7. For $f(x, y, z) = x^2 - y^2 + z^2$, find $f(-1, 2, 3)$ and $f(2, -1, 3)$.

8. For $f(x, y, z) = 2^x + 5zy - x$, find $f(0, 1, -3)$ and $f(1, 0, -3)$.

Applications

BUSINESS AND ECONOMICS

9. *Cost of Storage Equipment.* Consider the storage model in Example 3. For $100,000, a company buys a storage tank that has a capacity of 80,000 gallons. Later it replaces the tank with a new tank with double the capacity of the original. Estimate the cost of the new tank.

10. *Price–Earnings Ratio.* The *price–earnings ratio* of a stock is given by

$$R(P, E) = \frac{P}{E},$$

where P is the price per share of the stock and E is the earnings per share. Recently, the price per share of IBM stock was $92\frac{3}{16}$ and the earnings per share were $3.28. Find the price–earnings ratio. Give decimal notation to the nearest tenth.

11. *Yield of a Stock.* The *yield* of a stock is given by

$$Y(D, P) = \frac{D}{P},$$

where D is the dividends per share of a stock and P is the price per share. Recently, the price per share of Goodyear stock was $50\frac{7}{16}$ and the dividends per share were $1.20. Find the yield. Give percent notation to the nearest tenth of a percent.

LIFE AND PHYSICAL SCIENCES

12. *Poiseuille's Law.* The speed of blood in a vessel is given by

$$V(L, p, R, r, v) = \frac{p}{4Lv}(R^2 - r^2),$$

where R is the radius of the vessel, r is the distance of the blood from the center of the vessel, L is the length of the blood vessel, p is the pressure, and v is the viscosity. Find $V(1, 100, 0.0075, 0.0025, 0.05)$.

13. *Wind Speed of a Tornado.* Under certain conditions, the *wind speed S,* in miles per hour, of a tornado at a distance d from its center can be approximated by the function

$$S = \frac{aV}{0.51d^2},$$

where a is an atmospheric constant that depends on certain atmospheric conditions and V is the approximate volume of the tornado, in cubic feet. Approximate the wind speed 100 ft from the center of a tornado when its volume is $1,600,000$ ft^3 and $a = 0.78$.

SOCIAL SCIENCES

14. *Intelligence Quotient.* The *intelligence quotient* in psychology is given by

$$Q(m, c) = 100 \cdot \frac{m}{c},$$

where m is a person's mental age and c is his or her chronological, or actual, age. Find $Q(21, 20)$ and $Q(19, 20)$.

Synthesis

tw 15. Explain the difference between a function of two variables and a function of one variable.

tw 16. Find some examples of functions of several variables not considered in the text, even some that may not have formulas.

TECHNOLOGY CONNECTION

GENERAL INTEREST

Wind Chill Temperature. Because wind speed enhances the loss of heat from the skin, we feel colder when there is wind than when there is not. The *wind chill temperature* is what the temperature would have to be with no wind in order to give the same chilling effect. The wind chill temperature W is given by

$$W(v, T) = 91.4 - \frac{(10.45 + 6.68\sqrt{v} - 0.447v)(457 - 5T)}{110},$$

where T is the actual temperature as given by a thermometer, in degrees Fahrenheit, and v is the speed of the wind, in miles per hour. Find the wind chill temperature in each case. Round to the nearest one degree.

17. $T = 30°F$, $v = 25$ mph
18. $T = 20°F$, $v = 20$ mph
19. $T = 20°F$, $v = 40$ mph
20. $T = -10°F$, $v = 30$ mph

Use a 3D grapher to generate the graph of the function.

21. $f(x, y) = y^2$
22. $f(x, y) = x^2 + y^2$
23. $f(x, y) = (x^4 - 16x^2)e^{-y^2}$
24. $f(x, y) = 4(x^2 + y^2) - (x^2 + y^2)^2$
25. $f(x, y) = x^3 - 3xy^2$
26. $f(x, y) = \dfrac{1}{x^2 + 4y^2}$

7.2

Partial Derivatives

Finding Partial Derivatives

Consider the function f given by

$$z = f(x, y) = x^2y^3 + xy + 4y^2.$$

Suppose for the moment that we fix y at 3. Then

$$f(x, 3) = x^2(3^3) + x(3) + 4(3^2) = 27x^2 + 3x + 36.$$

Note that we now have a function of only one variable. Taking the first derivative with respect to x, we have

$$54x + 3.$$

In general, without replacing y with a specific number, let's consider y fixed. Then f becomes a function of x alone and we can calculate its derivative with respect to x. This derivative is called the *partial derivative of f with respect to x*. Notation for this partial derivative is

$$\frac{\partial f}{\partial x} \quad \text{or} \quad \frac{\partial z}{\partial x}.$$

Thus let's again consider the function

$$z = f(x, y) = x^2 y^3 + xy + 4y^2.$$

The color blue indicates the variable x when we fix y and treat it as a constant. The expressions y^3, y, and y^2 are then constants. We have

$$\frac{\partial f}{\partial x} = \frac{\partial z}{\partial x}$$
$$= 2xy^3 + y.$$

Similarly, we find $\partial f/\partial y$ or $\partial z/\partial y$ by fixing x (treating it as a constant) and calculating the derivative with respect to y. From

$$z = f(x, y) = x^2 y^3 + xy + 4y^2, \qquad \text{The color blue indicates the variable.}$$

we get

$$\frac{\partial f}{\partial y} = \frac{\partial z}{\partial y}$$
$$= 3x^2 y^2 + x + 8y.$$

A definition of partial derivatives is as follows.

Definition

For $z = f(x, y)$,

$$\frac{\partial z}{\partial x} = \lim_{h \to 0} \frac{f(x + h, y) - f(x, y)}{h},$$

$$\frac{\partial z}{\partial y} = \lim_{k \to 0} \frac{f(x, y + k) - f(x, y)}{k}.$$

We can find partial derivatives of functions of any number of variables.

Example 1   For $w = x^2 - xy + y^2 + 2yz + 2z^2 + z$, find

$$\frac{\partial w}{\partial x}, \quad \frac{\partial w}{\partial y}, \quad \text{and} \quad \frac{\partial w}{\partial z}.$$

Solution   In order to find $\partial w/\partial x$, we consider x the variable and the other letters the constants. From

$$w = x^2 - xy + y^2 + 2yz + 2z^2 + z,$$

we get

$$\frac{\partial w}{\partial x} = 2x - y.$$

From

$$w = x^2 - xy + y^2 + 2yz + 2z^2 + z,$$

we get

$$\frac{\partial w}{\partial y} = -x + 2y + 2z;$$

and from

$$w = x^2 - xy + y^2 + 2yz + 2z^2 + z,$$

we get

$$\frac{\partial w}{\partial z} = 2y + 4z + 1.$$

We will often make use of a simpler notation f_x for the partial derivative of f with respect to x and f_y for the partial derivative of f with respect to y. That is, if $z = f(x, y)$, then z_x represents the partial derivative of z with respect to x and z_y represents the partial derivative of z with respect to y.

Example 2   For $f(x, y) = 3x^2y + xy$, find f_x and f_y.

Solution   We have

$$f_x = 6xy + y,$$
$$f_y = 3x^2 + x.$$

For the function in the preceding example, let's evaluate f_x at $(2, -3)$:

$$f_x(2, -3) = 6 \cdot 2 \cdot (-3) + (-3)$$
$$= -39.$$

If we use the notation $\partial f / \partial x = 6xy + y$, where $f = 3x^2y + xy$, the value of the partial derivative at $(2, -3)$ is given by

$$\left. \frac{\partial f}{\partial x} \right|_{(2, -3)} = 6 \cdot 2 \cdot (-3) + (-3)$$
$$= -39,$$

but this notation is not as convenient as $f_x(2, -3)$.

TECHNOLOGY CONNECTION

Consider finding values of a partial derivative of $f(x, y) = 3x^3y + 2xy$ using a grapher that finds derivatives of functions of one variable. How can you find $f_x(-4, 1)$? Then how can you find $f_y(2, 6)$?

Example 3   For $f(x, y) = e^{xy} + y \ln x$, find f_x and f_y.

Solution

$$f_x = y \cdot e^{xy} + y \cdot \frac{1}{x}$$
$$= ye^{xy} + \frac{y}{x},$$
$$f_y = x \cdot e^{xy} + 1 \cdot \ln x$$
$$= xe^{xy} + \ln x$$

The Geometric Interpretation of Partial Derivatives

The *graph* of a function of two variables $z = f(x, y)$ is a surface S, which might have a graph similar to the one shown below, where each input pair (x, y) has only one output $z = f(x, y)$.

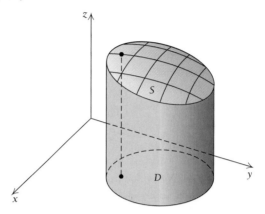

Now suppose that we hold x fixed, say, at the value a. The set of all points for which $x = a$ is a plane parallel to the yz-plane, so when x is fixed at a, y and z vary along the plane, as shown below. The plane in the figure cuts the surface in some curve C_1. The partial derivative f_y gives the slope of tangent lines to this curve.

Similarly, if we hold y fixed, say, at the value b, we obtain a curve C_2, as follows. The partial derivative f_x gives the slope of tangent lines to this curve.

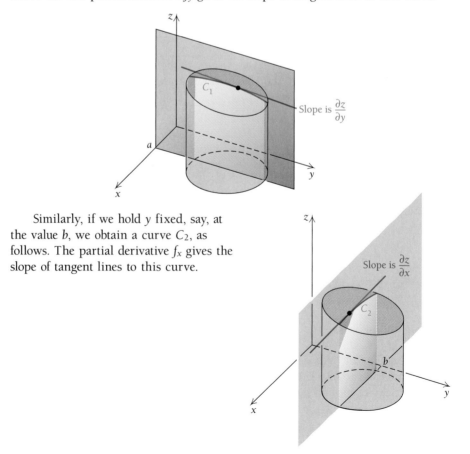

An Economics Application: The Cobb–Douglas Production Function

One model of production that is frequently considered in business and economics is the *Cobb–Douglas production function:*

$$p(x, y) = Ax^a y^{1-a}, \qquad A > 0 \text{ and } 0 < a < 1,$$

where p is the number of units produced with x units of labor and y units of capital. (Capital is the cost of machinery, buildings, tools, and other supplies.) The partial derivatives

$$\frac{\partial p}{\partial x} \quad \text{and} \quad \frac{\partial p}{\partial y}$$

are called, respectively, the *marginal productivity of labor* and the *marginal productivity of capital.*

Example 4   A cellular phone company has the following production function for a certain product:

$$p(x, y) = 50x^{2/3} y^{1/3}.$$

a) Find the production from 125 units of labor and 64 units of capital.

b) Find the marginal productivities.

c) Evaluate the marginal productivities at $x = 125$ and $y = 64$.

Solution

a) $p(125, 64) = 50(125)^{2/3}(64)^{1/3} = 50(25)(4) = 5000$ units

b) $\dfrac{\partial p}{\partial x} = 50\left(\dfrac{2}{3}\right)x^{-1/3}y^{1/3} = \dfrac{100y^{1/3}}{3x^{1/3}}, \quad \text{or} \quad \dfrac{100}{3}\left(\dfrac{y}{x}\right)^{1/3};$

$\dfrac{\partial p}{\partial y} = 50\left(\dfrac{1}{3}\right)x^{2/3}y^{-2/3} = \dfrac{50x^{2/3}}{3y^{2/3}}, \quad \text{or} \quad \dfrac{50}{3}\left(\dfrac{x}{y}\right)^{2/3}$

c) $\left.\dfrac{\partial p}{\partial x}\right|_{(125, 64)} = \dfrac{100(64)^{1/3}}{3(125)^{1/3}} = \dfrac{100(4)}{3(5)} = 26\dfrac{2}{3};$

$\left.\dfrac{\partial p}{\partial y}\right|_{(125, 64)} = \dfrac{50(125)^{2/3}}{3(64)^{2/3}} = \dfrac{50(25)}{3(16)} = 26\dfrac{1}{24}$

How can we interpret these marginal productivities? Suppose that the amount spent on capital is fixed at 64. Then if the amount of labor changes by one unit from 125, production will change by about 27 units. Suppose that the amount of labor is held fixed at 125. Then if the amount of capital spent changes by one unit from 64, production will change by about 26 units.

A Cobb–Douglas production function is consistent with the law of diminishing returns. That is, if one input (of either labor or capital) is held fixed while the other increases infinitely, then production will eventually increase at a decreasing rate. With such functions, it also turns out that if a certain maximum production is possible, then the expense of more labor, for example, will not prevent that maximum output from still being attainable.

EXERCISE SET 7.2

Find $\dfrac{\partial z}{\partial x}, \dfrac{\partial z}{\partial y}, \dfrac{\partial z}{\partial x}\Big|_{(-2,\,-3)}$ and $\dfrac{\partial z}{\partial y}\Big|_{(0,\,-5)}$

1. $z = 2x - 3xy$ 2. $z = (x - y)^3$

3. $z = 3x^2 - 2xy + y$ 4. $z = 2x^3 + 3xy - x$

Find f_x, f_y, $f_x(-2, 4)$, and $f_y(4, -3)$.

5. $f(x, y) = 2x - 3y$

6. $f(x, y) = 5x + 7y$

Find f_x, f_y, $f_x(-2, 1)$, and $f_y(-3, -2)$.

7. $f(x, y) = \sqrt{x^2 + y^2}$

8. $f(x, y) = \sqrt{x^2 - y^2}$

Find f_x and f_y.

9. $f(x, y) = e^{2x+3y}$

10. $f(x, y) = e^{3x-2y}$

11. $f(x, y) = e^{xy}$

12. $f(x, y) = e^{2xy}$

13. $f(x, y) = y \ln (x + y)$

14. $f(x, y) = x \ln (x + y)$

15. $f(x, y) = x \ln (xy)$

16. $f(x, y) = y \ln (xy)$

17. $f(x, y) = \dfrac{x}{y} - \dfrac{y}{x}$

18. $f(x, y) = \dfrac{x}{y} + \dfrac{y}{x}$

19. $f(x, y) = 3(2x + y - 5)^2$

20. $f(x, y) = 4(3x + y - 8)^2$

Find $\dfrac{\partial f}{\partial b}$ and $\dfrac{\partial f}{\partial m}$.

21. $f(b, m) = (m + b - 4)^2 + (2m + b - 5)^2 + (3m + b - 6)^2$

22. $f(b, m) = (m + b - 6)^2 + (2m + b - 8)^2 + (3m + b - 9)^2$

Find f_x, f_y, and f_λ.

23. $f(x, y, \lambda) = 3xy - \lambda(2x + y - 8)$

24. $f(x, y, \lambda) = 4xy - \lambda(3x - y + 7)$

25. $f(x, y, \lambda) = x^2 + y^2 - \lambda(10x + 2y - 4)$

26. $f(x, y, \lambda) = x^2 - y^2 - \lambda(4x - 7y - 10)$

Applications ....

BUSINESS AND ECONOMICS

27. *The Cobb–Douglas Model.* A publishing company has the following production function for a certain product:

$$p(x, y) = 1800x^{0.621}y^{0.379},$$

where p is the number of units produced with x units of labor and y units of capital.

a) Find the production from 2500 units of labor and 1700 units of capital.

b) Find the marginal productivities.

tw c) Interpret the meanings of the marginal productivities found in part (b).

d) Evaluate the marginal productivities at $x = 2500$ and $y = 1700$.

28. *The Cobb–Douglas Model.* A sporting goods company has the following production function for a certain product:

$$p(x, y) = 2400x^{2/5}y^{3/5},$$

where p is the number of units produced with x units of labor and y units of capital.

a) Find the production from 32 units of labor and 1024 units of capital.

b) Find the marginal productivities.

tw c) Interpret the meanings of the marginal productivities found in part (b).

d) Evaluate the marginal productivities at $x = 32$ and $y = 1024$.

LIFE AND PHYSICAL SCIENCES

Temperature–Humidity Heat Index. In the summer, humidity affects the actual temperature, making a person feel hotter due to a reduced heat loss from the skin caused by higher humidity. The *temperature–humidity index*, T_h, is what the temperature would have to be with no humidity in order to give the same heat effect. One index often used is given by

$$T_h = 1.98T - 1.09(1 - H)(T - 58) - 56.9,$$

where T is the air temperature, in degrees Fahrenheit, and H is the relative humidity, which is the ratio of the amount of water vapor in the air to the maximum amount of water vapor possible in the air at that temperature. H is usually expressed as a percentage. Find the temperature–humidity index in each case. Round to the nearest tenth of a degree.

29. $T = 85°$ and $H = 60\%$
30. $T = 90°$ and $H = 90\%$
31. $T = 90°$ and $H = 100\%$
32. $T = 78°$ and $H = 100\%$

tw 33. Find $\dfrac{\partial T_h}{\partial H}$ and interpret its meaning.

tw 34. Find $\dfrac{\partial T_h}{\partial T}$ and interpret its meaning.

SOCIAL SCIENCES

Reading Ease. The following formula is used by psychologists and educators to predict the *reading*

ease E of a passage of words:

$$E = 206.835 - 0.846w - 1.015s,$$

where w is the number of syllables in a 100-word section and s is the average number of words per sentence. Find the reading ease in each case.

35. $w = 146$ and $s = 5$
36. $w = 180$ and $s = 6$

37. Find $\dfrac{\partial E}{\partial w}$.

38. Find $\dfrac{\partial E}{\partial s}$.

Synthesis

Find f_x and f_t.

39. $f(x, t) = \dfrac{x^2 + t^2}{x^2 - t^2}$ 40. $f(x, t) = \dfrac{x^2 - t}{x^3 + t}$

41. $f(x, t) = \dfrac{2\sqrt{x} - 2\sqrt{t}}{1 + 2\sqrt{t}}$ 42. $f(x, t) = \sqrt[4]{x^3 t^5}$

43. $f(x, t) = 6x^{2/3} - 8x^{1/4}t^{1/2} - 12x^{-1/2}t^{3/2}$

44. $f(x, t) = \left(\dfrac{x^2 + t^2}{x^2 - t^2}\right)^5$

tw 45. Do some library research on the Cobb–Douglas production function. Can you find how it was developed?

tw 46. Explain the meaning of the first partial derivatives of a function of two variables in terms of slopes of tangent lines.

7.3

Higher-Order Partial Derivatives

Consider

$$z = f(x, y) = 3xy^2 + 2xy + x^2.$$

Then

$$\frac{\partial z}{\partial x} = \frac{\partial f}{\partial x} = 3y^2 + 2y + 2x.$$

Suppose that we continue and find the first partial derivative of $\partial z/\partial x$ with respect to y. This will be a **second-order partial derivative** of the original function z. Its notation is as follows:

$$\frac{\partial}{\partial y}\left(\frac{\partial z}{\partial x}\right) = \frac{\partial}{\partial y}\left(\frac{\partial f}{\partial x}\right)$$

$$= \frac{\partial^2 z}{\partial y\,\partial x}$$

$$= \frac{\partial^2 f}{\partial y\,\partial x}$$

$$= 6y + 2.$$

We could also denote the preceding partial derivative using the notation f_{xy}. Then

$$f_{xy} = 6y + 2.$$

Note that in the notation f_{xy}, x and y are in the order (left to right) in which the differentiation is done, but in

$$\frac{\partial^2 f}{\partial y\,\partial x},$$

the order of x and y is reversed. In both notations, the differentiation with respect to x is done first, followed by differentiation with respect to y.

Notation for the four second-order partial derivatives is as follows.

Definition

1. $\dfrac{\partial^2 z}{\partial x\,\partial x} = \dfrac{\partial^2 f}{\partial x\,\partial x} = \dfrac{\partial^2 z}{\partial x^2} = \dfrac{\partial^2 f}{\partial x^2} = f_{xx}$

    Take the partial with respect to x, and then with respect to x again.

2. $\dfrac{\partial^2 z}{\partial y\,\partial x} = \dfrac{\partial^2 f}{\partial y\,\partial x} = f_{xy}$

    Take the partial with respect to x, and then with respect to y.

3. $\dfrac{\partial^2 z}{\partial x\,\partial y} = \dfrac{\partial^2 f}{\partial x\,\partial y} = f_{yx}$

    Take the partial with respect to y, and then with respect to x.

4. $\dfrac{\partial^2 z}{\partial y\,\partial y} = \dfrac{\partial^2 f}{\partial y\,\partial y} = \dfrac{\partial^2 z}{\partial y^2} = \dfrac{\partial^2 f}{\partial y^2} = f_{yy}$

    Take the partial with respect to y, and then with respect to y again.

Example 1   For

$$z = f(x, y) = x^2 y^3 + x^4 y + xe^y,$$

find the four second-order partial derivatives.

Solution

a) $\dfrac{\partial^2 f}{\partial x^2} = f_{xx} = \dfrac{\partial}{\partial x}(2xy^3 + 4x^3y + e^y)$

$\qquad\qquad = 2y^3 + 12x^2y$

Differentiate twice with respect to x.

b) $\dfrac{\partial^2 f}{\partial y\,\partial x} = f_{xy} = \dfrac{\partial}{\partial y}(2xy^3 + 4x^3y + e^y)$

$\qquad\qquad = 6xy^2 + 4x^3 + e^y$

Differentiate with respect to x, and then with respect to y.

c) $\dfrac{\partial^2 f}{\partial x\,\partial y} = f_{yx} = \dfrac{\partial}{\partial x}(3x^2y^2 + x^4 + xe^y)$

$\qquad\qquad = 6xy^2 + 4x^3 + e^y$

Differentiate with respect to y, and then with respect to x.

d) $\dfrac{\partial^2 f}{\partial y^2} = f_{yy} = \dfrac{\partial}{\partial y}(3x^2y^2 + x^4 + xe^y)$

$\qquad\qquad = 6x^2y + xe^y$

Differentiate twice with respect to y.

We see by comparing parts (b) and (c) in Example 1 that

$$\frac{\partial^2 f}{\partial y\,\partial x} = \frac{\partial^2 f}{\partial x\,\partial y} \quad \text{and} \quad f_{xy} = f_{yx}.$$

Although this will be true for virtually all functions that we consider in this text, it is *not* true for all functions. Such a function is given in Exercise 19 of Exercise Set 7.3.

EXERCISE SET 7.3

Find the four second-order partial derivatives.

1. $f(x, y) = 3x^2 - xy + y$
2. $f(x, y) = 5x^2 + xy - x$
3. $f(x, y) = 3xy$
4. $f(x, y) = 4xy$
5. $f(x, y) = x^5y^4 + x^3y^2$
6. $f(x, y) = x^4y^3 - x^2y^3$

Find f_{xx}, f_{xy}, f_{yx}, and f_{yy}. (Remember, f_{yx} means to differentiate with respect to y, and then to x.)

7. $f(x, y) = 2x - 3y$ 8. $f(x, y) = 3x + 5y$
9. $f(x, y) = e^{2xy}$ 10. $f(x, y) = e^{xy}$
11. $f(x, y) = x + e^y$ 12. $f(x, y) = y - e^x$
13. $f(x, y) = y \ln x$ 14. $f(x, y) = x \ln y$

Synthesis ....

Find f_{xx}, f_{xy}, f_{yx}, and f_{yy}.

15. $f(x, y) = \dfrac{x}{y^2} - \dfrac{y}{x^2}$

16. $f(x, y) = \dfrac{xy}{x - y}$

17. Consider $f(x, y) = \ln(x^2 + y^2)$. Show that f is a solution to the partial differential equation

$$\frac{\partial^2 f}{\partial x^2} + \frac{\partial^2 f}{\partial y^2} = 0.$$

18. Consider $f(x, y) = x^3 - 5xy^2$. Show that f is a solution to the partial differential equation

$$xf_{xy} - f_y = 0.$$

19. Consider the function f defined as follows:

$$f(x, y) = \begin{cases} \dfrac{xy(x^2 - y^2)}{x^2 + y^2}, & \text{for } (x, y) \neq (0, 0), \\ 0, & \text{for } (x, y) = (0, 0). \end{cases}$$

a) Find $f_x(0, y)$ by evaluating the limit

$$\lim_{h \to 0} \frac{f(h, y) - f(0, y)}{h}.$$

b) Find $f_y(x, 0)$ by evaluating the limit

$$\lim_{h \to 0} \frac{f(x, h) - f(x, 0)}{h}.$$

c) Now find and compare $f_{yx}(0, 0)$ and $f_{xy}(0, 0)$.

20. For $f = [\ln (x^3 + e^y)]^5$, find f_{xx}, f_{xy}, f_{yx}, and f_{yy}.

7.4

Maximum–Minimum Problems

We will now find maximum and minimum values of functions of two variables.

Definition

A function f of two variables:

1. has a **relative maximum** at (a, b) if

   $$f(x, y) \leq f(a, b)$$

   for all points in a rectangular region containing (a, b);
2. has a **relative minimum** at (a, b) if

   $$f(x, y) \geq f(a, b)$$

   for all points in a rectangular region containing (a, b).

This definition is illustrated in Figs. 1 and 2. A relative maximum (minimum) may not be an "absolute" maximum (minimum), as illustrated in Fig. 3.

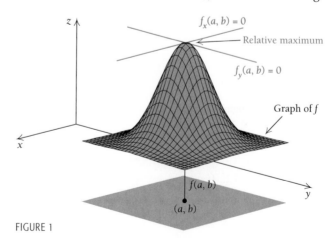

FIGURE 1

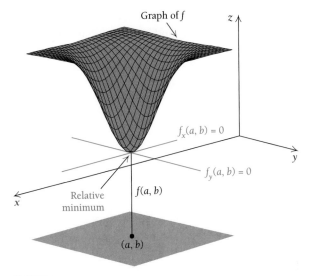

FIGURE 2

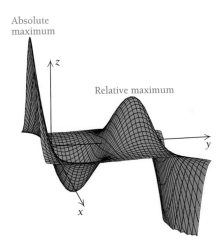

FIGURE 3

Determining Maximum and Minimum Values

Suppose that a function f assumes a relative maximum (or minimum) value at some point (a, b) inside its domain. We assume that f and its partial derivatives exist and are "continuous" inside its domain, though we will not take the space to define continuity. If we hold y fixed at the value b, then $f(x, b)$ is the output of a function f of one variable x, and the resulting function has its relative maximum value at $x = a$. Thus its derivative must be 0 there—that is, $f_x = 0$ at the point (a, b). Similarly, $f_y = 0$ at (a, b). The equations

$$f_x = 0 \quad \text{and} \quad f_y = 0 \tag{1}$$

are thus satisfied by the point (a, b) at which the relative maximum occurs. We call a point (a, b) at which both partial derivatives are 0 a *critical point*. This is comparable to the earlier definition for functions of one variable. Thus one strategy for finding relative maximum or minimum values is to solve the system of equations (1) above to find critical points. Just as for functions of one variable, this strategy does *not* guarantee that we will have a relative maximum or minimum value. We have argued only that *if f* has a maximum or minimum value at (a, b), *then* both its partial derivatives must be 0 at that point. Look back at Figs. 1 and 2. Then note Fig. 4, which illustrates a case in which the partial derivatives are 0 but the function does not have a relative maximum or minimum value at (a, b).

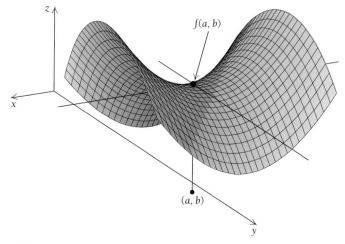

FIGURE 4

Where is the saddle point?

In Fig. 4, suppose that we fix y at a value b. Then $f(x, b)$, considered as the output of a function of one variable x, has a minimum at a, but f does not. Similarly, if we fix x at a, then $f(a, y)$, considered as the output of a function of one variable y, has a maximum at b, but f does not. The point $f(a, b)$ is called a **saddle point.** In other words, $f_x(a, b) = 0$ and $f_y(a, b) = 0$ [the point (a, b) is a critical point], but f does not attain a relative maximum or minimum value at (a, b). A saddle point for a function of two variables is comparable to a point of inflection (which is simultaneously a critical point) for a function of one variable.

A test for finding relative maximum and minimum values that involves the use of first- and second-order partial derivatives is stated below. We will not prove this theorem.

Theorem 1

The D-test

To find the relative maximum and minimum values of f:

1. Find f_x, f_y, f_{xx}, f_{yy}, and f_{xy}.
2. Solve the system of equations $f_x = 0, f_y = 0$. Let (a, b) represent a solution.
3. Evaluate D, where $D = f_{xx}(a, b) \cdot f_{yy}(a, b) - [f_{xy}(a, b)]^2$.

4. Then:

    a) f has a maximum at (a, b) if $D > 0$ and $f_{xx}(a, b) < 0$.
    b) f has a minimum at (a, b) if $D > 0$ and $f_{xx}(a, b) > 0$.
    c) f has neither a maximum nor a minimum at (a, b) if $D < 0$. The function has a **saddle point** at (a, b). See Fig. 4.
    d) This test is not applicable if $D = 0$.

The D-test is somewhat analogous to the Second Derivative Test (Section 3.2) for functions of one variable. Saddle points are analogous to critical points where concavity changes and that are not relative maximum or minimum values.

A relative maximum or minimum *may not be an absolute maximum or minimum value*. Tests for absolute maximum or minimum values are rather complicated. We will restrict our attention to finding *relative* maximum or minimum values. Fortunately, in most of our applications, relative maximum or minimum values turn out to be absolute as well.

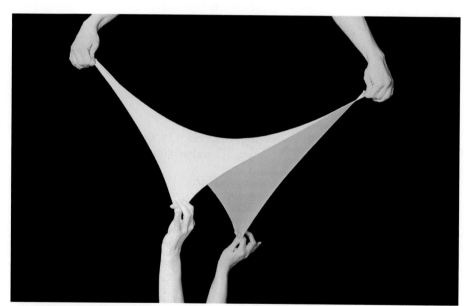

The shape of a perfect tent. To give a tent roof the maximum strength possible, designers draw the fabric into a series of three-dimensional shapes that, viewed in profile, resemble a horse's saddle and that mathematicians call an anticlastic curve. Two people with a stretchy piece of fabric such as Lycra Spandex can duplicate the shape, as shown above. One person pulls up and out on two diagonal corners; the other person pulls down and out on the other two corners. The opposing tensions draw each point of the fabric's surface into rigid equilibrium. The more pronounced the curve, the stiffer the surface.

Example 1   Find the relative maximum and minimum values of

$$f(x, y) = x^2 + xy + y^2 - 3x.$$

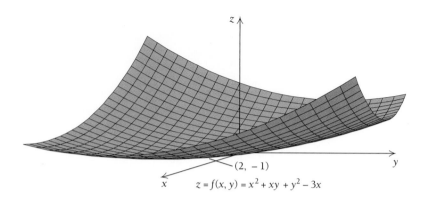

$$z = f(x, y) = x^2 + xy + y^2 - 3x$$

(2, −1)

Solution

1. Find f_x, f_y, f_{xx}, f_{yy}, and f_{xy}:

$$f_x = 2x + y - 3, \qquad f_y = x + 2y,$$
$$f_{xx} = 2; \qquad\qquad f_{yy} = 2;$$
$$f_{xy} = 1.$$

2. Solve the system of equations $f_x = 0, f_y = 0$:

$$2x + y - 3 = 0, \tag{1}$$
$$x + 2y = 0. \tag{2}$$

   Solving equation (2) for x, we get $x = -2y$. Substituting $-2y$ for x in equation (1) and solving, we get

$$2(-2y) + y - 3 = 0$$
$$-4y + y - 3 = 0$$
$$-3y = 3$$
$$y = -1.$$

   To find x when $y = -1$, we substitute -1 for y in either equation (1) or equation (2). We choose equation (2):

$$x + 2(-1) = 0$$
$$x = 2.$$

   Thus, $(2, -1)$ is the only critical point and $f(2, -1)$ is our candidate for a maximum or minimum value.

3. We must check to see whether $f(2, -1)$ is a maximum or minimum value:

$$D = f_{xx}(2, -1) \cdot f_{yy}(2, -1) - [f_{xy}(2, -1)]^2$$
$$= 2 \cdot 2 - [1]^2$$
$$= 3.$$

4. Thus, $D = 3$ and $f_{xx}(2, -1) = 2$. Since $D > 0$ and $f_{xx}(2, -1) > 0$, it follows that f has a relative minimum at $(2, -1)$ and that the minimum value is found as follows:

$$f(2, -1) = 2^2 + 2(-1) + (-1)^2 - 3 \cdot 2$$
$$= 4 - 2 + 1 - 6$$
$$= -3.$$

Example 2   Find the relative maximum and minimum values of

$$f(x, y) = xy - x^3 - y^2.$$

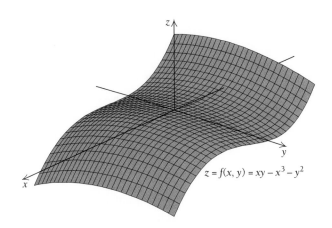

$z = f(x, y) = xy - x^3 - y^2$

Solution

1. Find $f_x, f_y, f_{xx}, f_{yy},$ and f_{xy}:

$$f_x = y - 3x^2, \qquad f_y = x - 2y,$$
$$f_{xx} = -6x; \qquad\quad f_{yy} = -2;$$
$$f_{xy} = 1.$$

2. Solve the system of equations $f_x = 0, f_y = 0$:

$$y - 3x^2 = 0, \tag{1}$$
$$x - 2y = 0. \tag{2}$$

Solving equation (1) for y, we get $y = 3x^2$. Substituting $3x^2$ for y in equation (2) and solving, we get

$$x - 2(3x^2) = 0$$
$$x - 6x^2 = 0$$
$$x(1 - 6x) = 0. \qquad \text{Factoring}$$

Setting each factor equal to 0 and solving, we have

$$x = 0 \quad or \quad 1 - 6x = 0$$
$$x = 0 \quad or \qquad\quad x = \tfrac{1}{6}.$$

To find y when $x = 0$, we substitute 0 for x in either equation (1) or equation (2). We choose equation (2):

$$0 - 2y = 0$$
$$-2y = 0$$
$$y = 0.$$

Thus, $(0, 0)$ is a critical point and $f(0, 0)$ is one candidate for a maximum or minimum value. To find the other, we substitute $\frac{1}{6}$ for x in either equation (1) or equation (2). We choose equation (2):

$$\tfrac{1}{6} - 2y = 0$$
$$-2y = -\tfrac{1}{6}$$
$$y = \tfrac{1}{12}.$$

Thus, $\left(\frac{1}{6}, \frac{1}{12}\right)$ is another critical point and $f\left(\frac{1}{6}, \frac{1}{12}\right)$ is another candidate for a maximum or minimum value.

3., 4. We must check both $(0, 0)$ and $\left(\frac{1}{6}, \frac{1}{12}\right)$ to see whether they yield maximum or minimum values:

$$\text{For } (0, 0): \quad D = f_{xx}(0, 0) \cdot f_{yy}(0, 0) - [\, f_{xy}(0, 0)]^2$$
$$= (-6 \cdot 0) \cdot (-2) - [1]^2$$
$$= -1.$$

Since $D < 0$, it follows that $f(0, 0)$ is neither a maximum nor a minimum value, but a saddle point.

$$\text{For } \left(\tfrac{1}{6}, \tfrac{1}{12}\right): \quad D = f_{xx}\!\left(\tfrac{1}{6}, \tfrac{1}{12}\right) \cdot f_{yy}\!\left(\tfrac{1}{6}, \tfrac{1}{12}\right) - \left[\, f_{xy}\!\left(\tfrac{1}{6}, \tfrac{1}{12}\right)\right]^2$$
$$= \left(-6 \cdot \tfrac{1}{6}\right) \cdot (-2) - [1]^2$$
$$= -1(-2) - 1$$
$$= 1.$$

Thus, $D = 1$ and $f_{xx}\!\left(\frac{1}{6}, \frac{1}{12}\right) = -1$. Since $D > 0$ and $f_{xx}\!\left(\frac{1}{6}, \frac{1}{12}\right) < 0$, it follows that f has a relative maximum at $\left(\frac{1}{6}, \frac{1}{12}\right)$ and that maximum value is found as follows:

$$f\!\left(\tfrac{1}{6}, \tfrac{1}{12}\right) = \tfrac{1}{6} \cdot \tfrac{1}{12} - \left(\tfrac{1}{6}\right)^3 - \left(\tfrac{1}{12}\right)^2$$
$$= \tfrac{1}{72} - \tfrac{1}{216} - \tfrac{1}{144} = \tfrac{1}{432}.$$

Example 3   *Business: Maximizing Profit.* A firm produces two kinds of golf ball, one that sells for $3 and one for $2. The total revenue, in thousands of dollars, from the sale of x thousand balls at $3 each and y thousand at $2 each is given by

$$R(x, y) = 3x + 2y.$$

The company determines that the total cost, in thousands of dollars, of producing x thousand of the $3 ball and y thousand of the $2 ball is given by

$$C(x, y) = 2x^2 - 2xy + y^2 - 9x + 6y + 7.$$

Find the amount of each type of ball that must be produced and sold in order to maximize profit.

Solution   The total profit $P(x, y)$ is given by

$$P(x, y) = R(x, y) - C(x, y)$$
$$= 3x + 2y - (2x^2 - 2xy + y^2 - 9x + 6y + 7)$$
$$P(x, y) = -2x^2 + 2xy - y^2 + 12x - 4y - 7.$$

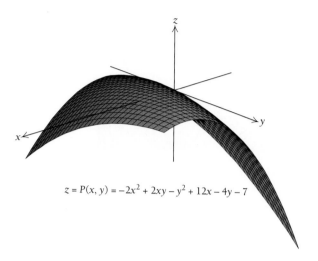

$$z = P(x, y) = -2x^2 + 2xy - y^2 + 12x - 4y - 7$$

1. Find P_x, P_y, P_{xx}, P_{yy}, and P_{xy}:

$$P_x = -4x + 2y + 12, \qquad P_y = 2x - 2y - 4,$$
$$P_{xx} = -4; \qquad\qquad P_{yy} = -2;$$
$$P_{xy} = 2.$$

2. Solve the system of equations $P_x = 0$, $P_y = 0$:

$$-4x + 2y + 12 = 0, \tag{1}$$
$$2x - 2y - 4 = 0. \tag{2}$$

Adding these equations, we get

$$-2x + 8 = 0.$$

Then

$$-2x = -8$$
$$x = 4.$$

To find y when $x = 4$, we substitute 4 for x in either equation (1) or equation (2). We choose equation (2):

$$2 \cdot 4 - 2y - 4 = 0$$
$$-2y + 4 = 0$$
$$-2y = -4$$
$$y = 2.$$

Thus, $(4, 2)$ is the only critical point and $P(4, 2)$ is a candidate for a maximum or minimum value.

3. We must check to see whether $P(4, 2)$ is a maximum or minimum value:

$$D = P_{xx}(4, 2) \cdot P_{yy}(4, 2) - [P_{xy}(4, 2)]^2$$
$$= (-4)(-2) - 2^2$$
$$= 4.$$

4. Thus, $D = 4$ and $P_{xx}(4, 2) = -4$. Since $D > 0$ and $P_{xx}(4, 2) < 0$, it follows that P has a relative maximum at $(4, 2)$. Thus in order to maximize profit, the company must produce and sell 4 thousand of the \$3 golf balls and 2 thousand of the \$2 golf balls.   ◄

EXERCISE SET 7.4

Find the relative maximum and minimum values.

1. $f(x, y) = x^2 + xy + y^2 - y$

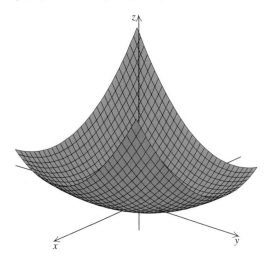

2. $f(x, y) = x^2 + xy + y^2 - 5y$

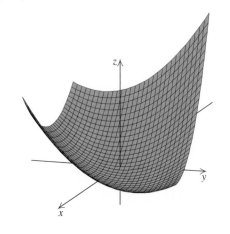

3. $f(x, y) = 2xy - x^3 - y^2$

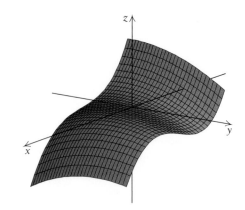

4. $f(x, y) = 4xy - x^3 - y^2$

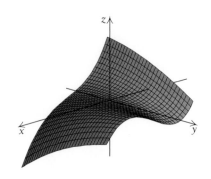

5. $f(x, y) = x^3 + y^3 - 3xy$
6. $f(x, y) = x^3 + y^3 - 6xy$
7. $f(x, y) = x^2 + y^2 - 2x + 4y - 2$
8. $f(x, y) = x^2 + 2xy + 2y^2 - 6y + 2$

9. $f(x, y) = x^2 + y^2 + 2x - 4y$

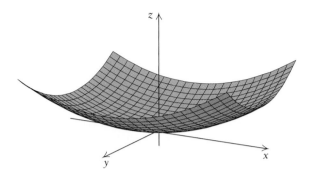

10. $f(x, y) = 4y + 6x - x^2 - y^2$

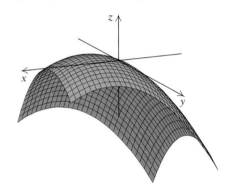

11. $f(x, y) = 4x^2 - y^2$

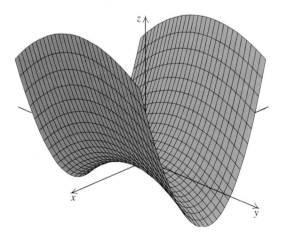

12. $f(x, y) = x^2 - y^2$

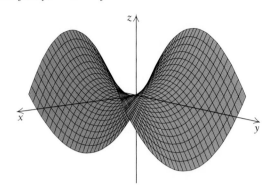

Applications

BUSINESS AND ECONOMICS

In these problems, assume that relative maximum and minimum values are absolute maximum and minimum values.

13. *Maximizing Profit.* A firm produces two kinds of radio, one that sells for $17 and the other for $21. The total revenue from the sale of x thousand radios at $17 each and y thousand at $21 each is given by

$$R(x, y) = 17x + 21y.$$

The company determines that the total cost, in thousands of dollars, of producing x thousand of the $17 radio and y thousand of the $21 radio is given by

$$C(x, y) = 4x^2 - 4xy + 2y^2 - 11x + 25y - 3.$$

Find the amount of each type of radio that must be produced and sold in order to maximize profit.

14. *Maximizing Profit.* A firm produces two kinds of baseball glove, one that sells for $18 and the other for $25. The total revenue from the sale of x thousand gloves at $18 each and y thousand at $25 each is given by

$$R(x, y) = 18x + 25y.$$

The company determines that the total cost, in thousands of dollars, of producing x thousand of the $18 glove and y thousand of the $25 glove is given by

$$C(x, y) = 4x^2 - 6xy + 3y^2 + 20x + 19y - 12.$$

Find the amount of each type of glove that must be produced and sold in order to maximize profit.

15. *Maximizing Profit.* A one-product company finds that its profit, in millions of dollars, is a function P given by

$$P(a, p) = 2ap + 80p - 15p^2 - \tfrac{1}{10}a^2p - 100,$$

where a is the amount spent on advertising, in millions of dollars, and p is the price charged per item of the product, in dollars. Find the maximum value of P and the values of a and p at which it is attained.

16. *Maximizing Profit.* A one-product company finds that its profit, in millions of dollars, is a function P given by

$$P(a, n) = -5a^2 - 3n^2 + 48a - 4n + 2an + 300,$$

where a is the amount spent on advertising, in millions of dollars, and n is the number of items sold, in thousands. Find the maximum value of P and the values of a and n at which it is attained.

17. *Minimizing the Cost of a Container.* A trash company is designing an open-top, rectangular container that will have a volume of 320 ft^3. The cost of making the bottom of the container is $5 per square foot, and the cost of the sides is $4 per square foot. Find the dimensions of the container that will minimize total cost. (*Hint*: Make a substitution using the formula for volume.)

18. *Two-Variable Revenue Maximization.* Boxowitz, Inc., a computer firm, markets two kinds of electronic calculator that compete with one another. Their demand functions are expressed by the following relationships:

$$q_1 = 78 - 6p_1 - 3p_2, \qquad (1)$$
$$q_2 = 66 - 3p_1 - 6p_2, \qquad (2)$$

where p_1 and p_2 are the price of each calculator, in multiples of $10, and q_1 and q_2 are the quantity of each calculator demanded, in hundreds of units.

a) Find a formula for the total-revenue function R in terms of the variables p_1 and p_2. [*Hint:* $R = p_1q_1 + p_2q_2$; then substitute expressions from equations (1) and (2) to find $R(p_1, p_2)$.]

b) What prices p_1 and p_2 should be charged for each product in order to maximize total revenue?

c) How many units will be demanded?

d) What is the maximum total revenue?

19. *Two-Variable Revenue Maximization.* Repeat Exercise 18, where

$$q_1 = 64 - 4p_1 - 2p_2$$

and

$$q_2 = 56 - 2p_1 - 4p_2.$$

20. *Temperature.* A flat metal plate is located on a coordinate plane. The temperature of the plate, in degrees Fahrenheit, at point (x, y) is given by

$$T(x, y) = x^2 + 2y^2 - 8x + 4y.$$

Find the minimum temperature and where it occurs. Is there a maximum temperature?

Synthesis

Find the relative maximum and minimum values and the saddle points.

21. $f(x, y) = e^x + e^y - e^{x+y}$

22. $f(x, y) = xy + \dfrac{2}{x} + \dfrac{4}{y}$

23. $f(x, y) = 2y^2 + x^2 - x^2y$

24. $S(b, m) = (m + b - 72)^2 + (2m + b - 73)^2 + (3m + b - 75)^2$

tw 25. Describe the D-test and how it is used.

tw 26. Explain the difference between a relative minimum and an absolute minimum of a function of two variables.

TECHNOLOGY CONNECTION

Use a 3D grapher to graph each of the following functions. Then estimate any relative extrema.

27. $f(x, y) = \dfrac{-5}{x^2 + 2y^2 + 1}$

28. $f(x, y) = x^3 + y^3 + 3xy$

29. $f(x, y) = \dfrac{3xy(x^2 - y^2)}{(x^2 + y^2)}$

30. $f(x, y) = \dfrac{y + x^2y^2 - 8x}{xy}$

7.5

➤ Find a regression line.
➤ Solve applied problems involving regression lines.

An Application: The Least-Squares Technique

The problem of fitting an equation to a set of data occurs frequently. We considered this in Section 1.6. Such an equation provides a model of the phenomena from which predictions can be made. For example, in business, one might want to predict future sales on the basis of past data. In ecology, one might want to predict future demands for natural gas on the basis of past need. Suppose that we are trying to find a linear equation

$$y = mx + b$$

to fit the data. To determine this equation is to determine the values of m and b. But how? Let's consider some factual data.

The graph shown in Fig. 1 appeared in a newspaper advertisement for the Indianapolis Life Insurance Company. It pertains to the total amount of life insurance in force in various years. The same data are compiled in the table following the graph.

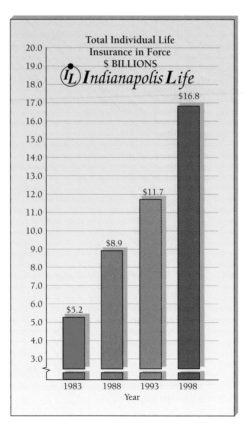

FIGURE 1

Year, x	1. 1983	2. 1988	3. 1993	4. 1998	5. 2003
Total Amount of Individual Life Insurance in Force (in billions), y	$5.2	$8.9	$11.7	$16.8	?

Suppose that we plot these points and try to draw a line through them that fits. Note that there are several ways in which this might be done (see Figs. 2 and 3). Each would give a different estimate of the total amount of insurance in force in 2003.

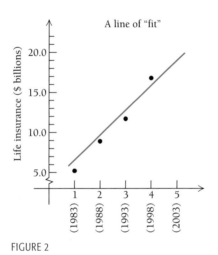

FIGURE 2

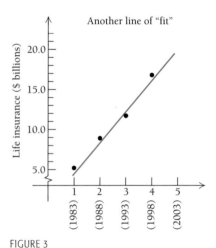

FIGURE 3

Note that the time is given in increments of five years, making computations easier. Consider the data points (1, 5.2), (2, 8.9), (3, 11.7), and (4, 16.8), as plotted in Fig. 4.

We will try to fit these data with a line

$$y = mx + b$$

by determining the values of m and b. Note the y-errors, or y-deviations, $y_1 - 5.2$, $y_2 - 8.9$, $y_3 - 11.7$, and $y_4 - 16.8$ between the observed points (1, 5.2), (2, 8.9), (3, 11.7), and (4, 16.8) and the points (1, y_1), (2, y_2), (3, y_3), and (4, y_4) on the line. We would like, somehow, to minimize these deviations in order to have a good fit. One way of minimizing the deviations is based on the *least-squares assumption*.

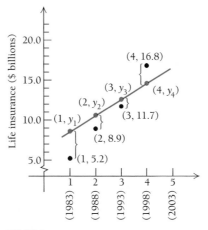

FIGURE 4

The Least-Squares Assumption

The line of best fit is the line for which the sum of the squares of the y-deviations is a minimum. This is called the **regression line.**

Using the least-squares assumption for the life insurance data, we would minimize

$$(y_1 - 5.2)^2 + (y_2 - 8.9)^2 + (y_3 - 11.7)^2 + (y_4 - 16.8)^2, \tag{1}$$

and since the points $(1, y_1)$, $(2, y_2)$, $(3, y_3)$, and $(4, y_4)$ must be solutions of $y = mx + b$, it follows that

$$y_1 = m(1) + b = m + b,$$
$$y_2 = m(2) + b = 2m + b,$$
$$y_3 = m(3) + b = 3m + b,$$
$$y_4 = m(4) + b = 4m + b.$$

Substituting $m + b$ for y_1, $2m + b$ for y_2, $3m + b$ for y_3, and $4m + b$ for y_4 in equation (1), we have

$$(m + b - 5.2)^2 + (2m + b - 8.9)^2 + (3m + b - 11.7)^2 + (4m + b - 16.8)^2. \tag{2}$$

Thus, to find the regression line for the given set of data, we must find the values of m and b that minimize the function S given by the sum in equation (2).

To apply the D-test, we first find the partial derivatives $\partial S/\partial b$ and $\partial S/\partial m$:

$$\frac{\partial S}{\partial b} = 2(m + b - 5.2) + 2(2m + b - 8.9) + 2(3m + b - 11.7)$$
$$+ 2(4m + b - 16.8)$$
$$= 20m + 8b - 85.2,$$

and

$$\frac{\partial S}{\partial m} = 2(m + b - 5.2) + 2(2m + b - 8.9)2 + 2(3m + b - 11.7)3 \\ + 2(4m + b - 16.8)4 \\ = 60m + 20b - 250.6.$$

We set these derivatives equal to 0 and solve the resulting system:

$$20m + 8b - 85.2 = 0, \qquad 5m + 2b = 21.3,$$
$$\quad\quad\quad\quad\quad\quad\quad\quad or$$
$$60m + 20b - 250.6 = 0; \qquad 15m + 5b = 62.65.$$

The solution of this system is

$$b = 1.25, \qquad m = 3.76.$$

We leave it to the student to complete the D-test to verify that $(1.25, 3.76)$ does, in fact, yield the minimum of S. We need not bother to compute $S(1.25, 3.76)$.

The values of m and b are all we need to determine $y = mx + b$. The regression line is

$$y = 3.76x + 1.25.$$

We see the graph of the "best-fit" regression line together with the data in Fig. 5. Compare it to Figs. 1 and 2.

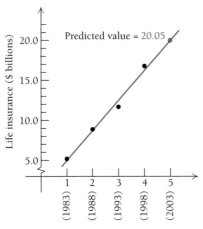

FIGURE 5

We can extrapolate from the data to predict the total amount of life insurance in force in 2003:

$$y = 3.76(5) + 1.25 = \$20.05.$$

Thus the total amount of life insurance in force in 2003 will be about $20.05 billion.

The method of least squares is a statistical process illustrated here with only four data points in order to simplify the explanation. Most statistical researchers would warn that many more than four data points should be used to get a "good" regression line. Furthermore, making predictions too far in the future from any linear model may not be valid. It can be done, but the further into the future the prediction is made, the more dubious one should be about the prediction.

*The Regression Line for an Arbitrary Collection of Data Points (c_1, d_1), (c_2, d_2), . . . , (c_n, d_n)

Look again at the regression line

$$y = 3.76x + 1.25$$

for the data points $(1, 5.2)$, $(2, 8.9)$, $(3, 11.7)$, and $(4, 16.8)$. Let's consider the arithmetic averages, or means, of the x-coordinates, denoted $\bar{x}$, and the y-coordinates, denoted $\bar{y}$:

$$\bar{x} = \frac{1 + 2 + 3 + 4}{4} = 2.5,$$

$$\bar{y} = \frac{5.2 + 8.9 + 11.7 + 16.8}{4} = 10.65.$$

It turns out that the point $(\bar{x}, \bar{y})$, or $(2.5, 10.65)$, is on the regression line since

$$10.65 = 3.76(2.5) + 1.25.$$

Thus the regression line is

$$y - \bar{y} = m(x - \bar{x}),$$

or

$$y - 10.65 = m(x - 2.5).$$

All that remains, in general, is to determine m.

Suppose that we want to find the regression line for an arbitrary number of points (c_1, d_1), $(c_2, d_2), \ldots, (c_n, d_n)$. To do so, we find the values m and b that minimize the function S given by

$$S(b, m) = (y_1 - d_1)^2 + (y_2 - d_2)^2 + \cdots + (y_n - d_n)^2$$

$$= \sum_{i=1}^{n} (y_i - d_i)^2,$$

where $y_i = mc_i + b$.

Using a procedure like the one we used earlier to minimize S, we can show that $y = mx + b$ takes the form

$$y - \bar{y} = m(x - \bar{x}),$$

where

$$\bar{x} = \frac{\sum_{i=1}^{n} c_i}{n}, \qquad \bar{y} = \frac{\sum_{i=1}^{n} d_i}{n},$$

---

*This part is considered optional and can be omitted without loss of continuity.

and

$$m = \frac{\sum_{i=1}^{n} (c_i - \bar{x})(d_i - \bar{y})}{\sum_{i=1}^{n} (c_i - \bar{x})^2}.$$

Let's see how this works out for the individual life-insurance example used previously.

c_i	d_i	$c_i - \bar{x}$	$(c_i - \bar{x})^2$	$(d_i - \bar{y})$	$(c_i - \bar{x})(d_i - \bar{y})$
1	5.2	-1.5	2.25	-5.45	8.175
2	8.9	-0.5	0.25	-1.75	0.875
3	11.7	0.5	0.25	1.05	0.525
4	16.8	1.5	2.25	6.15	9.225

$$\sum_{i=1}^{4} c_i = 10 \qquad \sum_{i=1}^{4} d_i = 42.6 \qquad \sum_{i=1}^{4} (c_i - \bar{x})^2 = 5 \qquad \sum_{i=1}^{4} (c_i - \bar{x})(d_i - \bar{y}) = 18.8$$

$$\bar{x} = 2.5 \qquad \bar{y} = 10.65 \qquad\qquad\qquad m = \frac{18.8}{5} = 3.76$$

Thus the regression line is

$$y - 10.65 = 3.76(x - 2.5),$$

which simplifies to

$$y = 3.76x + 1.25.$$

TECHNOLOGY CONNECTION

As we have seen in Section 1.6 and in other parts of the book, graphers can perform linear regression, as well as other kinds of regression such as quadratic, exponential, and logarithmic. Use such a grapher to fit a linear equation to the life-insurance data.

With some graphers, you will also obtain a number r, called the **coefficient of correlation.** Although we cannot develop that concept in detail in this text, keep in mind that r is used to describe the strength of the linear relationship between x and y. The closer r is to 1, the better the correlation.

For the life-insurance data, $r \approx 0.993$, which indicates a fairly good linear relationship. Keep in mind that a high linear correlation does not necessarily indicate a "cause-and-effect" connection between the variables.

All of the following exercises can be done with a grapher if your instructor so directs. The grapher can also be used to check your work.

Applications
BUSINESS AND ECONOMICS

1. *Sales of Sport Utility Vehicles.* Sales of sport utility vehicles (SUVs) have experienced a steady growth since 1993. Consider the data in the following table.

Years, x, Since 1993	Sales of SUVs, y (in millions)
0. 1993	5.1
1. 1994	8.8
2. 1995	10.0
3. 1996	11.1
4. 1997	12.6

Source: Polk

a) Find the regression line $y = mx + b$.
b) Use the regression line to predict sales in 2005 and in 2010.

2. *Medicare and Medicaid Payments.* The following table lists total health-care payments in billions of dollars, by Medicare, which is paid mainly to the elderly, and Medicaid, which is paid mainly to people unable to pay their medical bills.

Years, x, Since 1993	Health Care Payments, y (in billions)
0. 1993	$268.7
1. 1994	296.9
2. 1995	328.9
3. 1996	344.3
4. 1997	377.1
5. 1998	405.7

Source: Health-care Financing Administration, U.S. Department of Health and Human Services

a) Find the regression line $y = mx + b$.
b) Use the regression line to predict total health-care payments in 2005 and in 2010.

LIFE AND PHYSICAL SCIENCES

3. *Life Expectancy of Women.* Consider the data below showing the average life expectancy of women in various years.

Year, x	Life Expectancy of Women, y (in years)
1. 1950	70.9
2. 1960	73.2
3. 1970	74.8
4. 1980	77.5
5. 1990	78.6

a) Find the regression line $y = mx + b$.
b) Use the regression line to predict the life expectancy of women in 2005 and in 2010.

4. *Life Expectancy of Men.* Consider the data below showing the average life expectancy of men in various years.

Year, x	Life Expectancy of Men, y (in years)
1. 1950	65.3
2. 1960	66.6
3. 1970	67.1
4. 1980	69.9
5. 1990	71.8

a) Find the regression line $y = mx + b$.
b) Use the regression line to predict the life expectancy of men in 2000 and in 2010.

GENERAL INTEREST

5. *Grade Predictions.* A professor wants to predict students' final examination scores on the basis of their midterm test scores. An equation was determined on the basis of data on the scores of three students who took the same course with the same instructor the previous semester (see the following table).

Midterm Score, x	Final Exam Score, y
70%	75%
60	62
85	89

a) Find the regression line $y = mx + b$. (*Hint:* The y-deviations are $70m + b - 75$, $60m + b - 62$, and so on.)
b) The midterm score of a student was 81. Use the regression line to predict the student's final exam score.

6. *Predicting the World Record in the High Jump.* On July 29, 1989, Javier Sotomayor of Cuba set an astounding world record of 8 ft in the high jump. It has been established that most world records in track and field can be modeled by a linear function. The following table shows world records for various years.

Year, x (Use the actual year for x.)	World Record in High Jump, y (in inches)
1912 (George Horme)	78.0
1956 (Charles Dumas)	84.5
1973 (Dwight Stones)	90.5
1989 (Javier Sotomayer)	96.0

a) Find the regression line $y = mx + b$.
b) Use the regression line to predict the world record in the high jump in 2000 and in 2050.
tw c) Does your answer in part (b) for 2050 seem realistic? Explain why extrapolating so far into the future could be a problem.

tw 7. Explain the concept of linear regression to a friend.

tw 8. Discuss the idea of linear regression with an expert in your major. Write a brief report.

TECHNOLOGY   CONNECTION

9. *General Interest: Predicting the World Record in the One-Mile Run.*

   a) Find the regression line $y = mx + b$ that fits the set of data in the table at right. $\left(Hint:$ Convert each time to decimal notation; for example, $4{:}24.5 = 4\frac{24.5}{60} = 4.4083.\right)$

   b) Use the regression line to predict the world record in the mile in 1998 and in 2004.

   c) In August 1993, Noureddine Morceli set a new world record of 3:44.39 for the mile. How does this compare with what can be predicted by the regression?

Year, x (Use the actual year for x.)	World Record in Mile (min:sec), y
1875 (Walter Slade)	4:24.5
1894 (Fred Bacon)	4:18.2
1923 (Paavo Nurmi)	4:10.4
1937 (Sidney Wooderson)	4:06.4
1942 (Gunder Haegg)	4:06.2
1945 (Gunder Haegg)	4:01.4
1954 (Roger Bannister)	3:59.4
1964 (Peter Snell)	3:54.4
1967 (Jim Ryun)	3:51.1
1975 (John Walker)	3:49.4
1979 (Sebastian Coe)	3:49.0
1980 (Steve Ovett)	3:48.8
1985 (Steve Cram)	3:46.31

7.6

OBJECTIVES

➤ Find maximum and minimum values using Lagrange multipliers.

➤ Solve applied problems involving Lagrange multipliers.

Constrained Maximum and Minimum Values: Lagrange Multipliers

Before we proceed in detail, let's return to a problem we considered in Chapter 3.

Problem   A hobby store has 20 ft of fencing to fence off a rectangular electric-train area in one corner of its display room. The two sides up against the wall require no fence. What dimensions of the rectangle will maximize the area?

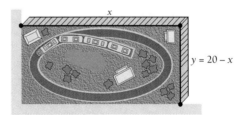

We maximize the function

$$A = xy$$

subject to the condition, or *constraint*, $x + y = 20$. Note that A is a function of two variables.

When we solved this earlier, we first solved the constraint for y:

$$y = 20 - x.$$

We then substituted $20 - x$ for y to obtain

$$A(x, y) = x(20 - x)$$
$$= 20x - x^2,$$

which is a function of one variable. Next, we found a maximum value using Maximum–Minimum Principle 1 (see Section 3.5). By itself, the function of two variables

$$A(x, y) = xy$$

has no maximum value. This can be checked using the D-test. With the constraint $x + y = 20$, however, the function does have a maximum. We see this in the following graph.

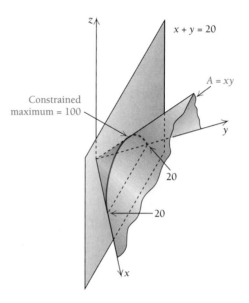

It may be quite difficult to solve a constraint for one variable. The procedure outlined below allows us to proceed without doing so.

The Method of Lagrange Multipliers

To find a maximum or minimum value of a function $f(x, y)$ subject to the constraint $g(x, y) = 0$:

1. Form a new function:

$$F(x, y, \lambda) = f(x, y) - \lambda g(x, y).$$

2. Find the first partial derivatives F_x, F_y, and F_λ.

3. Solve the system

$$F_x = 0, \qquad F_y = 0, \quad \text{and} \quad F_\lambda = 0.$$

Let (a, b, λ) represent a solution of this system. We still must determine whether (a, b, λ) yields a maximum or minimum of the function f, but we will assume such a maximum or minimum for each of the problems considered in this text.

The method of Lagrange multipliers can be extended to functions of three (or more) variables.

The variable λ (lambda) is called a **Lagrange multiplier.** We first illustrate the method of Lagrange multipliers by resolving the hobby-store problem.

Example 1   Find the maximum value of

$$A(x, y) = xy$$

subject to the constraint $x + y = 20$.

Solution

1. We form the new function F given by

$$F(x, y, \lambda) = xy - \lambda \cdot (x + y - 20).$$

Note that we first had to express $x + y = 20$ as $x + y - 20 = 0$.

2. We find the first partial derivatives:

$$F_x = y - \lambda,$$
$$F_y = x - \lambda,$$
$$F_\lambda = -(x + y - 20).$$

3. We set these derivatives equal to 0 and solve the resulting system:

$$y - \lambda = 0, \tag{1}$$
$$x - \lambda = 0, \tag{2}$$
$$-(x + y - 20) = 0, \quad \text{or} \quad x + y - 20 = 0. \tag{3}$$

From equations (1) and (2), it follows that

$$x = y = \lambda.$$

Substituting λ for x and y in equation (3), we get

$$\lambda + \lambda - 20 = 0$$
$$2\lambda = 20$$
$$\lambda = 10.$$

Thus, $x = \lambda = 10$ and $y = \lambda = 10$. The maximum value of A subject to the constraint occurs at (10, 10) and is

$$A(10, 10) = 10 \cdot 10$$
$$= 100.$$

Example 2   Find the maximum value of

$$f(x, y) = 3xy$$

subject to the constraint

$$2x + y = 8.$$

(*Note:* f might be interpreted, for example, as a production function with budget constraint $2x + y = 8$.)

Solution

1. We form the new function F given by

$$F(x, y, \lambda) = 3xy - \lambda(2x + y - 8).$$

Note that we first had to express $2x + y = 8$ as $2x + y - 8 = 0$.

2. We find the first partial derivatives:

$$F_x = 3y - 2\lambda,$$
$$F_y = 3x - \lambda,$$
$$F_\lambda = -(2x + y - 8).$$

3. We set these derivatives equal to 0 and solve the resulting system:

$$3y - 2\lambda = 0, \tag{1}$$
$$3x - \lambda = 0, \tag{2}$$
$$-(2x + y - 8) = 0, \quad \text{or} \quad 2x + y - 8 = 0. \tag{3}$$

Solving equation (1) for y, we get

$$y = \frac{2}{3}\lambda.$$

Solving equation (2) for x, we get

$$x = \frac{\lambda}{3}.$$

Substituting $(2/3)\lambda$ for y and $(\lambda/3)$ for x in equation (3), we get

$$2\left(\frac{\lambda}{3}\right) + \left(\frac{2}{3}\lambda\right) - 8 = 0$$

$$\frac{4}{3}\lambda = 8$$

$$\lambda = \frac{3}{4} \cdot 8 = 6.$$

Then

$$x = \frac{\lambda}{3} = \frac{6}{3} = 2 \quad \text{and} \quad y = \frac{2}{3}\lambda = \frac{2}{3} \cdot 6 = 4.$$

The maximum value of f subject to the constraint occurs at $(2, 4)$ and is

$$f(2, 4) = 3 \cdot 2 \cdot 4 = 24.$$

Example 3 *Business: The Beverage-Can Problem (12 oz).* The standard beverage can has a volume of 12 oz, or 21.66 in^3. What dimensions yield the minimum surface area? Find the minimum surface area.

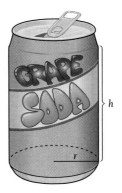

Solution We want to minimize the function s given by

$$s(h, r) = 2\pi rh + 2\pi r^2$$

subject to the volume constraint

$$\pi r^2 h = 21.66, \quad \text{or} \quad \pi r^2 h - 21.66 = 0.$$

Note that s does not have a minimum without the constraint.

1. We form the new function S given by

$$S(h, r, \lambda) = 2\pi rh + 2\pi r^2 - \lambda(\pi r^2 h - 21.66).$$

2. We find the first partial derivatives:

$$\frac{\partial S}{\partial h} = 2\pi r - \lambda\pi r^2,$$

$$\frac{\partial S}{\partial r} = 2\pi h + 4\pi r - 2\lambda\pi rh,$$

$$\frac{\partial S}{\partial \lambda} = -(\pi r^2 h - 21.66).$$

3. We set these derivatives equal to 0 and solve the resulting system:

$$2\pi r - \lambda\pi r^2 = 0, \tag{1}$$

$$2\pi h + 4\pi r - 2\lambda\pi rh = 0, \tag{2}$$

$$-(\pi r^2 h - 21.66) = 0, \quad \text{or} \quad \pi r^2 h - 21.66 = 0. \tag{3}$$

Note that we can solve equation (1) for r:

$$\pi r(2 - \lambda r) = 0$$

$$\pi r = 0 \quad or \quad 2 - \lambda r = 0$$

$$r = 0 \quad or \quad r = \frac{2}{\lambda}.$$

Since $r = 0$ cannot be a solution to the original problem, we continue by substituting $2/\lambda$ for r in equation (2):

$$2\pi h + 4\pi \cdot \frac{2}{\lambda} - 2\lambda\pi \cdot \frac{2}{\lambda} \cdot h = 0$$

$$2\pi h + \frac{8\pi}{\lambda} - 4\pi h = 0$$

$$\frac{8\pi}{\lambda} - 2\pi h = 0$$

$$-2\pi h = -\frac{8\pi}{\lambda},$$

so

$$h = \frac{4}{\lambda}.$$

Since $h = 4/\lambda$ and $r = 2/\lambda$, it follows that $h = 2r$. Substituting $2r$ for h in equation (3) yields

$$\pi r^2(2r) - 21.66 = 0$$

$$2\pi r^3 - 21.66 = 0$$

$$2\pi r^3 = 21.66$$

$$\pi r^3 = 10.83$$

$$r^3 = \frac{10.83}{\pi}$$

$$r = \sqrt[3]{\frac{10.83}{\pi}} \approx 1.51 \text{ in.}$$

Thus when $r = 1.51$ in., $h = 3.02$ in. The surface area is a minimum and is approximately

$$2\pi(1.51)(3.02) + 2\pi(1.51)^2, \quad \text{or about} \quad 42.98 \text{ in}^2. \qquad \blacktriangleleft$$

The actual dimensions of a standard-sized 12-oz beverage can are $r = 1.25$ in. and $h = 4.875$ in. A natural question after studying Example 3 is, "Why don't beverage companies make cans using the dimensions found in that example?" To do this at this time would mean an enormous cost in retooling. New can-making machines would have to be purchased at a cost of millions. New beverage-filling machines would have to be purchased. Vending machines would no longer be the correct size. A partial response to the desire to save aluminum has been found in recycling and

in manufacturing cans with a rippled effect at the top. These cans require less aluminum. As a result of many engineering ideas, the amount of aluminum required to make 1000 cans has been reduced from 36.5 lb to 28.1 lb. The consumer is actually a very important factor in the shape of the can. Market research has shown that a can with the dimensions found in Example 3 is not as comfortable to hold and might not be accepted by consumers.

EXERCISE SET 7.6

Find the maximum value of f subject to the given constraint.

1. $f(x, y) = xy; \quad 2x + y = 8$
2. $f(x, y) = 2xy; \quad 4x + y = 16$
3. $f(x, y) = 4 - x^2 - y^2; \quad x + 2y = 10$
4. $f(x, y) = 3 - x^2 - y^2; \quad x + 6y = 37$

Find the minimum value of f subject to the given constraint.

5. $f(x, y) = x^2 + y^2; \quad 2x + y = 10$
6. $f(x, y) = x^2 + y^2; \quad x + 4y = 17$
7. $f(x, y) = 2y^2 - 6x^2; \quad 2x + y = 4$
8. $f(x, y) = 2x^2 + y^2 - xy; \quad x + y = 8$
9. $f(x, y, z) = x^2 + y^2 + z^2; \quad y + 2x - z = 3$
10. $f(x, y, z) = x^2 + y^2 + z^2; \quad x + y + z = 1$

Use the method of Lagrange multipliers to solve each of the following.

11. Of all numbers whose sum is 70, find the two that have the maximum product.
12. Of all numbers whose sum is 50, find the two that have the maximum product.
13. Of all numbers whose difference is 6, find the two that have the minimum product.
14. Of all numbers whose difference is 4, find the two that have the minimum product.

Applications ....
BUSINESS AND ECONOMICS

15. *Maximizing Typing Area.* A standard piece of typing paper has a perimeter of 39 in. Find the dimensions of the paper that will give the most typing area, subject to the perimeter constraint of 39 in. What is its area? Does the standard $8\frac{1}{2}$-in. $\times$ 11-in. paper have maximum area?

16. *Maximizing Room Area.* A carpenter is building a rectangular room with a fixed perimeter of 80 ft. What are the dimensions of the largest room that can be built? What is its area?

17. *Minimizing Surface Area.* An oil drum of standard size has a volume of 200 gal, or 27 ft^3. What dimensions yield the minimum surface area? Find the minimum surface area.

Do these drums appear to be made in such a way as to minimize surface area?

18. *Juice-Can Problem.* A standard-sized juice can has a volume of 99 in³. What dimensions yield the minimum surface area? Find the minimum surface area.

19. *Maximizing Total Sales.* The total sales S of a one-product firm are given by

    $$S(L, M) = ML - L^2,$$

    where M is the cost of materials and L is the cost of labor. Find the maximum value of this function subject to the budget constraint

    $$M + L = 80.$$

20. *Maximizing Total Sales.* The total sales S of a one-product firm are given by

    $$S(L, M) = 2ML - L^2,$$

    where M is the cost of materials and L is the cost of labor. Find the maximum value of this function subject to the budget constraint

    $$M + L = 60.$$

21. *Minimizing Construction Costs.* A company is planning to construct a warehouse whose cubic footage is to be 252,000 ft³. Construction costs per square foot are estimated to be as follows:

    Walls: $3.00
    Floor: $4.00
    Ceiling: $3.00

    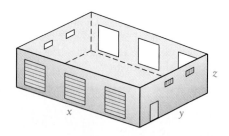

    a) The total cost of the building is a function $C(x, y, z)$, where x is the length, y is the width, and z is the height. Find a formula for $C(x, y, z)$.
    b) What dimensions of the building will minimize the total cost? What is the minimum cost?

22. *Minimizing the Costs of Container Construction.* A container company is going to construct a shipping container of volume 12 ft³ with a square

bottom and top. The cost of the top and the sides is $2 per square foot and for the bottom is $3 per square foot. What dimensions will minimize the cost of the container?

23. *Minimizing Total Cost.* A product can be made entirely on either machine A or machine B, or both. The nature of the machines makes their cost functions differ:

    $$\text{Machine A:} \quad C(x) = 10 + \frac{x^2}{6},$$

    $$\text{Machine B:} \quad C(y) = 200 + \frac{y^3}{9}.$$

    Total cost is given by $C(x, y) = C(x) + C(y)$. How many units should be made on each machine in order to minimize total costs if $x + y = 10{,}100$ units are required?

Synthesis

Find the indicated maximum or minimum values of f subject to the given constraint.

24. Minimum: $f(x, y) = xy;$ $x^2 + y^2 = 4$
25. Minimum: $f(x, y) = 2x^2 + y^2 + 2xy + 3x + 2y;$ $y^2 = x + 1$
26. Maximum: $f(x, y, z) = x + y + z;$ $x^2 + y^2 + z^2 = 1$
27. Maximum: $f(x, y, z) = x^2y^2z^2;$ $x^2 + y^2 + z^2 = 1$
28. Maximum: $f(x, y, z) = x + 2y - 2z;$ $x^2 + y^2 + z^2 = 4$
29. Maximum: $f(x, y, z, t) = x + y + z + t;$ $x^2 + y^2 + z^2 + t^2 = 1$
30. Minimum: $f(x, y, z) = x^2 + y^2 + z^2;$ $x - 2y + 5z = 1$

31. *Economics: The Law of Equimarginal Productivity.* Suppose that $p(x, y)$ represents the production of a two-product firm. We give no formula for p. The company produces x items of the first product at a cost of c_1 each and y items of the second product at a cost of c_2 each. The budget constraint B is a constant given by

    $$B = c_1 x + c_2 y.$$

    Find the value of λ using the Lagrange-multiplier method in terms of p_x, p_y, c_1, and c_2. The resulting equation is called the *Law of Equimarginal Productivity.*

32. *Business: Maximizing Production.* A computer company has the following Cobb–Douglas production function for a certain product:

$$p(x, y) = 800x^{3/4}y^{1/4},$$

where x is the labor, measured in dollars, and y is the capital, measured in dollars. Suppose that a company can make a total investment in labor and capital of $1,000,000. How should it allocate the investment between labor and capital in order to maximize production?

tw 33. Write a brief report on the life and work of the mathematician Joseph Louis Lagrange (1736–1813).

tw 34. Discuss the difference between solving a problem using Lagrange multipliers and using the method of Section 7.4. Describe the difference graphically.

 TECHNOLOGY CONNECTION

35.–42. Use a 3D grapher to graph both equations in each of Exercises 1–8. Then visually check the results that you found analytically.

7.7

OBJECTIVES

> Evaluate a multiple integral.

Multiple Integration

The following is an example of a *double integral:*

$$\int_3^6 \int_{-1}^2 10xy^2 \, dx \, dy, \quad \text{or} \quad \int_3^6 \left(\int_{-1}^2 10xy^2 \, dx \right) dy.$$

Evaluating a double integral is somewhat similar to "undoing" a second partial derivative. We first evaluate the inside x-integral, treating y as a constant:

$$\int_{-1}^2 10xy^2 \, dx = 10y^2 \left[\frac{x^2}{2} \right]_{-1}^2 = 5y^2[x^2]_{-1}^2 = 5y^2[2^2 - (-1)^2] = 15y^2.$$

Color indicates the variable. All else is constant.

Then we evaluate the outside y-integral:

$$\begin{aligned}
\int_3^6 15y^2 \, dy &= 15 \left[\frac{y^3}{3} \right]_3^6 \\
&= 5[y^3]_3^6 \\
&= 5(6^3 - 3^3) \\
&= 945.
\end{aligned}$$

More precisely, the above is called a **double iterated integral.** The word "iterate" means "to do again."

If the dx and dy and the limits of integration are interchanged, as follows,

$$\int_{-1}^2 \int_3^6 10xy^2 \, dy \, dx,$$

we first evaluate the inside y-integral, treating x as a constant:

$$\int_3^6 10xy^2 \, dy = 10x\left[\frac{y^3}{3}\right]_3^6$$

$$= \frac{10x}{3}\left[y^3\right]_3^6$$

$$= \frac{10}{3}x(6^3 - 3^3)$$

$$= 630x.$$

Then we evaluate the outside x-integral:

$$\int_{-1}^2 630x \, dx = 630\left[\frac{x^2}{2}\right]_{-1}^2$$

$$= 315[x^2]_{-1}^2$$

$$= 315[2^2 - (-1)^2]$$

$$= 945.$$

Note that we get the same result. This is not always true, but will be for the types of function that we consider.

Sometimes variables occur as limits of integration.

Example 1   Evaluate

$$\int_0^1 \int_{x^2}^x xy^2 \, dy \, dx.$$

Solution   We first evaluate the y-integral, treating x as a constant:

$$\int_{x^2}^x xy^2 \, dy = x\left[\frac{y^3}{3}\right]_{x^2}^x$$

$$= \frac{1}{3}x[x^3 - (x^2)^3]$$

$$= \frac{1}{3}(x^4 - x^7).$$

Then we evaluate the outside integral:

$$\frac{1}{3}\int_0^1 (x^4 - x^7) \, dx = \frac{1}{3}\left[\frac{x^5}{5} - \frac{x^8}{8}\right]_0^1$$

$$= \frac{1}{3}\left[\left(\frac{1^5}{5} - \frac{1^8}{8}\right) - \left(\frac{0^5}{5} - \frac{0^8}{8}\right)\right] = \frac{1}{40}.$$

Thus,

$$\int_0^1 \int_{x^2}^x xy^2 \, dy \, dx = \frac{1}{40}.$$

The Geometric Interpretation of Multiple Integrals

Suppose that the region G in the xy-plane is bounded by the functions $y_1 = g(x)$ and $y_2 = h(x)$ and the lines $x_1 = a$ and $x_2 = b$. We want the volume V of the solid above G and under the surface $z = f(x, y)$. We can think of the solid as composed of many vertical columns, one of which is shown in Fig. 1. The volume of one such column can be thought of as $l \cdot w \cdot h$, or $z \cdot \Delta y \cdot \Delta x$. Integrating such columns in the y-direction, we obtain

$$\left[\int_{y_1}^{y_2} z \, dy \right] dx,$$

which can be pictured as a "slab," or slice. Then integrating such slices in the x-direction, we obtain the entire volume:

$$V = \int_a^b \left[\int_{y_1}^{y_2} z \, dy \right] dx,$$

or

$$V = \int_a^b \int_{g(x)}^{h(x)} z \, dy \, dx,$$

where $z = f(x, y)$.

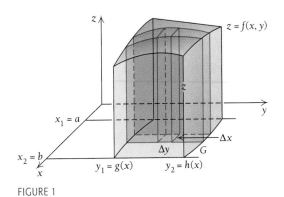

FIGURE 1

In Example 1, the region of integration G is the plane region between the graphs of $y = x^2$ and $y = x$, as shown in Figs. 2 and 3.

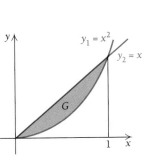

FIGURE 2

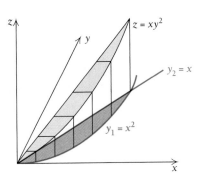

FIGURE 3

When we evaluated the double integral in Example 1, we found the volume of the solid based on G and capped by the surface $z = xy^2$, as shown in Fig. 2.

An Application to Probability

Suppose that we throw a dart at a region R in a plane. We assume that the dart lands on a point (x, y) in R (Fig. 4). We can think of (x, y) as a continuous random variable whose coordinates assume all values in some region R. A function f is said to be a **joint probability density function** if

$$f(x, y) \geq 0 \quad \text{for all } (x, y) \text{ in } R$$

and

$$\int \int_R f(x, y) \, dx \, dy = 1,$$

where $\int \int_R$ refers to the double integral evaluated over the region R.

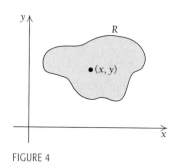

FIGURE 4   FIGURE 5

Suppose that we want to know the probability that the dart hits a point (x, y) in a rectangular subregion G of R, where G is the set of points for which $a \leq x \leq b$ and $c \leq y \leq d$ (Fig. 5). This would be given by

$$\int \int_G f(x, y) \, dx \, dy = \int_c^d \int_a^b f(x, y) \, dx \, dy.$$

EXERCISE SET 7.7

Evaluate.

1. $\displaystyle\int_0^1 \int_0^1 2y \, dx \, dy$

2. $\displaystyle\int_0^1 \int_0^1 2x \, dx \, dy$

3. $\displaystyle\int_{-1}^1 \int_x^1 xy \, dy \, dx$

4. $\displaystyle\int_{-1}^1 \int_x^2 (x + y) \, dy \, dx$

5. $\displaystyle\int_0^1 \int_{-1}^3 (x + y) \, dy \, dx$

6. $\displaystyle\int_0^1 \int_{-1}^1 (x + y) \, dy \, dx$

7. $\displaystyle\int_0^1 \int_{x^2}^x (x + y) \, dy \, dx$

8. $\displaystyle\int_0^1 \int_{-1}^x (x^2 + y^2) \, dy \, dx$

9. $\displaystyle\int_0^2 \int_0^x (x + y^2) \, dy \, dx$ 　　　10. $\displaystyle\int_1^3 \int_0^x 2e^{x^2} \, dy \, dx$

11. Find the volume of the solid capped by the surface $z = 1 - y - x^2$ over the region bounded above and below by $y = 1 - x^2$ and $y = 0$ and left and right by $x = 0$ and $x = 1$, by evaluating the integral

$$\int_0^1 \int_0^{1-x^2} (1 - y - x^2) \, dy \, dx.$$

12. Find the volume of the solid capped by the surface $z = x + y$ over the region bounded above and below by $y = 1 - x$ and $y = 0$ and left and right by $x = 0$ and $x = 1$, by evaluating the integral

$$\int_0^1 \int_0^{1-x} (x + y) \, dy \, dx.$$

Suppose that a continuous random variable has a joint probability density function given by

$$f(x, y) = x^2 + \tfrac{1}{3}xy,$$
$$0 \le x \le 1, \quad 0 \le y \le 2.$$

13. Find

$$\int_0^2 \int_0^1 f(x, y) \, dx \, dy.$$

14. Find the probability that a point (x, y) is in the region bounded by $0 \le x \le \tfrac{1}{2}, 1 \le y \le 2$, by evaluating the integral

$$\int_1^2 \int_0^{1/2} f(x, y) \, dx \, dy.$$

Synthesis ....

A *triple iterated integral* such as

$$\int_r^s \int_c^d \int_a^b f(x, y, z) \, dx \, dy \, dz$$

is evaluated in much the same way as the double iterated integral. We first evaluate the inside x-integral, treating y and z as constants. Then we evaluate the middle y-integral, treating z as a constant. Finally, we evaluate the outside z-integral. Evaluate these triple integrals.

15. $\displaystyle\int_0^1 \int_1^3 \int_{-1}^2 (2x + 3y - z) \, dx \, dy \, dz$

16. $\displaystyle\int_0^2 \int_1^4 \int_{-1}^2 (8x - 2y + z) \, dx \, dy \, dz$

17. $\displaystyle\int_0^1 \int_0^{1-x} \int_0^{2-x} xyz \, dz \, dy \, dx$

18. $\displaystyle\int_0^2 \int_{2-y}^{6-2y} \int_0^{\sqrt{4-y^2}} z \, dz \, dx \, dy$

tw 19. Describe the geometric meaning of the multiple integral of a function of two variables.

 TECHNOLOGY CONNECTION

20. Use a grapher that does multiple integration to evaluate some double integrals found in this exercise set.

CHAPTER 7 SUMMARY AND REVIEW

Terms to Know

Function of two variables, p. 496
Function of several variables, p. 497
Partial derivative, p. 503
Second-order partial derivative, p. 509

Relative maximum, p. 512
Relative minimum, p. 512
Saddle point, p. 514
D-test, p. 515

Least-squares technique, p. 523
Method of Lagrange multipliers, p. 532
Multiple integration, p. 539
Double iterated integral, p. 539

Review Exercises

These review exercises are for test preparation. They can also be used as a lengthened practice test. Answers are at the back of the book. The answers also contain bracketed section references, which tell you where to restudy if your answer is incorrect.

Given $f(x, y) = e^y + 3xy^3 + 2y$, find each of the following.

1. $f(2, 0)$

2. f_x

3. f_y

4. f_{xy}

5. f_{yx}

6. f_{xx}

7. f_{yy}

Given $z = x^2 \ln y + y^4$, find each of the following.

8. $\dfrac{\partial z}{\partial x}$

9. $\dfrac{\partial z}{\partial y}$

10. $\dfrac{\partial^2 z}{\partial y\, \partial x}$

11. $\dfrac{\partial^2 z}{\partial x\, \partial y}$

12. $\dfrac{\partial^2 z}{\partial x^2}$

13. $\dfrac{\partial^2 z}{\partial y^2}$

Find the relative maximum and minimum values.

14. $f(x, y) = x^3 - 6xy + y^2 + 6x + 3y - \frac{1}{5}$

15. $f(x, y) = x^2 - xy + y^2 - 2x + 4y$

16. $f(x, y) = 3x - 6y - x^2 - y^2$

17. $f(x, y) = x^4 + y^4 + 4x - 32y + 80$

18. Consider the data in the following table regarding the average salary of an NFL player over a recent six-year period.

Year, x	Average Salary of an NFL Player, y (in millions)
1. 1991	$0.43
2. 1992	0.52
3. 1993	0.65
4. 1994	0.80
5. 1995	0.90
6. 1996	1.00

a) Find the regression line $y = mx + b$.
b) Use the regression line to predict the average NFL player salary in 1997 and in 2000.

19. Consider the data in the following table regarding enrollment in colleges and universities during a recent three-year period.

Year, x	Enrollment, y (in millions)
1	7.2
2	8.0
3	8.4

a) Find the regression line $y = mx + b$.
b) Use the regression line to predict enrollment in the fourth year.

20. Find the minimum value of

$$f(x, y) = x^2 - 2xy + 2y^2 + 20$$

subject to the constraint $2x - 6y = 15$.

21. Find the maximum value of

$$f(x, y) = 6xy$$

subject to the constraint $2x + y = 20$.

Evaluate.

22. $\displaystyle\int_0^1 \int_{x^2}^{3x} (x^3 + 2y)\, dy\, dx$

23. $\displaystyle\int_0^1 \int_{x^2}^{x} (x - y)\, dy\, dx$

Synthesis

24. Evaluate

$$\int_0^2 \int_{1-2x}^{1-x} \int_0^{\sqrt{2-x^2}} z\, dz\, dy\, dx.$$

25. *Business: Minimizing Surface Area.* Suppose that beverages could be packaged in either a cylindrical container or a rectangular container with a square top and bottom. Each container is designed to be of minimum surface area for its shape. If we assume a volume of 26 in^3, which container would have the smaller surface area?

 TECHNOLOGY CONNECTION

26. Use a 3D grapher to graph

$$f(x, y) = x^2 + 4y^2.$$

CHAPTER 7 TEST

Given $f(x, y) = e^x + 2x^3y + y$, find each of the following.

1. $f(-1, 2)$　　**2.** $\dfrac{\partial f}{\partial x}$　　**3.** $\dfrac{\partial f}{\partial y}$

4. $\dfrac{\partial^2 f}{\partial x^2}$　　**5.** $\dfrac{\partial^2 f}{\partial x\, \partial y}$　　**6.** $\dfrac{\partial^2 f}{\partial y\, \partial x}$

7. $\dfrac{\partial^2 f}{\partial y^2}$

Find the relative maximum and minimum values.

8. $f(x, y) = x^2 - xy + y^3 - x$

9. $f(x, y) = y^2 - x^2$

10. *Business: Predicting Total Sales.* Consider the data in the following table regarding the total sales of a company during the first three years of operation.

Year, x	Sales, y (in millions)
1	$10
2	15
3	19

a) Find the regression line $y = mx + b$.
b) Use the regression line to predict sales in the fourth year.

11. Find the maximum value of

$$f(x, y) = 6xy - 4x^2 - 3y^2$$

subject to the constraint $x + 3y = 19$.

12. Evaluate

$$\int_0^2 \int_1^x (x^2 - y)\, dy\, dx.$$

Synthesis ....

13. *Business: Maximizing Production.* An appliance company has the following Cobb–Douglas production function for a certain product,

$$p(x, y) = 50x^{2/3}y^{1/3},$$

where x is labor, measured in dollars, and y is capital, measured in dollars. Suppose that a company can make a total investment in labor and capital of $600,000. How should it allocate the investment between labor and capital in order to maximize production?

14. Find f_x and f_t:

$$f(x, t) = \frac{x^2 - 2t}{x^3 + 2t}.$$

 TECHNOLOGY CONNECTION

15. Use a 3D grapher to graph

$$f(x, y) = x - \tfrac{1}{2}y^2 - \tfrac{1}{3}x^3.$$

Minimizing Employees' Travel Time in a Building

For a business in which employees spend considerable time moving between offices, designing a building to minimize travel time can reap enormous financial savings.

In multilevel building design, one consideration is travel time between the most remote points in a rectangular building with a square base. We will make use of Lagrange multipliers to design such a building.

Let's assume that each floor has a square grid of hallways, as shown in the figure below. Suppose that you are standing at the most remote point P in the top northeast corner of such a building with 12 floors. How long will it take to reach the southwest corner on the first floor—that is, point Q?

Let's call the time t. We find a formula for t in two steps:

1. You are to go from the twelfth floor to the first floor. This is a move in a vertical direction.

2. You need to go across the first floor. This is a move in a horizontal direction.

The vertical time is h, the height of the top floor from the ground, divided by a, the speed at which you can travel in a vertical direction (elevator speed). Thus vertical time is given by h/a.

The horizontal time is the time it takes to go across the first level, by way of the square grid of hallways (from R to Q above). If each floor is a square with side of length k, then the distance from R to Q is $2k$. If the walking speed is b, then the horizontal time is given by $2k/b$. Thus the time it will take to go from P to Q is a function of two variables, h and k, given by

$$t(h, k) = \text{Vertical time} + \text{Horizontal time}$$
$$= \frac{h}{a} + \frac{2k}{b}.$$

What happens if we must choose between two (or more) building plans with the same floor area, but with different dimensions? Will the travel time be the same? Or will it be different for the two buildings? First of all, what is the total floor area of a given building? Suppose that the building has n floors, each a square of side k. Then the total floor area is given by

$$A = nk^2.$$

Note that the area of the roof is not included.

If h is the height of the top of the building to the ground and c is the height of each floor, then $n = h/c$. (For simplicity, we are ignoring the thickness of each floor.) Thus,

$$A = \frac{h}{c}k^2.$$

Let's return to the problem of two buildings with the same total floor area, but with different dimensions, and see what happens to $t(h, k)$.

Exercises

1. Use the **TABLE** feature on your grapher, if you have one, or a spreadsheet software program. For each case below, let the elevator speed $a = 10$ ft/sec, the walking speed $b = 4$ ft/sec, and the height of each floor $c = 15$ ft. Each case in the table covers two situations, though the floor area stays the same for a particular case. Complete the table at the top of the next column.

2. Do different dimensions, with a fixed floor area, yield different travel times?

3. Do you think there are values of h and k for a building with a given floor area that will minimize travel time? Try fixing the floor

Case	Building	n	k	A	h	$t(h, k)$
1	B1	2	40	3200	30	23
	B2	3	32.66	3200	45	20.83
2	B1	2	60	7200		
	B2	3	48.99			
3	B1	4	40			
	B2	5	35.777			
4	B1	5	60			
	B2	10	42.426			
5	B1	5	150			
	B2	10	106.066			
6	B1	10	40			
	B2	17	30.679			
7	B1	10	80			
	B2	17	61.357			
8	B1	17	40			
	B2	26	32.344			
9	B1	17	50			
	B2	26	40.43			
10	B1	26	77			
	B2	50	55.525			

area at 40,000 ft^2 and explore the results of different dimensions. What do we need to use to know for sure?

Find the dimensions of a rectangular building with a square base that will minimize travel time t between the most remote points in the building described as follows. Each floor has a square grid of hallways. The height of the top floor from the ground is h, and the length of a side of each floor is k. The elevator speed is 10 ft/sec and the average speed of a person walking is 4 ft/sec. The total floor area of the building is 40,000 ft^2. The height of each floor is 8 ft.

4. Use the information given to find a formula for the function $t(h, k)$.

5. Find a formula for the constraint.

6. Use the method of Lagrange multipliers to find the dimensions of the building that will minimize travel time t between the most remote points in the building. Use a 3D grapher to graph both equations in Exercises 4 and 5. Then visually check the results you found analytically.

7. Find a general solution in terms of a, b, c, and A.

Cumulative Review

1. Write an equation of the line with slope -4 and containing the point $(-7, 1)$.

2. For $f(x) = x^2 - 5$, find $f(x + h)$.

3. a) Graph:

$$f(x) = \begin{cases} 5 - x, & \text{for } x \neq 2, \\ -3, & \text{for } x = 2. \end{cases}$$

   b) Find $\lim_{x \to 2} f(x)$.
   c) Find $f(2)$.
   d) Is f continuous at 2?

Find the limit, if it exists.

4. $\lim_{x \to -4} \dfrac{x^2 - 16}{x + 4}$

5. $\lim_{x \to 1} \sqrt{x^3 + 8}$

6. $\lim_{x \to 3} \dfrac{4}{x - 3}$

7. $\lim_{x \to \infty} \dfrac{12x - 7}{3x + 2}$

8. $\lim_{h \to 0} \dfrac{3x^2 + h}{x - 2h}$

9. $\lim_{x \to \infty} \dfrac{8x^3 + 5x^2 - 7}{16x^5 - 4x^3 + 9}$

Differentiate.

10. $y = -9x + 3$

11. $y = x^2 - 7x + 3$

12. $y = x^{1/4}$

13. $f(x) = x^{-6}$

14. $f(x) = (x - 3)(x + 1)^5$

15. $f(x) = \dfrac{x^3 - 1}{x^5}$

16. $y = \dfrac{e^x + x}{e^x}$

17. $y = \ln (x^2 + 5)$

18. $y = e^{\ln x}$

19. $y = e^{3x} + x^2$

20. $y = e^{\sqrt{x} - 3}$

21. $f(x) = \ln (e^x - 4)$

22. For $y = x^2 - \dfrac{2}{x}$, find d^2y/dx^2.

23. Differentiate implicitly to find dy/dx if
$$x^3 + x/y = 7.$$

24. Find an equation of the tangent line to the graph of $y = e^x - x^2 - 3$ at the point $(0, -2)$.

Find the relative extrema of the function. List your answer in terms of an ordered pair. Then sketch a graph of the function.

25. $f(x) = x^3 - 3x + 1$

26. $f(x) = 2x^2 - x^4 - 3$

27. $f(x) = \dfrac{8x}{x^2 + 1}$

28. $f(x) = \dfrac{8}{x^2 - 4}$

Find the absolute maximum and minimum values, if they exist, over the indicated interval. If no interval is indicated, use the real line.

29. $f(x) = 3x^2 - 6x - 4$

30. $f(x) = -5x + 1$

31. $f(x) = \frac{1}{3}x^3 - x^2 - 3x + 5;\quad [-2, 0]$

32. *Business: Maximizing Profit.* For a certain product, the total-revenue and total-cost functions are given by

$$R(x) = 4x^2 + 11x + 110,$$
$$C(x) = 4.2x^2 + 5x + 10.$$

Find the number of units that must be produced and sold in order to maximize profit.

33. *Business: Minimizing Inventory Costs.* An appliance store sells 450 pocket radios each year. It costs $4 to store one radio for one year. To order radios, there is a fixed cost of $1 plus $0.75 for each radio. How many times per year should the store reorder radios, and in what lot size, in order to minimize inventory costs?

34. For $f(x) = 3x^2 - 7$, $x = 5$, and $\Delta x = 0.1$, find Δy and $f'(x)\,\Delta x$.

35. *Business: Exponential Growth.* A national frozen yogurt firm is experiencing a growth pattern of 10% per year in the number N of franchises that it owns; that is,

$$\frac{dN}{dt} = 0.1N,$$

where N is the number of franchises and t is the time, in years, from 1996.

   a) Given that there were 8000 franchises in 1996, find the solution of the equation, assuming $N_0 = 8000$ and $k = 0.1$.
   b) How many franchises will there be in 2004?
   c) What is the doubling time of the number of franchises?

36. *Economics: Elasticity of Demand.* Consider the demand function

$$x = D(p) = 240 - 20p.$$

   a) Find the elasticity.
   b) Find the elasticity at $p = \$2$, stating whether the demand is elastic or inelastic.
   c) Find the elasticity at $p = \$9$, stating whether the demand is elastic or inelastic.
   d) At a price of $2, will a small increase in price cause total revenue to increase or decrease?
   e) Find the value of p for which the total revenue is a maximum.

Evaluate.

37. $\displaystyle\int 3x^5\,dx$

38. $\displaystyle\int_{-1}^{0} (2e^x + 1)\,dx$

39. $\displaystyle\int \frac{x}{(7 - 3x)^2}\,dx$ (Use Table 1.)

40. $\displaystyle\int x^3 e^{x^4}\,dx$ (Use substitution. Do not use Table 1.)

41. $\displaystyle\int (x + 3)\ln x\,dx$

42. $\displaystyle\int \frac{260}{x}\,dx$

43. $\displaystyle\int_0^1 6\sqrt{x}\,dx$

44. Find the area under the graph of $y = x^2 + 3x$ over the interval $[1, 5]$.

45. *Business: Present Value.* Find the present value of $250,000 due in 30 yr at 8%, compounded continuously.

46. *Business: Accumulated Present Value.* Find the accumulated present value of an investment over a 50-yr period in which there is a continuous money flow of $4500 and the current interest rate is 7.2%, compounded continuously.

47. Determine whether the following improper integral is convergent or divergent, and calculate its value if convergent:

$$\int_3^\infty \frac{1}{x^7}\,dx.$$

48. Given the probability density function

$$f(x) = \frac{3}{2x^2} \quad \text{over } [1, 3],$$

find $E(x)$.

49. Let x be a continuous random variable that is normally distributed with mean $\mu = 3$ and standard deviation $\sigma = 5$. Using Table 2, find $P(-2 \le x \le 8)$.

50. *Economics: Supply and Demand.* Given the demand and supply functions

$$p = D(x) = (x - 20)^2$$

and

$$p = S(x) = x^2 + 10x + 50,$$

find the equilibrium point and the consumer's surplus.

51. Find the volume of the solid of revolution generated by rotating the region under the graph of

$$y = e^{-x} \quad \text{from } x = 0 \text{ to } x = 5$$

about the x-axis.

52. Solve the differential equation $dy/dx = xy$.

53. *Business: Effect of Advertising.* In an advertising experiment, it was determined that

$$\frac{dP}{dt} = k(L - P),$$

where P is the percentage of people in the city who bought the product after the ad was run t times, and $L = 100\%$, or 1.

a) Express $P(t)$ in terms of $L = 1$.
b) It was determined that $P(10) = 70\%$. Use this to determine k. Round your answer to the nearest hundredth.
c) Rewrite $P(t)$ in terms of k using the value of k found in part (b).
d) Use the equation in part (c) to find $P(20)$.

Given $f(x, y) = e^y + 4x^2y^3 + 3x$, find each of the following.

54. f_x 55. f_{yy}

56. Find the relative maximum and minimum values of $f(x, y) = 8x^2 - y^2$.

57. Maximize $f(x, y) = 4x + 2y - x^2 - y^2 + 4$, subject to the constraint $x + 2y = 9$.

58. Evaluate

$$\int_0^3 \int_{-1}^2 e^x \, dy \, dx.$$

Appendix

OBJECTIVES

➤ Manipulate exponential expressions.
➤ Multiply and factor algebraic expressions.
➤ Solve equations, inequalities, and applied problems.

Review of Basic Algebra

This appendix covers most of the algebraic topics essential to a study of calculus. It might be used in conjunction with Chapter 1 or as the need for certain skills arises throughout the book.

Exponential Notation

Let's review the meaning of an expression

$$a^n,$$

where a is any real number and n is an integer; that is, n is a number in the set $\{\ldots, -3, -2, -1, 0, 1, 2, 3, \ldots\}$. The number a is called the **base** and n is called the **exponent**. When n is greater than 1, then

$$a^n = \underbrace{a \cdot a \cdot a \cdots a}_{n \text{ factors}}.$$

In other words, a^n is the product of n **factors**, each of which is a.

Example 1 Express without exponents.

a) $4^3 = 4 \cdot 4 \cdot 4 = 64$

b) $(-2)^5 = (-2)(-2)(-2)(-2)(-2) = -32$

c) $(-2)^4 = (-2)(-2)(-2)(-2) = 16$

d) $-2^4 = -(2^4) = -(2)(2)(2)(2) = -16$

e) $(1.08)^2 = 1.08 \times 1.08 = 1.1664$

f) $\left(\dfrac{1}{2}\right)^3 = \dfrac{1}{2} \cdot \dfrac{1}{2} \cdot \dfrac{1}{2} = \dfrac{1}{8}$

We define an exponent of 1 as follows:

$$a^1 = a, \quad \text{for any real number } a.$$

In other words, any real number to the first power is that number itself.
We define an exponent of 0 as follows:

$$a^0 = 1, \quad \text{for any nonzero real number } a.$$

That is, any nonzero real number a to the zero power is 1.

Example 2   Express without exponents.

a) $(-2x)^0 = 1$ b) $(-2x)^1 = -2x$

c) $\left(\dfrac{1}{2}\right)^0 = 1$ d) $e^0 = 1$

e) $e^1 = e$ f) $\left(\dfrac{1}{2}\right)^1 = \dfrac{1}{2}$

The meaning of a negative integer as an exponent is as follows:

$$a^{-n} = \frac{1}{a^n}, \quad \text{for any nonzero real number } a.$$

That is, any nonzero real number a to the $-n$ power is the reciprocal of a^n.

Example 3   Express without negative exponents.

a) $2^{-5} = \dfrac{1}{2^5} = \dfrac{1}{2 \cdot 2 \cdot 2 \cdot 2 \cdot 2} = \dfrac{1}{32}$

b) $10^{-3} = \dfrac{1}{10^3} = \dfrac{1}{10 \cdot 10 \cdot 10} = \dfrac{1}{1000}$, or 0.001

c) $\left(\dfrac{1}{4}\right)^{-2} = \dfrac{1}{\left(\dfrac{1}{4}\right)^2} = \dfrac{1}{\dfrac{1}{4} \cdot \dfrac{1}{4}} = \dfrac{1}{\dfrac{1}{16}} = 1 \cdot \dfrac{16}{1} = 16$

d) $x^{-5} = \dfrac{1}{x^5}$

e) $e^{-k} = \dfrac{1}{e^k}$

f) $t^{-1} = \dfrac{1}{t^1} = \dfrac{1}{t}$

Properties of Exponents

Note the following:

$$b^5 \cdot b^{-3} = (b \cdot b \cdot b \cdot b \cdot b) \cdot \frac{1}{b \cdot b \cdot b}$$

$$= \frac{b \cdot b \cdot b}{b \cdot b \cdot b} \cdot b \cdot b$$

$$= 1 \cdot b \cdot b = b^2.$$

We could have obtained the same result by adding the exponents. This is true in general.

Theorem 1

For any nonzero real number a and any integers n and m,

$$a^n \cdot a^m = a^{n+m}.$$

(To multiply when the bases are the same, add the exponents.)

Example 4 Multiply.

a) $x^5 \cdot x^6 = x^{5+6} = x^{11}$

b) $x^{-5} \cdot x^6 = x^{-5+6} = x$

c) $2x^{-3} \cdot 5x^{-4} = 10x^{-3+(-4)} = 10x^{-7}$, or $\dfrac{10}{x^7}$

d) $r^2 \cdot r = r^{2+1} = r^3$

Note the following:

$$b^5 \div b^2 = \frac{b^5}{b^2} = \frac{b \cdot b \cdot b \cdot b \cdot b}{b \cdot b}$$

$$= \frac{b \cdot b}{b \cdot b} \cdot b \cdot b \cdot b$$

$$= 1 \cdot b \cdot b \cdot b = b^3.$$

We could have obtained the same result by subtracting the exponents. This is true in general.

Theorem 2

For any nonzero real number a and any integers n and m,

$$\frac{a^n}{a^m} = a^{n-m}.$$

(To divide when the bases are the same, subtract the exponent in the denominator from the exponent in the numerator.)

Example 5   Divide.

a) $\dfrac{a^3}{a^2} = a^{3-2} = a^1 = a$

b) $\dfrac{x^7}{x^7} = x^{7-7} = x^0 = 1$

c) $\dfrac{e^3}{e^{-4}} = e^{3-(-4)} = e^{3+4} = e^7$

d) $\dfrac{e^{-4}}{e^{-1}} = e^{-4-(-1)} = e^{-4+1} = e^{-3}$, or $\dfrac{1}{e^3}$

Note the following:

$$(b^2)^3 = b^2 \cdot b^2 \cdot b^2 = b^{2+2+2} = b^6.$$

We could have obtained the same result by multiplying the exponents.

Theorem 3

For any nonzero real numbers a and b, and any integers n and m,

$$(a^n)^m = a^{nm}, \qquad (ab)^n = a^n b^n, \quad \text{and} \quad \left(\dfrac{a}{b}\right)^n = \dfrac{a^n}{b^n}.$$

Example 6   Simplify.

a) $(x^{-2})^3 = x^{-2 \cdot 3} = x^{-6}$, or $\dfrac{1}{x^6}$

b) $(e^x)^2 = e^{2x}$

c) $(2x^4 y^{-5} z^3)^{-3} = 2^{-3}(x^4)^{-3}(y^{-5})^{-3}(z^3)^{-3}$

$$= \dfrac{1}{2^3} x^{-12} y^{15} z^{-9}, \text{ or } \dfrac{y^{15}}{8x^{12} z^9}$$

d) $\left(\dfrac{x^2}{p^4 q^5}\right)^3 = \dfrac{(x^2)^3}{(p^4 q^5)^3} = \dfrac{x^6}{(p^4)^3(q^5)^3} = \dfrac{x^6}{p^{12} q^{15}}$

Multiplication

The distributive laws are important in multiplying. These laws are as follows.

The Distributive Laws

For any numbers A, B, and C.

$$A(B + C) = AB + AC \quad \text{and} \quad A(B - C) = AB - AC.$$

Example 7 Multiply.

a) $3(x - 5) = 3 \cdot x - 3 \cdot 5 = 3x - 15$

b) $P(1 + i) = P \cdot 1 + P \cdot i = P + Pi$

c) $(x - 5)(x + 3) = (x - 5)x + (x - 5)3$
$$= x \cdot x - 5x + 3x - 5 \cdot 3$$
$$= x^2 - 2x - 15$$

d) $(a + b)(a + b) = (a + b)a + (a + b)b$
$$= a \cdot a + ba + ab + b \cdot b$$
$$= a^2 + 2ab + b^2$$

The following formulas, which are obtained using the distributive laws, are also useful in multiplying.

$$(A + B)^2 = A^2 + 2AB + B^2, \qquad\qquad (1)$$
$$(A - B)^2 = A^2 - 2AB + B^2 \qquad\qquad (2)$$
$$(A - B)(A + B) = A^2 - B^2 \qquad\qquad (3)$$

Example 8 Multiply.

a) $(x + h)^2 = x^2 + 2xh + h^2$

b) $(2x - t)^2 = (2x)^2 - 2(2x)t + t^2 = 4x^2 - 4xt + t^2$

c) $(3c + d)(3c - d) = (3c)^2 - d^2 = 9c^2 - d^2$

Factoring

Factoring is the reverse of multiplication. That is, to factor an expression, we find an equivalent expression that is a product. Always remember to look first for a common factor.

Example 9 Factor.

a) $P + Pi = P \cdot 1 + P \cdot i = P(1 + i)$ We used a distributive law.

b) $2xh + h^2 = h(2x + h)$

c) $x^2 - 6xy + 9y^2 = (x - 3y)^2$

d) $x^2 - 5x - 14 = (x - 7)(x + 2)$ We looked for factors of -14 whose sum is -5.

e) $6x^2 + 7x - 5 = (2x - 1)(3x + 5)$ We first considered ways of factoring the first coefficient—for example, $(2x\ )(3x\ )$. Then we looked for factors of -5 such that when we multiply, we obtain the given expression.

f) $x^2 - 9t^2 = (x - 3t)(x + 3t)$ We used the formula $(A - B)(A + B) = A^2 - B^2$.

Some expressions with four terms can be factored by first looking for a common binomial factor. This is called **factoring by grouping**.

Example 10   Factor.

a) $t^3 + 6t^2 - 2t - 12 = t^2(t + 6) - 2(t + 6)$ Factoring the first two terms and then the second two terms

$$= (t^2 - 2)(t + 6)$$ Factoring out the common binomial factor, $t + 6$

b) $x^3 - 7x^2 - 4x + 28 = x^2(x - 7) - 4(x - 7)$ Factoring the first two terms and then the second two terms

$$= (x^2 - 4)(x - 7)$$ Factoring out the common binomial factor $x - 7$

$$= (x - 2)(x + 2)(x - 7)$$ ◄

Solving Equations

Basic to the solution of many equations are the Addition Principle and the Multiplication Principle. We can add (or subtract) the same number on both sides of an equation and obtain an equivalent equation; that is, a new equation that has the same solutions as the original equation. We can also multiply (or divide) by a nonzero number on both sides of an equation and obtain an equivalent equation.

The Addition Principle

For any real numbers a, b, and c,

$$a = b \quad \text{is equivalent to} \quad a + c = b + c.$$

The Multiplication Principle

For any real numbers a, b, and c,

$$a = b \quad \text{is equivalent to} \quad a \cdot c = b \cdot c.$$

When solving an equation, we use these equation-solving principles and other properties of real numbers to get the variable alone on one side. Then it is easy to determine the solution.

Example 11   Solve: $-\frac{5}{6}x + 10 = \frac{1}{2}x + 2$.

Solution   We first multiply by 6 on both sides to clear the fractions:

$$6\left(-\frac{5}{6}x + 10\right) = 6\left(\frac{1}{2}x + 2\right)$$ Using the Multiplication Principle

$$6\left(-\frac{5}{6}x\right) + 6 \cdot 10 = 6\left(\frac{1}{2}x\right) + 6 \cdot 2$$ Using the Distributive Law

$$-5x + 60 = 3x + 12$$ Simplifying

$$60 = 8x + 12$$ Using the Addition Principle: We add $5x$ on both sides.

$$48 = 8x$$ Adding -12 on both sides

$$\frac{1}{8} \cdot 48 = \frac{1}{8} \cdot 8x$$ Multiplying by $\frac{1}{8}$ on both sides

$$6 = x.$$

The variable is now alone, and we see that 6 is the solution. We can check by substituting 6 into the original equation.

The third principle for solving equations is the *Principle of Zero Products*.

> ## The Principle of Zero Products
> For any numbers a and b, if $ab = 0$, then $a = 0$ or $b = 0$; and if $a = 0$ or $b = 0$, then $ab = 0$.

An equation being solved by this principle must have a 0 on one side and a product on the other. The solutions are then obtained by setting each factor equal to 0 and solving the resulting equations.

Example 12   Solve: $3x(x - 2)(5x + 4) = 0$.

Solution   We have

$$3x(x - 2)(5x + 4) = 0$$

$3x = 0$ *or* $x - 2 = 0$ *or* $5x + 4 = 0$ Using the Principle of Zero Products

$\frac{1}{3} \cdot 3x = \frac{1}{3} \cdot 0$ *or* $x = 2$ *or* $5x = -4$ Solving each separately

$x = 0$ *or* $x = 2$ *or* $x = -\frac{4}{5}$.

The solutions are 0, 2, and $-\frac{4}{5}$.

Note that the Principle of Zero Products can be applied *only* when a product is 0. For example, although we may know that $ab = 8$, we *do not know* that $a = 8$ or $b = 8$.

Example 13   Solve: $4x^3 = x$.

Solution   We have

$$4x^3 = x$$
$$4x^3 - x = 0 \quad \text{Adding } -x$$
$$x(4x^2 - 1) = 0$$
$$x(2x - 1)(2x + 1) = 0 \quad \text{Factoring}$$

$x = 0$ *or* $2x - 1 = 0$ *or* $2x + 1 = 0$ Using the Principle of Zero Products

$x = 0$ *or* $2x = 1$ *or* $2x = -1$

$x = 0$ *or* $x = \frac{1}{2}$ *or* $x = -\frac{1}{2}$.

The solutions are 0, $\frac{1}{2}$, and $-\frac{1}{2}$.

Rational Equations

Expressions like the following are **polynomials in one variable:**

$$x^2 - 4, \qquad x^3 + 7x^2 - 8x + 9, \qquad t - 19.$$

The **least common multiple, LCM,** of two polynomials is found by factoring and using each factor the greatest number of times that it occurs in any one factorization.

Example 14 Find the LCM: $x^2 + 2x + 1$, $5x^2 - 5x$, and $x^2 - 1$.

Solution

$$\left. \begin{array}{l} x^2 + 2x + 1 = (x + 1)(x + 1); \\ 5x^2 - 5x = 5x(x - 1); \\ x^2 - 1 = (x + 1)(x - 1) \end{array} \right\} \quad \text{Factoring}$$

$$\text{LCM} = 5x(x + 1)(x + 1)(x - 1) \qquad \blacktriangleleft$$

A **rational expression** is a ratio of polynomials. Each of the following is a rational expression:

$$\frac{x^2 - 6x + 9}{x^2 - 4}, \quad \frac{x - 2}{x - 3}, \quad \frac{a + 7}{a^2 - 16}, \quad \frac{5}{5t - 15}.$$

A **rational equation** is an equation containing one or more rational expressions. Here are some examples:

$$\frac{2}{3} - \frac{5}{6} = \frac{1}{x}, \qquad x + \frac{6}{x} = 5, \qquad \frac{2x}{x - 3} - \frac{6}{x} = \frac{18}{x^2 - 3x}.$$

To solve a rational equation, we first clear the equation of fractions by multiplying on both sides by the LCM of all the denominators. The resulting equation might have solutions that are *not* solutions of the original equation. Thus we must check all possible solutions in the original equation.

Example 15 Solve: $\dfrac{2x}{x - 3} - \dfrac{6}{x} = \dfrac{18}{x^2 - 3x}$.

Solution The LCM of the denominators is $x(x - 3)$. We multiply by $x(x - 3)$.

$$x(x - 3)\left(\frac{2x}{x - 3} - \frac{6}{x} \right) = x(x - 3)\left(\frac{18}{x^2 - 3x} \right) \qquad \begin{array}{l}\text{Multiplying by the}\\ \text{LCM on both sides}\end{array}$$

$$x(x - 3) \cdot \frac{2x}{x - 3} - x(x - 3) \cdot \frac{6}{x} = x(x - 3)\left(\frac{18}{x^2 - 3x} \right) \qquad \begin{array}{l}\text{Multiplying to}\\ \text{remove parentheses}\end{array}$$

$$2x^2 - 6(x - 3) = 18 \qquad \text{Simplifying}$$

$$2x^2 - 6x + 18 = 18$$

$$2x^2 - 6x = 0$$

$$2x(x - 3) = 0$$

$$2x = 0 \quad or \quad x - 3 = 0$$

$$x = 0 \quad or \qquad x = 3$$

The numbers 0 and 3 are possible solutions. We look at the original equation and see that each makes a denominator 0. We can also carry out a check, as follows.

Check:

For 0:

$$\frac{2x}{x - 3} - \frac{6}{x} = \frac{18}{x^2 - 3x}$$

$$\frac{2(0)}{0 - 3} - \frac{6}{0} \; ? \; \frac{18}{0^2 - 3(0)}$$

$$0 - \frac{6}{0} \; \bigg| \; \frac{18}{0} \qquad \text{UNDEFINED;}$$
$$\qquad\qquad\qquad \text{FALSE}$$

For 3:

$$\frac{2x}{x - 3} - \frac{6}{x} = \frac{18}{x^2 - 3x}$$

$$\frac{2(3)}{3 - 3} - \frac{6}{3} \; ? \; \frac{18}{3^2 - 3(3)}$$

$$\frac{6}{0} - 2 \; \bigg| \; \frac{18}{0} \qquad \text{UNDEFINED;}$$
$$\qquad\qquad\qquad \text{FALSE}$$

The equation has *no solution*.

Example 16 Solve: $\dfrac{x^2}{x - 2} = \dfrac{4}{x - 2}$.

Solution   The LCM of the denominators is $x - 2$. We multiply by $x - 2$.

$$(x - 2) \cdot \frac{x^2}{x - 2} = (x - 2) \cdot \frac{4}{x - 2}$$

$$x^2 = 4 \qquad \text{Simplifying}$$

$$x^2 - 4 = 0$$

$$(x + 2)(x - 2) = 0$$

$$x = -2 \quad or \quad x = 2 \qquad \text{Using the Principle of Zero Products}$$

Check:

For 2:

$$\frac{x^2}{x - 2} = \frac{4}{x - 2}$$

$$\frac{2^2}{2 - 2} \; ? \; \frac{4}{2 - 2}$$

$$\frac{4}{0} \; \bigg| \; \frac{4}{0} \qquad \text{UNDEFINED;}$$
$$\qquad\qquad \text{FALSE}$$

For -2:

$$\frac{x^2}{x - 2} = \frac{4}{x - 2}$$

$$\frac{(-2)^2}{-2 - 2} \; ? \; \frac{4}{-2 - 2}$$

$$\frac{4}{-4} \; \bigg| \; \frac{4}{-4}$$
$$-1 \; \bigg| \; -1 \qquad \text{TRUE}$$

The number -2 is a solution, but 2 is not (it results in division by 0).

Solving Inequalities

Two inequalities are **equivalent** if they have the same solutions. For example, the inequalities $x > 4$ and $4 < x$ are equivalent. Principles for solving inequalities are similar to those for solving equations. We can add the same number on both sides of an inequality. We can also multiply on both sides by the same nonzero number, but if that number is negative, we must reverse the inequality sign. The following are the inequality-solving principles.

> ## The Inequality-Solving Principles
> For any real numbers a, b, and c,
>
> $$a < b \quad \text{is equivalent to} \quad a + c < b + c.$$
>
> For any real numbers a, b, and any *positive* number c,
>
> $$a < b \quad \text{is equivalent to} \quad ac < bc.$$
>
> For any real numbers a, b, and any *negative* number c,
>
> $$a < b \quad \text{is equivalent to} \quad ac > bc.$$
>
> Similar statements hold for $\leq$ and $\geq$.

Example 17 Solve: $17 - 8x \geq 5x - 4$.

Solution We have

$$17 - 8x \geq 5x - 4$$
$$-8x \geq 5x - 21 \qquad \text{Adding } -17$$
$$-13x \geq -21 \qquad \text{Adding } -5x$$
$$-\tfrac{1}{13}(-13x) \leq -\tfrac{1}{13}(-21) \qquad \text{Multiplying by } -\tfrac{1}{13} \text{ and } \textit{reversing} \text{ the inequality sign}$$
$$x \leq \tfrac{21}{13}.$$

Any number less than or equal to $\frac{21}{13}$ is a solution.

Applications

To solve applied problems, we first translate to mathematical language, usually an equation. Then we solve the equation and check to see whether the solution to the equation is a solution to the problem.

Example 18 *Life Science: Weight Gain.* After a 5% gain in weight, an animal weighs 693 lb. What was its original weight?

Solution We first translate to an equation:

$$\underbrace{(Original\ weight)}_{w} + \underbrace{5\%}_{+\ 5\%}\underbrace{(Original\ weight)}_{w} = 693$$
$$= 693.$$

Now we solve the equation:

$$w + 5\%w = 693$$
$$1 \cdot w + 0.05w = 693$$
$$(1 + 0.05)w = 693$$
$$1.05w = 693$$
$$w = \frac{693}{1.05} = 660.$$

Check: $660 + 5\% \times 660 = 660 + 0.05 \times 660 = 660 + 33 = 693.$

The original weight of the animal was 660 lb.

Example 19 *Business: Total Sales.* Raggs, Ltd., a clothing firm, determines that its total revenue, in dollars, from the sale of x suits is given by

$$200x + 50.$$

Determine the number of suits that the firm must sell to ensure that its total revenue will be more than $70,050.

Solution We translate to an inequality and solve:

$$200x + 50 > 70,050$$
$$200x > 70,000 \qquad \text{Adding} -50$$
$$x > 350. \qquad \text{Multiplying by } \tfrac{1}{200}$$

Thus the company's total revenue will exceed $70,050 when it sells more than 350 suits.

EXERCISE SET A

Express without exponents.

1. 5^3
2. 7^2
3. $(-7)^2$
4. $(-5)^3$
5. $(1.01)^2$
6. $(1.01)^3$
7. $\left(\dfrac{1}{2}\right)^4$
8. $\left(\dfrac{1}{4}\right)^3$
9. $(6x)^0$
10. $(6x)^1$
11. t^1
12. t^0
13. $\left(\dfrac{1}{3}\right)^0$
14. $\left(\dfrac{1}{3}\right)^1$

Express without negative exponents.

15. 3^{-2}
16. 4^{-2}
17. $\left(\dfrac{1}{2}\right)^{-3}$
18. $\left(\dfrac{1}{2}\right)^{-2}$
19. 10^{-1}
20. 10^{-4}
21. e^{-b}
22. t^{-k}
23. b^{-1}
24. h^{-1}

Multiply.

25. $x^2 \cdot x^3$
26. $t^3 \cdot t^4$
27. $x^{-7} \cdot x$
28. $x^5 \cdot x$
29. $5x^2 \cdot 7x^3$
30. $4t^3 \cdot 2t^4$
31. $x^{-4} \cdot x^7 \cdot x$
32. $x^{-3} \cdot x \cdot x^3$
33. $e^{-t} \cdot e^t$
34. $e^k \cdot e^{-k}$

Divide.

35. $\dfrac{x^5}{x^2}$
36. $\dfrac{x^7}{x^3}$
37. $\dfrac{x^2}{x^5}$
38. $\dfrac{x^3}{x^7}$
39. $\dfrac{e^k}{e^k}$
40. $\dfrac{t^k}{t^k}$

41. $\dfrac{e^t}{e^4}$
42. $\dfrac{e^k}{e^3}$
43. $\dfrac{t^6}{t^{-8}}$
44. $\dfrac{t^5}{t^{-7}}$
45. $\dfrac{t^{-9}}{t^{-11}}$
46. $\dfrac{t^{-11}}{t^{-7}}$
47. $\dfrac{ab(a^2b)^3}{ab^{-1}}$
48. $\dfrac{x^2y^3(xy^3)^2}{x^{-3}y^2}$

Simplify.

49. $(t^{-2})^3$
50. $(t^{-3})^4$
51. $(e^x)^4$
52. $(e^x)^5$
53. $(2x^2y^4)^3$
54. $(2x^2y^4)^5$
55. $(3x^{-2}y^{-5}z^4)^{-4}$
56. $(5x^3y^{-7}z^{-5})^{-3}$
57. $(-3x^{-8}y^7z^2)^2$
58. $(-5x^4y^{-5}z^{-3})^4$
59. $\left(\dfrac{cd^3}{2q^2}\right)^4$
60. $\left(\dfrac{4x^2y}{a^3b^3}\right)^3$

Multiply.

61. $5(x - 7)$
62. $x(1 + t)$
63. $(x - 5)(x - 2)$
64. $(x - 4)(x - 3)$
65. $(a - b)(a^2 + ab + b^2)$
66. $(x^2 - xy + y^2)(x + y)$
67. $(2x + 5)(x - 1)$
68. $(3x + 4)(x - 1)$
69. $(a - 2)(a + 2)$
70. $(3x - 1)(3x + 1)$
71. $(5x + 2)(5x - 2)$
72. $(t - 1)(t + 1)$
73. $(a - h)^2$
74. $(a + h)^2$
75. $(5x + t)^2$
76. $(7a - c)^2$

77. $5x(x^2 + 3)^2$

78. $-3x^2(x^2 - 4)(x^2 + 4)$

Use the following equation (equation 5) for Exercises 79–81.

$$(x + h)^3 = (x + h)(x + h)^2$$
$$= (x + h)(x^2 + 2xh + h^2)$$
$$= (x + h)x^2 + (x + h)2xh + (x + h)h^2$$
$$= x^3 + x^2h + 2x^2h + 2xh^2 + xh^2 + h^3$$
$$= x^3 + 3x^2h + 3xh^2 + h^3 \quad (5)$$

79. $(a + b)^3$

80. $(a - b)^3$

81. $(x - 5)^3$

82. $(2x + 3)^3$

Factor.

83. $x - xt$

84. $x + xh$

85. $x^2 + 6xy + 9y^2$

86. $x^2 - 10xy + 25y^2$

87. $x^2 - 2x - 15$

88. $x^2 + 8x + 15$

89. $x^2 - x - 20$

90. $x^2 - 9x - 10$

91. $49x^2 - t^2$

92. $9x^2 - b^2$

93. $36t^2 - 16m^2$

94. $25y^2 - 9z^2$

95. $a^3b - 16ab^3$

96. $2x^4 - 32$

97. $a^8 - b^8$

98. $36y^2 + 12y - 35$

99. $10a^2x - 40b^2x$

100. $x^3y - 25xy^3$

101. $2 - 32x^4$

102. $2xy^2 - 50x$

103. $9x^2 + 17x - 2$

104. $6x^2 - 23x + 20$

105. $x^3 + 8$
    (Hint: See Exercise 66.)

106. $a^3 - 27b^3$
    (Hint: See Exercise 65.)

107. $y^3 - 64t^3$

108. $m^3 + 1000p^3$

109. $3x^3 - 6x^2 - x + 2$

110. $5y^3 + 2y^2 - 10y - 4$

111. $x^3 - 5x^2 - 9x + 45$

112. $t^3 + 3t^2 - 25t - 75$

Solve.

113. $-7x + 10 = 5x - 11$

114. $-8x + 9 = 4x - 70$

115. $5x - 17 - 2x = 6x - 1 - x$

116. $5x - 2 + 3x = 2x + 6 - 4x$

117. $x + 0.8x = 216$

118. $x + 0.5x = 210$

119. $x + 0.08x = 216$

120. $x + 0.05x = 210$

121. $2x(x + 3)(5x - 4) = 0$

122. $7x(x - 2)(2x + 3) = 0$

123. $x^2 + 1 = 2x + 1$

124. $2t^2 = 9 + t^2$

125. $t^2 - 2t = t$

126. $6x - x^2 = x$

127. $6x - x^2 = -x$

128. $2x - x^2 = -x$

129. $9x^3 = x$

130. $16x^3 = x$

131. $(x - 3)^2 = x^2 + 2x + 1$

132. $(x - 5)^2 = x^2 + x + 3$

133. $\dfrac{4x}{x + 5} + \dfrac{20}{x} = \dfrac{100}{x^2 + 5x}$

134. $\dfrac{x}{x + 1} + \dfrac{3x + 5}{x^2 + 4x + 3} = \dfrac{2}{x + 3}$

135. $\dfrac{50}{x} - \dfrac{50}{x - 2} = \dfrac{4}{x}$

136. $\dfrac{60}{x} = \dfrac{60}{x - 5} + \dfrac{2}{x}$

137. $0 = 2x - \dfrac{250}{x^2}$

138. $5 - \dfrac{35}{x^2} = 0$

139. $3 - x \le 4x + 7$

140. $x + 6 \le 5x - 6$

141. $5x - 5 + x > 2 - 6x - 8$

142. $3x - 3 + 3x > 1 - 7x - 9$

143. $-7x < 4$

144. $-5x \ge 6$

145. $5x + 2x \le -21$

146. $9x + 3x \ge -24$

147. $2x - 7 < 5x - 9$

148. $10x - 3 \ge 13x - 8$

149. $8x - 9 < 3x - 11$

150. $11x - 2 \ge 15x - 7$

151. $8 < 3x + 2 < 14$

152. $2 < 5x - 8 \le 12$

153. $3 \le 4x - 3 \le 19$

154. $9 \le 5x + 3 < 19$

155. $-7 \le 5x - 2 \le 12$

156. $-11 \le 2x - 1 < -5$

Applications
BUSINESS AND ECONOMICS

157. *Investment Increase.* An investment is made at $8\frac{1}{2}\%$, compounded annually. It grows to $705.25 at the end of 1 year. How much was invested originally?

158. *Investment Increase.* An investment is made at 7%, compounded annually. It grows to $856 at the end of 1 year. How much was invested originally?

159. *Total Revenue.* A firm determines that the total revenue, in dollars, from the sale of x units of a product is

$$3x + 1000.$$

Determine the number of units that must be sold so that its total revenue will be more than $22,000.

160. *Total Revenue.* A firm determines that the total revenue, in dollars, from the sale of x units of a product is

$$5x + 1000.$$

Determine the number of units that must be sold so that its total revenue will be more than $22,000.

LIFE AND PHYSICAL SCIENCES

161. *Weight Gain.* After a 6% gain in weight, an animal weighs 508.8 lb. What was its original weight?

162. *Weight Gain.* After a 7% gain in weight, an animal weighs 363.8 lb. What was its original weight?

SOCIAL SCIENCES

163. *Population Increase.* After a 2% increase, the population of a city is 826,200. What was the former population?

164. *Population Increase.* After a 3% increase, the population of a city is 741,600. What was the former population?

GENERAL INTEREST

165. *Grade Average.* To get a B in a course, a student's average must be greater than or equal to 80% (at least 80%) and less than 90%. On the first three tests, the student scores 78%, 90%, and 92%. Determine the scores on the fourth test that will guarantee a B.

166. *Grade Average.* To get a C in a course, a student's average must be greater than or equal to 70% and less than 80%. On the first three tests, the student scores 65%, 83%, and 82%. Determine the scores on the fourth test that will guarantee a C.

Tables

TABLE 1
INTEGRATION FORMULAS

(Whenever $\ln X$ is used, it is assumed that $X > 0$.)

1. $\displaystyle\int x^n \, dx = \frac{x^{n+1}}{n+1} + C, n \neq -1$

2. $\displaystyle\int \frac{dx}{x} = \ln x + C$

3. $\displaystyle\int u \, dv = uv - \int v \, du$

4. $\displaystyle\int e^x \, dx = e^x + C$

5. $\displaystyle\int e^{ax} \, dx = \frac{1}{a} \cdot e^{ax} + C$

6. $\displaystyle\int xe^{ax} \, dx = \frac{1}{a^2} \cdot e^{ax}(ax - 1) + C$

7. $\displaystyle\int x^n e^{ax} \, dx = \frac{x^n e^{ax}}{a} - \frac{n}{a} \int x^{n-1} e^{ax} \, dx + C$

8. $\displaystyle\int \ln x \, dx = x \ln x - x + C$

9. $\displaystyle\int (\ln x)^n \, dx = x(\ln x)^n - n \int (\ln x)^{n-1} \, dx + C, n \neq -1$

10. $\displaystyle\int x^n \ln x \, dx = x^{n+1} \left[\frac{\ln x}{n+1} - \frac{1}{(n+1)^2} \right] + C, n \neq -1$

11. $\displaystyle\int a^x \, dx = \frac{a^x}{\ln a} + C, a > 0, a \neq 1$

12. $\int \dfrac{1}{\sqrt{x^2 + a^2}}\, dx = \ln\left(x + \sqrt{x^2 + a^2}\right) + C$

13. $\int \dfrac{1}{\sqrt{x^2 - a^2}}\, dx = \ln\left(x + \sqrt{x^2 - a^2}\right) + C$

14. $\int \dfrac{1}{x^2 - a^2}\, dx = \dfrac{1}{2a} \ln\left(\dfrac{x - a}{x + a}\right) + C$

15. $\int \dfrac{1}{a^2 - x^2}\, dx = \dfrac{1}{2a} \ln\left(\dfrac{a + x}{a - x}\right) + C$

16. $\int \dfrac{1}{x\sqrt{a^2 + x^2}}\, dx = -\dfrac{1}{a} \ln\left(\dfrac{a + \sqrt{a^2 + x^2}}{x}\right) + C$

17. $\int \dfrac{1}{x\sqrt{a^2 - x^2}}\, dx = -\dfrac{1}{a} \ln\left(\dfrac{a + \sqrt{a^2 - x^2}}{x}\right) + C,\; 0 < x < a$

18. $\int \dfrac{x}{ax + b}\, dx = \dfrac{b}{a^2} + \dfrac{x}{a} - \dfrac{b}{a^2} \ln(ax + b) + C$

19. $\int \dfrac{x}{(ax + b)^2}\, dx = \dfrac{b}{a^2(ax + b)} + \dfrac{1}{a^2} \ln(ax + b) + C$

20. $\int \dfrac{1}{x(ax + b)}\, dx = \dfrac{1}{b} \ln\left(\dfrac{x}{ax + b}\right) + C$

21. $\int \dfrac{1}{x(ax + b)^2}\, dx = \dfrac{1}{b(ax + b)} + \dfrac{1}{b^2} \ln\left(\dfrac{x}{ax + b}\right) + C$

22. $\int \sqrt{x^2 \pm a^2}\, dx = \tfrac{1}{2}\left[x\sqrt{x^2 \pm a^2} \pm a^2 \ln\left(x + \sqrt{x^2 \pm a^2}\right)\right] + C$

23. $\int x\sqrt{a + bx}\, dx = \dfrac{2}{15b^2}(3bx - 2a)(a + bx)^{3/2} + C$

24. $\int x^2\sqrt{a + bx}\, dx = \dfrac{2}{105b^3}(15b^2x^2 - 12abx + 8a^2)(a + bx)^{3/2} + C$

25. $\int \dfrac{x\, dx}{\sqrt{a + bx}} = \dfrac{2}{3b^2}(bx - 2a)\sqrt{a + bx} + C$

26. $\int \dfrac{x^2\, dx}{\sqrt{a + bx}} = \dfrac{2}{15b^3}(3b^2x^2 - 4abx + 8a^2)\sqrt{a + bx} + C$

TABLE 2
AREAS FOR A STANDARD NORMAL DISTRIBUTION

Entries in the table represent area under the curve between $t = 0$ and a positive value of t. Because of the symmetry of the curve, area under the curve between $t = 0$ and a negative value of t would be found in a similar manner.

Area = Probability

$$= P(0 \le x \le t)$$

$$= \int_0^t \frac{1}{\sqrt{2\pi}} e^{-x^2/2} \, dx$$

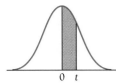

t	0.00	0.01	0.02	0.03	0.04	0.05	0.06	0.07	0.08	0.09
0.0	.0000	.0040	.0080	.0120	.0160	.0199	.0239	.0279	.0319	.0359
0.1	.0398	.0438	.0478	.0517	.0557	.0596	.0636	.0675	.0714	.0753
0.2	.0793	.0832	.0871	.0910	.0948	.0987	.1026	.1064	.1103	.1141
0.3	.1179	.1217	.1255	.1293	.1331	.1368	.1406	.1443	.1480	.1517
0.4	.1554	.1591	.1628	.1664	.1700	.1736	.1772	.1808	.1844	.1879
0.5	.1915	.1950	.1985	.2019	.2054	.2088	.2123	.2157	.2190	.2224
0.6	.2257	.2291	.2324	.2357	.2389	.2422	.2454	.2486	.2517	.2549
0.7	.2580	.2611	.2642	.2673	.2704	.2734	.2764	.2794	.2823	.2852
0.8	.2881	.2910	.2939	.2967	.2995	.3023	.3051	.3078	.3106	.3133
0.9	.3159	.3186	.3212	.3238	.3264	.3289	.3315	.3340	.3365	.3389
1.0	.3413	.3438	.3461	.3485	.3508	.3531	.3554	.3577	.3599	.3621
1.1	.3643	.3665	.3686	.3708	.3729	.3749	.3770	.3790	.3810	.3830
1.2	.3849	.3869	.3888	.3907	.3925	.3944	.3962	.3980	.3997	.4015
1.3	.4032	.4049	.4066	.4082	.4099	.4115	.4131	.4147	.4162	.4177
1.4	.4192	.4207	.4222	.4236	.4251	.4265	.4279	.4292	.4306	.4319
1.5	.4332	.4345	.4357	.4370	.4382	.4394	.4406	.4418	.4429	.4441
1.6	.4452	.4463	.4474	.4484	.4495	.4505	.4515	.4525	.4535	.4545
1.7	.4554	.4564	.4573	.4582	.4591	.4599	.4608	.4616	.4625	.4633
1.8	.4641	.4649	.4656	.4664	.4671	.4678	.4686	.4693	.4699	.4706
1.9	.4713	.4719	.4726	.4732	.4738	.4744	.4750	.4756	.4761	.4767
2.0	.4772	.4778	.4783	.4788	.4793	.4798	.4803	.4808	.4812	.4817
2.1	.4821	.4826	.4830	.4834	.4838	.4842	.4846	.4850	.4854	.4857
2.2	.4861	.4864	.4868	.4871	.4875	.4878	.4881	.4884	.4887	.4890
2.3	.4893	.4896	.4898	.4901	.4904	.4906	.4909	.4911	.4913	.4916
2.4	.4918	.4920	.4922	.4925	.4927	.4929	.4931	.4932	.4934	.4936
2.5	.4938	.4940	.4941	.4943	.4945	.4946	.4948	.4949	.4951	.4952
2.6	.4953	.4955	.4956	.4957	.4959	.4960	.4961	.4962	.4963	.4964
2.7	.4965	.4966	.4967	.4968	.4969	.4970	.4971	.4972	.4973	.4974
2.8	.4974	.4975	.4976	.4977	.4977	.4978	.4979	.4979	.4980	.4981
2.9	.4981	.4982	.4982	.4983	.4984	.4984	.4985	.4985	.4986	.4986
3.0	.4987	.4987	.4987	.4988	.4988	.4989	.4989	.4989	.4990	.4990

Answers

Technology Connection, p. 7

1.–20. Left to the student.

Exercise Set 1.1, p. 13

1.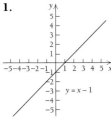

3.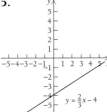

5.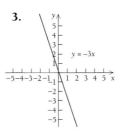

7.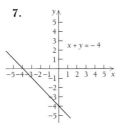

9.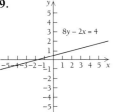

11.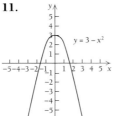

13.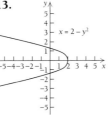

15.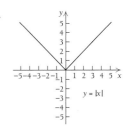

17. (a) $1060; **(b)** $1060.90; **(c)** $1061.36;
(d) $1061.83, assuming 365 days in a year; **(e)** $1061.84
19. $578.70   **21. (a)** 325 per 100,000; **(b)** 67 and 88;
(c) approximately 79; **(d)** tw

23.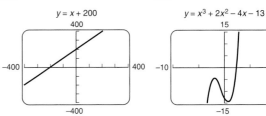

25.

27.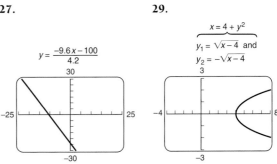

29.

Technology Connection, p. 18

1. 951; 42,701   **2.** 21,813

Technology Connection, p. 19

1.–2. Left to the student.

Technology Connection, p. 21

1. 6; 3.99; 150; $-1.\overline{5}$, or $-\frac{14}{9}$ **2.** -21.3; -18.39; -117.3; $3.2\overline{5}$, or $\frac{293}{90}$ **3.** -75; -65.466; -420.6; $1.6\overline{8}$, or $\frac{76}{45}$

Technology Connection, p. 23

1.–3. Left to the student.

Technology Connection, p. 25

1. $(-2.87234, -8.336082)$, $(-2.234043, 1.0202267)$, $(0, 1)$, $(0.95744681, -2.909538)$, $(2.5531915, 4.8777535)$; answers may vary

Exercise Set 1.2, p. 26

1. Yes   **3.** No   **5.** Yes   **7.** Yes   **9.** Yes   **11.** No

13. (a)

Inputs	Outputs
4.1	11.2
4.01	11.02
4.001	11.002
4	11

(b) $f(5) = 13, f(-1) = 1, f(k) = 2k + 3,$ $f(1 + t) = 2t + 5, f(x + h) = 2x + 2h + 3$
15. $g(-1) = -2, g(0) = -3, g(1) = -2, g(5) = 22,$ $g(u) = u^2 - 3, g(a + h) = a^2 + 2ah + h^2 - 3,$ $g(1 - h) = h^2 - 2h - 2$

17. (a) $f(4) = \dfrac{1}{49}, f(-3)$ does not exist, $f(0) = \dfrac{1}{9},$

$f(a) = \dfrac{1}{(a + 3)^2}, f(t + 1) = \dfrac{1}{(t + 4)^2},$

$f(t + 3) = \dfrac{1}{(t + 6)^2}, f(x + h) = \dfrac{1}{(x + h + 3)^2}$ **(b)** Take

an input, square it, add six times the input, add 9, and then take the reciprocal of the result.

19.

21.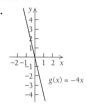

23.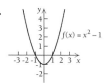

25.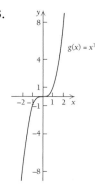

27. Yes   **29.** Yes   **31.** No   **33.** No   **35.** Yes   **37.** Yes
39. (a) 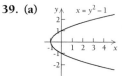   **(b)** no

41. $f(x + h) = x^2 + 2xh + h^2 - 3x - 3h$
43. 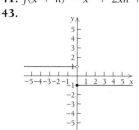   **45.**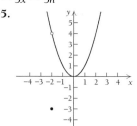

47. $R(10) = \$2050, R(100) = \$20{,}050$ **49. (a)** \$40 million;   **(b)** week 6;   **(c)** week 3   **51. (a)** Yes, each event has exactly one scale of impact numbers.
(b) events; scale of impact numbers

53. $y = \pm\sqrt{\dfrac{x + 5}{2}}$; not a function

55. $y = 2\sqrt[3]{x}$; a function   **57.** tw
59.   **61.** Left to the student.

X	Y1	
-3	.6	
-2	ERROR	
-1	-1	
0	-.75	
1	-1	
2	ERROR	
3	.6	
X = -3		

Technology Connection, p. 34

1. Domain = all real numbers; range = $[-7, \infty)$
2. Domain = all real numbers; range = all real numbers
3. Domain = $(-\infty, 0) \cup (0, \infty)$; range = $(-\infty, 0) \cup (0, \infty)$
4. Domain = all real numbers; range = $[-8, \infty)$
5. Domain = $[-4, \infty)$; range = $[0, \infty)$

6. Domain $= [-3, 3]$; range $= [0, 3]$
7. Domain $= [-3, 3]$; range $= [-3, 0]$
8. Domain $=$ all real numbers; range $=$ all real numbers

Exercise Set 1.3, p. 37

1. $(0, 5)$ **3.** $[-9, -4)$ **5.** $[x, x + h]$ **7.** (p, ∞)
9. $[-3, 3]$ **11.** $[-14, -11]$ **13.** $(-\infty, -4]$
15. (a) 1;   **(b)** $\{-3, -1, 1, 3, 5\}$;   **(c)** 3;
(d) $\{-1, 0, 1, 2\ 3\}$ **17. (a)** 4;
(b) $\{-5, -3, 1, 2, 3, 4, 5\}$;   **(c)** $-5, -3, 4$;
(d) $\{-3, 2, 4, 5\}$ **19. (a)** $2\frac{2}{3}$;   **(b)** $[-2, 3]$;   **(c)** $1\frac{3}{4}$;
(d) $[1, 5]$ **21. (a)** -2;   **(b)** $[-4, 2]$;   **(c)** -2;
(d) $[-3, 3]$ **23. (a)** 3;   **(b)** $[-3, 3]$;   **(c)** $-1\frac{2}{5}, 1\frac{2}{5}$;
(d) $[-5, 4]$ **25. (a)** -1;   **(b)** $[-6, 5]$;   **(c)** $-4, 0, 3$;
(d) $[-2, 2]$ **27.** $\{x \mid x$ is a real number *and* $x \neq 5\}$, or
$(-\infty, 5) \cup (5, \infty)$ **29.** All real numbers
31. All real numbers   **33.** $\left\{x \mid x \text{ is a real number } and \right.$
$\left. x \neq -\frac{4}{3}\right\}$, or $\left(-\infty, -\frac{4}{3}\right) \cup \left(-\frac{4}{3}, \infty\right)$
35. All real numbers   **37.** $\left\{x \mid x \text{ is a real number } and \right.$
$\left. x \neq \frac{7}{4}\right\}$, or $\left(-\infty, \frac{7}{4}\right) \cup \left(\frac{7}{4}, \infty\right)$ **39.** $\{x \mid x$ is a real number
and $x \neq -4\}$, or $(-\infty, -4) \cup (-4, \infty)$
41. All real numbers   **43.** All real numbers
45. $\{x \mid x$ is a real number *and* $x \neq 2\}$, or
$(-\infty, 2) \cup (2, \infty)$ **47.** All real numbers   **49.** $\{x \mid x$ is
a real number *and* $x \neq 2\}$, or $(-\infty, 2) \cup (2, \infty)$
51. $f(-1) = -8, f(0) = 0, f(1) = -2$
53. (a) $A(i) = 10,000\left(1 + \dfrac{i}{2}\right)^{16}$;   **(b)** all positive real
numbers   **55.** tw   **57.** tw
59. Exercise 29: all real numbers; Exercise 32: $[3, \infty)$;
Exercise 44: all real numbers; Exercise 47: $[1, \infty)$;
Exercise 48: $[0, \infty)$

Technology Connection, p. 41

1. The line will slant up from left to right, will intersect
the y-axis at $(0, 1)$, and will be steeper than $y = 10x + 1$.
2. The line will slant up from left to right, will pass
through the origin, and will be less steep than $y = \frac{2}{31}x$.
3. The line will slant down from left to right, will pass
through the origin, and will be steeper than $y = -10x$.
4. The line will slant down from left to right, will
intersect the y-axis at $(0, -1)$, and will be less steep than
$y = -\frac{5}{32}x - 1$.

Technology Connection, p. 42

1. The graph of y_2 is a shift 3 units up of the graph of y_1,
and y_2 has y-intercept $(0, 3)$. The graph of y_3 is a shift
4 units down of the graph of y_1, and y_3 has y-intercept

$(0, -4)$. The graph of $y = x - 5$ is a shift 5 units down
of the graph of $y = x$, and $y = x - 5$ has y-intercept
$(0, -5)$. All lines are parallel.
2. For any x-value, the y_2-value is 3 more than the
y_1-value and the y_3-value is 4 less than the y_1-value.

Exercise Set 1.4, p. 51

1. 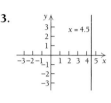   **3.**

5. $m = -3$,   **7.** $m = 0.5$,
y-intercept: $(0, 0)$ y-intercept: $(0, 0)$

9. $m = -2$,   **11.** $m = -1$,
y-intercept: $(0, 3)$ y-intercept: $(0, -2)$

13. $m = -2$, y-intercept: $(0, 2)$ **15.** $m = -1$,
y-intercept: $\left(0, -\frac{5}{2}\right)$ **17.** $y + 5 = -5(x - 1)$, or
$y = -5x$ **19.** $y - 3 = -2(x - 2)$, or $y = -2x + 7$
21. $y = \frac{1}{2}x - 6$ **23.** $y = 3$ **25.** $\frac{3}{2}$ **27.** $-\frac{3}{34}$
29. No slope   **31.** 0   **33.** 3   **35.** 2
37. $y - 1 = \frac{3}{2}(x + 2)$, or $y + 2 = \frac{3}{2}(x + 4)$, or $y = \frac{3}{2}x + 4$
39. $y - \frac{1}{2} = -\frac{3}{34}\left(x - \frac{2}{5}\right)$, or $y - \frac{4}{5} = -\frac{3}{34}(x + 3)$, or
$y = -\frac{3}{34}x + \frac{91}{170}$ **41.** $x = 3$ **43.** $y = 3$
45. $y = 3x$ **47.** $y = 2x + 3$ **49.** $\frac{2}{25}$, or 8%
51. 3.5%   **53.** $100 per year
55. (a) $A = P + 8\%P = P + 0.08P = 1.08P$;   **(b)** $108;
(c) $240   **57. (a)** $C(x) = 20x + 100,000$;   **(b)** $R(x) = 45x$;
(c) $P(x) = R(x) - C(x) = 25x - 100,000$;   **(d)** $3,650,000,
a profit;   **(e)** 4000   **59. (a)** $C(x) = x + 250$;
(b) $R(x) = 10x$. The student charges $10 per lawn.
(c) 28   **61. (a)** $R = 4.17T$;   **(b)** $R = 25.02$
63. (a) $B(W) = 0.025W$;   **(b)** $B = 2.5\%W$. The weight of
the brain is 2.5% of the body weight.   **(c)** 3 lb

65. (a) $D(0°) = 115$ ft, $D(-20°) = 75$ ft, $D(10°) = 135$ ft, $D(32°) = 179$ ft;   **(b)** tw
67. (a) 145.78 cm;   **(b)** 142.98 cm
69. (a) $A(0) = 19.7$, $A(1) = 19.78$, $A(10) = 20.5$, $A(30) = 22.1$, $A(50) = 23.7$;   **(b)** 23.94;
(c)

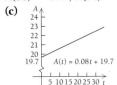

71. tw
73. Left to the student. Answers may vary.

Technology Connection, p. 57

1.–2. Left to the student.

Technology Connection, p. 59

1. (a) (2, 0) and (4, 0);   **(b)** 2, 4;   **(c)** The kx-coordinates of the x-intercepts of $f(x) = x^2 - 6x + 8$ are the solutions of $x^2 - 6x + 8 = 0$.

Technology Connection, p. 61

1. $-5, 2$ **2.** $-4, 6$ **3.** $-2, 1$ **4.** $-1.414214, 0, 1.414214$ **5.** $0, 700$ **6.** $-2.079356, 0.46295543, 3.1164004$ **7.** $-3.095574, -0.6460838, 0.64608382, 3.0955736$ **8.** $-1, 1$ **9.** $-2, -1.414214, 1, 1.414214$ **10.** $-3, -1, 2, 3$ **11.** $-0.3874259, 1.7207592$ **12.** 6.13293

Technology Connection, p. 64

1.–8. Left to the student.
9. If we use ZOOM 4 (ZDecimal) and $[-9.4, 9.4, -3.1, 9.3]$, the "hole" will be evident.

Technology Connection, p. 65

1.–2. Left to the student.

Technology Connection, p. 67

1. Domain $= [0, \infty)$; range $= [0, \infty)$
2. Domain $= [-2, \infty)$; range $= [0, \infty)$
3. Domain $= (-\infty, \infty)$; range $= (-\infty, \infty)$
4. Domain $= (-\infty, \infty)$; range $= (-\infty, \infty)$
5. Domain $= [1, \infty)$; range $= [0, \infty)$
6. Domain $= (-\infty, \infty)$; range $= (-\infty, \infty)$
7. Domain $= [-3, \infty)$; range $= (-\infty, 5]$
8. Domain $= (-\infty, \infty)$; range $= (-\infty, \infty)$
9. Domain $= (-\infty, \infty)$; range $= [0, \infty)$
10. Domain $= [0, \infty)$; range $= [0, \infty)$
11. Incorrect   **12.** Correct

Technology Connection, p. 71

1. ($14, 266)

Exercise Set 1.5, p. 71

1.

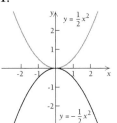

3.

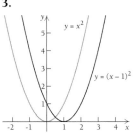

5.

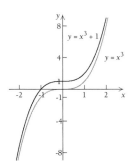

7.

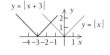

9.

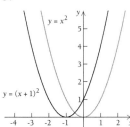

11.

13.

15.

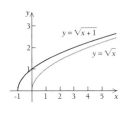

17.

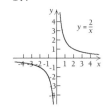

19.

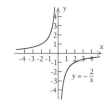

21.

23.

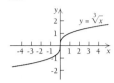

25.

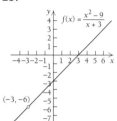

27.

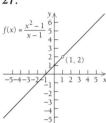

29. $1 \pm \sqrt{3}$ **31.** $-3 \pm \sqrt{10}$ **33.** $\dfrac{1 \pm \sqrt{2}}{2}$

35. $\dfrac{-4 \pm \sqrt{10}}{3}$ **37.** $x^{3/2}$ **39.** $a^{3/5}$ **41.** $t^{1/7}$

43. $t^{-4/3}$ **45.** $t^{-1/2}$ **47.** $(x^2 + 7)^{-1/2}$ **49.** $\sqrt[5]{x}$

51. $\sqrt[3]{y^2}$ **53.** $\dfrac{1}{\sqrt[5]{t^2}}$ **55.** $\dfrac{1}{\sqrt[3]{b}}$ **57.** $\dfrac{1}{\sqrt[6]{e^{17}}}$

59. $\dfrac{1}{\sqrt{x^2 - 3}}$ **61.** 27   **63.** 16   **65.** 8   **67.** All

real numbers except 5   **69.** All real numbers except 2, 3
71. $\left[-\dfrac{4}{5}, \infty \right)$ **73.** ($50, 500$)   **75.** ($5, 1$)
77. ($1, 4$)   **79.** ($2, 3$)   **81.** \$12.05   **83.** 84; 220; 364
85. (a) 90,109; 99,130; 114,595;
(b)

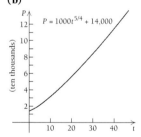

87. ▮tw   **89.** 2.359   **91.** 1.5, 9.5   **93.** -2.646,
2.646   **95.** None   **97.** -10.153, -1.871, -0.821,
-0.303, 0.098, 0.535, 1.219, 3.297

Technology Connection, p. 79

1. (a) $y = 2.7x + 31.4$;   **(b)**

(c) 93.5;   **(d)** ▮tw

Technology Connection, p. 81

1. (a) $y = -0.00005x^4 + 0.0376x^3 - 3.4792x^2 + 105.8108x - 916.6895$. The quartic fits better than the quadratic and the same as the cubic.
(b) Most live births to women occur between the ages of 15 and 45. Since there cannot be a negative number of live births and since live births do not increase for women older than 45, we choose [15, 45] as the domain of this function.
2. (a) $y = -52.3651x^2 + 4533.4594x - 49,405.1373$;
(b)

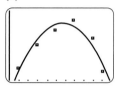

(c) $y = -1.4734x^3 + 133.5946x^2 - 2669.8377x + 34,218.4235$;
(d)

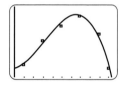

(e) $y = -0.0560x^4 + 7.9980x^3 - 436.1840x^2 + 11,627.8376x - 90,625.0001$;   **(f)**

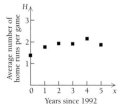

(g) quartic   **(h)** \$30,555; \$48,592   **3. (a)** Left to the
student;   **(b)** $y = 93.2857x^2 - 1336x + 5460.8286$;
(c) 3162 per 100,000, 743 per 100,000, 1429 per 100,000

Exercise Set 1.6, p. 83

1. Linear   **3.** Polynomial   **5.** Quadratic $(a > 0)$
7. (a) , yes;

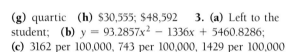

(b) $H(x) = 0.0275x + 1.6925$;

(c) 1.91 in 2000, 2.05 in 2005
9. (a) $S(x) = 0.8x + 1.7$, using the years 1994 and 1998;
(b)

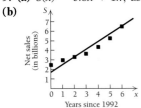

(c) $8.1 billion in 2000, $12.1 billion in 2005
11. (a) $N(x) = 65.6x^2 - 143.1x + 131.5$;   **(b)** $3185 million in 2002, $14,636 million in 2010;   **(c)** tw
13. (a) $N(x) = 0.1875x^2 - 33.75x + 1750$;   **(b)** 531
15. tw
17. (a) $H(x) = 0.114x + 1.52$;   **(b)** 2.43 in 2000, 3.00 in 2005;   **(c)** tw
19. (a) $y = 24.675x^2 - 69.435x + 123.315$; $1147 million in 2002, $5329 million in 2010;   **(c)** tw

Summary and Review: Chapter 1, p. 86

1. [1.1] **(a)** 100 per 1000;   **(b)** 20 and 30;   **(c)** tw
2. [1.1] $1212.75   **3.** [1.1] $5017.60   **4.** [1.3] $(-6, 1]$
5. [1.3] $(0, \infty)$ **6.** [1.2] 13   **7.** [1.2] $2h^2 + 3h + 4$
8. [1.2] 3   **9.** [1.2] 36   **10.** [1.2] $h^2 - 2h + 1$
11. [1,2] 9
12. [1.1]   **13.** [1.2]

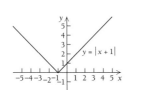   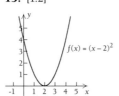

14. [1.2]

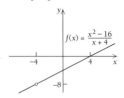

15. [1.2] No   **16.** [1.2] Yes   **17.** [1.2] Yes
18. [1.3] **(a)** 1.2;   **(b)** $[-3, 4]$;   **(c)** -3;   **(d)** $[-1, 2]$
19. [1.4]   **20.** [1.4]

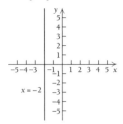   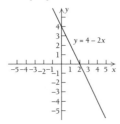

21. [1.4] $y = -\frac{7}{11}x + \frac{6}{11}$, or
$7x + 11y = 6$ **22.** [1.4] $y = 8x + 7$
23. [1.4] $m = -\frac{1}{6}$; y-intercept: $(0, 3)$
24. [1.4] 75 pages per day   **25.** [1.4] $-\frac{20}{3}$ m/sec
26. [1.4] 72 lb   **27.** [1.4] **(a)** $R(x) = 28x$;
(b) $C(x) = 16x + 80,000$;   **(c)** $P(x) = 12x - 80,000$;
(d) $16,000 profit;   **(e)** 6667   **28.** [1.5] $\sqrt[6]{y}$
29. [1.5] $x^{3/20}$ **30.** [1.5] 125   **31.** [1.5] $(-\infty, 3]$
32. [1.5] All real numbers except 1 and -1
33. [1.2]

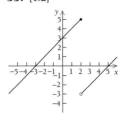

34. [1.6] **(a)** $G(x) = \frac{9}{7}x + \frac{437}{7}$;
(b) 86, 95   **35.** [1.6] **(a)** $P(x) = 0.0018x^2 + 0.0046x + 0.29$;   **(b)** 52%, 96%;   **(c)** tw   **36.** [1.5] $\frac{1}{32}$
37. [1.2], [1.3], [1.5];

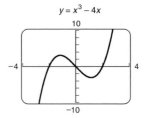

$-2, 0, 2$; domain $= (-\infty, \infty)$; range $= (-\infty, \infty)$
38. [1.2], [1.3], [1.5];

$$y = \sqrt[3]{|9 - x^2|} - 1$$

$-3.162, -2.828, 2.828, 3.162$; domain $= (-\infty, \infty)$;
range $= [-1, \infty)$
39. [1.5] $(-1.9, 0.8), (-0.3, 1), (2.1, 0.7)$
40. [1.6] **(a)** $y = 0.6255x + 75.4766$;   **(b)** 87, 91;
(c) tw   **41.** [1.6] **(a)** $y = 0.0021x^2 - 0.0002x + 0.2958$;   **(b)** 51%, 99%;   **(c)** tw

Test: Chapter 1, p. 89

1. [1.1] **(a)** 1150 minutes/month;   **(b)** 62;   **(c)** tw
2. [1.1] $920   **3.** [1.2] **(a)** 5;   **(b)** $x^2 + 2xh + h^2 - 4$

4. [1.4] $m = -3$; y-intercept: $(0, 2)$
5. [1.4] $y + 5 = \frac{1}{4}(x - 8)$, or $y = \frac{1}{4}x - 7$
6. [1.4] $m = 6$ **7.** [1.4] $-\$700$ per year
8. [1.4] $\frac{1}{2}$ lb per bag   **9.** [1.4] $F(W) = \frac{2}{3}W$
10. [1.4] **(a)** $C(x) = 0.5x + 10,000$;   **(b)** $R(x) = 7.5x$;
(c) $P(x) = 7x - 10,000$;   **(d)** 1429   **11.** [1.5] $(\$3, 16)$
12. [1.2] Yes   **13.** [1.2] No   **14.** [1.3] **(a)** -4;
(b) $(-\infty, \infty)$;   **(c)** $-3, 3$;   **(d)** $[-5, \infty)$
15. [1.2]
16. [1.5] $t^{-1/2}$
17. [1.5] $\dfrac{1}{\sqrt[5]{t^3}}$
18. [1.5]

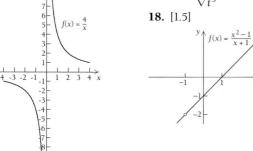

19. [1.5] All real numbers except $2, -7$
20. [1.3] $[-2, \infty)$ **21.** [1.3] $[c, d]$
22. [1.2]

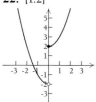

23. [1.6] **(a)** $M(r) = \frac{1}{5}r + 160$;   **(b)** 172, 175
24. [1.6] **(a)**

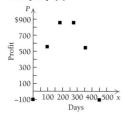

(b) yes;   **(c)** $P(x) = -0.02x^2 + 9x - 100$;
(d) $\$912.50$;   **(e)** tw
25. [1.5] **(a)** 525;   **(b)** $\$4.31$

26. [1.2], [1.3], [1.5]

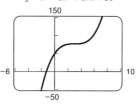

-1.25; domain $= (-\infty, \infty)$; range $= (-\infty, \infty)$
27. [1.2], [1.3], [1.5]

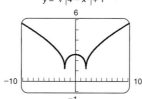

none; domain $= (-\infty, \infty)$; range $= [1, \infty)$
28. [1.5] $(-1.2, 2.4)$ **29.** [1.6] **(a)** $M(r) = 0.2x + 160$;
(b) 172, 175;   **(c)** tw
30. [1.6] **(a)** $P(x) = -0.0200x^2 + 8.9919x - 95.8571$;
(b) $\$914.38$;   **(c)** tw
31. [1.6] **(a)** $y = 37.5761x + 294.4774$,
$y = -0.5925x^2 + 74.6068x - 117.7247$, $y = 0.0220x^3 - 2.6042x^2 + 125.7143x - 439.6475$, $y = 0.0028x^4 - 0.3240x^3 + 11.4571x^2 - 88.5121x + 507.8387$;
(b) left to the student;   **(c)** tw ;   **(d)** tw

Chapter 2 ....

Technology Connection, p. 98

1. 5   **2.** -4 **3.** 327, 456.95, 475.24, 492.1, 493.81,
494, 494.19, 495.9, 513.24, 573.9, 685.17   **4.** 494
5. -1

Exercise Set 2.1, p. 106

1. No   **3.** Yes   **5. (a)** -1, 2 does not exist;   **(b)** -1;
(c) no;   **(d)** 3;   **(e)** 3;   **(f)** yes   **7. (a)** 2;   **(b)** 2;
(c) yes;   **(d)** 0;   **(e)** 0;   **(f)** yes   **9. (a)** 3;   **(b)** 3;
(c) 3;   **(d)** 2;   **(e)** no;   **(f)** yes   **11. (a)** F;   **(b)** T;
(c) F;   **(d)** F;   **(e)** F;   **(f)** T;   **(g)** T;   **(h)** F;   **(i)** F;
(j) T   **13.** No; yes; no; yes   **15.** 33¢, 55¢, does not
exist   **17.** 77¢   **19.** $\$20,000$, $\$28,000$, does not exist.
21. No, yes   **23.** tw   **25.** 100, 30, does not exist
27. No, yes   **29.** tw   **31. (a)** Does not exist;
(b) 2   **33.** tw

Technology Connection, p. 111

1. 53   **2.** 2.8284, or $2\sqrt{2}$ **3.** 0.16666, or $\frac{1}{6}$

Exercise Set 2.2, p. 113

1. -2 **3.** Does not exist   **5.** 11   **7.** -10 **9.** $-\frac{5}{2}$
11. 5   **13.** 2   **15.** 3   **17.** Does not exist   **19.** $\frac{2}{7}$

21. $\frac{5}{4}$ **23.** $6x^2$ **25.** $\dfrac{-2}{x^3}$

27. **(a)** 1;   **(b)** 1;   **(c)** 1;   **(d)** 1;   **(e)** yes, no

29. Does not exist   **31.** 6   **33.** -0.2887, or $-\dfrac{1}{2\sqrt{3}}$

35. 0.75   **37.** 0.25

Technology Connection, p. 117

1.–2. Left to the student.

Exercise Set 2.3, p. 121

1. **(a)** $7(2x + h)$;   **(b)** 70, 63, 56.7, 56.07
3. **(a)** $-7(2x + h)$;   **(b)** $-70, -63, -56.7, -56.07$
5. **(a)** $7(3x^2 + 3xh + h^2)$;   **(b)** 532, 427, 344.47, 336.8407
7. **(a)** $\dfrac{-5}{x(x + h)}$;   **(b)** $-0.2083, -0.25, -0.3049, -0.3117$
9. **(a)** -2;   **(b)** all -2 **11.** **(a)** $2x + h - 1$;   **(b)** 9,
8, 7.1, 7.01   **13.** **(a)** 70, 39, 29, 23 pleasure units/unit of
product;   **(b)** tw   **15.** **(a)** $300,093.99;   **(b)** $299,100;
(c) $993.99;   **(d)** $993.99;   **(e)** tw   **17.** **(a)** $0.581
million/year; **(b)** $0.781 million/year;   **(c)** tw
19. **(a)** 1.25 words/minute, 1.25 words/minute, 0.625
words/minute, 0 words/minute; 0 words/minute;
(b) tw   **21.** **(a)** 144 ft;   **(b)** 400 ft;   **(c)** 128 ft/sec
23. **(a)** 125 million people/year for each;   **(b)** no, tw ;
(c) A: 290 million people/year, -40 million people/year,
-50 million people/year, 300 million people/year; B:
125 million people/year in all intervals;   **(d)** A, tw
25. tw   **27.** $2ax + b + ah$

29. $\dfrac{1}{\sqrt{x + h} + \sqrt{x}}$ **31.** $\dfrac{-2x - h}{x^2(x + h)^2}$

33. $\dfrac{1}{(x + 1)(x + 1 + h)}$

Technology Connection, p. 133

1. $f'(x) = -\dfrac{3}{x^2}; f'(-2) = -\dfrac{3}{4}; f'\left(-\dfrac{1}{2}\right) = -12$
2. $y = -\dfrac{3}{4}x - 3; y = -12x - 12$

Technology Connection, p. 136

1.–4. Graphs are left to the student.   **1.** 60, 26, 0, -80
2. $-96, -11, 24, -11, -56$ **3.** $-36, 0, 12, 0, -43.47$
4. Does not exist, 0.41, 1.81, 2.00, 1.15, does not exist
5. tw

Exercise Set 2.4, p. 137

1. **(a), (b)**

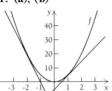

(c) $f'(x) = 10x$;
(d) $-20, 0, 10$

3. **(a), (b)**

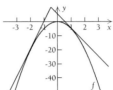

(c) $f'(x) = -10x$;
(d) $20, 0, -10$

5. **(a), (b)**

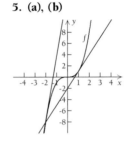

(c) $f'(x) = 3x^2$;
(d) $12, 0, 3$

7. **(a), (b)** All the tangent
lines are identical to the
original function.

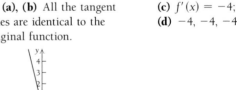

(c) $f'(x) = 2$;
(d) $2, 2, 2$

9. **(a), (b)** All the tangent
lines are identical to the
original function.

(c) $f'(x) = -4$;
(d) $-4, -4, -4$

11. (a), (b)

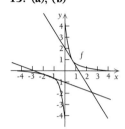

(c) $f'(x) = 2x + 1$;
(d) $-3, 1, 3$

13. (a), (b)

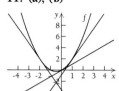

(c) $f'(x) = \dfrac{-1}{x^2}$;

(d) $-\dfrac{1}{4}$, does not exist, -1

15. $f'(x) = m$ **17.** $y = 6x - 9$; $y = -2x - 1$;
$y = 20x - 100$ **19.** $y = -5x + 10$; $y = -5x - 10$;
$y = -0.0005x + 0.1$ **21.** $y = 2x + 5$; $y = 4$;
$y = -10x + 29$ **23.** $x_0, x_3, x_4, x_6, x_{12}$

25. 1, 2, 3, 4, and so on   **27.** tw   **29.** $\dfrac{-2}{x^3}$

31. $\dfrac{1}{(1 + x)^2}$ **33.** $x = -3$

35.–39. Left to the student.

Technology Connection, p. 141

1.–4. Graphs are left to the student.   **1.** 152, -76,
$-100, -180$ **2.** $-96, -11, 24, -11, -56$ **3.** $-36, 0$,
12, 0, -43.47 **4.** Does not exist, 0.41, 1.81, 2.00, 1.15,
does not exist

Technology Connection, p. 146

1. 1, 11   **2.** 0, 4   **3.** 2   **4.** $-1.4, 1.4$

Technology Connection, p. 147

1. 2

Exercise Set 2.5, p. 148

1. $7x^6$ **3.** 0   **5.** $600x^{149}$ **7.** $3x^2 + 6x$ **9.** $\dfrac{4}{\sqrt{x}}$

11. $0.07x^{-0.93}$ **13.** $\dfrac{2}{5\sqrt[5]{x}}$ **15.** $\dfrac{-3}{x^4}$ **17.** $6x - 8$

19. $\dfrac{1}{4\sqrt[4]{x^3}} + \dfrac{1}{x^2}$ **21.** $1.6x^{1.5}$ **23.** $\dfrac{-5}{x^2} - 1$ **25.** 4

27. 4   **29.** x^3 **31.** $-0.02x - 0.5$

33. $-2x^{-5/3} + \dfrac{3}{4}x^{-1/4} + \dfrac{6}{5}x^{1/5} - 24x^{-4}$ **35.** $-\dfrac{2}{x^2} - \dfrac{1}{2}$

37. $-16x^{-2} + 24x^{-4} - 4x^{-5}$

39. $\dfrac{1}{2}x^{-1/2} + \dfrac{1}{3}x^{-2/3} - \dfrac{1}{4}x^{-3/4} + \dfrac{1}{5}x^{-4/5}$

41. $y = 10x - 15$; $y = x + 3$; $y = -2x + 1$

43. $(0, 0)$ **45.** $(0, 0)$ **47.** $\left(\dfrac{5}{6}, \dfrac{23}{12}\right)$ **49.** $(-25, 76.25)$

51. There are none.   **53.** The tangent line is horizontal
at all points on the graph.   **55.** $\left(\dfrac{5}{3}, \dfrac{148}{27}\right), (-1, -4)$

57. $(\sqrt{3}, 2 - 2\sqrt{3}), (-\sqrt{3}, 2 + 2\sqrt{3})$ **59.** $\left(\dfrac{19}{2}, \dfrac{399}{4}\right)$

61. $(60, 150)$ **63.** $\left(-2 + \sqrt{3}, \dfrac{4}{3} - \sqrt{3}\right),$
$\left(-2 - \sqrt{3}, \dfrac{4}{3} + \sqrt{3}\right)$ **65.** $(0, -4), \left(\sqrt{\dfrac{2}{3}}, -\dfrac{40}{9}\right),$
$\left(-\sqrt{\dfrac{2}{3}}, -\dfrac{40}{9}\right)$ **67.** $2x - 1$ **69.** $2x + 1$

71. $3x^2 - \dfrac{1}{x^2}$ **73.** $-192x^2$ **75.** $\dfrac{2}{3\sqrt[3]{x^2}}$

77. $3x^2 + 6x + 3$

79. $\dfrac{F(x + h) - F(x)}{h}$

$$= \dfrac{[f(x + h) - g(x + h)] - [f(x) - g(x)]}{h}$$

$$= \dfrac{f(x + h) - f(x)}{h} - \dfrac{g(x + h) - g(x)}{h}.$$

As h approaches 0, the two terms on the right approach
$f'(x)$ and $g'(x)$, respectively, so their difference
approaches $f'(x) - g'(x)$. Thus, $F'(x) = f'(x) - g'(x)$.
81. tw
83. $(-0.692, -2.639),$
$(0.692, -4.761)$ **85.** $(-0.346, -2.191),$
$(1.929, 2.358)$

$y = 1.6x^3 - 2.3x - 3.7$

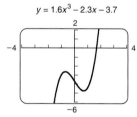

$y = \dfrac{5x^2 + 8x - 3}{3x^2 + 2}$

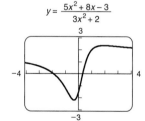

87.

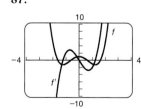

89.

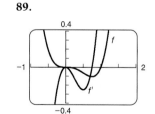

91.

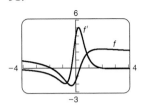

Technology Connection, p. 154

1. Graphs are left to the student.
$P(x) = -0.5x^2 + 40x - 3$; $R'(x) = 50 - x$, $C'(x) = 10$,
$P'(x) = 40 - x$; \$1200, \$403, \$797; \$10, \$10, \$0;
marginal cost

Exercise Set 2.6, p. 155

1. (a) $v(t) = 3t^2 + 1$;   **(b)** $a(t) = 6t$;
(c) $v(4) = 49$ ft/sec, $a(4) = 24$ ft/sec^2
3. (a) $P(x) = -0.001x^2 + 3.8x - 60$;
(b) $R(100) = \$500$, $C(100) = \$190$, $P(100) = \$310$;
(c) $R'(x) = 5$, $C'(x) = 0.002x + 1.2$,
$P'(x) = -0.002x + 3.8$;   **(d)** $R'(100) = \$5$ per unit,
$C'(100) = \$1.40$ per unit, $P'(100) = \$3.60$ per
unit;   **(e)** tw   **5. (a)** $N'(a) = -2a + 300$;   **(b)** 2906;
(c) 280 units/thousand dollars;   **(d)** tw
7. (a) 630 units per month, 980 units per month,
1430 units per month, 630 units per month;
(b) $M'(t) = -4t + 100$;   **(c)** 80, 60, 0, -80;   **(d)** tw
9. (a) $\dfrac{dD}{dp} = -\dfrac{1}{2\sqrt{p}}$;   **(b)** 95;   **(c)** $-\dfrac{1}{10}$ dollars per unit;
(d) tw   **11. (a)** $\dfrac{dC}{dr} = 2\pi$;   **(b)** tw
13. (a) $T'(t) = -0.2t + 1.2$;   **(b)** $100.175°$;
(c) 0.9 degree/day;   **(d)** tw
15. (a) $\dfrac{dT}{dW} = 1.31W^{0.31}$;   **(b)** tw
17. (a) $\dfrac{dR}{dQ} = kQ - Q^2$;   **(b)** tw
19. (a) $A'(t) = 0.08$;   **(b)** tw   **21.** tw
23.   **25.**

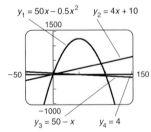

Technology Connection, p. 161

1. (a) Incorrect;   **(b)** incorrect;   **(c)** correct
2.–5. Left to the student.

Technology Connection, p. 162

1. $R(p) = p(200 - p)$, $R'(p) = 200 - 2p$; graphs are left
to the student.

Exercise Set 2.7, p. 163

1. $11x^{10}$ **3.** $\dfrac{1}{x^2}$ **5.** $3x^2$

7. $(8x^5 - 3x^2 + 20)\left(32x^3 - \dfrac{3}{2\sqrt{x}}\right) +$
$(40x^4 - 6x)(8x^4 - 3\sqrt{x})$ **9.** $300 - 2x$
11. $(4\sqrt{x} - 6)(3x^2 - 2) + (2x^{-1/2})(x^3 - 2x + 4)$
13. $2x + 6$ **15.** $6x^5 - 32x^3 + 32x$
17. $(5x^{-3})(4x^3 - 15x^2 + 10) -$
$15x^{-4}(x^4 - 5x^3 + 10x - 2)$, or $5 - 100x^{-3} + 30x^{-4}$
19. $\left(x + \dfrac{2}{x}\right)(2x) + \left(1 - \dfrac{2}{x^2}\right)(x^2 - 3)$, or
$3x^2 + \dfrac{6}{x^2} - 1$ **21.** $\dfrac{300}{(300 - x)^2}$ **23.** $\dfrac{17}{(2x + 5)^2}$
25. $\dfrac{-x^4 - 3x^2 - 2x}{(x^3 - 1)^2}$ **27.** $\dfrac{1}{(1 - x)^2}$ **29.** $\dfrac{2}{(x + 1)^2}$
31. $\dfrac{-1}{(x - 3)^2}$ **33.** $\dfrac{-2x^2 + 6x + 2}{(x^2 + 1)^2}$ **35.** $\dfrac{-18x + 35}{x^8}$
37. $\dfrac{4\sqrt{x} - \frac{1}{2}x^{-1/2}(4x + 3)}{x}$, or $\dfrac{4x - 3}{2x^{3/2}}$
39.–75. Left to the student.   **77.** $y = 2$; $y = \frac{1}{2}x + 2$
79. (a) $D'(p) = \dfrac{-2978}{(10p + 11)^2}$;   **(b)** $-\dfrac{2978}{2601}$
81. (a) $R(p) = p(400 - p) = 400p - p^2$;
(b) $R'(p) = 400 - 2p$ **83. (a)** $R(p) = 4000 + 3p$;
(b) $R'(p) = 3$ **85.** $A'(x) = \dfrac{xC'(x) - C(x)}{x^2}$
87. (a) $T'(t) = \dfrac{4 - 4t^2}{(t^2 + 1)^2}$;   **(b)** $T(2) = 100.2°F$;
(c) $-0.48°F/\text{hour}$ **89.** $\dfrac{5x^3 - 30x^2\sqrt{x}}{2\sqrt{x}(\sqrt{x} - 5)^2}$
91. $\dfrac{-3(1 + 2v)}{(1 + v + v^2)^2}$ **93.** $\dfrac{2t^3 - t^2 + 1}{(1 - t + t^2 - t^3)^2}$
95. $\dfrac{5x^3 + 15x^2 + 2}{2x\sqrt{x}}$
97. $[x(9x^2 + 6) + (3x^3 + 6x - 2)](3x^4 + 7) +$
$12x^4(3x^3 + 6x - 2)$ **99.** $\dfrac{6t^2(t^5 + 3)}{(t^3 + 1)^2} + \dfrac{5t^4(t^3 - 1)}{t^3 + 1}$
101.
$$\dfrac{\{(x^7 - 2x^6 + 9)[(2x^2 + 3)(12x^2 - 7) + 4x(4x^3 - 7x + 2)] - (7x^6 - 12x^5)(2x^2 + 3)(4x^3 - 7x + 2)\}}{(x^7 - 2x^6 + 9)^2}$$
103. tw

105. No points at which the tangent line is horizontal.

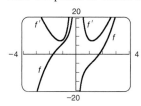

107. $(-0.2, -0.75), (0.2, 0.75)$ **109.** $(-1, -2), (1, 2)$

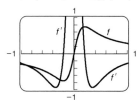   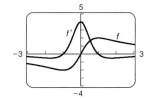

Exercise Set 2.8, p. 172

1. $-55(1 - x)^{54}$ **3.** $\dfrac{4}{\sqrt{1 + 8x}}$ **5.** $\dfrac{3x}{\sqrt{3x^2 - 4}}$

7. $-240x(3x^2 - 6)^{-41}$ **9.** $\sqrt{2x + 3} + \dfrac{x}{\sqrt{2x + 3}}$, or

$\dfrac{3(x + 1)}{\sqrt{2x + 3}}$ **11.** $2x\sqrt{x - 1} + \dfrac{x^2}{2\sqrt{x - 1}}$, or $\dfrac{5x^2 - 4x}{2\sqrt{x - 1}}$

13. $\dfrac{-6}{(3x + 8)^3}$ **15.** $(1 + x^3)^2(-3x^2 - 12x^5)$, or

$-3x^2(1 + x^3)^2(1 + 4x^3)$ **17.** $4x - 400$

19. $2(x + 6)^9(x - 5)^3(7x - 13)$

21. $4(x - 4)^7(3 - x)^3(10 - 3x)$ **23.** $4(2x - 3)^2(3 - 8x)$

25. $\left(\dfrac{1 - x}{1 + x}\right)^{-1/2} \cdot \dfrac{-1}{(x + 1)^2}$ **27.** $\left(\dfrac{3x - 1}{5x + 2}\right)^3 \cdot \dfrac{44}{(5x + 2)^2}$

29. $\frac{1}{3}(x^4 + 3x^2)^{-2/3}(4x^3 + 6x)$

31. $100(2x^3 - 3x^2 + 4x + 1)^{99}(6x^2 - 6x + 4)$

33. $\left(\dfrac{2x + 3}{5x - 1}\right)^{-5} \cdot \dfrac{68}{(5x - 1)^2}$

35. $\left(\dfrac{x^2 + 1}{x^2 - 1}\right)^{-1/2} \cdot \dfrac{-2x}{(x^2 - 1)^2}$

37. $\dfrac{(2x + 3)^3(-6x - 61)}{(3x - 2)^6}$

39. $16(2x + 1)^{-1/3}(3x - 4)^{5/4} + 45(2x + 1)^{2/3}(3x - 4)^{1/4}$,
or $(2x + 1)^{-1/3}(3x - 4)^{1/4}(138x - 19)$ **41.** $\frac{1}{2}u^{-1/2}, 2x,$

$\dfrac{x}{\sqrt{x^2 - 1}}$ **43.** $50u^{49}, 12x^2 - 4x,$

$50(4x^3 - 2x^2)^{49}(12x^2 - 4x)$ **45.** $2u + 1, 3x^2 - 2,$
$(2x^3 - 4x + 1)(3x^2 - 2)$ **47.** $y = \frac{5}{4}x + \frac{3}{4}$

49. (a) $\dfrac{2x - 3x^2}{(1 + x)^6}$;   **(b)** $\dfrac{2x - 3x^2}{(1 + x)^6}$;   **(c)** same

51. $12x^2 - 12x + 5, 6x^2 + 3$ **53.** $\dfrac{16}{x^2} - 1, \dfrac{2}{4x^2 - 1}$

55. $x^4 - 2x^2 + 2, x^4 + 2x^2$ **57.** $f(x) = x^5,$

$g(x) = 3x^2 - 7$ **59.** $f(x) = \dfrac{x + 1}{x - 1}, g(x) = x^3$

61. -216 **63.** $-4(-11)^{-2/3} \approx -0.8087$

65. (a) $C'(x) = \dfrac{1500x^2}{\sqrt{x^3 + 2}}, C'(10) = 4738.68$;   **(b)** tw

67. (a) $P'(x) = \dfrac{2000x}{\sqrt{x^2 + 3}} - \dfrac{1500x^2}{\sqrt{x^3 + 2}}$;   **(b)** tw

69. (a) $\$3000(1 + i)^2$;   **(b)** tw

71. (a) $D(t) = \dfrac{80{,}000}{1.6t + 9}$;   **(b)** $D'(t) = \dfrac{-128{,}000}{(1.6t + 9)^2}$;

(c) $\dfrac{-128{,}000}{28{,}561}$, or about -4.48

73. $\dfrac{x^2 - 2}{\sqrt[3]{(x^3 - 6x + 1)^2}}$ **75.** $\dfrac{x - 2}{2(x - 1)^{3/2}}$

77. $\dfrac{-4(1 + 2v)^3}{v^5}$ **79.** $\dfrac{1}{\sqrt{1 - x^2}(1 - x)}$, or

$\dfrac{\sqrt{1 - x^2}}{(1 - x^2)(1 - x)}$ **81.** $3\left(\dfrac{x^2 - x - 1}{x^2 + 1}\right)^2 \cdot \dfrac{x^2 + 4x - 1}{(x^2 + 1)^2}$,

or $\dfrac{3(x^2 - x - 1)^2(x^2 + 4x - 1)}{(x^2 + 1)^4}$

83. $\dfrac{1}{\sqrt{t}(1 + \sqrt{t})^2}$ **85.** tw

87. $(-2.14476, -7.728), (2.14476, 7.728)$

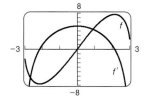

89.

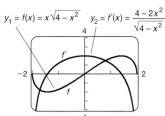

Exercise Set 2.9, p. 176

1. 0 **3.** $-\dfrac{2}{x^3}$ **5.** $-\dfrac{3}{16}x^{-7/4}$ **7.** $12x^2 + \dfrac{8}{x^3}$ **9.** $\dfrac{12}{x^5}$

11. $n(n - 1)x^{n-2}$ **13.** $12x^2 - 2$

15. $-\frac{1}{4}(x - 1)^{-3/2}$, or $\dfrac{-1}{4\sqrt{(x - 1)^3}}$ **17.** $2a$

19. $86(x^2 - 8x)^{41}(85x^2 - 680x + 1344)$
21. $200x^2(x^4 - 4x^2)^{48}(199x^4 - 798x^2 + 792)$
23. $-\frac{2}{9}x^{-4/3}$ **25.** $-\frac{3}{16}(x - 8)^{-5/4}$ **27.** $6x^{-4} + 24x^{-5}$
29. 24 **31.** $720x$ **33.** $20(x^2 - 5)^8[19x^2 - 5]$
35. $a(t) = 6t + 2$ **37.** $P''(t) = 200,000$
39. $y' = -x^{-2} - 2x^{-3}, \, y'' = 2x^{-3} + 6x^{-4},$
$y''' = -6x^{-4} - 24x^{-5}$ **41.** $y' = \dfrac{1 + 2x^2}{\sqrt{1 + x^2}},$
$y'' = \dfrac{2x^3 + 3x}{(1 + x^2)^{3/2}}, \, y''' = \dfrac{3}{(1 + x^2)^{5/2}}$ **43.** $y' = \dfrac{11}{(2x + 3)^2},$
$y'' = \dfrac{-44}{(2x + 3)^3}, \, y''' = \dfrac{264}{(2x + 3)^4}$ **45.** $y' = \dfrac{x - 2}{2(x - 1)^{3/2}},$
$y'' = \dfrac{4 - x}{4(x - 1)^{5/2}}, \, y''' = \dfrac{3(x - 6)}{8(x - 1)^{7/2}}$ **47.** $\dfrac{2}{(x - 1)^3}$
49. $\dfrac{3}{(x + 2)^2}, \dfrac{-6}{(x + 2)^3}, \dfrac{18}{(x + 2)^4}, \dfrac{-72}{(x + 2)^5}, \dfrac{360}{(x + 2)^6}$
51. **53.**

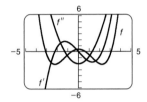

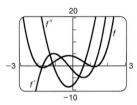

Summary and Review: Chapter 2, p. 178

1. [2.1] **(a)** $-11, -10.5, -10.1, -10.01, -10.001,$
$-10.0001; -9, -9.5, -9.9, -9.99, -9.999, -9.9999;$
(b) $-10, -10, -10$ **2.** [2.1] -10 **3.** [2.1] -10
4. [2.1], [2.2] -4 **5.** [2.1], [2.2] 10 **6.** [2.2] -10
7. [2.2] 5 **8.** [2.1] No **9.** [2.1] Yes **10.** [2.1] -4
11. [2.1] -4 **12.** [2.1] Yes **13.** [2.1] Does not exist
14. [2.1] -2 **15.** [2.1] No **16.** [2.3] 1
17. [2.3] -3 **18.** [2.3] $4x + 2h$ **19.** [2.4], [2.5]
$y = x - 1$ **20.** [2.5] $(4, 5)$ **21.** [2.5] $(5, -108)$
22. [2.5] $20x^4$ **23.** [2.5] $x^{-2/3}$ **24.** [2.5] $64x^{-9}$
25. [2.5] $6x^{-3/5}$ **26.** [2.5] $0.7x^6 - 12x^3 - 3x^2$
27. [2.5] $x^5 + 32x^3 - 5$ **28.** [2.7] $2x$
29. [2.7] $\dfrac{-x^2 + 16x + 8}{(8 - x)^2}$
30. [2.8] $2(5 - x)(2x - 1)^4(26 - 7x)$
31. [2.8] $35x^4(x^5 - 2)^6$
32. [2.8] $3x^2(4x + 3)^{-1/4} + 2x(4x + 3)^{3/4}$, or
$x(4x + 3)^{-1/4}(11x + 6)$ **33.** [2.9] $240x^{-6}$
34. [2.9] $840x^3$ **35.** [2.6], [2.9] **(a)** $1 + 4t^3$; **(b)** $12t^2$;
(c) 33, 48 **36.** [2.6] **(a)** $-8x^2 + 47x + 10$; **(b)** $800,
$3050, -$2250$; **(c)** $40, 16x - 7, -16x + 47$; **(d)** $40
per unit, $313 per unit, $-$273 per unit
37. [2.6] **(a)** $100t$; **(b)** $30,000$; **(c)** 2000 per year
38. [2.8] $4x^2 - 4x + 6, -2x^2 - 9$

39. [2.8] $\dfrac{-9x^4 - 4x^3 + 9x + 2}{2\sqrt{1 + 3x}(1 + x^3)^2}$ **40.** [2.2] -0.25
41. [2.2] 0.1667, or $\frac{1}{6}$
42. [2.5] $(-1.7137, 37.445), (0, 0), (1.7137, -37.445)$

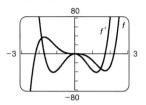

Test: Chapter 2, p. 180

1. [2.1] **(a)** $11, 11.7, 11.9, 11.99, 11.999, 11.9999; 13, 12.5,$
$12.1, 12.01, 12.001, 12.0001;$ **(b)** $12, 12, 12$ **2.** [2.1] 12
3. [2.1] 12 **4.** [2.1] Does not exist. **5.** [2.1] 0
6. [2.1] Does not exist **7.** [2.1] 2 **8.** [2.1] 4
9. [2.1] 1 **10.** [2.1] 1 **11.** [2.1] 1 **12.** [2.1] Yes
13. [2.1] No **14.** [2.1] Does not exist **15.** [2.1] 1
16. [2.1] No **17.** [2.1] 3 **18.** [2.1] 3 **19.** [2.1] Yes
20. [2.1], [2.2] 6 **21.** [2.1], [2.2] $\frac{1}{2}$
22. [2.1], [2.2] Does not exist **23.** [2.3] $3(2x + h)$
24. [2.4] $y = \frac{3}{4}x + 2$ **25.** [2.5] $(0, 0), (2, -4)$
26. [2.5] $84x^{83}$ **27.** [2.5] $\dfrac{5}{\sqrt{x}}$ **28.** [2.5] $\dfrac{10}{x^2}$
29. [2.5] $\frac{5}{4}x^{1/4}$, or $\frac{5}{4}\sqrt[4]{x}$ **30.** [2.5] $-x + 0.61$
31. [2.5] $x^2 - 2x + 2$ **32.** [2.5], [2.7] $\dfrac{-6x + 20}{x^5}$
33. [2.7] $\dfrac{5}{(5 - x)^2}$
34. [2.8] $(x + 3)^3(7 - x)^4(-9x + 13)$
35. [2.8] $-5(x^5 - 4x^3 + x)^{-6}(5x^4 - 12x^2 + 1)$
36. [2.8] $\sqrt{x^2 + 5} + \dfrac{x^2}{\sqrt{x^2 + 5}}$, or $\dfrac{2x^2 + 5}{\sqrt{x^2 + 5}}$
37. [2.9] $24x$
38. [2.6] **(a)** $P(x) = -0.001x^2 + 48.8x - 60$;
(b) $R(10) = $500, C(10) = $72.10,$
$P(10) = 427.90; **(c)** $R'(x) = 50, C'(x) = 0.002x + 1.2,$
$P'(x) = -0.002x + 48.8$; **(d)** $R'(10) = 50 per unit;
$C'(10) = 1.22 per unit; $P'(10) = 48.78 per unit
39. [2.6] **(a)** $\dfrac{dM}{dt} = -0.003t^2 + 0.2t$; **(b)** 9;
(c) 1.7 words/minute **40.** [2.8] $x^3 + x^6,$
$x^3 + 3x^4 + 3x^5 + x^6$
41. [2.7], [2.8] $-2\left(\dfrac{1 + 3x}{1 - 3x}\right)^{1/3} + \left(\dfrac{1 - 3x}{1 + 3x}\right)^{2/3}$
42. [2.2] 12

43. [2.4] (1.0836, 25.1029), (2.9503, 8.6247)

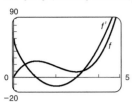

44. [2.4] $\frac{1}{2}$

Chapter 3

Technology Connection, p. 195

1. Relative maximum: $(-1, 19)$; relative minimum: $(2, -8)$

Technology Connection, p. 197

1. Relative minimum: $(0, 0)$ **2.** Relative maximum: $(1.67, 2.36)$ **3.** Relative maximum: $(-1, 2.17)$; relative minimum: $(2, -2.33)$ **4.** Relative minima: $(-3.20, -0.01)$ and $(3.20, -0.01)$; relative maximum: $(0, 22)$

Technology Connection, p. 199

1. Left to the student.   **2.** There is a vertical asymptote at $x = 1$. The derivative does not exist at $x = 1$.
3. Relative maximum: $(1, 2)$

Exercise Set 3.1, p. 199

1. Relative minimum at $(2, 1)$

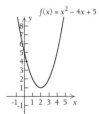

3. Relative maximum at $\left(\frac{1}{2}, \frac{21}{4}\right)$

5. Relative minimum at $(-1, -2)$

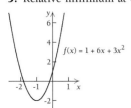

7. Relative minimum at $(1, 1)$; relative maximum at $\left(-\frac{1}{3}, \frac{59}{27}\right)$

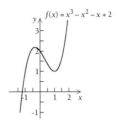

9. Relative maximum at $(-1, 8)$; relative minimum at $(1, 4)$

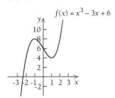

11. Relative minimum at $(0, 0)$; relative maximum at $(1, 1)$

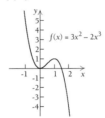

13. None

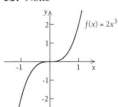

15. Relative maximum at $(0, 10)$; relative minimum at $(4, -22)$

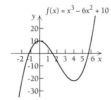

17. Relative maximum at $\left(\frac{3}{4}, \frac{27}{256}\right)$

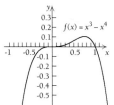

59. Relative maximum at (200, 86.6)

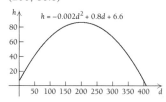

61. tw

19. Relative minima at $(-2, -13)$ and $(2, -13)$; relative maximum at $(0, 3)$

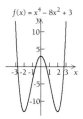

63. Relative minima at $(-5, 425)$ and $(4, -304)$; relative maximum at $(-2, 560)$

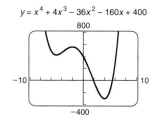

21. Relative maximum at $(0, 1)$

23. Relative minimum at $(0, -8)$

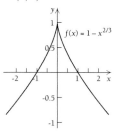

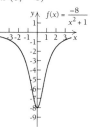

65. Relative maximum at $(2.12, 4.5)$; relative minimum at $(-2.12, -4.5)$

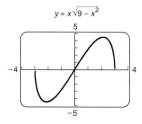

25. Relative minimum at $(-1, -2)$; relative maximum at $(1, 2)$

27. None

67. tw

Technology Connection, p. 214

1. Relative minimum at $(1, -1)$; inflection points at $(0, 0)$, $(0.553, -0.512)$, $(1.447, -0.512)$, and $(2, 0)$

Exercise Set 3.2, p. 214

1. Relative maximum at $(0, 2)$

3. Relative minimum at $\left(-\frac{1}{2}, -\frac{5}{4}\right)$

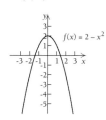

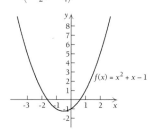

29.–55. Left to the student.
57. Relative maximum at $(150, 22{,}506)$

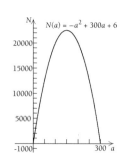

5. Relative maximum at $\left(\frac{3}{8}, -\frac{7}{16}\right)$

7. Relative maximum at $(-2, 72)$; relative minimum at $(3, -53)$

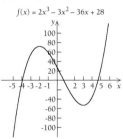

9. Relative maximum at $\left(-\frac{1}{2}, 1\right)$; relative minimum at $\left(\frac{1}{2}, -\frac{1}{3}\right)$

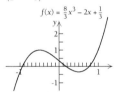

11. Relative minimum at $(0, -4)$; relative maximum at $(2, 0)$

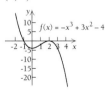

13. Relative minima at $(0, 0)$ and $(3, -27)$; relative maximum at $(1, 5)$

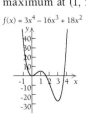

15. Relative minimum at $(-1, 0)$

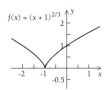

17. Relative minima at $(-\sqrt{3}, -9)$ and $(\sqrt{3}, -9)$; relative maximum at $(0, 0)$

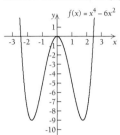

19. Relative maximum at $\left(-\frac{2}{3}, \frac{121}{27}\right)$; relative minimum at $(2, -5)$

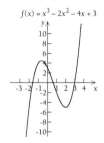

21. Relative minimum at $(-1, -1)$

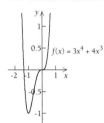

23. Relative maximum at $(-5, 400)$; relative minimum at $(9, -972)$

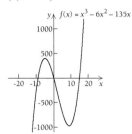

25. Relative minimum at $\left(-1, -\frac{1}{2}\right)$; relative maximum at $\left(1, \frac{1}{2}\right)$

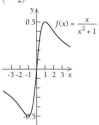

27. Relative maximum at $(0, 3)$

29. None

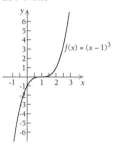

31. Relative minima at $(0, 0)$ and $(1, 0)$; relative maximum at $\left(\frac{1}{2}, \frac{1}{16}\right)$

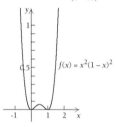

33. Relative minimum at $(-2, -64)$; relative maximum at $(2, 64)$

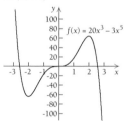

35. Relative minimum at $(-\sqrt{2}, -2)$; relative maximum at $(\sqrt{2}, 2)$

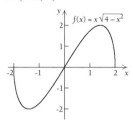

37. None   **39.** $(0, 1)$ **41.** $\left(\frac{1}{2}, \frac{1}{6}\right)$

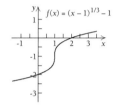

43.–83. Left to the student.   **85.**

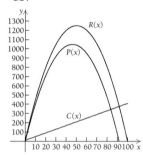

87. An object of radius $13\frac{1}{3}$ mm   **89.** tw   **91.** tw
93. Relative maximum at $(1, 1)$; relative minimum at $(0, 0)$

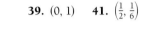

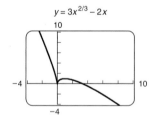

95. Relative maximum at $(0, 0)$; relative minimum at $(0.8, -1.11)$

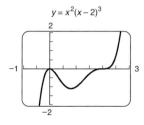

97. Relative minimum at $\left(\frac{1}{4}, -\frac{1}{4}\right)$

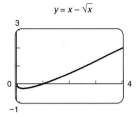

$y = x - \sqrt{x}$

99. tw

Technology Connection, p. 219

1. 2   **2.** 2

Technology Connection, p. 220

1. $\frac{2}{3}$ **2.** 0   **3.** ∞ **4.** 1

Technology Connection, p. 223

1. $x = -7$ and $x = 4$ **2.** $x = 0$, $x = 3$, and $x = -2$

Technology Connection, p. 224

1. $y = 0$ **2.** $y = 3$ **3.** $y = 0$ **4.** $y = \frac{1}{2}$

Technology Connection, p. 225

1. $y = 3x - 1$ **2.** $y = 5x$

Technology Connection, p. 226

1. x-intercepts: $(0, 0)$, $(3, 0)$, $(-5, 0)$; y-intercept: $(0, 0)$
2. x-intercepts: $(0, 0)$, $(-3, 0)$, $(1, 0)$; y-intercept $(0, 0)$

Exercise Set 3.3, p. 232

1. $\frac{2}{5}$ **3.** 5   **5.** $\frac{1}{2}$ **7.** $\frac{2}{3}$ **9.** 0   **11.** ∞ **13.** 0
15. ∞
17.

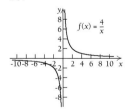

$f(x) = \frac{4}{x}$

19.

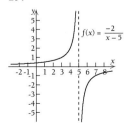

$f(x) = \frac{-2}{x-5}$

21.

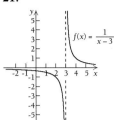

$f(x) = \frac{1}{x-3}$

23.

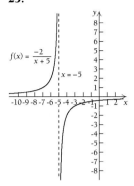

$f(x) = \frac{-2}{x+5}$ $x = -5$

25.

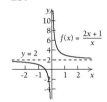

$f(x) = \frac{2x+1}{x}$ $y = 2$

27.

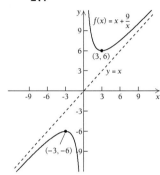

$f(x) = x + \frac{9}{x}$ $(3, 6)$ $y = x$ $(-3, -6)$

29.

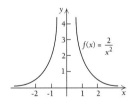

$f(x) = \frac{2}{x^2}$

31.

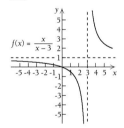

$f(x) = \frac{x}{x-3}$

33.

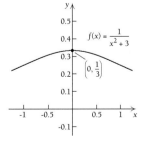

$f(x) = \frac{1}{x^2+3}$ $\left(0, \frac{1}{3}\right)$

35.

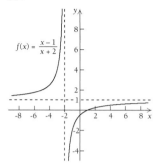

$f(x) = \dfrac{x-1}{x+2}$

37.

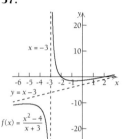

$x = -3$

$y = x - 3$

$f(x) = \dfrac{x^2 - 4}{x + 3}$

39.

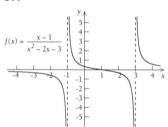

$f(x) = \dfrac{x-1}{x^2 - 2x - 3}$

41.

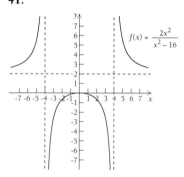

$f(x) = \dfrac{2x^2}{x^2 - 16}$

43.

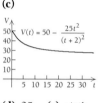

$f(x) = \dfrac{1}{x^2 - 1}$

45.

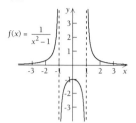

$f(x) = \dfrac{x^2 + 1}{x}$

$y = x$

47. (a) $50, $37.24, $32.64, $26.37;
(b) maximum = 50 at $x = 0$;
(c)

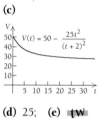

$V(t) = 50 - \dfrac{25t^2}{(t+2)^2}$

(d) 25; **(e)** tw

49. (a) $480, $600, $2400, $4800;
(b) ∞; **(c)** tw
(d)

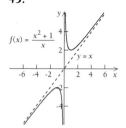

$C(p) = \dfrac{48{,}000}{100 - p}$

(e) tw

51. (a) 4.00, 4.50, 5.14, 6.00, 7.20, 9.00, 12.00, 18.00, 36.00, 54.00, 108.00; **(b)** ∞; **(c)** tw   **53.** tw
55. Does not exist   **57.** $-\infty$ **59.** $\frac{3}{2}$ **61.** $-\infty$
63.   **65.**

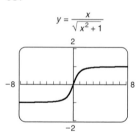

$y = \dfrac{x}{\sqrt{x^2 + 1}}$

$y = \dfrac{x^3 + 2x^2 - 15x}{x^2 - 5x - 14}$

67.   **69.** tw

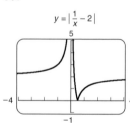

$y = \left| \dfrac{1}{x} - 2 \right|$

Technology Connection, p. 238

1. Minimum = -8 at $x = -2$, maximum = 2.18 at $x = -0.3$; minimum = 1 at $x = -1$ and $x = 1$, maximum = 4 at $x = 2$

Technology Connection, p. 240

1. Minimum $= -4$ at $x = 2$; no absolute maximum

Technology Connection, p. 244

1. Minimum $= 6.32$ at $x = 0.32$; no absolute maximum

Exercise Set 3.4, p. 245

1. **(a)** 41 mph; **(b)** 80 mph; **(c)** 13.5 mpg;
(d) 16.5 mpg; **(e)** about 22%
3. Maximum $= 5\frac{1}{4}$ at $x = \frac{1}{2}$; minimum $= 3$ at $x = 2$
5. Maximum $= 4$ at $x = 2$; minimum $= 1$ at $x = 1$
7. Maximum $= \frac{59}{27}$ at $x = -\frac{1}{3}$; minimum $= 1$ at $x = -1$
9. Maximum $= 1$ at $x = 1$; minimum $= -5$ at $x = -1$
11. Maximum $= 15$ at $x = -2$; minimum $= -13$ at $x = 5$
13. Maximum $=$ minimum $= -5$ for all x in $[-1, 1]$
15. Maximum $= 4$ at $x = -1$; minimum $= -12$ at $x = 3$
17. Maximum $= 3\frac{1}{5}$ at $x = -\frac{1}{5}$; minimum $= -48$ at $x = 3$
19. Minimum $= -4$ at $x = 2$; maximum $= 50$ at $x = 5$
21. Minimum $= -110$ at $x = -5$; maximum $= 2$ at $x = -1$
23. Maximum $= 513$ at $x = -8$; minimum $= -511$ at $x = 8$
25. Maximum $= 17$ at $x = 1$; minimum $= -15$ at $x = -3$
27. Maximum $= 32$ at $x = -2$; minimum $= -\frac{27}{16}$ at $x = \frac{3}{2}$
29. Maximum $= 13$ at $x = -2$ and $x = 2$; minimum $= 4$ at $x = -1$ and $x = 1$
31. Minimum $= -5$ at $x = -3$; maximum $= -1$ at $x = 5$
33. Minimum $= 2$ at $x = 1$; maximum $= 20\frac{1}{20}$ at $x = 20$
35. Maximum $= \frac{4}{5}$ at $x = -2$ and $x = 2$; minimum $= 0$ at $x = 0$
37. Minimum $= -1$ at $x = -2$; maximum $= 3$ at $x = 26$
39.–47. Left to the student.
49. Maximum $= 1225$ at $x = 35$
51. Minimum $= 200$ at $x = 10$
53. Maximum $= \frac{1}{3}$ at $x = \frac{1}{2}$ **55.** Maximum $= \frac{289}{4}$ at $x = \frac{17}{2}$
57. Maximum $= 2\sqrt{3}$ at $x = -\sqrt{3}$; minimum $= -2\sqrt{3}$ at $x = \sqrt{3}$
59. Maximum $= 5700$ at $x = 2400$
61. Minimum $= -55\frac{1}{3}$ at $x = 1$
63. Maximum $= 2000$ at $x = 20$; minimum $= 0$ at $x = 0$ and $x = 30$
65. Minimum $= 24$ at $x = 6$
67. Minimum $= 108$ at $x = 6$

69. Maximum $= 3$ at $x = -1$; minimum $= -\frac{3}{8}$ at $x = \frac{1}{2}$
71. Maximum $= 2$ at $x = 8$; minimum $= 0$ at $x = 0$
73. None
75. Maximum $= -1$ at $x = 1$; minimum $= -5$ at $x = -1$
77. None
79. Minimum $= 0$ at $x = 0$; maximum $= 1$ at $x = -1$ and $x = 1$ **81.** None
83. Maximum $= -\frac{10}{3} + 2\sqrt{3}$ at $x = 2 - \sqrt{3}$; minimum $= -\frac{10}{3} - 2\sqrt{3}$ at $x = 2 + \sqrt{3}$
85. Minimum $= -1$ at $x = -1$ and $x = 1$
87.–95. Left to the student.
97. Maximum $= 1430$ units per month at $t = 25$ years of service
99. Maximum $= \$1500$ at $x = 3$ **101.** 61.64 mph
103. Maximum $= 3\sqrt{6}$ at $x = 3$; minimum $= -2$ at $x = -2$
105. 7   **107.** tw
109. Minimum $= -17$ at $x = 3$; maximum $= 10$ at $x = 0$ and $x = 4$
111. Minimum $= 0$ at $x = -1$ and $x = 1$
113. **(a)** $y = x + 8.857$; 15.857 millimeters;
(b) $y = 0.117x^4 - 1.520x^3 + 6.193x^2 - 7.018x + 10$; 23.888 mm, 7.62 mm

Technology Connection, p. 250

1. 20, 0; 16, 64; 13.5, 87.75; 12, 96; 10, 100; 8, 96; 6.8, 89.76; 0, 0   **2.** Left to the student.   **3.** 100 at $x = 10$

Technology Connection, p. 252

1. 4, 0; 3.5, 3.5; 3, 12; 2.5, 22.5; 2, 32; 1.7, 35.972; 1.5, 37.5; 1, 36; 0.6, 27.744; 0.5, 24.5; 0, 0   **2.** Left to the student.   **3.** 38 at $x = 5\frac{1}{3}$

Technology Connection, p. 259

1. Left to the student.   **2.** $23,500 at $x = 100$

Exercise Set 3.5, p. 261

1. 25 and 25; maximum product $= 625$ **3.** No; $Q = x(50 - x)$ has no minimum   **5.** 2 and -2; minimum product $= -4$ **7.** $x = \frac{1}{2}$, $y = \sqrt{\frac{1}{2}}$; maximum $= \frac{1}{4}$ **9.** $x = 10$, $y = 10$; minimum $= 200$
11. $x = 2$, $y = \frac{32}{3}$; maximum $= \frac{64}{3}$ **13.** 30 yd by 60 yd; maximum area $= 1800$ yd^2 **15.** 13.5 ft by 13.5 ft; 182.25 ft^2 **17.** 20 in. by 20 in. by 5 in.; maximum $= 2000$ in^3 **19.** 5 in. by 5 in. by 2.5 in.; minimum $= 75$ in^2 **21.** 46 units; maximum profit $= \$1048$ **23.** 70 units; maximum profit $= \$19$
25. Approximately 1667 units; maximum profit $\approx \$5481$
27. **(a)** $R(x) = 150x - 0.5x^2$;

(b) $P(x) = -0.75x^2 + 150x - 4000$;   **(c)** 100;   **(d)** \$3500;
(e) \$100   **29.** \$5.75, 72,500 (Will the stadium hold that
many?)   **31.** 25   **33.** 4 ft by 4 ft by 20 ft   **35.** 9%
37. Reorder 5 times per year; lot size = 20
39. Reorder 12 times per year; lot size = 30
41. Reorder about 13 times per year; lot size = 28

43. 14 in. by 14 in. by 28 in.   **45.** $x = y = \dfrac{24}{4 + \pi}$ ft

47. $\sqrt[3]{\dfrac{1}{10}}$ **49.** Reorder $\dfrac{Q}{\sqrt{2bQ/a}}$ times per year; lot

size $= \sqrt{2bQ/a}$ **51.** Minimum at $x = \dfrac{24\pi}{\pi + 4} \approx 10.56$ in.,

$24 - x = \dfrac{96}{\pi + 4} \approx 13.45$ in. There is no maximum if the
string is to be cut. One would interpret the maximum to
be at the endpoint, with the string uncut and used to form
a circle.   **53.** S should be about 4.25 miles downshore

from A.   **55. (a)** $C'(x) = 8 + \dfrac{3x^2}{100}$;

(b) $A(x) = 8 + \dfrac{20}{x} + \dfrac{x^2}{100}$;   **(c)** $A'(x) = \dfrac{x}{50} - \dfrac{20}{x^2}$;

(d) minimum = 11 at $x_0 = 10$, $C'(10) = 11$;
(e) $A(10) = 11$, $C'(10) = 11$ **57.** Minimum $= 6 - 4\sqrt{2}$
at $x = 2 - \sqrt{2}$ and $y = -1 + \sqrt{2}$ **59.** tw

Exercise Set 3.6, p. 271

1. $\Delta y = 0.0401$; $f'(x)\,\Delta x = 0.04$ **3.** 0.2816, 0.28
5. -0.556, -1 **7.** 6, 6   **9.** $\Delta C = \$2.01$, $C'(70) = \$2$
11. $\Delta R = \$2$, $R'(70) = \$2$
13. (a) $P(x) = -0.01x^2 + 1.4x - 30$;   **(b)** $\Delta P = -\$0.01$,
$P'(70) = 0$ **15.** 4.375   **17.** 10.1   **19.** 2.167
21. $dy = 9x^2(2x^3 + 1)^{1/2}\,dx$ **23.** $dy = \frac{1}{5}(x + 27)^{-4/5}\,dx$
25. $dy = (4x^3 - 6x^2 + 10x + 3)\,dx$ **27.** $dy = 3.1$

29. -0.01 **31.** 100   **33.** 2.512 cm^3 **35.** $\dfrac{5}{\pi} \approx 1.6$ ft

37. tw

Exercise Set 3.7, p. 277

1. $\dfrac{1 - y}{x + 2}$; $-\dfrac{1}{9}$ **3.** $-\dfrac{x}{y}$; $-\dfrac{1}{\sqrt{3}}$ **5.** $\dfrac{6x^2 - 2xy}{x^2 - 3y^2}$; $-\dfrac{36}{23}$

7. $-\dfrac{y}{x}$ **9.** $\dfrac{x}{y}$ **11.** $\dfrac{3x^2}{5y^4}$ **13.** $\dfrac{-3xy^2 - 2y}{3x + 4x^2y}$

15. $\dfrac{dp}{dx} = \dfrac{-2}{2p + 1}$ **17.** $\dfrac{dp}{dx} = -\dfrac{p + 4}{x + 3}$ **19.** $-\dfrac{3}{4}$

21. \$400 per day, \$80 per day, \$320 per day   **23.** \$16 per
day, \$8 per day, \$8 per day   **25.** 0.1728π cm^3/day $\approx$

0.54 cm^3/day   **27. (a)** $1000R \cdot \dfrac{dR}{dt}$;

(b) -0.01125 mm^3/min   **29.** 65 mph   **31.** $-\dfrac{\sqrt{y}}{\sqrt{x}}$

33. $\dfrac{2}{3y^2(x + 1)^2}$ **35.** $-\dfrac{9}{4}\sqrt[3]{y}\sqrt{x}$ **37.** $\dfrac{dy}{dx} = \dfrac{1 + y}{2 - x}$,

$\dfrac{d^2y}{dx^2} = \dfrac{2 + 2y}{(2 - x)^2}$ **39.** $\dfrac{dy}{dx} = \dfrac{x}{y}$, $\dfrac{d^2y}{dx^2} = \dfrac{y^2 - x^2}{y^3}$

41. tw

43.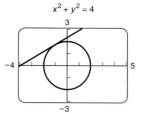

$x^2 + y^2 = 4$

45.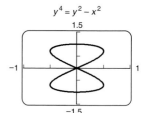

$y^4 = y^2 - x^2$

47.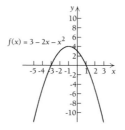

$y^2 = x^3$

Summary and Review: Chapter 3, p. 279

1. [3.1], [3.2] Relative maximum at $(-1, 4)$

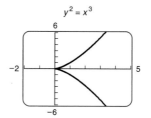

$f(x) = 3 - 2x - x^2$

2. [3.1], [3.2] Relative maximum at $(0, 3)$; relative
minima at $(-1, 2)$ and $(1, 2)$

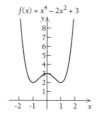

$f(x) = x^4 - 2x^2 + 3$

3. [3.1], [3.2] Relative maximum at $(-1, 4)$; relative minimum at $(1, -4)$

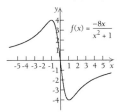

4. [3.1], [3.2] None

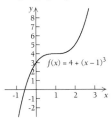

5. [3.1], [3.2] Relative maximum at $(-1, 4)$; relative minimum at $\left(\frac{1}{3}, \frac{76}{27}\right)$

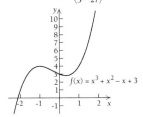

6. [3.1], [3.2] Relative minimum at $(0, 0)$

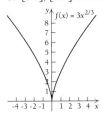

7. [3.1], [3.2] Relative maximum at $(-1, 17)$; relative minimum at $(2, -10)$

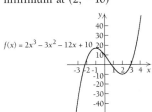

8. [3.1], [3.2] Relative maximum at $(-1, 4)$; relative minimum at $(1, 0)$

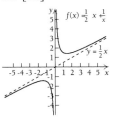

9. [3.3]

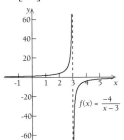

10. [3.3]

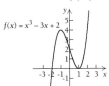

11. [3.3]

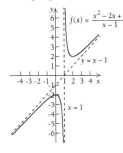

12. [3.3]

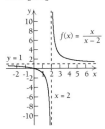

13. [3.4] None   **14.** [3.4] Maximum $= 32$ at $x = -2$; minimum $= -17$ at $x = 5$ **15.** [3.4] Maximum $= 19$ at $x = -1$; minimum $= -35$ at $x = 2$
16. [3.4] Maximum $= -2$ at $x = 1$; minimum $= -12$ at $x = -1$ **17.** [3.4] Minimum $= 3$ at $x = -1$

18. [3.4] None   **19.** [3.4] Maximum $= 163$ at $x = -3$; minimum $= -1$ at $x = -1$ **20.** [3.4] Minimum $= 10$ at $x = 1$ **21.** [3.4] Maximum $= 256$ at $x = 4$; minimum $= 0$ at $x = 0$ and $x = 5$
22. [3.4] Maximum $= \frac{53}{4}$ at $x = \frac{5}{2}$ **23.** [3.5] 30, 30
24. [3.5] $Q = -1$ at $x = -1$, $y = -1$ **25.** [3.5] 30 units; $451 profit   **26.** [3.5] 10 ft $\times$ 10 ft $\times$ 25 ft; $1500
27. [3.5] 12 times; 30 bicycles   **28.** [3.6] $\Delta y = -10.875$; $f'(x)\,\Delta x = -13$ **29.** [3.6] **(a)** $dy = (3x^2 - 1)\,dx$;

(b) $dy = 0.11$ **30.** [3.6] 8.3125   **31.** [3.7] $\dfrac{2x^2 + 3y}{-2y^2 - 3x}$;

$\dfrac{4}{5}$ **32.** [3.7] $-1\frac{3}{4}$ ft/sec   **33.** [3.7] $600 per day, $450 per day, $150 per day   **34.** [3.4] Minimum $= 0$ at $x = 3$

35. [3.7] $\dfrac{3x^5 - 4x^3 - 12xy^2}{12x^2y + 4y^3 - 3y^5}$ **36.** [3.1], [3.2] Relative minimum at $(-9, -9477)$; relative maximum at $(0, 0)$; relative minimum at $(15, -37{,}125)$ **37.** [3.1], [3.2] Relative maximum at $(-1.714, 37.445)$; relative minimum at $(1.714, -37.445)$ **38.** [3.1], [3.2] Relative minima at $(-3, -1)$ and $(3, -1)$; relative maximum at $(0, 1.08)$
39. (a) Linear: $y = 6.998187602x - 124.6183581$; quadratic:
$y = 0.0439274846x^2 + 2.881202838x - 53.51475166$;
cubic: $y = -0.0033441547x^3 + 0.4795643605x^2$
$\qquad - 11.35931622x + 5.276985809$;
quartic: $y = -0.00005539834x^4 + 0.0067192294x^3$
$\qquad - 0.0996735857x^2 - 0.8409991942x$
$\qquad - 0.246072967$;
(b) tw ;   **(c)** tw ;   **(d)** absolute maximum at $(78.650443, 456.75983)$; the maximum value of about 457 per 100,000 women occurs at approximately age 79.

Test: Chapter 3, p. 281

1. [3.1], [3.2] Relative minimum at $(2, -9)$

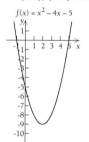

2. [3.1], [3.2] Relative minima at $(-1, -1)$ and $(1, -1)$; relative maximum at $(0, 1)$

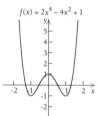

3. [3.1], [3.2] Relative minimum at $(2, -4)$

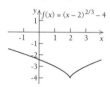

4. [3.1], [3.2] Relative maximum at $(0, 4)$

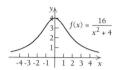

5. [3.1], [3.2] Relative maximum at $(-1, 2)$; relative minimum at $\left(\dfrac{1}{3}, \dfrac{22}{27}\right)$

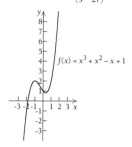

6. [3.1], [3.2] Relative minimum at $(-1, 2)$; relative maximum at $(1, 6)$

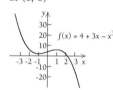

7. [3.1], [3.2] None

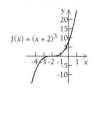

8. [3.1], [3.2] Relative minimum at $\left(-\dfrac{3}{\sqrt{2}}, -\dfrac{9}{2}\right)$; relative maximum at $\left(\dfrac{3}{\sqrt{2}}, \dfrac{9}{2}\right)$

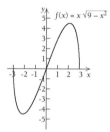

9. [3.3]

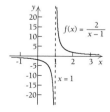

10. [3.3]

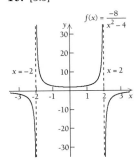

11. [3.3]

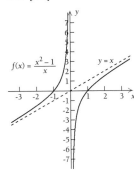

12. [3.3]

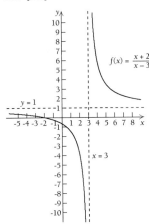

13. [3.4] Maximum $= 9$ at $x = 3$ **14.** [3.4] Maximum $= 2$
at $x = -1$; minimum $= -1$ at $x = -2$
15. [3.4] Maximum $= 28.49$ at $x = 4.3$
16. [3.4] Maximum $= 7$ at $x = -1$; minimum $= 3$ at
$x = 1$ **17.** [3.4] None   **18.** [3.4] Minimum $= -\frac{13}{12}$ at
$x = \frac{1}{6}$ **19.** [3.4] Minimum $= 48$ at $x = 4$
20. [3.5] 4 and -4 **21.** [3.5] $x = 5$, $y = -5$;
minimum $= 50$ **22.** [3.5] Maximum profit $= \$24,980$;
500 units   **23.** [3.5] 40 in. by 40 in. by 10 in.;
maximum volume $= 16,000$ in^3 **24.** [3.5] 35 times at
lot size 35   **25.** [3.6] $\Delta y = 1.01$, $f'(x)\,\Delta x = 1$

26. [3.6] 10.2   **27.** [3.7] **(a)** $\dfrac{x\,dx}{\sqrt{x^2+3}}$;   **(b)** 0.0075593

28. [3.7] $-\dfrac{x^2}{y^2}$, $-\dfrac{1}{4}$ **29.** [3.7] -0.96 ft/sec

30. [3.4] Maximum $= \frac{1}{3}\cdot 2^{2/3}$ at $x = \sqrt[3]{2}$;
minimum $= 0$ at $x = 0$

31. [3.5] **(a)** $A(x) = \dfrac{C(x)}{x} = 100 + \dfrac{100}{\sqrt{x}} + \dfrac{\sqrt{x}}{100}$;

(b) minimum $= 102$ at $x = 10,000$
32. [3.1], [3.2] Relative maximum at $(1.09, 25.1)$; relative
minimum at $(2.97, 8.6)$

33. (a) Linear: $y = -0.7707142857x + 12691.60714$,
quadratic:
$y = -0.9998904762x^2 + 299.1964286x + 192.9761905$,
cubic: $y = 0.000084x^3 - 1.037690476x^2 + 303.3964286x$
$\qquad + 129.9761905$,
quartic: $y = -0.000001966061x^4 + 0.0012636364x^3$
$\qquad\qquad - 1.256063636x^2 + 315.8247403x$
$\qquad\qquad + 66.78138528$;
(b) tw ;   **(c)** tw ;   **(d)** 22,450; \$149,000

Chapter 4

Technology Connection, p. 287
1. 156.993   **2.** 16.242   **3.** 0.064   **4.** 0.000114

Exercise Set 4.1, p. 298
1.

3.

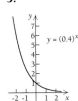

5.

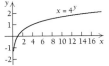

7. $3e^{3x}$
9. $-10e^{-2x}$

11. e^{-x} **13.** $-7e^x$ **15.** e^{2x} **17.** $x^3 e^x(x+4)$
19. $\dfrac{e^x(x-4)}{x^5}$ **21.** $(-2x+7)e^{-x^2+7x}$ **23.** $-xe^{-x^2/2}$
25. $\dfrac{e^{\sqrt{x-7}}}{2\sqrt{x-7}}$ **27.** $\dfrac{e^x}{2\sqrt{e^x-1}}$
29. $(1-2x)e^{-2x} - e^{-x} + 3x^2$ **31.** e^{-x} **33.** ke^{-kx}

35.

37.

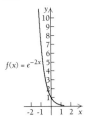

39.

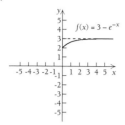

41.–45. Left to the student.
47. $y = x + 1$ **49.** Left to the student.
51. (a) $C'(t) = 50e^{-t}$;   **(b)** \$50 million;
(c) \$0.916 million;   **(d)** tw
53. (a) Approximately 335, 280, 173;
(b)

(c) $D'(p) = -1.44e^{-0.003p}$;   **(d)** tw
55. (a) 0 ppm, 3.7 ppm, 5.4 ppm, 4.48 ppm, 0.05 ppm;
(b)

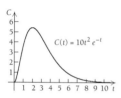

(c) $C'(t) = 10te^{-t}(2 - t)$ **(d)** $C \approx 5.4$ ppm at $t = 2$ hr;
(e) tw   **57.** $15e^{3x}(e^{3x} + 1)^4$ **59.** $-e^{-t} - 3e^{3t}$
61. $\dfrac{e^x(x - 1)^2}{(x^2 + 1)^2}$ **63.** $\dfrac{e^{\sqrt{x}}}{2\sqrt{x}} + \dfrac{1}{2}\sqrt{e^x}$
65. $\dfrac{1}{2}e^{x/2}\left[\dfrac{x}{\sqrt{x - 1}}\right]$ **67.** $\dfrac{4}{(e^x + e^{-x})^2}$
69. 2, 2.25, 2.48832, 2.59374, 2.71692
71. Maximum $= 4e^{-2} \approx 0.54$ at $x = 2$ **73.** tw
75. Relative minimum at $(0, 0)$; relative maximum at $(2, 0.54)$

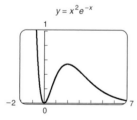

77. See Example 9. The graphs of $f(x)$, $f'(x)$, and $f''(x)$ are all the graph of $y = e^x$.

79.

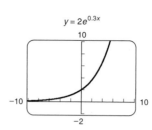

81.

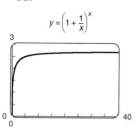

Technology Connection, p. 301
1.–4. Left to the student.

Technology Connection, p. 304
1. 1000, 5, 0.699, 3

Technology Connection, p. 307
1. 6.9   **2.** -4.1 **3.** 74.9   **4.** 46.2   **5.** 38.7

Exercise Set 4.2, p. 314
1. $2^3 = 8$ **3.** $8^{1/3} = 2$ **5.** $a^J = K$ **7.** $b^v = T$
9. $\log_e b = M$ **11.** $\log_{10} 100 = 2$ **13.** $\log_{10} 0.1 = -1$
15. $\log_M V = p$ **17.** 2.708   **19.** -1.609 **21.** 2.609
23. 2.9957   **25.** -1.3863 **27.** 4   **29.** 8.681690
31. -4.006334 **33.** 8.999619   **35.** $t \approx 4.6$

37. $t \approx 4.1$ **39.** $t \approx 2.3$ **41.** $t \approx 141$ **43.** $-\dfrac{6}{x}$
45. $x^3(1 + 4 \ln x) - x$ **47.** $\dfrac{1 - 4 \ln x}{x^5}$ **49.** $\dfrac{1}{x}$
51. $\dfrac{10x}{5x^2 - 7}$ **53.** $\dfrac{1}{x \ln 4x}$ **55.** $\dfrac{x^2 + 7}{x(x^2 - 7)}$
57. $e^x\left(\dfrac{1}{x} + \ln x\right)$ **59.** $\dfrac{e^x}{e^x + 1}$ **61.** $\dfrac{2 \ln x}{x}$
63. (a) 1000;   **(b)** $N'(a) = \dfrac{200}{a}$, $N'(10) = 20$;
(c) minimum $= 1000$ at $a = 1$;   **(d)** tw   **65.** 58 days
67. (a) \$58.69, \$78.00;   **(b)** $63.80e^{-1.1t}$;   **(c)** 2.7;
(d) tw   **69. (a)** 18.1%, 69.9%;   **(b)** $P'(t) = 20e^{-0.2t}$;
(c) 11.5;   **(d)** tw   **71. (a)** 68%;   **(b)** 36%;   **(c)** 3.6%;
(d) 5%;   **(e)** $-\dfrac{20}{t + 1}$;   **(f)** maximum $= 68\%$ at $t = 0$;
(g) tw   **73. (a)** 2.37 ft/sec;   **(b)** 3.37 ft/sec;
(c) $v'(p) = \dfrac{0.37}{p}$;   **(d)** tw   **75.** $\dfrac{-4(\ln x)^{-5}}{x}$
77. $\dfrac{15t^2}{t^3 + 1}$ **79.** $\dfrac{4[\ln (x + 5)]^3}{x + 5}$
81. $\dfrac{5t^4 - 3t^2 + 6t}{(t^3 + 3)(t^2 - 1)}$

83. $\dfrac{24x + 25}{8x^2 + 5x}$ **85.** $\dfrac{2(1 - \ln t^2)}{t^3}$

87. $x^n \ln x$ **89.** $\dfrac{1}{\sqrt{1 + t^2}}$ **91.** $\dfrac{3}{x \ln x}$

93. 1   **95.** $t = \dfrac{\ln P - \ln P_0}{k}$ **97.** Let $a = \ln x$; then

$e^a = x$, so $\log x = \log e^a = a \log e$. Then

$a = \ln x = \dfrac{\log x}{\log e} \approx \dfrac{\log x}{0.4343} \approx 2.3026 \log x$.   **99.** **tw**

101. $\sqrt[e]{e} \approx 1.444667861$; $\sqrt[e]{e} > \sqrt[x]{x}$ for any $x > 0$ such that $x \ne e$ **103.** ∞

105.

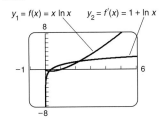

$y_1 = f(x) = x \ln x$ $y_2 = f'(x) = 1 + \ln x$

107.

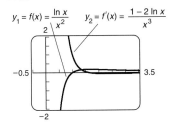

$y_1 = f(x) = \dfrac{\ln x}{x^2}$ $y_2 = f'(x) = \dfrac{1 - 2 \ln x}{x^3}$

109. $-\dfrac{1}{2e} \approx -0.184$

Technology Connection, p. 323

1. 6.5322 billion   **2.** 7.0644 billion   **3.** 8.2626 billion
4. 13.2204 billion
5. $y = 6172.367867(1.059990837)^x$, $P(t) = 6172.4e^{0.058t}$,
5.8%   **6.** \$11,716; \$26,485; \$113,647

Exercise Set 4.3, p. 327

1. $Q(t) = Q_0 e^{kt}$ **3. (a)** $N(t) = 50e^{0.1t}$;   **(b)** 369;
(c) 6.9 yr   **5. (a)** $P(t) = P_0 e^{0.065t}$;   **(b)** \$1067.16,
\$1138.83;   **(c)** 10.7 yr   **7.** 6.9%   **9.** 6.9 yr (2006)
11. 11.2 yr; \$102,256.88   **13.** \$7500; 8.3 yr
15 (a) $k = 0.161602$; $V(t) = \$84,000e^{0.161602t}$;
(b) \$271,283,000;   **(c)** 4.3 yr;   **(d)** 58 yr
17. (a) $P(t) = 0.386e^{0.198t}$;   **(b)** \$0.852 billion, \$1.039
billion;   **(c)** 8.31 yr;   **(d)** 3.5 yr;   **(e)** **tw**
19. (a) $y = 0.1097854248(1.442743433)^x$,
$P(t) = 0.1e^{0.367t}$;   **(b)** 17.0 million, 154.1 million;
(c) 20.7 yr (since 1990);   **(d)** 1.9 yr
21. Approximately \$237,000,000,000,000

23. $k = 0.154876$, or 15.4876%; approximately \$3,053,274;
approximately \$10,540,475   **25. (a)** 2%;   **(b)** 3.8%,
7.0%, 21.6%, 50.2%, 93.1%, 98.03%;
(c) $P'(t) = \dfrac{6.37e^{-0.13t}}{(1 + 49e^{-0.13t})^2}$;
(d)

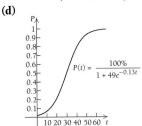

$P(t) = \dfrac{100\%}{1 + 49e^{-0.13t}}$

27. 19.8 yr   **29.** 10%   **31.** 24.8 yr
33. (a) $P(t) = 241e^{0.009t}$;   **(b)** 273 million;   **(c)** 77 yr
35. 0.20%   **37. (a)** Approximately 2.2%;   **(b)** **tw**
39. (a) 400, 520, 1214, 2059, 2396, 2478;
(b) $P'(t) = \dfrac{4200e^{-0.32t}}{(1 + 5.25e^{-0.32t})^2}$;
(c)

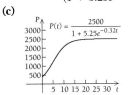

$P(t) = \dfrac{2500}{1 + 5.25e^{-0.32t}}$

41. (a) 0%, 33.0%, 55.1%, 69.9%, 86.5%, 99.2%, 99.8%;
(b) $P'(t) = 0.4e^{-0.4t}$;
(c)

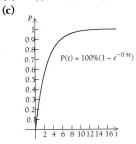

$P(t) = 100\%(1 - e^{-0.4t})$

43. 7.573%   **45.** 9%   **47.** $T_3 = \dfrac{\ln 3}{k}$ **49.** Answers
depend on particular data.   **51.** $\dfrac{\ln 2}{24} \approx 2.9\%$ per hour
53. **tw**   **55.** **tw**

Exercise Set 4.4, p. 340

1. \$15,059.71   **3.** \$35,957.75
5. 6 stereos at \$403.73 each   **7.** \$27,194.82
9. (a) $y = (34001.78697)(0.6702977719)^x$,
$V(t) = 34001.78697e^{-0.4t}$;   **(b)** \$2067, \$623;   **(c)** 8.8 yr;
(d) 1.7 yr;   **(e)** **tw**   **11.** (f)   **13.** (b)   **15.** (c)

17. 3.2% per year   **19.** 2.8% per year
21. 0.003% per year   **23.** 9.9 g   **25.** 19,109 yr
27. 7604 yr   **29. (a)** $A = A_0 e^{-kt}$;   **(b)** 9 hr
31. (a) 0.8%;   **(b)** 78.7% W_0
33. (a) 25%I_0, 6%I_0, 1.5%I_0;   **(b)** 0.00008%I_0
35. (a) 25°;   **(b)** $k = 0.05$;   **(c)** 84.2°;
(d) 32 min;   **(e)** tw   **37.** 7 P.M.
39. (a) $k = 0.0162649$, $P(t) = 453,000 e^{-0.0162649t}$, where
t is the number of years since 1970;   **(b)** 244,161;
(c) 801 yr, 173 million
41. (a) 11 watts;   **(b)** 173 days;   **(c)** 402 days;
(d) 50 watts;   **(e)** tw
43. (a) $y = 84.94353992 - 0.5412834098 \ln x$;
(b) 83.8%, 83.7%, 83.2%, 83.0%;   **(c)** 230 months;
(d) tw   **45.** ($166.16, 292)   **47.** tw

Exercise Set 4.5, p. 348

1. $e^{6.4378}$ **3.** $e^{12.238}$ **5.** $e^{k \cdot \ln 4}$ **7.** $e^{kT \cdot \ln 8}$
9. $(\ln 6)6^x$ **11.** $(\ln 10)10^x$ **13.** $(6.2)^x[x \ln 6.2 + 1]$
15. $10^x x^2[x \ln 10 + 3]$ **17.** $\dfrac{1}{\ln 4} \cdot \dfrac{1}{x}$
19. $\dfrac{2}{\ln 10} \cdot \dfrac{1}{x}$ **21.** $\dfrac{1}{\ln 10} \cdot \dfrac{1}{x}$
23. $x^2 \left[\dfrac{1}{\ln 8} + 3 \log_8 x \right]$ **25. (a)** $-250,000(\ln 4)(\frac{1}{4})^t$;
(b) tw   **27.** 5   **29. (a)** $I = 10^7 \cdot I_0$;   **(b)** $I = 10^8 \cdot I_0$;
(c) the intensity in (b) is 10 times that in part (a);
(d) $\dfrac{dI}{dR} = (I_0 \cdot \ln 10) \cdot 10^R$;   **(e)** tw
31. (a) $\dfrac{1}{\ln 10} \cdot \dfrac{1}{I}$;   **(b)** tw   **33. (a)** $\dfrac{m}{\ln 10} \cdot \dfrac{1}{x}$;
(b) tw   **35.** $2(\ln 3)3^{2x}$ **37.** $(\ln x + 1)x^x$
39. $x^{e^x} e^x \left(\ln x + \dfrac{1}{x} \right)$ **41.** $\dfrac{1}{\ln a} \cdot \dfrac{f'(x)}{f(x)}$ **43.** tw

Technology Connection, p. 352

1. $E(p) = \dfrac{p}{300 - p}$, $R(p) = p(300 - p)$ **2.** Left to the
student.   **3.** $150; maximum revenue $22,500

Exercise Set 4.6, p. 354

1. (a) $E(p) = \dfrac{p}{400 - p}$;   **(b)** $E(125) = \dfrac{5}{11}$, inelastic;
(c) $p = $200   **3. (a)** $E(p) = \dfrac{p}{50 - p}$;   **(b)** $E(46) = \dfrac{23}{2}$,
elastic;   **(c)** $p = $25   **5. (a)** $E(p) = 1$, for all $p > 0$;
(b) $E(50) = 1$, unit elasticity;   **(c)** Total revenue $=$
$R(p) = 400$, for all $p > 0$. It has 400 as a maximum for

all $p > 0$.   **7. (a)** $E(p) = \dfrac{p}{1000 - 2p}$;   **(b)** $E(400) = 2$,
elastic;   **(c)** $p = \$\dfrac{1000}{3} \approx \$333.33$ **9. (a)** $E(p) = \dfrac{p}{4}$;
(b) $E(10) = \dfrac{5}{2}$, elastic;   **(c)** $p = \$4$
11. (a) $E(p) = \dfrac{2p}{p + 3}$;   **(b)** $E(1) = \dfrac{1}{2}$, inelastic;
(c) $p = \$3$ **13. (a)** $E(p) = \dfrac{25p}{967 - 25p}$;
(b) $p = 19.34¢$;   **(c)** $p > 19.34¢$;   **(d)** $p < 19.34¢$;
(e) $p = 19.34¢$;   **(f)** decrease
15. (a) $E(p) = \dfrac{3p^3}{2(200 - p^3)}$;   **(b)** $\dfrac{81}{346} \approx 0.234$;
(c) increases it   **17. (a)** $E(p) = n$;
(b) no, E is a constant n;   **(c)** only when $n = 1$
19. $E(p) = -pL'(p)$ **21.** tw

Summary and Review: Chapter 4, p. 356

1. [4.2] $\dfrac{1}{x}$ **2.** [4.1] e^x **3.** [4.2] $\dfrac{4x^3}{x^4 + 5}$
4. [4.1] $\dfrac{e^{2\sqrt{x}}}{\sqrt{x}}$ **5.** [4.2] $\dfrac{6}{x}$ **6.** [4.1] $4e^{4x} + 4x^3$
7. [4.2] $\dfrac{1 - 3 \ln x}{x^4}$ **8.** [4.1], [4.2] $e^{x^2} \left(\dfrac{1}{x} + 2x \ln 4x \right)$
9. [4.1], [4.2] $4e^{4x} - \dfrac{1}{x}$ **10.** [4.2] $8x^7 - \dfrac{8}{x}$
11. [4.1], [4.2] $\dfrac{1 - x}{e^x}$ **12.** [4.2] 6.9300
13. [4.2] -3.2698 **14.** [4.2] 8.7601   **15.** [4.2] 3.2698
16. [4.2] 2.54995   **17.** [4.2] -3.6602
18. [4.3] $Q(t) = Q_0 e^{kt}$ **19.** [4.3] 4.3%
20. [4.3] 10.2 yr   **21.** [4.3] **(a)** $k = 0.050991$,
$C(t) = \$4.65 e^{0.050991t}$;   **(b)** $32.28, $46.13
22. [4.3] **(a)** $N(t) = 60 e^{0.12t}$;   **(b)** 123;   **(c)** 5.8 yr
23. [4.4] 5.3 yr   **24.** [4.4] 18.2% per day
25. [4.4] **(a)** $A(t) = 800 e^{-0.07t}$;   **(b)** 197 g;   **(c)** 9.9 days
26. [4.3] **(a)** 0.5034, 0.7534, 0.9698, 0.9991, 0.9999;
(b) $p'(t) = 0.7 e^{-0.7t}$;   **(c)** tw ;
(d)

27. [4.4] $34,735.26   **28.** [4.5] $(\ln 3)3^x$

29. [4.5] $\dfrac{1}{\ln 15} \cdot \dfrac{1}{x}$ **30.** [4.6] **(a)** $E(p) = \dfrac{2p}{p+4}$;

(b) $E(\$1) = \frac{2}{5}$, inelastic;   **(c)** $E(\$12) = \frac{3}{2}$, elastic;

(d) decrease;   **(e)** $p = \$4$ **31.** [4.1] $\dfrac{-8}{(e^{2x} - e^{-2x})^2}$

32. [4.2] $-\dfrac{1}{1024e}$

33. [4.1]   **34.** [4.1] 0

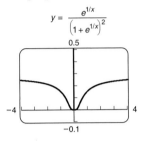

$$y = \dfrac{e^{1/x}}{\left(1 + e^{1/x}\right)^2}$$

35. [4.3] **(a)** $y = (0.979050351)(2.120505653)^x$,
$P(t) = 0.979e^{0.752x}$;   **(b)** \$36,562 billion,
\$67,441,854 billion;   **(c)** 4.9 yr;   **(d)** 0.9 yr

Test: Chapter 4, p. 358

1. [4.1] e^x **2.** [4.2] $\dfrac{1}{x}$ **3.** [4.1] $-2xe^{-x^2}$

4. [4.2] $\dfrac{1}{x}$ **5.** [4.1] $e^x - 15x^2$

6. [4.1], [4.2] $3e^x\left(\dfrac{1}{x} + \ln x\right)$

7. [4.1], [4.2] $\dfrac{e^x - 3x^2}{e^x - x^3}$

8. [4.1], [4.2] $\dfrac{\frac{1}{x} - \ln x}{e^x}$, or $\dfrac{1 - x \ln x}{xe^x}$

9. [4.2] 1.0674   **10.** [4.2] 0.5554   **11.** [4.2] 0.4057
12. [4.3] $M(t) = M_0 e^{kt}$ **13.** [4.3] 17.3% per hour
14. [4.3] 10 yr   **15.** [4.3] **(a)** 0.0330326;
$C(t) = \$0.54e^{0.0330326t}$;   **(b)** \$3.79, \$5.28
16. [4.3] **(a)** $A(t) = 3e^{-0.1t}$;   **(b)** 1.1 cc;   **(c)** 6.9 hr
17. [4.4] 63 days   **18.** [4.4] 0.000069% per year
19. [4.4] **(a)** 4%;   **(b)** 5.2%, 14.5%, 40.7%, 73.5%, 91.8%,
99.5%, 99.9%;   **(c)** $P'(t) = \dfrac{6.72e^{-0.28t}}{(1 + 24e^{-0.28t})^2}$;

(d) tw;   **(e)**

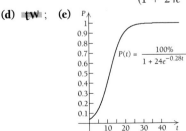

$$P(t) = \dfrac{100\%}{1 + 24e^{-0.28t}}$$

20. [4.4] \$28,503.49   **21.** [4.5] $(\ln 20)20^x$
22. [4.5] $\dfrac{1}{\ln 20} \cdot \dfrac{1}{x}$ **23.** [4.6] **(a)** $E(p) = \dfrac{p}{5}$;

(b) $E(\$3) = \frac{3}{5}$, inelastic;   **(c)** $E(\$18) = \frac{18}{5}$, elastic;
(d) increase;   **(e)** $p = \$5$ **24.** [4.2] $(\ln x)^2$
25. [4.2] Maximum $= 256/e^4 \approx 4.69$ at $x = 4$;
minimum $= 0$ at $x = 0$
26. [4.1]   **27.** [4.1] 0

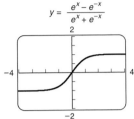

$$y = \dfrac{e^x - e^{-x}}{e^x + e^{-x}}$$

28. [4.3] **(a)** $y = 706278.2683(1.07455664)^x$,
$y = 706{,}278e^{0.072x}$;   **(b)** \$1,451,001; \$6,124,236;
(c) 101 yr;   **(d)** 9.6 yr

Chapter 5

Exercise Set 5.1, p. 371

1. $\dfrac{x^7}{7} + C$ **3.** $2x + C$ **5.** $\dfrac{4}{5}x^{5/4} + C$

7. $\dfrac{x^3}{3} + \dfrac{x^2}{2} - x + C$ **9.** $\dfrac{t^3}{3} - t^2 + 3t + C$

11. $\dfrac{5}{8}e^{8x} + C$ **13.** $\dfrac{x^4}{4} - \dfrac{7}{15}x^{15/7} + C$

15. $1000 \ln x + C$ **17.** $-x^{-1} + C$

19. $\dfrac{2}{3}x^{3/2} + C$ **21.** $-18x^{1/3} + C$ **23.** $-4e^{-2x} + C$

25. $\dfrac{1}{3}x^3 - x^{3/2} - 3x^{-1/3} + C$

27. $f(x) = \dfrac{x^2}{2} - 3x + 13$ **29.** $f(x) = \dfrac{x^3}{3} - 4x + 7$

31. $C(x) = \dfrac{x^4}{4} - x^2 + 100$ **33.** **(a)** $R(x) = \dfrac{x^3}{3} - 3x$;

(b) tw   **35.** $D(p) = \dfrac{4000}{p} + 3$

37. **(a)** $E(t) = 30t - 5t^2 + 32$;   **(b)** $E(3) = 77\%$,
$E(5) = 57\%$ **39.** $s(t) = t^3 + 4$ **41.** $v(t) = 2t^2 + 20$
43. $s(t) = -\frac{1}{3}t^3 + 3t^2 + 6t + 10$
45. $s(t) = -16t^2 + v_0 t + s_0$ **47.** $\frac{1}{4}$ mi
49. **(a)** $M(t) = -0.001t^3 + 0.1t^2$;   **(b)** about 6
51. $f(t) = \dfrac{t^{\sqrt{3}+1}}{\sqrt{3}+1} + 8$ **53.** $\dfrac{x^6}{6} - \dfrac{2}{5}x^5 + \dfrac{1}{4}x^4 + C$

55. $\frac{2}{5}t^{5/2} + 4t^{3/2} + 18\sqrt{t} + C$

57. $\frac{t^4}{4} + t^3 + \frac{3}{2}t^2 + t + C$, or $\frac{(t+1)^4}{4} + C$

59. $\frac{b}{a}e^{ax} + C$ **61.** $\frac{12}{7}x^{7/3} + C$

63. $\frac{t^3}{3} - t^2 + 4t + C$ **65.** tw

Technology Connection, p. 384

1. 0 **2.** 13.75 **3.** 0.535 **4.** 27.972 **5.** -260

Exercise Set 5.2, p. 385

1. 8 **3.** 8 **5.** $41\frac{2}{3}$ **7.** $\frac{1}{4}$ **9.** $10\frac{2}{3}$

11. $e^3 - 1 \approx 19.086$ **13.** $\ln 3 \approx 1.0986$ **15.** 0; the area above the x-axis is equal to the area below the x-axis. **17.** $-\frac{4}{15}$; the area is below the x-axis. **19.–35.** Left to the student.

37. $e^b - e^a$ **39.** $b^3 - a^3$ **41.** $\frac{e^2}{2} + \frac{1}{2}$ **43.** $\frac{2}{3}$

45. $-\frac{1275}{34}$ **47.** 4 **49.** $9\frac{5}{6}$ **51.** 12 **53.** $e^5 - \frac{1}{e}$

55. tw **57.** $3\frac{1}{2}$ **59.** $359\frac{7}{15}$ **61.** $6\frac{3}{4}$ **63.** $\frac{307}{6}$

65. $\frac{15}{4}$ **67.** 8 **69.** 12 **71.** tw **73.** 6.25

75. 529.36 **77.** 10.60 **79.** 6.28 **81.** $885.\overline{3}$

83. 1.51

Technology Connection, p. 395

1. $\frac{16}{5}$, or $3\frac{1}{5}$; although the values of $f(x)$ increase from 0 to 16 over $[0, 2]$, we would not expect the average value to be 8, because we see from the graph that $f(x)$ is less than 8 over more than half the interval.

Exercise Set 5.3, p. 396

1. (a) 1.4914; (b) 1.1418; (c) 0.8571
3. An antiderivative, velocity **5.** An antiderivative, energy used in time t **7.** An antiderivative, total revenue **9.** An antiderivative, amount of drug in blood **11.** An antiderivative, number of words memorized in time t **13.** 0 **15.** $e - 1$ **17.** $\frac{4}{3}$ **19.** 13

21. $\frac{1}{n+1}$ **23.** (a) \$2948.26; (b) \$2913.90;

(c) $k = 6.9$, so it will be on the 7th day **25.** \$6300
27. \$68,676,000 **29.** (a) 90 words per minute;
(b) 96 words per minute at $t = 1$; **(c)** 70 words per minute **31.** 90 **33.** (a) 2,253,169; (b) 21.8 days
35. (a) $\frac{95}{3} \approx 31.7°$; (b) $-10°$; (c) 46.25° **37.** 7

39. 256.8 million **41.** (a) 100 after 10 hr; (b) $33\frac{1}{3}$
43. 47.5 **45.** 5.333 **47.** 2.159

Exercise Set 5.4, p. 406

1. $\frac{1}{4}$ **3.** $\frac{9}{2}$ **5.** $\frac{125}{6}$ **7.** $\frac{9}{2}$ **9.** $\frac{1}{6}$ **11.** $\frac{125}{3}$ **13.** $\frac{32}{3}$

15. 3 **17.** 128 **19.** 9 **21.** $17\frac{1}{3}$

23. (a) B; (b) 2 **25.** $\frac{(e-1)^2}{e}$, or ≈ 1.086 **27.** $\frac{2}{3}$

29. 96 **31.** $\frac{\pi p R^4}{8Lv}$ **33.** 24.961 **35.** 16.7083

37. (a)

(b) $a = -1.8623, b = 0, c = 1.4594$; **(c)** 64.5239; **(d)** 17.683

Technology Connection, p. 412

1. $62.\overline{03}$

Exercise Set 5.5, p. 412

1. $\ln(7 + x^3) + C$ **3.** $\frac{1}{4}e^{4x} + C$ **5.** $2e^{x/2} + C$

7. $\frac{1}{4}e^{x^4} + C$ **9.** $-\frac{1}{3}e^{-t^3} + C$ **11.** $\frac{(\ln 4x)^2}{2} + C$

13. $\ln(1 + x) + C$ **15.** $-\ln(4 - x) + C$

17. $\frac{1}{24}(t^3 - 1)^8 + C$ **19.** $\frac{1}{8}(x^4 + x^3 + x^2)^8 + C$

21. $\ln(4 + e^x) + C$

23. $\frac{1}{4}(\ln x^2)^2 + C$, or $(\ln x)^2 + C$

25. $\ln(\ln x) + C$ **27.** $\frac{2}{3a}(ax + b)^{3/2} + C$

29. $\frac{b}{a}e^{ax} + C$ **31.** $\frac{-1}{4(1 + x^3)^4} + C$

33. $-\frac{21}{8}(4 - x^2)^{4/3} + C$ **35.** $e - 1$ **37.** $\frac{21}{4}$

39. $\ln 4 - \ln 2 = \ln \frac{4}{2} = \ln 2$ **41.** $\ln 19$ **43.** $1 - \frac{1}{e^b}$

45. $1 - \frac{1}{e^{mb}}$ **47.** $\frac{208}{3}$ **49.** $\frac{1640}{6561}$ **51.** $\frac{315}{8}$

53. Left to the student. **55.** (a) $P(T) = 2000T^2 + [-250{,}000(e^{-0.1T} - 1)] - 250{,}000$, or $2000T^2 - 250{,}000e^{-0.1T}$; (b) $P(10) = \$108{,}030$

57. (a) $\frac{100{,}000}{0.025}(e^{2.475} - 1) \approx 43{,}526{,}828$;

(b) $\frac{100{,}000}{0.025}(e^{2.475} - e^2) \approx 17{,}970{,}604$

59. $\frac{128}{3}$ **61.** $-\frac{5}{12}(1 - 4x^2)^{3/2} + C$ **63.** $-\frac{1}{3}e^{-x^3} + C$

65. $-e^{1/t} + C$ **67.** $-\frac{1}{3}(\ln x)^{-3} + C$

69. $\frac{2}{9}(x^3 + 1)^{3/2} + C$ **71.** $\frac{3}{4}(x^2 - 6x)^{2/3} + C$

73. $t + \dfrac{1}{t + 1} + C$ **75.** $x + 2 \ln (x + 1) + C$

77. $-\dfrac{1}{n - 1} (\ln x)^{-n+1} + C$ **79.** $\ln (e^x + e^{-x}) + C$

81. $\ln [\ln (\ln x)] + C$ **83.** $\dfrac{9(7x^2 + 9)^{n+1}}{14(n + 1)} + C$

85. tw

Technology Connection, p. 419

1. 1.9407

Exercise Set 5.6, p. 421

1. $xe^{5x} - \dfrac{1}{5}e^{5x} + C$ **3.** $\dfrac{1}{2}x^6 + C$

5. $\dfrac{x}{2}e^{2x} - \dfrac{1}{4}e^{2x} + C$ **7.** $-\dfrac{x}{2}e^{-2x} - \dfrac{1}{4}e^{-2x} + C$

9. $\dfrac{x^3}{3}\ln x - \dfrac{x^3}{9} + C$ **11.** $\dfrac{x^2}{2}\ln x^2 - \dfrac{x^2}{2} + C$, or

$x^2 \ln x - \dfrac{x^2}{2} + C$ **13.** $(x + 3) \ln (x + 3) - x + C$.

Let $u = \ln (x + 3)$, $dv = dx$, and choose $v = x + 3$ for an antiderivative of v.

15. $\left(\dfrac{x^2}{2} + 2x\right) \ln x - \dfrac{x^2}{4} - 2x + C$

17. $\left(\dfrac{x^2}{2} - x\right) \ln x - \dfrac{x^2}{4} + x + C$

19. $\frac{2}{3}x(x + 2)^{3/2} - \frac{4}{15}(x + 2)^{5/2} + C$

21. $\dfrac{x^4}{4}\ln 2x - \dfrac{x^4}{16} + C$ **23.** $x^2 e^x - 2xe^x + 2e^x + C$

25. $\frac{1}{2}x^2 e^{2x} + \frac{1}{4}e^{2x} - \frac{1}{2}xe^{2x} + C$

27. $e^{-2x}\left(-\frac{1}{2}x^3 - \frac{3}{4}x^2 - \frac{3}{4}x - \frac{3}{8}\right) + C$

29. $e^{3x}\left[\frac{1}{3}(x^4 + 1) - \frac{4}{9}x^3 + \frac{4}{9}x^2 - \frac{8}{27}x + \frac{8}{81}\right] + C$

31. $\frac{8}{3}\ln 2 - \frac{7}{9}$ **33.** $9 \ln 9 - 5 \ln 5 - 4$ **35.** 1

37. $C(x) = \frac{8}{3}x(x + 3)^{3/2} - \frac{16}{15}(x + 3)^{5/2}$

39. (a) $10[e^{-T}(-T - 1) + 1]$; **(b)** ≈ 9.084

41. $\frac{2}{9}x^{3/2}[3 \ln x - 2] + C$ **43.** $\dfrac{e^t}{t + 1} + C$

45. $2\sqrt{x} \ln x - 4\sqrt{x} + C$

47. $(13t^2 - 48)\left[\frac{5}{16}(4t + 7)^{4/5}\right] - t\left[\frac{325}{288}(4t + 7)^{9/5}\right] + \frac{1625}{16,128}(4t + 7)^{14/5} + C$ **49.** Let $u = x^n$ and $dv = e^x \, dx$. Then $du = nx^{n-1} \, dx$ and $v = e^x$. Then use integration by parts.   **51.** tw   **53.** 355,986

Exercise Set 5.7, p. 426

1. $\dfrac{1}{9}e^{-3x}(-3x - 1) + C$ **3.** $\dfrac{5^x}{\ln 5} + C$

5. $\dfrac{1}{8}\ln\left(\dfrac{4 + x}{4 - x}\right) + C$ **7.** $5 - x - 5 \ln (5 - x) + C$

9. $\dfrac{1}{5(5 - x)} + \dfrac{1}{25}\ln\left(\dfrac{x}{5 - x}\right) + C$

11. $(\ln 3)x + x \ln x - x + C$

13. $\dfrac{x^4 e^{5x}}{5} - \dfrac{4}{25}x^3 e^{5x} + \dfrac{12}{125}x^2 e^{5x} - \dfrac{24}{3125}e^{5x}(5x - 1) + C$

15. $x^4\left(\frac{1}{4}\ln x - \frac{1}{16}\right) + C$ **17.** $\ln (x + \sqrt{x^2 + 7}) + C$

19. $\dfrac{2}{5 - 7x} + \dfrac{2}{5}\ln\left(\dfrac{x}{5 - 7x}\right) + C$

21. $-\dfrac{5}{4}\ln\left(\dfrac{x - \frac{1}{2}}{x + \frac{1}{2}}\right) + C$

23. $m\sqrt{m^2 + 4} + 4 \ln (m + \sqrt{m^2 + 4}) + C$

25. $\dfrac{5 \ln x}{2x^2} + \dfrac{5}{4x^2} + C$

27. $x^3 e^x - 3x^2 e^x + 6xe^x - 6e^x + C$

29. $S(p) = 100\left[\dfrac{20}{20 - p} + \ln (20 - p)\right]$

31. $-4 \ln\left(\dfrac{x}{3x - 2}\right) + C$

33. $\dfrac{1}{-2(x - 2)} + \dfrac{1}{4}\ln\left(\dfrac{x}{x - 2}\right) + C$

35. $\dfrac{-3}{(e^{-x} - 3)} + \ln (e^{-x} - 3) + C$

Summary and Review: Chapter 5, p. 428

1. [5.1] $\frac{8}{5}x^5 + C$ **2.** [5.1] $3e^x + 2x + C$

3. [5.1] $t^3 + \frac{7}{2}t^2 + \ln t + C$ **4.** [5.2] 9   **5.** [5.2] $\frac{62}{3}$

6. [5.3] Total number of words written in t minutes; an antiderivative   **7.** [5.3] Total amount of sales at the end of t days; an antiderivative   **8.** [5.2] $\frac{1}{6}(b^6 - a^6)$

9. [5.2] $-\frac{2}{5}$ **10.** [5.2] $e - \frac{1}{2}$ **11.** [5.2] $3 \ln 3$

12. [5.2] 0   **13.** [5.2] Negative   **14.** [5.2] Positive

15. [5.4] $\frac{27}{2}$ **16.** [5.5] $\frac{1}{4}e^{x^4} + C$

17. [5.5] $\ln (4t^6 + 3) + C$ **18.** [5.5] $\frac{1}{4}(\ln 4x)^2 + C$

19. [5.5] $-\frac{2}{3}e^{-3x} + C$ **20.** [5.6] $e^{3x}\left(x - \frac{1}{3}\right) + C$

21. [5.6] $x \ln x^7 - 7x + C$ **22.** [5.6] $x^3 \ln x - \dfrac{x^3}{3} + C$

23. [5.7] $\dfrac{1}{14}\ln\left(\dfrac{7 + x}{7 - x}\right) + C$

24. [5.7] $\frac{1}{5}x^2 e^{5x} - \frac{2}{25}xe^{5x} + \frac{2}{125}e^{5x} + C$

25. [5.7] $\dfrac{1}{49} + \dfrac{x}{7} - \dfrac{1}{49} \ln (7x + 1) + C$

26. [5.7] $\ln (x + \sqrt{x^2 - 36}) + C$

27. [5.7] $x^7 \left(\dfrac{\ln x}{7} - \dfrac{1}{49} \right) + C$

28. [5.7] $\dfrac{1}{64} e^{8x}(8x - 1) + C$ **29.** [5.3] 3.667

30. [5.3] $\frac{1}{2}(11 - e^{-2})$ **31.** [5.3] 80

32. [5.5] $e^{12} - 1$, or about \$162,754

33. [5.6] $e^{0.1x}(10x^3 - 300x^2 + 6000x - 60,000) + C$

34. [5.5] $\ln (4t^3 + 7) + C$

35. [5.7] $\frac{2}{75}(5x - 8)\sqrt{4 + 5x} + C$ **36.** [5.5] $e^{x^5} + C$

37. [5.5] $\ln (x + 9) + C$ **38.** [5.5] $\frac{1}{96}(t^8 + 3)^{12} + C$

39. [5.6] $x \ln 7x - x + C$ **40.** [5.6] $\dfrac{x^2}{2} \ln 8x - \dfrac{x^2}{4} + C$

41. [5.5], [5.7] $\frac{1}{10}[\ln (t^5 + 3)]^2 + C$

42. [5.5] $-\frac{1}{2} \ln (1 + 2e^{-x}) + C$

43. [5.5], [5.6], [5.7] $(\ln \sqrt{x})^2 + C$

44. [5.6], [5.7] $\dfrac{x^{92}}{92} \left(\ln x - \dfrac{1}{92} \right) + C$

45. [5.6] $(x - 3) \ln (x - 3) - (x - 4) \ln (x - 4) + C$

46. [5.5] $-\dfrac{1}{3(\ln x)^3} + C$ **47.** [5.4] 1.333

Test: Chapter 5, p. 429

1. [5.1] $x + C$ **2.** [5.1] $200x^5 + C$

3. [5.1] $e^x + \ln x + \frac{8}{11}x^{11/8} + C$ **4.** [5.2] $\frac{1}{6}$

5. [5.2] $4 \ln 3$ **6.** [5.3] An antiderivative, total number of words typed in t minutes   **7.** [5.2] 12

8. [5.2] $-\dfrac{1}{2} \left(\dfrac{1}{e^2} - 1 \right)$ **9.** [5.2] $\ln b - \ln a$, or $\ln \dfrac{b}{a}$

10. [5.2] Positive   **11.** [5.5] $\ln (x + 8) + C$

12. [5.5] $-2e^{-0.5x} + C$ **13.** [5.5] $\frac{1}{40}(t^4 + 1)^{10} + C$

14. [5.6] $\dfrac{x}{5} e^{5x} - \dfrac{e^{5x}}{25} + C$ **15.** [5.6] $\dfrac{x^4}{4} \ln x^4 - \dfrac{x^4}{4} + C$

16. [5.7] $\dfrac{2^x}{\ln 2} + C$ **17.** [5.7] $\dfrac{1}{7} \ln \left(\dfrac{x}{7 - x} \right) + C$

18. [5.3] 6   **19.** [5.4] $\dfrac{1}{3}$ **20.** [5.3] 95

21. [5.3] \$49,000   **22.** [5.3] 94 words

23. [5.7] $\dfrac{1}{10} \ln \left(\dfrac{x}{10 - x} \right) + C$

24. [5.6] $e^x(x^5 - 5x^4 + 20x^3 - 60x^2 + 120x - 120) + C$

25. [5.5] $\frac{1}{6}e^{x^6} + C$ **26.** [5.6] $\frac{2}{3}x^{3/2} \ln x - \frac{4}{9}x^{3/2} + C$

27. [5.6], [5.7] $\frac{1}{15}(3x^2 - 8)(x^2 + 4)^{3/2} + C$

28. [5.7] $\dfrac{1}{16} \ln \left(\dfrac{8 + x}{8 - x} \right) + C$

29. [5.6] $e^{0.1x}(10x^4 - 400x^3 + 12,000x^2 - 240,000x +$

$2,400,000) + C$ **30.** [5.6] $\dfrac{x^2}{2} \ln 13x - \dfrac{x^2}{4} + C$

31. [5.5] $\dfrac{(\ln x)^4}{4} - \dfrac{4}{3}(\ln x)^3 + 5 \ln x + C$

32. [5.5] $(x + 3) \ln (x + 3) - (x + 5) \ln (x + 5) + C$

33. [5.7] $\frac{3}{10}(8x^3 + 10)(5x - 4)^{2/3} - \frac{108}{125}x^2(5x - 4)^{5/3} +$ $\frac{81}{625}x(5x - 4)^{8/3} - \frac{243}{34,375}(5x - 4)^{11/3} + C$ **34.** [5.4] 8

Chapter 6

Technology Connection, p. 434

1. (2, \$9)

Exercise Set 6.1, p. 439

1. (a) (6, \$5);   (b) \$15;   (c) \$9   **3.** (a) (1, \$9);
(b) \$3.33;   (c) \$1.67   **5.** (a) (3, \$9);   (b) \$36;   (c) \$18
7. (a) (50, \$500);   (b) \$12,500;   (c) \$6250
9. (a) (2, \$3);   (b) \$2;   (c) \$0.35   **11.** (a) (5, \$0.61);
(b) \$86.36;   (c) \$2.45   **13.** tw
15. (a) (1, \$9);
(b)

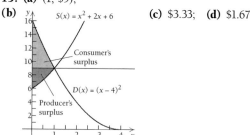

(c) \$3.33;   (d) \$1.67

17. (a) Linear;   (b) $y = -2.5x + 22.5$;   (c) \$45;
(d) \$24.20

Technology Connection, p. 441

1. $f(0) = \$1000, f(1) = \$1083.29, f(2) = \$1173.51,$
$f(3) = \$1271.25$

Technology Connection, p. 441

1. \$3390.61

Exercise Set 6.2, p. 446

1. \$131   **3.** \$5610.72   **5.** \$340,754.12   **7.** \$949.94
9. \$259.37   **11.** \$8264.94   **13.** \$29,676.18
15. \$17,802.91   **17.** \$966,112.17
19. 3,674,343,000 tons   **21.** After 38 yr (2028)
23. 19.994 lb   **25.** 112,891.1 million barrels
27. \$2,383,120   **29.** \$125.30   **31.** tw
33. The area and the amount of continuous money flow are the same. Both are 5610.72.

Technology Connection, p. 449

1. 1

Technology Connection, p. 449

1. ∞

Exercise Set 6.3, p. 452

1. Convergent, $\frac{1}{3}$ **3.** Divergent   **5.** Convergent, 1
7. Convergent, $\frac{1}{2}$ **9.** Divergent   **11.** Convergent, 5
13. Divergent   **15.** Divergent   **17.** Divergent
19. Convergent, 1   **21.** $\frac{1}{2}$ **23.** 1   **25.** $45,000
27. $6250   **29.** $4500   **31.** 33,333 lb
33. Divergent   **35.** Convergent, 2   **37.** Convergent, $\frac{1}{2}$
39. $\frac{1}{k^2}$; the total dose of the drug   **41.** tw

43.

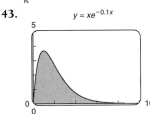
$y = xe^{-0.1x}$

45. 1.25

Exercise Set 6.4, p. 464

1. $\int_0^1 2x\,dx = [x^2]_0^1 = 1^2 - 0^2 = 1$

3. $\int_4^7 \frac{1}{3}\,dx = \left[\frac{1}{3}x\right]_4^7 = \frac{1}{3}(7-4) = 1$

5. $\int_1^3 \frac{3}{26}x^2\,dx = \left[\frac{3}{26}\cdot\frac{x^3}{3}\right]_1^3 = \frac{1}{26}(3^3 - 1^3) = 1$

7. $\int_1^e \frac{1}{x}\,dx = [\ln x]_1^e = \ln e - \ln 1 = 1 - 0 = 1$

9. $\int_{-1}^1 \frac{3}{2}x^2\,dx = \left[\frac{3}{2}\cdot\frac{1}{3}x^3\right]_{-1}^1$

$\qquad = \frac{1}{2}(1^3 - (-1)^3)$

$\qquad = \frac{1}{2}(1 + 1) = 1$

11. $\int_0^\infty 3e^{-3x}\,dx = \lim_{b\to\infty} \int_0^b 3e^{-3x}\,dx$

$\qquad = \lim_{b\to\infty} \left[\frac{3}{-3}e^{-3x}\right]_0^b$

$\qquad = \lim_{b\to\infty} [-e^{-3x}]_0^b$

$\qquad = \lim_{b\to\infty} [-e^{-3b} - (-e^{-3\cdot 0})]$

$\qquad = \lim_{b\to\infty} \left(1 - \frac{1}{e^{3b}}\right) = 1$

13. $\frac{1}{4}$; $f(x) = \frac{1}{4}x$ **15.** $\frac{3}{2}$; $f(x) = \frac{3}{2}x^2$ **17.** $\frac{1}{5}$; $f(x) = \frac{1}{5}$
19. $\frac{1}{2}$; $f(x) = \frac{2-x}{2}$ **21.** $\frac{1}{\ln 3}$; $f(x) = \frac{1}{x\ln 3}$
23. $\frac{1}{e^3 - 1}$; $f(x) = \frac{e^x}{e^3 - 1}$ **25.** (a) $\frac{8}{25}$;   (b) tw
27. $\frac{1}{2}$ **29.** 0.3297   **31.** 0.99995   **33.** 0.3935
35. 0.9502   **37.** $b = \sqrt[4]{4}$, or $\sqrt{2}$ **39.** tw
41.–51. Left to the student.

Technology Connection, p. 475

1. (a) 0.63, or 63%;   **(b)** 0.023, or 2.3%
2. (a) 13.4%;   **(b)** 64.5%

Exercise Set 6.5, p. 475

1. $\mu = E(x) = \frac{7}{2}$, $E(x^2) = 13$, $\sigma^2 = \frac{3}{4}$, $\sigma = \frac{1}{2}\sqrt{3}$
3. $\mu = E(x) = 2$, $E(x^2) = \frac{9}{2}$, $\sigma^2 = \frac{1}{2}$, $\sigma = \sqrt{\frac{1}{2}}$
5. $\mu = E(x) = \frac{14}{9}$, $E(x^2) = \frac{5}{2}$, $\sigma^2 = \frac{13}{162}$, $\sigma = \sqrt{\frac{13}{162}} = \frac{1}{9}\sqrt{\frac{13}{2}}$
7. $\mu = E(x) = -\frac{5}{4}$, $E(x^2) = \frac{11}{5}$, $\sigma^2 = \frac{51}{80}$, $\sigma = \sqrt{\frac{51}{80}} = \frac{1}{4}\sqrt{\frac{51}{5}}$
9. $\mu = E(x) = \dfrac{2}{\ln 3}$, $E(x^2) = \dfrac{4}{\ln 3}$, $\sigma^2 = \dfrac{4\ln 3 - 4}{(\ln 3)^2}$,
$\sigma = \dfrac{2}{\ln 3}\sqrt{\ln 3 - 1}$ **11.** 0.4964   **13.** 0.3665
15. 0.6442   **17.** 0.0078   **19.** 0.1716   **21.** 0.0013
23. (a) 0.6826;   **(b)** 68.26%   **25.** 0.2898   **27.** 0.4514
29.–45. Left to the student.   **47.** 0.62%   **49.** 0.1115
51. $\mu = E(x) = \dfrac{a+b}{2}$, $E(x^2) = \dfrac{b^3 - a^3}{3(b-a)}$, or
$\dfrac{b^2 + ba + a^2}{3}$, $\sigma^2 = \dfrac{(b-a)^2}{12}$,
$\sigma = \dfrac{b-a}{2\sqrt{3}}$ **53.** $\sqrt{2}$ **55.** $\dfrac{\ln 2}{k}$
57. 5.801 oz   **59.** tw

Exercise Set 6.6, p. 479

1. $\dfrac{\pi}{3}$ **3.** $\dfrac{7\pi}{3}$ **5.** $\dfrac{\pi}{2}(e^{10} - e^{-4})$ **7.** $\dfrac{2\pi}{3}$ **9.** $\pi \ln 3$
11. 32π **13.** $\dfrac{32\pi}{5}$ **15.** 56π **17.** $\dfrac{32\pi}{3}$ **19.** $2\pi e^3$
21. π

Exercise Set 6.7, p. 488

1. $y = x^4 + C$; $y = x^4 + 3$, $y = x^4$, $y = x^4 - 796$;
answers may vary   **3.** $y = \frac{1}{2}e^{2x} + \frac{1}{2}x^2 + C$;
$y = \frac{1}{2}e^{2x} + \frac{1}{2}x^2 - 5$, $y = \frac{1}{2}e^{2x} + \frac{1}{2}x^2 + 7$;
$y = \frac{1}{2}e^{2x} + \frac{1}{2}x^2$; answers may vary

5. $y = 3 \ln x - \frac{1}{3}x^3 + \frac{1}{6}x^6 + C$;
$y = 3 \ln x - \frac{1}{3}x^3 + \frac{1}{6}x^6 - 15$,
$y = 3 \ln x - \frac{1}{3}x^3 + \frac{1}{6}x^6 - 7$,
$y = 3 \ln x - \frac{1}{3}x^3 + \frac{1}{6}x^6$; answers may vary
7. $y = \frac{1}{3}x^3 + x^2 - 3x + 4$　**9.** $f(x) = \frac{3}{5}x^{5/3} - \frac{1}{2}x^2 - \frac{61}{10}$
11. $y'' = \frac{1}{x}$. Then

$$\begin{array}{c|c} y'' - \dfrac{1}{x} = 0 & \\ \hline \dfrac{1}{x} - \dfrac{1}{x} & 0 \\ 0 & \end{array}$$

13. $y' = 4e^x + 3xe^x$, $y'' = 7e^x + 3xe^x$. Then

$$\begin{array}{c|c} y'' - 2y' + y = 0 & \\ \hline (7e^x + 3xe^x) - 2(4e^x + 3xe^x) + (e^x + 3xe^x) & 0 \\ 7e^x + 3xe^x - 8e^x - 6xe^x + e^x + 3xe^x & \\ 0 & \end{array}$$

15. $y = C_1 e^{x^4}$, where $C_1 = e^C$　**17.** $y = \sqrt[3]{\frac{5}{2}x^2 + C}$
19. $y = \sqrt{2x^2 + C_1}$, $y = -\sqrt{2x^2 + C_1}$, where $C_1 = 2C$
21. $y = \sqrt{6x + C_1}$, $y = -\sqrt{6x + C_1}$, where $C_1 = 2C$
23. $y = -3 + 8e^{x^2/2}$　**25.** $y = \sqrt[3]{15x - 3}$
27. $y = C_1 e^{3x}$, where $C_1 = e^C$　**29.** $P = C_1 e^{2t}$, where $C_1 = e^C$　**31.** $f(x) = \ln x - 2x^2 + \frac{2}{3}x^{3/2} + 9$
33. $C(x) = 2.6x - 0.01x^2 + 120$,
$A(x) = 2.6 - 0.01x + \dfrac{120}{x}$
35. (a) $P(C) = \dfrac{400}{(C + 3)^{1/2}} - 40$;　**(b)** \$97
37. (a) $R = k \cdot \ln (S + 1) + C$;　**(b)** $R = k \cdot \ln (S + 1)$;
(c) no units, no pleasure from them　**39.** $x = 200 - p$
41. $x = \dfrac{C_1}{p^n}$　**43.** $R = C_1 \cdot S^k$, where $C_1 = e^C$
45.

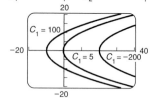

$y_1 = \sqrt{10x + C_1}$, $y_2 = -\sqrt{10x + C_1}$

---

11. [6.3] Convergent, 1　**12.** [6.3] Divergent
13. [6.3] Convergent, $\dfrac{1}{2}$　**14.** [6.4] $\dfrac{8}{3}$; $f(x) = \dfrac{8}{3x^3}$
15. [6.4] 0.6　**16.** [6.5] $\frac{3}{5}$　**17.** [6.5] $\frac{3}{4}$　**18.** [6.5] $\frac{3}{4}$
19. [6.5] $\frac{3}{80}$　**20.** [6.5] $\dfrac{\sqrt{15}}{20}$　**21.** [6.5] 0.4678
22. [6.5] 0.8827　**23.** [6.5] 0.1002　**24.** [6.5] 0.5000
25. [6.5] 0.0918　**26.** [6.6] $\dfrac{127\pi}{7}$　**27.** [6.6] $\dfrac{\pi}{6}$
28. [6.7] $y = C_1 e^{x^{11}}$, where $C_1 = e^C$
29. [6.7] $y = \pm\sqrt{4x + C_1}$, where $C_1 = 2C$
30. [6.7] $y = 5e^{4x}$　**31.** [6.7] $v = \sqrt[3]{15t + 19}$
32. [6.7] $y = \pm\sqrt{3x^2 + C_1}$, where $C_1 = 2C$
33. [6.7] $y = 8 - C_1 e^{-x^2/2}$, where $C_1 = e^{-C}$
34. [6.7] $x = 100 - p$
35. [6.7] $V = \$36.37 - \$6.37e^{-kt}$
36. [6.4] $c = \sqrt[9]{4.5}$　**37.** [6.3] Convergent, $-\frac{1}{5}$
38. [6.3] Convergent, 3　**39.** [6.3] 1

Test: Chapter 6, p. 491

1. [6.1] (3, \$16)　**2.** [6.1] \$45　**3.** [6.1] \$22.50
4. [6.2] \$29,192　**5.** [6.2] \$344.66　**6.** [6.2] 37,121,870
thousand tons　**7.** [6.2] 39 yr　**8.** [6.2] \$2465.97
9. [6.2] \$30,717.71　**10.** [6.2] \$34,545.45
11. [6.3] Convergent, $\dfrac{1}{4}$　**12.** [6.3] Divergent
13. [6.4] $\dfrac{1}{4}$; $f(x) = \dfrac{x^3}{4}$　**14.** [6.4] 0.8647
15. [6.5] $E(x) = \dfrac{13}{6}$　**16.** [6.5] $E(x^2) = 5$
17. [6.5] $\mu = \dfrac{13}{6}$　**18.** [6.5] $\sigma^2 = \dfrac{11}{36}$
19. [6.5] $\sigma = \dfrac{\sqrt{11}}{6}$　**20.** [6.5] 0.4332
21. [6.5] 0.4420　**22.** [6.5] 0.9071　**23.** [6.5] 0.4207
24. [6.6] $\pi \ln 5$　**25.** [6.6] $\dfrac{5\pi}{2}$　**26.** [6.7] $y = C_1 e^{x^8}$,
where $C_1 = e^C$　**27.** [6.7] $y = \sqrt{18x + C_1}$,
$y = -\sqrt{18x + C_1}$, where $C_1 = 2C$　**28.** [6.7] $y = 11e^{6t}$
29. [6.7] $y = 5 - C_1 e^{-x^3/3}$, where $C_1 = e^{-C}$
30. [6.7] $v = \pm\sqrt[4]{8t + C_1}$, where $C_1 = 4C$
31. [6.7] $y = C_1 e^{4x + x^2/2}$, where $C_1 = e^C$
32. [6.7] $x = \dfrac{C_1}{p^4}$　**33.** [6.7] **(a)** $V(t) = 36(1 - e^{-kt})$;
(b) $k = 0.12$;　**(c)** $V(t) = 36(1 - e^{-0.12t})$;
(d) $V(12) \approx \$27.47$;　**(e)** $t \approx 14.9$ months
34. [6.4] $b = \sqrt[6]{6}$　**35.** [6.3] Convergent, $-\frac{1}{4}$
36. [6.3] 3.14

Summary and Review: Chapter 6, p. 490

1. [6.1] (2, \$16)　**2.** [6.1] \$18.67　**3.** [6.1] \$5.33
4. [6.2] \$83,560.54　**5.** [6.2] \$251.50　**6.** [6.2] 282,732
thousand tons　**7.** [6.2] 183 yr　**8.** [6.2] \$247.88
9. [6.2] \$44,517　**10.** [6.2], [6.3] \$53,333

Chapter 7 ....

Technology Connection, p. 498

1. 8,161,517,629

Exercise Set 7.1, p. 501

1. $f(0, -2) = 0, f(2, 3) = -8, f(10, -5) = 200$
3. $f(0, -2) = 1, f(-2, 1) = -13\frac{8}{9}, f(2, 1) = 23$
5. $f(e, 2) = \ln e + 2^3 = 1 + 8 = 9, f(e^2, 4) = 66,$
$f(e^3, 5) = 128$ **7.** $f(-1, 2, 3) = 6, f(2, -1, 3) = 12$
9. \$151,571.66   **11.** 2.4%   **13.** 244.7 mph   **15.**
17. $0°$ **19.** $-22°$
21.

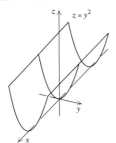

$z = y^2$

23.

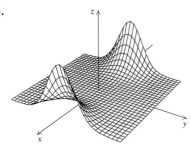

$z = (x^4 - 16x^2)e^{-y^2}$

25.

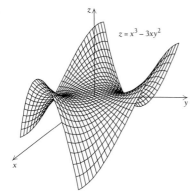

$z = x^3 - 3xy^2$

Exercise Set 7.2, p. 508

1. $\frac{\partial z}{\partial x} = 2 - 3y, \frac{\partial z}{\partial y} = -3x, \frac{\partial z}{\partial x}\Big|_{(-2, -3)} = 11, \frac{\partial z}{\partial y}\Big|_{(0, -5)} = 0$

3. $\frac{\partial z}{\partial x} = 6x - 2y, \frac{\partial z}{\partial y} = -2x + 1, \frac{\partial z}{\partial x}\Big|_{(-2, -3)} = -6,$

$\frac{\partial z}{\partial y}\Big|_{(0, -5)} = 1$ **5.** $f_x = 2, f_y = -3, f_x(-2, 4) = 2,$

$f_y(4, -3) = -3$ **7.** $f_x = \frac{x}{\sqrt{x^2 + y^2}}, f_y = \frac{y}{\sqrt{x^2 + y^2}},$

$f_x(-2, 1) = \frac{-2}{\sqrt{5}}, f_y(-3, -2) = \frac{-2}{\sqrt{13}}$ **9.** $f_x = 2e^{2x+3y},$

$f_y = 3e^{2x+3y}$ **11.** $f_x = ye^{xy}, f_y = xe^{xy}$

13. $f_x = \frac{y}{x + y}, f_y = \frac{y}{x + y} + \ln(x + y)$

15. $f_x = 1 + \ln xy, f_y = \frac{x}{y}$

17. $f_x = \frac{1}{y} + \frac{y}{x^2}, f_y = -\frac{x}{y^2} - \frac{1}{x}$
19. $f_x = 12(2x + y - 5), f_y = 6(2x + y - 5)$
21. $\frac{\partial f}{\partial b} = 12m + 6b - 30, \frac{\partial f}{\partial m} = 28m + 12b - 64$
23. $f_x = 3y - 2\lambda, f_y = 3x - \lambda, f_\lambda = -(2x + y - 8)$
25. $f_x = 2x - 10\lambda, f_y = 2y - 2\lambda, f_\lambda = -(10x + 2y - 4)$

27. (a) 3,888,064 units;   **(b)** $\frac{\partial p}{\partial x} = 1117.8\left(\frac{y}{x}\right)^{0.379}$

$\frac{\partial p}{\partial y} = 682.2\left(\frac{x}{y}\right)^{0.621}$;   **(c)** ;   **(d)** 965.8, 866.8

29. $99.6°$ **31.** $121.3°$ **33.**   **35.** 78.244

37. -0.846 **39.** $f_x = \frac{-4xt^2}{(x^2 - t^2)^2}, f_t = \frac{4x^2t}{(x^2 - t^2)^2}$

41. $f_x = \frac{1}{\sqrt{x}(1 + 2\sqrt{t})}, f_t = \frac{-1 - 2\sqrt{x}}{\sqrt{t}(1 + 2\sqrt{t})^2}$

43. $f_x = 4x^{-1/3} - 2x^{-3/4}t^{1/2} + 6x^{-3/2}t^{3/2},$
$f_t = -4x^{1/4}t^{-1/2} - 18x^{-1/2}t^{1/2}$ **45.**

Exercise Set 7.3, p. 511

1. $\frac{\partial^2 f}{\partial x^2} = 6, \frac{\partial^2 f}{\partial y \partial x} = \frac{\partial^2 f}{\partial x \partial y} = -1, \frac{\partial^2 f}{\partial y^2} = 0$ **3.** $\frac{\partial^2 f}{\partial x^2} = 0,$

$\frac{\partial^2 f}{\partial y \partial x} = \frac{\partial^2 f}{\partial x \partial y} = 3, \frac{\partial^2 f}{\partial y^2} = 0$ **5.** $\frac{\partial^2 f}{\partial x^2} = 20x^3y^4 + 6xy^2,$

$\frac{\partial^2 f}{\partial y \partial x} = \frac{\partial^2 f}{\partial x \partial y} = 20x^4y^3 + 6x^2y, \frac{\partial^2 f}{\partial y^2} = 12x^5y^2 + 2x^3$
7. $f_{xx} = 0, f_{yx} = 0, f_{xy} = 0, f_{yy} = 0$ **9.** $f_{xx} = 4y^2e^{2xy},$
$f_{yx} = f_{xy} = 4xye^{2xy} + 2e^{2xy}, f_{yy} = 4x^2e^{2xy}$ **11.** $f_{xx} = 0,$
$f_{yx} = f_{xy} = 0, f_{yy} = e^y$ **13.** $f_{xx} = -\frac{y}{x^2}, f_{yx} = f_{xy} = \frac{1}{x},$
$f_{yy} = 0$ **15.** $f_{xx} = \frac{-6y}{x^4}, f_{yx} = f_{xy} = \frac{2(y^3 - x^3)}{x^3y^3}, f_{yy} = \frac{6x}{y^4}$
17. $\frac{\partial^2 f}{\partial x^2} = \frac{2y^2 - 2x^2}{(x^2 + y^2)^2}, \frac{\partial^2 f}{\partial y^2} = \frac{2x^2 - 2y^2}{(x^2 + y^2)^2}$, so the sum is 0

19. (a) $-y$; **(b)** x; **(c)** $f_{yx}(0, 0) = 1, f_{xy}(0, 0) = -1$; so $f_{yx}(0, 0) \neq f_{xy}(0, 0)$

Exercise Set 7.4, p. 520

1. Relative minimum $= -\frac{1}{3}$ at $\left(-\frac{1}{3}, \frac{2}{3}\right)$

3. Relative maximum $= \frac{4}{27}$ at $\left(\frac{2}{3}, \frac{2}{3}\right)$

5. Relative minimum $= -1$ at $(1, 1)$

7. Relative minimum $= -7$ at $(1, -2)$

9. Relative minimum $= -5$ at $(-1, 2)$ **11.** None

13. 6 thousand of the \$17 radio and 5 thousand of the \$21 radio **15.** Maximum of $P = 35$ (million dollars) when $a = 10$ (million dollars) and $p = \$3$

17. The dimensions of the bottom are 8 ft by 8 ft, and the height is 5 ft.

19. (a) $R(p_1, p_2) = 64p_1 - 4p_1^2 - 4p_1p_2 + 56p_2 - 4p_2^2$; **(b)** $p_1 = 6(\$60), p_2 = 4(\$40)$; **(c)** $q_1 = 32$ (hundreds), $q_2 = 28$ (hundreds); **(d)** \$304,000 **21.** None; saddle point at $(0, 0)$ **23.** Minimum $= 0$ at $(0, 0)$; saddle points at $(2, 1)$ and $(-2, 1)$ **25.** tw

27. Relative minimum $= -5$ at $(0, 0)$ **29.** None

Exercise Set 7.5, p. 529

1. (a) $y = 1.73x + 6.06$; **(b)** 26.82 million; 35.47 million

3. (a) $y = 1.97x + 69.09$; **(b)** 80.91; 82.88

5. (a) $y = 1.068421x - 1.236842$; **(b)** 85.3 **7.** tw

9. (a) $y = -0.005925x + 15.54734$; **(b)** 3:42.54; 3:40.40; **(c)** 3:44.32

Exercise Set 7.6, p. 537

1. Maximum $= 8$ at $(2, 4)$ **3.** Maximum $= -16$ at $(2, 4)$ **5.** Minimum $= 20$ at $(4, 2)$

7. Minimum $= -96$ at $(8, -12)$ **9.** Minimum $= \frac{3}{2}$ at $\left(1, \frac{1}{2}, -\frac{1}{2}\right)$ **11.** 35 and 35 **13.** 3 and -3

15. $9\frac{3}{4}$ in., $9\frac{3}{4}$ in.; $95\frac{1}{16}$ in^2; no **17.** $r = \sqrt[3]{\dfrac{27}{2\pi}} \approx 1.6$ ft;

$h = 2 \cdot r \approx 3.2$ ft; minimum surface area ≈ 45.97 ft^2

19. Maximum of $S = 800$ at $L = 20, M = 60$

21. (a) $C(x, y, z) = 7xy + 6yz + 6xz$; **(b)** $x = 60$ ft, $y = 60$ ft, $z = 70$ ft; \$75,600 **23.** 10,000 units of A, 100 units of B **25.** Minimum $= -\frac{155}{128}$ at $\left(-\frac{7}{16}, -\frac{3}{4}\right)$

27. Maximum $= \dfrac{1}{27}$ at $\left(\pm\dfrac{1}{\sqrt{3}}, \pm\dfrac{1}{\sqrt{3}}, \pm\dfrac{1}{\sqrt{3}}\right)$

29. Maximum $= 2$ at $\left(\dfrac{1}{2}, \dfrac{1}{2}, \dfrac{1}{2}, \dfrac{1}{2}\right)$

31. $\lambda = \dfrac{p_x}{c_1} = \dfrac{p_y}{c_2}$ **33.** tw **35.–41.** Left to the student.

Exercise Set 7.7, p. 542

1. 1 **3.** 0 **5.** 6 **7.** $\frac{3}{20}$ **9.** 4 **11.** $\frac{4}{15}$ **13.** 1

15. 39 **17.** $\frac{13}{240}$ **19.** tw

Summary and Review: Chapter 7, p. 543

1. [7.1] 1 **2.** [7.2] $3y^3$ **3.** [7.2] $e^y + 9xy^2 + 2$

4. [7.3] $9y^2$ **5.** [7.3] $9y^2$ **6.** [7.3] 0

7. [7.3] $e^y + 18xy$ **8.** [7.2] $2x \ln y$ **9.** [7.2] $\dfrac{x^2}{y} + 4y^3$

10. [7.3] $\dfrac{2x}{y}$ **11.** [7.3] $\dfrac{2x}{y}$ **12.** [7.3] $2 \ln y$

13. [7.3] $-\dfrac{x^2}{y^2} + 12y^2$ **14.** [7.4] Minimum $= -\dfrac{549}{20}$ at $\left(5, \dfrac{27}{2}\right)$ **15.** [7.4] Minimum $= -4$ at $(0, -2)$

16. [7.4] Maximum $= \dfrac{45}{4}$ at $\left(\dfrac{3}{2}, -3\right)$

17. [7.4] Minimum $= 29$ at $(-1, 2)$

18. [7.5] **(a)** $y = 0.118x + 0.303$; **(b)** \$1.13 million; \$1.48 million **19.** [7.5] **(a)** $y = 0.6x + 6.7$; **(b)** 9.1 million **20.** [7.6] Minimum $= \dfrac{125}{4}$ at $\left(-\dfrac{3}{2}, -3\right)$

21. [7.6] Maximum $= 300$ at $(5, 10)$ **22.** [7.7] $\frac{97}{30}$

23. [7.7] $\frac{1}{60}$ **24.** [7.7] 0

25. [7.4], [7.6] The cylindrical container

26. [7.1]

$z = x^2 + 4y^2$

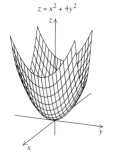

Test: Chapter 7, p. 545

1. [7.1] $e^{-1} - 2$ **2.** [7.2] $e^x + 6x^2y$ **3.** [7.2] $2x^3 + 1$

4. [7.3] $e^x + 12xy$ **5.** [7.3] $6x^2$ **6.** [7.3] $6x^2$

7. [7.3] 0 **8.** [7.4] Minimum $= -\dfrac{7}{16}$ at $\left(\dfrac{3}{4}, \dfrac{1}{2}\right)$

9. [7.4] None **10.** [7.5] **(a)** $y = \dfrac{9}{2}x + \dfrac{17}{3}$; **(b)** \$23.67 million **11.** [7.6] Maximum $= -19$ at $(4, 5)$

12. [7.7] 1 **13.** [7.4] \$400,000 for labor, \$200,000 for capital **14.** [7.2] $f_x = \dfrac{-x^4 + 4xt + 6x^2t}{(x^3 + 2t)^2}$,

$f_t = \dfrac{-2x^2(x + 1)}{(x^3 + 2t)^2}$

15. [7.1]

$$z = x - \tfrac{1}{2}y^2 - \tfrac{1}{3}x^3$$

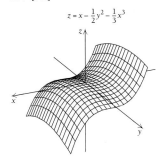

Cumulative Review, p. 549

1. [1.4] $y = -4x - 27$ **2.** [1.2] $x^2 + 2xh + h^2 - 5$
3. [2.2] **(a)**   **(b)** 3;   **(c)** -3;   **(d)** no

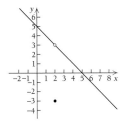

4. [2.2] -8 **5.** [2.2] 3   **6.** [2.2] Does not exist
7. [3.3] 4   **8.** [2.2] $3x$ **9.** [3.3] 0   **10.** [2.5] -9
11. [2.5] $2x - 7$ **12.** [2.5] $\tfrac{1}{4}x^{-3/4}$ **13.** [2.5] $-6x^{-7}$
14. [2.8] $(x - 3)5(x + 1)^4 + (x + 1)^5$, or
$2(x + 1)^4(3x - 7)$ **15.** [2.7] $\dfrac{5 - 2x^3}{x^6}$

16. [4.1] $\dfrac{1 - x}{e^x}$ **17.** [4.2] $\dfrac{2x}{x^2 + 5}$ **18.** [4.2] 1

19. [4.1] $3e^{3x} + 2x$ **20.** [4.1] $\dfrac{e^{\sqrt{x-3}}}{2\sqrt{x - 3}}$

21. [4.2] $\dfrac{e^x}{e^x - 4}$ **22.** [2.9] $2 - 4x^{-3}$

23. [3.7] $3xy^2 + \dfrac{y}{x}$ **24.** [2.4] $y = x - 2$

25. [3.1], [3.2] Relative   **26.** [3.1], [3.2] Relative
maximum at $(-1, 3)$;   maxima at $(-1, -2)$ and
relative minimum at $(1, -2)$; relative minimum
$(1, -1)$ at $(0, -3)$

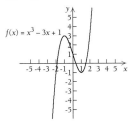   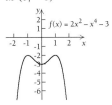

27. [3.1], [3.2] Relative
minimum at $(-1, -4)$;
relative maximum at $(1, 4)$

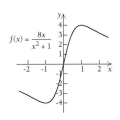

28. [3.1] Relative maximum at $(0, -2)$. For graph, see
Example 7 in Section 3.3.   **29.** [3.4] Minimum $= -7$ at
$x = 1$ **30.** [3.4] None   **31.** [3.4] Maximum $= 6\tfrac{2}{3}$ at
$x = -1$; minimum $= 4\tfrac{1}{3}$ at $x = -2$ **32.** [3.5] 15
33. [3.5] 30 times; lot size 15   **34.** [3.6] $\Delta y = 3.03$;
$f'(x)\Delta x = 3$ **35.** [4.3] **(a)** $N(t) = 8000e^{0.1t}$;

(b) 17,804;   **(c)** 6.9 yr   **36.** [4.6] **(a)** $E(p) = \dfrac{p}{12 - p}$;

(b) $E(2) = \tfrac{1}{5}$, inelastic;   **(c)** $E(9) = 3$; elastic;

(d) increase;   **(e)** $p = \$6$ **37.** [5.1] $\dfrac{1}{2}x^6 + C$

38. [5.2] $3 - \dfrac{2}{e}$

39. [5.7] $\dfrac{7}{9(7 - 3x)} + \dfrac{1}{9}\ln(7 - 3x) + C$

40. [5.5] $\dfrac{1}{4}e^{x^4} + C$

41. [5.1], [5.7] $\left(\dfrac{x^2}{2} + 3x\right)\ln x - \dfrac{x^2}{4} - 3x + C$

42. [5.1] $260\ln x + C$ **43.** [5.2] 4   **44.** [5.2] $\tfrac{232}{3}$
45. [6.2] \$22,679.49   **46.** [6.2] \$60,792.27   **47.** [6.3]
Convergent, $\tfrac{1}{4374}$ **48.** [6.4] $\tfrac{3}{2}\ln 3$ **49.** [6.5] 0.6826

50. [6.1] (169, \$7); \$751.33   **51.** [6.6] $-\dfrac{\pi}{2}\left(\dfrac{1}{e^{10}} - 1\right)$

52. [6.7] $y = C_1 e^{x^2/2}$, where $C_1 = e^C$
53. [4.3] **(a)** $P(t) = 1 - e^{-kt}$;   **(b)** $k = 0.12$;
(c) $P(t) = 1 - e^{-0.12t}$;   **(d)** $P(20) = 0.9093$, or 90.93%
54. [7.2] $8xy^3 + 3$ **55.** [7.3] $e^y + 24x^2y$
56. [7.4] None   **57.** [7.6] Maximum $= 4$ at (3, 3)
58. [7.7] $3(c^3 - 1)$

Appendix

Exercise Set A, p. 563

1. $5 \cdot 5 \cdot 5$, or 125   **2.** $7 \cdot 7$, or 49   **3.** $(-7)(-7)$,
or 49   **4.** $(-5)(-5)(-5)$, or -125 **5.** 1.0201
6. 1.030301   **7.** $\tfrac{1}{16}$ **8.** $\tfrac{1}{64}$ **9.** 1   **10.** $6x$ **11.** t

12. 1   **13.** 1   **14.** $\tfrac{1}{3}$ **15.** $\dfrac{1}{3^2}$, or $\dfrac{1}{9}$ **16.** $\dfrac{1}{4^2}$, or $\dfrac{1}{16}$

17. 8   **18.** 4   **19.** 0.1   **20.** 0.0001   **21.** $\dfrac{1}{e^b}$

22. $\dfrac{1}{t^k}$ **23.** $\dfrac{1}{b}$ **24.** $\dfrac{1}{h}$ **25.** x^5 **26.** t^7

27. x^{-6}, or $\dfrac{1}{x^6}$ **28.** x^6 **29.** $35x^5$ **30.** $8t^7$

31. x^4 **32.** x **33.** 1 **34.** 1 **35.** x^3 **36.** x^4

37. x^{-3}, or $\dfrac{1}{x^3}$ **38.** x^{-4}, or $\dfrac{1}{x^4}$ **39.** 1 **40.** 1

41. e^{t-4} **42.** c^{k-3} **43.** t^{14} **44.** t^{12} **45.** t^2

46. t^{-4}, or $\dfrac{1}{t^4}$ **47.** a^6b^5 **48.** x^7y^7 **49.** t^{-6}, or $\dfrac{1}{t^6}$

50. t^{-12}, or $\dfrac{1}{t^{12}}$ **51.** e^{4x} **52.** e^{5x} **53.** $8x^6y^{12}$

54. $32x^{10}y^{20}$ **55.** $\dfrac{1}{81}x^8y^{20}z^{-16}$, or $\dfrac{x^8y^{20}}{81z^{16}}$

56. $\dfrac{1}{125}x^{-9}y^{21}z^{15}$, or $\dfrac{y^{21}z^{15}}{125x^9}$ **57.** $9x^{-16}y^{14}z^4$, or $\dfrac{9y^{14}z^4}{x^{16}}$

58. $625x^{16}y^{-20}z^{-12}$, or $\dfrac{625x^{16}}{y^{20}z^{12}}$ **59.** $\dfrac{c^4d^{12}}{16q^8}$

60. $\dfrac{64x^6y^3}{a^9b^9}$ **61.** $5x-35$ **62.** $x+xt$

63. $x^2-7x+10$ **64.** $x^2-7x+12$
65. a^3-b^3 **66.** x^3+y^3 **67.** $2x^2+3x-5$
68. $3x^2+x-4$ **69.** a^2-4 **70.** $9x^2-1$
71. $25x^2-4$ **72.** t^2-1 **73.** $a^2-2ah+h^2$
74. $a^2+2ah+h^2$ **75.** $25x^2+10xt+t^2$
76. $49a^2-14ac+c^2$ **77.** $5x^5+30x^3+45x$
78. $-3x^6+48x^2$ **79.** $a^3+3a^2b+3ab^2+b^3$
80. $a^3-3a^2b+3ab^2-b^3$
81. $x^3-15x^2+75x-125$
82. $8x^3+36x^2+54x+27$ **83.** $x(1-t)$
84. $x(1+h)$ **85.** $(x+3y)^2$ **86.** $(x-5y)^2$
87. $(x-5)(x+3)$ **88.** $(x+5)(x+3)$
89. $(x-5)(x+4)$ **90.** $(x-10)(x+1)$
91. $(7x-t)(7x+t)$ **92.** $(3x-b)(3x+b)$

93. $4(3t-2m)(3t+2m)$ **94.** $(5y-3z)(5y+3z)$
95. $ab(a+4b)(a-4b)$ **96.** $2(x^2+4)(x+2)(x-2)$
97. $(a^4+b^4)(a^2+b^2)(a+b)(a-b)$
98. $(6y-5)(6y+7)$ **99.** $10x(a+2b)(a-2b)$
100. $xy(x+5y)(x-5y)$
101. $2(1+4x^2)(1+2x)(1-2x)$
102. $2x(y+5)(y-5)$ **103.** $(9x-1)(x+2)$
104. $(3x-4)(2x-5)$ **105.** $(x+2)(x^2-2x+4)$
106. $(a-3b)(a^2+3ab+9b^2)$
107. $(y-4t)(y^2+4yt+16t^2)$
108. $(m+10p)(m^2-10mp+100p^2)$
109. $(3x^2-1)(x-2)$ **110.** $(y^2-2)(5y+2)$
111. $(x-3)(x+3)(x-5)$
112. $(t-5)(t+5)(t+3)$ **113.** $\frac{7}{4}$ **114.** $\frac{79}{12}$
115. -8 **116.** $\frac{4}{5}$ **117.** 120 **118.** 140 **119.** 200
120. 200 **121.** $0,-3,\frac{4}{5}$ **122.** $0,2,-\frac{3}{2}$ **123.** $0,2$
124. $3,-3$ **125.** $0,3$ **126.** $0,5$ **127.** $0,7$
128. $0,3$ **129.** $0,\frac{1}{3},-\frac{1}{3}$ **130.** $0,\frac{1}{4},-\frac{1}{4}$ **131.** 1
132. 2 **133.** No solution   **134.** No solution
135. -23 **136.** -145 **137.** 5 **138.** $-\sqrt{7},\sqrt{7}$
139. $x\geq-\frac{4}{5}$ **140.** $x\geq3$ **141.** $x>-\frac{1}{12}$
142. $x>-\frac{5}{13}$ **143.** $x>-\frac{4}{7}$ **144.** $x\leq-\frac{6}{5}$
145. $x\leq-3$ **146.** $x\geq-2$ **147.** $x>\frac{2}{3}$
148. $x\leq\frac{5}{3}$ **149.** $x<-\frac{2}{5}$ **150.** $x\leq\frac{5}{4}$
151. $2<x<4$ **152.** $2<x\leq4$ **153.** $\frac{3}{2}\leq x\leq\frac{11}{2}$
154. $\frac{6}{5}\leq x<\frac{16}{5}$ **155.** $-1\leq x\leq\frac{14}{5}$
156. $-5\leq x<-2$ **157.** \$650   **158.** \$800
159. More than 7000 units
160. More than 4200 units   **161.** 480 lb
162. 340 lb   **163.** 810,000   **164.** 720,000
165. $60\%\leq x<100\%$ **166.** $50\%\leq x<90\%$

Index